INSTRUMENTATION, CONTROLS AND AUTOMATION IN THE POWER INDUSTRY
VOLUME 34

Proceedings of the Thirty-Fourth
Power Instrumentation Symposium
(1st Annual ISA/EPRI Joint Controls and Automation Conference)

June 3-5, 1991

Don CeSar Hotel
St. Petersburg Beach, Florida

ISA Program Chairman
Paul L. Kenny
Forney International Inc.

EPRI Program Chairman
Arvind Agarwal
EPRI

General Chairman
Ed Good
Florida Power Corp.

Hosted By
ISA Tampa Bay Section

PUBLICATION POLICY

Technical papers may not be reproduced in any form without written permission from the Instrument Society of America. The Society reserves the exclusive right of publication in its periodicals of all papers presented at the annual ISA Conference, at ISA Symposia, and at meetings co-sponsored by ISA when the Society acts as publisher. Papers not selected for such publication will be released to authors upon request.

In any event, following oral presentation, other publications are urged to publish up to 300-word excerpts of ISA papers, provided credits are given to the author, the meeting, and the Society using the Society's name in full, rather than simply "ISA."

Reprints of articles in this publication are available on a custom printing basis at reasonable prices in quantities of 50 or more.

Permission to photocopy for internal or personal use, or the internal or personal use of specific clients, is granted by Instrument Society of America for libraries and other users registered with the Copyright Clearance Center (CCC), provided that the fee of $.50 per page is paid directly to CCC, 21 Congress Street, Salem, Ma 01970. Payment should include the fee code information at the top of the first page of papers copied which appears in the form of:

0074-056X/91/[First Page]-[Last Page]/$0 + .50pp
[ISSN/Year of Publication/Page Numbers of Article/Fee for Copying] ($.50 Per Page)

Special requests should be addressed to:

Director of Publications Services
Instrument Society of America
P.O. Box 12277
Research Triangle Park, NC 27709

For further information concerning publications policy and reprint quotations, contact:

Publications Department
Instrument Society of America
67 Alexander Drive
P.O. Box 12277
Research Triangle Park, NC 27709
Phone: (919) 549-8411
Telex: 802 540 ISA DURM

ISSN: 0074-056X

ISBN: 1-55617-331-8

Library of Congress Catalog Card Number 62-52679

Printed in U.S.A.
By
Technical Communication Services

FOREWORD

In this period of rapid technological change, education and training are critical success factors. The ISA Power Industry Division's annual symposium, held over the past 34 years, has provided a forum where power industry colleagues could obtain knowledge and information about the latest innovations in power instrumentation, automation and control.

The year 1991 will be remembered as the year EPRI joined together with the ISA Power Division to present an outstanding group of technical papers at the First Annual ISA/EPRI Joint Controls and Automation Conference. All papers met the theme for the conference namely "Innovative Instrumentation, Controls, and Automation Techniques for the Power Generation Industry" and covered a myriad of applications ranging from nuclear to conventional fossil to co-generation plants involving nuclear, conventional BTG, and combined cycle equipment applications. EPRI and the ISA Power Division hope that these proceedings will broaden your horizon and provide you with a better understanding of current instrumentation and control principles.

I wish to thank the Authors, Session Developers, and other speakers who along with Roger Hull, the ISA Review Chairman and all of his reviewers made this years program a success. The overall symposium success was due to the hard work put forth by the General Chairman, Ed Good; the EPRI Program Chairman, Arvind Agarwal; and two (2) members of the host committee, namely Howard Cuddy and Les Miller. Carol Good and Pam Miller graciously organized and conducted a very good "spouses" program. Without their contribution, a successful Symposium would not have been possible and their efforts are appreciated.

EPRI and the ISA Power Division invite you to come to next year's Symposium, ISA 35th and the 2nd Annual Joint ISA/EPRI Controls and Automation Conference, to be held June 1-3, 1992 in Kansas City.

Paul L. Kenny
1991 ISA Program Chairman
Forney International, Inc.

LIST OF REVIEWERS

JUNE 1991

POWER SYMPOSIUM

Q.B. Chou - Ontario Hydro
George Huber Jr. - Gilbert Commonwealth
Mort Binstock - Westinghouse
Duane Hopkins - Southwestern Public Service
Charles Moore - Stone & Webster
Tom Russell - Honeywell
Steven Stultz - Baltimore Gas & Electric
A.H. Chen - Black & Veatch
Gordon McFarland - ABB Power Automation
J.M. Vavrek - Sargent and Lundy
Daniel Lee - Bailey Controls
Doug Kyr - Bailey Controls
Jimmy Williams - Utility Engineering Corp.
Paul Kenny - Forney
Joseph Tana - EBASCO
Dave Farrell - Honeywell
Don Olson - American Ref-Fuel
Eric Schaefer - The Foxboro Company
Roger Hull - Westinghouse
George Zdzieborski - United Eng. & Const.
W.E. Brown - Philadelphia Electric
Greg Essig - Bailey Controls
James Bayless III - Black & Veatch
Len Gruber - Westinghouse
Robert Bowser - ETSI
Cindy Whitaker - Baltimore Gas & Electric
Richard Givan - Sargent and Lundy
W.T. Crowe - Gulf States Utilities
Don Spriggs - TBI/Bailey Controls
Ronald Johnson - Sargent and Lundy
Robert Hubby - Leeds & Northrup
Ron DeVies - TU Electric
G.M. Hedge - Florida Power Corp.
Lawrence Karl - Westinghouse
David Hamme - Gilbert/Commonwealth
Karan Vasudeva - CRS Sirrine Engineers
John Poelma - Nevada Power
Gary Remmers - Fluor Daniel
Harold Sternberg - Bailey Controls
Robert Szczerbicki-Gilbert Commonwealth
James Flaherty - Allegheny Power
Leonard Sterle - Allen-Bradley Inc.
Ben Fassett - The Foxboro Company
Cyrus Taft - Southern Co. Services
Phil Fredricks - United Eng. & Const.
Charlie Menton - Westinghouse
James Rovnak - Stone & Webster
Jack Gabor - Westinghouse
Marjorie Widmeyer - WPPS
Bernie Traylor - Sandwell
Richard Roop - Hoosier Energy
P.Y. Keskar - Sandwell
Wayne Holland - Southern Co. Services
W.H. Radke - Souther Co. Services
David Tennant - Georgia Power
G.D. Debban - Bailey Controls
Duane Cao - EBASCO
Robert Larsen - Honeywell
Frank Ryan - Leeds & Northrup
Rafael Bossio - EBASCO
Frank Short - Houston Lighting & Power

LUNCHEON ADDRESSES

ISA/EPRI JOINT CONTROLS AND AUTOMATION CONFERENCE
ST. PETERSBURG BEACH
JUNE 3 & 4, 1991

June 3rd Abstract of Presentation by:
John A. Hancock
Senior V.P., Power Operations
Florida Power Corporation

In order to properly address, resolve and implement the changes to meet the requirements of the future, in the area of electric energy, a partnership of the major players is required. These major players are the electric utilities, the architect engineers, the manufacturers and EPRI.

There is a strong probability that utilities will be asked in the future to, in addition to providing reliable, quality, and quantity power, play strong roles in the solution of potential future problems in areas not conventionally connected to power production. Potable water supply through desalination might very well be one of the areas where electric utilities will be called to be the key players. The partnership needs to start gearing their efforts towards addressing these potential future needs.

Other areas, such as Demand Side Management, Alternative generation technologies such as Solar, Fuel Cells, Advanced Nuclear, Coal Gasification, Hydrogen Production, Electric Vehicles, Battery Development, Superconductivity, etc. require the partnership's strong interest and support.

June 4, 1991 Speaker was:
Dr. Jack Critchfield
Chief Executive Officer
Florida Progress Corporation

Dr. Critchfield's talk dealt with the subject of Education at all levels. He emphasized the need for industry and all of us as individuals to take a more active role in America's educational programs.

Dr. Critchfield was an excellent speaker and his message was outstanding and well received.

TABLE OF CONTENTS

0074-056X/91/001-028/$0 + .50pp

FOSSIL POWER PLANT AUTOMATION: ISSUES AND FUTURE TRENDS

S. Murthy Divakaruni, Program Manager, Controls & Diagnostics
George Touchton, Program Manager, Fossil Plant Operations

Present at:
ISA/EPRI Joint Controls and Automation Conference
St. Petersburg, Florida
June 3-5, 1991

INTRODUCTION

Computer information systems will continue to revolutionize the utility business in the decade ahead and help the electric power industry to focus on managing generation assets to seek maximum returns on their capital investment and to improve bottom-line profitability.

Major issues facing the utilities today are: plant aging, financial burdens on new plant construction, pressures from PUCs to meet performance standards, competition from IPPs and other utilities, and acid rain legislation compliance. This business climate, which is unlike any previously seen by regulated monopolies, demands utility managers to focus on improving plant efficiency and availability, and on generating the lowest cost electricity in the next 10-20 years with the most appropriate combination of old and new generation. The solution to all these concerns, as always, revolves around technology and trained personnel.

The utility instrumentation and control (I&C) engineers today perform many system integration functions and this new role is only likely to expand in the future as more and more computer systems are installed in power plant control rooms. Integration of diagnostic and performance monitoring systems will become a common element of utility control upgrade projects. Many new computer-based decision support and maintenance systems will emerge.

This paper elaborates on issues facing the industry and seeks to address how new computer-based control and automation technologies resulting from recent microprocessor evolution, can improve fossil plant operations and maintenance. This in turn can assist utilities to emerge stronger from the challenges ahead. Many presentations at the first ISA/EPRI co-sponsored conference are targeted towards improving the use of computer and control systems in the fossil and nuclear power plants and we believe this to be the right forum to share our ideas.

THE UTILITY CHALLENGE

Maintaining profitability in the face of increased environmental legislation and performance standards will be the major challenge of the electric utilities in the years ahead. According to the National Energy Strategy (ref. 1), the most abundant fuel, coal, will continue to be the primary fuel for new capacity additions at an average rate of 30 GW/yr between 1990-2000. (This does not take into account the 124,000 MW of current capacity which will be reaching 30-40 years of plant life and requiring retirement or upgrade.) However, as shown in Figure 1, nuclear power and renewable energy will become more viable sources for capacity additions beyond the year 2000. Two major points are worth noting before we elaborate on issues facing the power industry:

- The utilities have to commit to new large fossil construction projects in the next 2-3 years in order to allow for a two year licensing and five year construction period if they are to meet additional capacity requirements by the turn of the century. The new power plants will be more efficient, thus requiring integrated plant performance and diagnostic monitoring systems with control systems as well as seeking methods to reduce O&M and capital costs.

- For the nuclear energy resurgence to occur, a single step licensing process must evolve and the embedded technology must be proven. Fault-tolerant protection and control systems with established software validation and verification (V&V) practices will help the process. A close working relationship with the fossil power industry and demonstration of new control technologies in operating fossil plants will facilitate the process further.

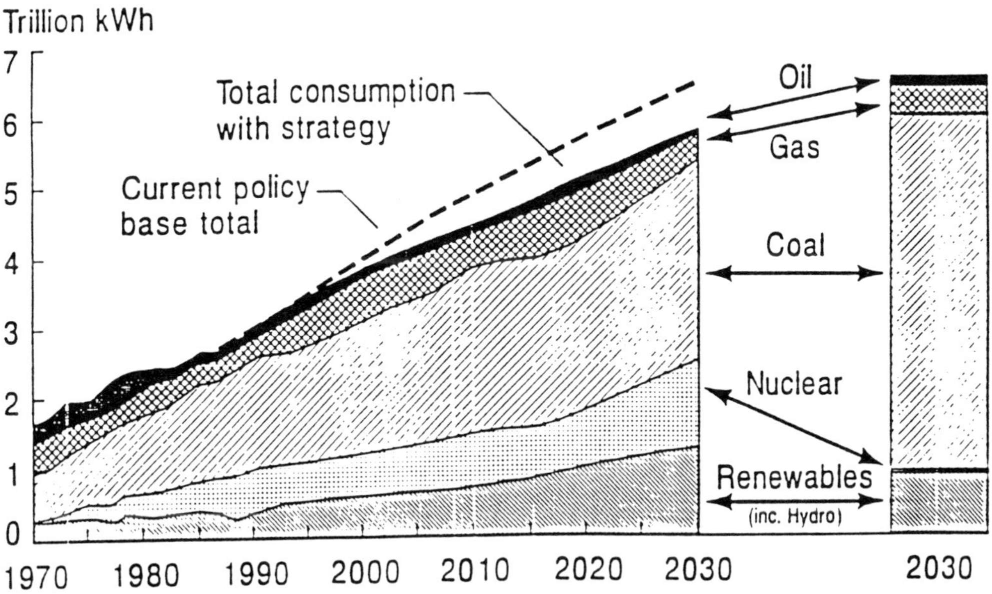

Figure 1: National Energy Strategy Electric Generation

The fossil power industry can also benefit immensely from the nuclear power industry experience, especially in addressing the human factors related issues in the control rooms and in upgrading them with full CRT-based systems. The fossil industry has not effectively addressed personnel related problems in the power plants very effectively, which in turn has contributed to equipment unavailability. The introduction of computer-based operator support systems and new digital systems may give rise to more errors if human factors are not properly addressed. Human factors related experiences gained in the nuclear industry with the safety parameter display system (SPDS) will be valuable for the fossil I&C engineers.

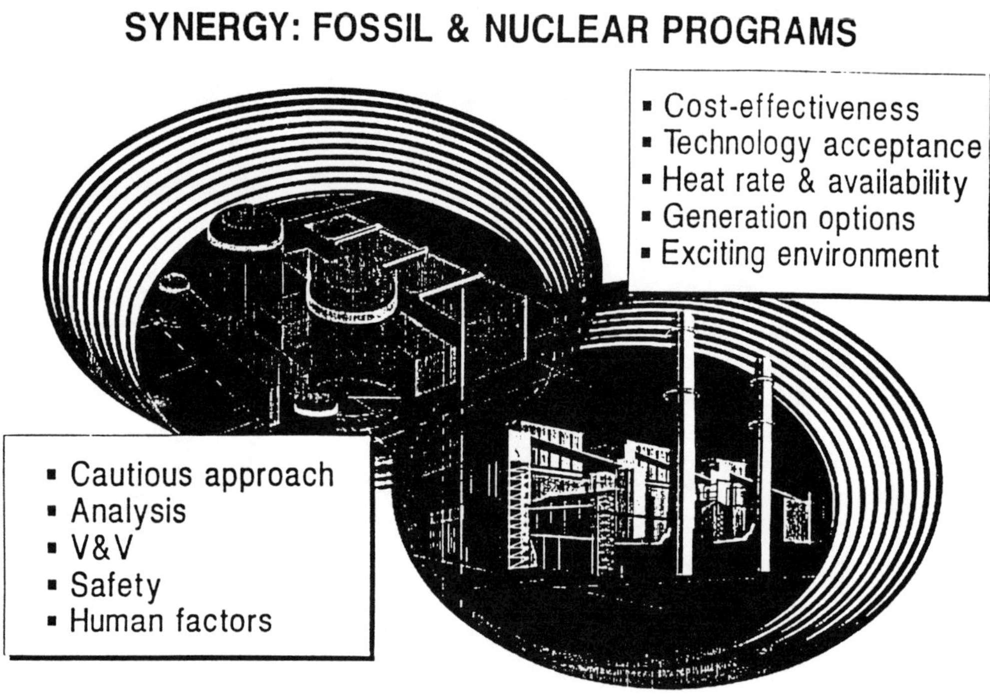

Figure 2: **Synergy between nuclear and fossil plants**

State of the Fossil Power Industry

It is important to step back and see what is happening in the fossil power generation industry to further assess the opportunity for I&C and automation engineers to help their utilities meet the challenges in this decade.

Aging Power Plants

By the year 2000, more than 50% of the U.S. fossil generating stations will be 30 years old, requiring upgrades in life optimization programs (Figure 3). Integration of new computer-based life optimization software, as well as a myriad of other diagnostic programs with distributed digital control hardware platforms, are necessary for utility engineers and operators to fully benefit from the enormous investment in the control retrofit projects and to keep the units running (Ref. 2). Figure 4 shows very clearly that the average power plant forced outage time is increasing from the conventional 60 days duration to over 100 days, and most of it can be attributed to aging plant systems and components. Without a well-established life optimization strategy, integrated with a powerful distributed controls platform and a training program, it will be difficult for utilities to maintain an aging fleet of fossil units at high availability (Ref. 3).

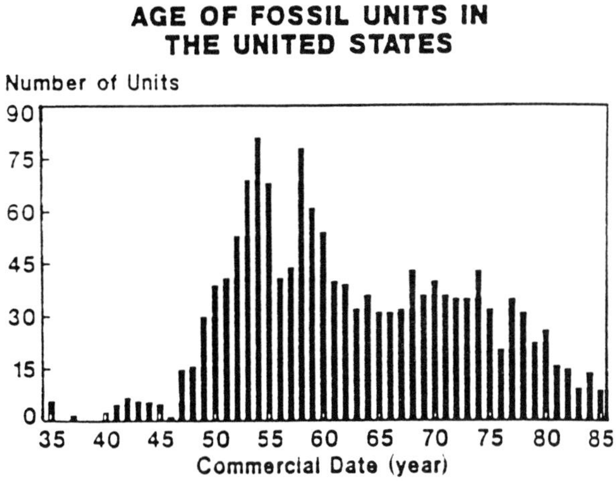

Figure 3: Aging distribution chart of the 1200 fossil units greater than 50 MW shows that by the mid 1990s the vast majority will be over 20 years old. For maximum benefit from control retrofits, utilities must plan on using computational horsepower of the digital systems to help operators assess remaining life of critical power plant components.

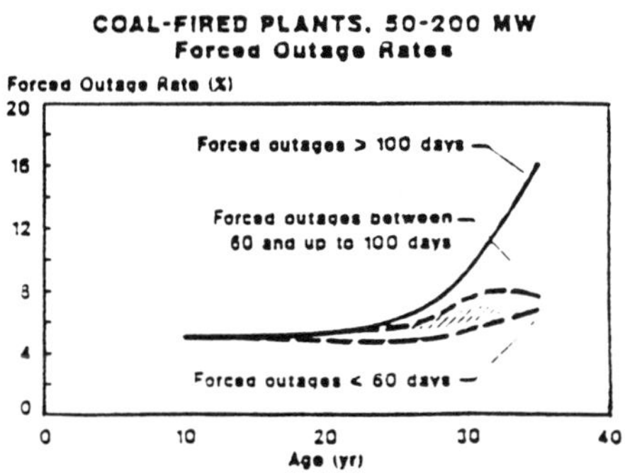

Figure 4: The average forced outage time is increasing. Without an integrated controls, diagnostics and performance monitoring program, it will be difficult to maintain high availability and low heat rates.

Environmental Compliance

In addition, the recent Clean Air Act Amendment lays out new emission standards requiring a 10-million-ton SO_2 and a 2-million-ton NO_x reduction by the year 2000 from existing facilities (Ref. 4 and 5). Scrubber operations, especially in the 21 states shown in Figure 5, will require training of operators to maintain the emission limits, and to some extent, even generate revenues for the operating company by meeting the limits (Ref. 6). Consequently, the role of continuous emission monitoring (CEM) system has also become very important and is expanding as more and more stack pollutants are measured in the power plants today. In addition to opacity, SO_2, NO_x and or CO_2, new constituents like hydrogen chloride, CO and NH_3 are being measured depending on the region and the type of the power plants. An August 1989 article in Power Magazine summarizes current problems with CEMs and recommends ways to integrate CEMs with the process control equipment (Ref. 7). The on-line coal analyzers, scrubber controls and CEM perform complementary functions and the industry must find ways to integrate them with the distributed digital controls effectively, even considering closed loop control of constituents such as NH_3, O_2 where possible. Some examples of emission monitoring systems integration with process control functions are given in figures 6 and 7 (Ref. 7). This needs to evolve as an inherent part of the overall fossil plant training program (Ref. 8).

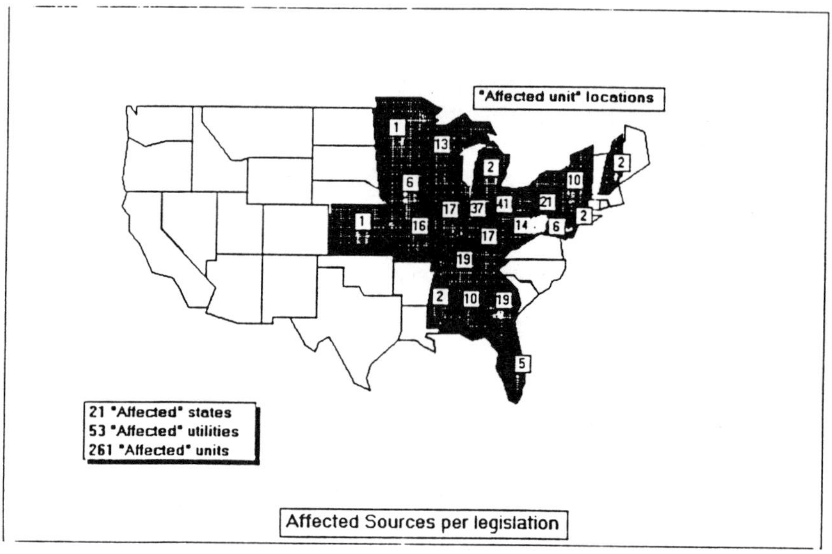

Figure 5: **Most Midwestern and Eastern utilities are affected by the recent Clean Air Act Amendment. Operator training to meet environmental control limits should be an integral part of a comprehensive utility training program.**

Manpower Needs and Training

By the time the new efficient power plants come on-line at the end of this decade, we will see a new generation of operators working in the control room, operators who have grown up with computers at home. The same will be true in the case of a new breed of engineers who would be experts at working with computers.

As power plants become more complex, automation will play a larger role. The power plant of the future may, if coal-based, resemble more of a "coal refinery"

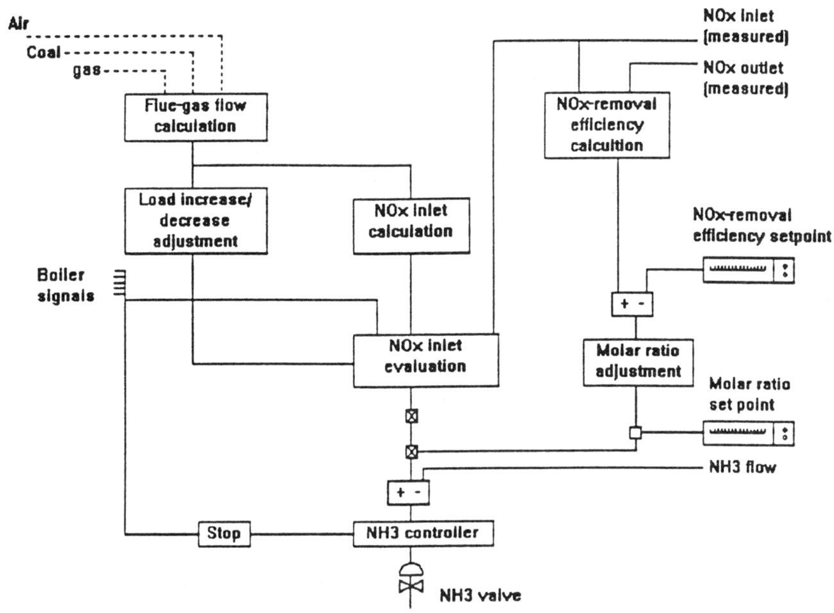

Figure 6: SCR control system can be interfaced with plant process control systems (Ref. 7).

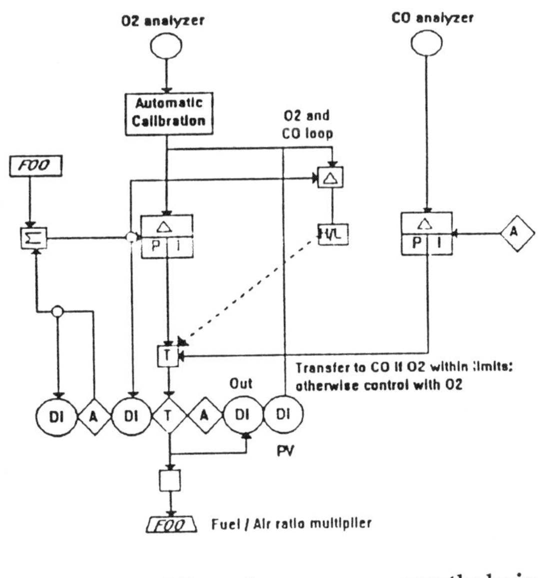

Figure 7: On-line coal analyzers and CO analyzers are currently being incorporated into boiler control schemes. Closed-loop control that optimizes combustion efficiency may be achieved in the future (Ref. 7).

where besides electricity, products will include heat, liquid and gaseous fuel, sulfur products, and a variety of salvaged metals and compounds. Simulation of such processes is one way to effectively verify the design concepts and to train and upgrade plant personnel, and some innovative steps have already been taken in this direction including the use of interactive "intelligent training" for such facilities as coal gasification plants (Ref. 5). Capturing knowledge in a way that can be used for manpower upgrading can be extended into expert systems that embody the skills and experience of key, long-term personnel. One concern that has often been expressed is that automation is a bar to understanding, the implication being that as we move to more computer controlled operation of power plants (a necessity as the process becomes more complex), the human in charge is less cognizant of the physical principles behind plant operation. This is an undesirable situation that could lead to poor decisions in emergency situation. So training, via simulators, expert systems or intelligent tutors, will be an essential growth activity for most forward-looking utilities (Ref. 5, 6).

Efficiency Improvements

Plant performance standards are a growing aspect of utility/PUC interactions, and most utilities see heat rate improvement as a major consideration in producing low cost electricity. There are many incentives for improving heat rates: lower heat rates mean reduced fuel consumption and reduced emissions. For example, just 1% improvement in heat rate can save the U.S. utility industry over $350 million per year in fuel costs, and due to less fuel burned per unit output, emissions are reduced, especially SO2, NOx and particulates. A 3% reduction in heat rate of existing coal-fired plants has the potential of reducing U.S. SO2 emissions by nearly half a million tons per year (Ref. 9). A knowledge of what parameters affect heat rate and the ability to precisely control them are both extremely important. Development of accurate measuring devices are critical to measuring these parameters. For example, good flame monitoring devices and consumption optimization methods will provide better means to control the burners in the power plants. Automation of parameters effecting controllable losses and intelligent on-line diagnosis of causes for heat rate deviations may produce significant benefits to the industry.

Cyclic Duty Operation

Although this issue is less important than a few years back, cyclic duty operation of existing fossil units poses challenging problems to control system designers. If the control systems are not designed or tuned to handle transient or cyclic operation of fossil plants, availability and heat rates are severely impacted, and can result in severe damage of equipment in extreme cases. On-line stress monitors and dynamic simulators can further help assess the condition of the plant and to estimate in real-time start up costs and life expenditure costs. New control algorithms that take into account the stresses in thick-walled components can be implemented automatically to schedule the ramp rates during the startup. EPRI's low-cost cycling advisor software is currently being evaluated for on-line application at a host utility (Ref. 3).

Variable pressure operation improves cycling of fossil power plants. Further, it lowers the heat rate and reduces turbine stresses at part load. EPRI's variable pressure operation program covers both sub- and supercritical units, and is aimed at working with host utilities to demonstrate the cycling guidelines for variable pressure operation in existing units (Ref. 9, 10). Cycling controllers being developed by some U.S. control suppliers can be set to maintain the rate of change of controlled parameters at a desired level. Instead of being hardwired into the controller set up, the rate is a variable whose value can change continuously in run time according to user demands (Ref. 11). These types of new controllers will facilitate cyclic duty operation of the fossil units and yet maintain the heat rates within the allowable operational limits of the major plant equipment.

EPRI R&D PROGRAM GOALS

EPRI's fossil power plant R&D objectives and strategic targets for both new and existing plants and are summarized below in Table 1. EPRI is developing the technology required to meet these targets by the year 2000 and believes these goals can be realized through the application of improved designs, new components and materials, improved equipment monitoring and control, and application of computer-based information systems and decision support methods (Ref. 12). The key elements that are of significance to this audience are illustrated in Figure 8.

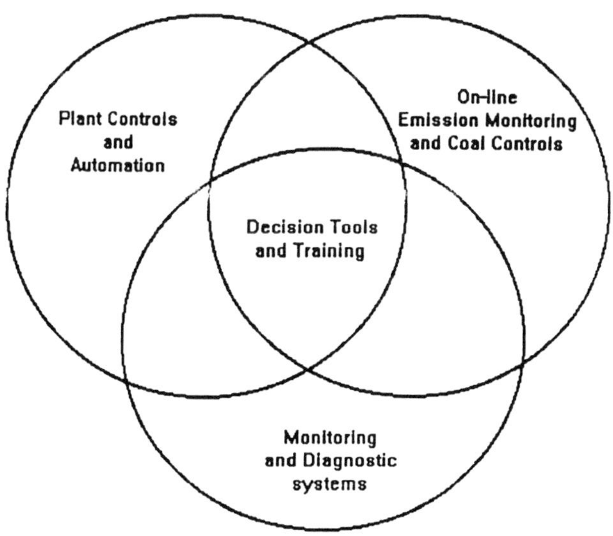

Figure 8: Key elements for integrated fossil plant automation

Table 1: EPRI's R&D Goals for Fossil Power Plants

Objective:	Develop and demonstrate advanced fossil technologies (coal, oil, and gas) that are cost-competitive, environmentally sound and promise significant energy efficiency improvements.
Strategic Target:	By 1995, produce tools to help utilities confidently specify and build 2300°F-CT combined cycle plants (<7200 Btu/kWh heat rate) and high efficiency state-of-the-art steam plants (<8500 Btu/kWh). By 2003 provide technology to build advanced, fossil-fired combined cycle units with <6500 Btu/kWh heat rate, 95% (gas) and 88% (coal) availability, 24-month (coal) construction time, and plant costs <8500/kW (coal), while meeting all environmental regulations.
Objective:	Improve the efficiency, availability, safety and cost of existing fossil facilities (coal, oil and gas).
Strategic Target:	Produce tools by 1995 to improve fossil steam plant heat rate and availability by 2-5% over 1990 levels and to reduce non-fuel O&M cost by 25%; and sustain those improvements through the year 2000. Demonstrate in at least 10 utilities. Produce tools by 1999 to cost effectively increase existing combustion turbine starting reliability to 97% and time between planned outages by 100%. Demonstrate in at least 5 utilities.
Objective:	Develop strategies for enhancing overall business efficiency.
Strategic Target:	By 1995, produce tools for optimum utilization of fossil plants in the overall utility business plan. Complete demonstrations at 5 utilities of strategic decision tools for asset management (including plant staffing, maintenance budgets, and environmental compliance options) which factor in heat rate and availability goals, plant remaining life, long-term fuel contracts, environmental dispatch, emission allowance trading, new generation plans, and purchased power options.

The goals of the Fossil Plant Operations program is to harness the power of microprocessor evolution and to apply computer control and information technology to enhance the overall productivity in the fossil plants.

COMPUTER POWER FOR FOSSIL PLANTS

Individual Computer Power Takes Off

The speed of individual computers has increased exponentially since their introduction, as Figure 9 shows. This is true of all computers from supercomputers to low priced personal computers. According to this figure, the rate of growth is forecast to slow after the turn of the century – a reasonable project based on fixed technology. However, many in the industry feel that this growth in processor power will accelerate, not slow, as new developments overcome each potential barrier. For example, the projections shown are for serial processors, but today parallel processing is emerging to push back the frontiers of speed and power. So end-users may prudently plan for hardware capabilities which will meet and exceed their needs. Additionally, the phenomenal growth in computing power has unveiled a whole new universe of technology needs, opportunities, and barriers.

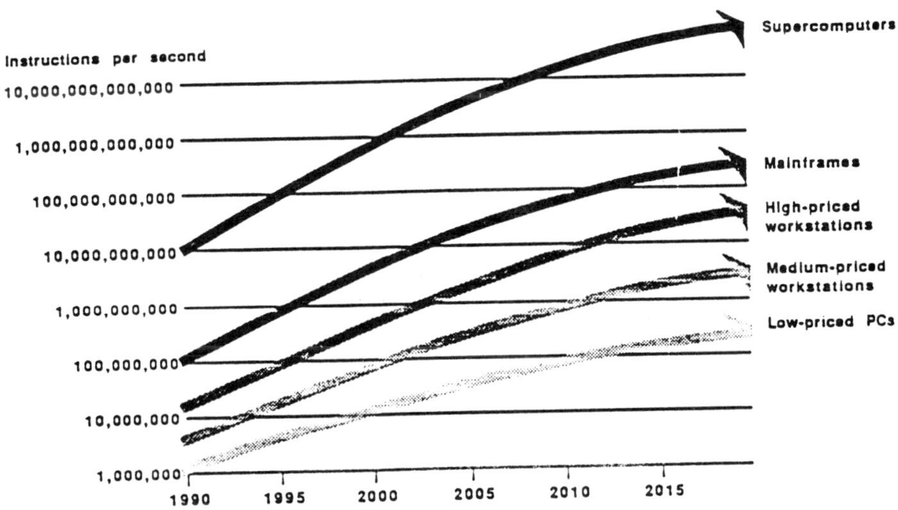

Figure 9: **The processing speed of personal computers, workstations, mainframes, and supercomputers has grown exponentially and is expected to continue to grow for the next three decades.**

Computers have affected the way the utilities do business in every facet of their operations. Automated computer-aided engineering and design (CAE/CAD) tools, have already resulted in productivity increases as high as 100 to 1 and engineers are now able to build logic ladder diagrams using data received from programmable controllers in a matter of two weeks instead of six months time. In another situation, a CAD/CAE software has decreased the time between receiving an order and releasing it for manufacture of a sootblower by a factor of five (Ref. 13). The key to such productivity gains cited in the literature is the creation and use of central database.

The sophistication and progress in power plant application software development is very encouraging. As the utilities gain more experience in using computer software for both on-line and off-line applications and as the benefits of this become significant , we will see a further explosion in application software development. The distributed digital control system hardware, which is mostly being used for data acquisition and digital control of PID loops, will use these sophisticated application packages to operate and maintain power plants. The systems to be integrated with control platforms will include on-line coal-quality software and analyzers, diagnostic vibration and performance monitors, incremental heat rate dispatch algorithms, what-if analyzers and expert systems, new instrumentation and closed-loop control of emission monitors. The real gain in productivity will be realized through inherent predictive software routines for controlling the parameters that effect the plant performance throughout the load range. New technologies such as the fuzzy logic and neural networking schemes and knowledge-based systems to perform traditional control functions will penetrate the utility operations in the decade ahead (Ref. 14-19).

Collaborative R&D with Vendors and Utilities

EPRI has invested millions of dollars in research and development of heat rate improvement, life optimization, monitoring and diagnostics, environmental control systems and improved coal-fired power plant technologies. Most of these technologies have demonstrated benefits at many host utility sites during the R&D phase of the projects, and vendors and architect engineers often use and refer to these technologies while working with member utilities.

The control system hardware suppliers simultaneously are also working on integrating diagnostic and performance monitoring systems as more and more utilities are requiring integrated systems as a part of planned life optimization or control retrofit programs. EPRI is working with most major vendors to develop "value-added" enhancements to the control system algorithms. EPRI is beginning to get into contracting arrangements with the suppliers to allow its member utilities to have an access to EPRI software through the distributed control platform. EPRI member utilities may expect to have an EPRI "advisory station" in the control room in the future, as a part of vendor supplied hardware providing performance updates, equipment diagnosis and recommended corrective actions to control room operators. Utility invested R&D results in the latest technology to improve plant performance, availability or plant operations will find their way to a place where it counts most and will stimulate vendor R&D in return. Collaborative R&D with vendors and host utilities is a win/win/win proposition for all the three parties as the vendors benefit by improving performance of their product lines and yet keep the proprietary data in tact; host utilities benefit by installing the latest technology to meet plant performance goals and utilize the distributed controls horsepower; and EPRI benefits by enhanced technology transfer through established vendor marketing channels and demonstrating EPRI membership value where it counts most viz., at the power plants. The combined effort on the part of these parties to define the product and its benefits is likely to reap rewards in terms of using the information technology for power industry applications. Figure 10 shows some of the products EPRI is developing with the vendors.

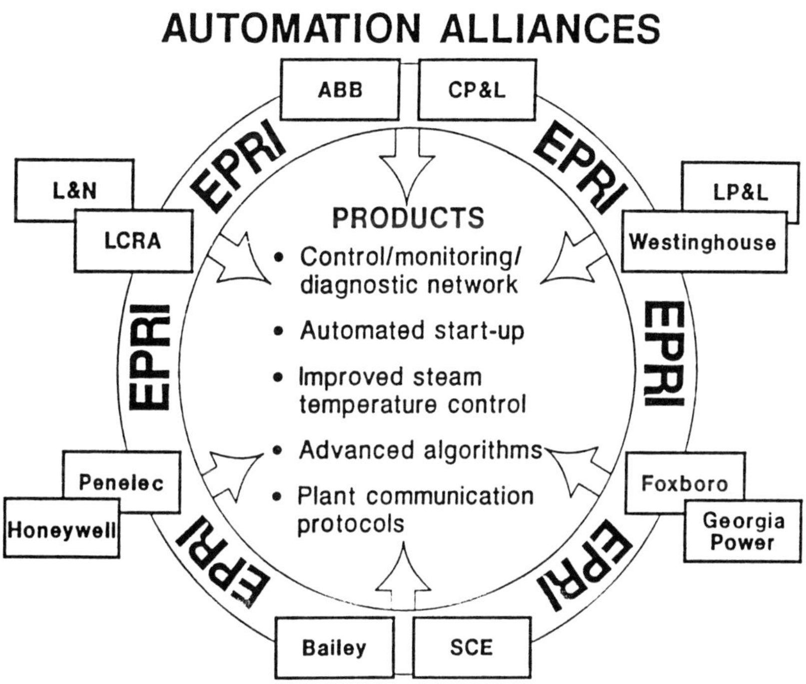

Figure 10: EPRI is forming alliances with control suppliers and utilities to jointly develop 'value - added' products.

COMPUTER-INTEGRATED FOSSIL PLANT AUTOMATION

Primarily, distributed computing techniques have significant impact on fossil plant operations in five areas: 1) controls and automation enhancements, 2) diagnostic and monitoring systems, 3) emission control systems, 4) decision-support methods and 5) personnel training and advanced technology transfer. EPRI has an established major R&D program in each of these areas and many projects are currently on-going in these technical disciplines (Ref. 20).

Controls Enhancement and Automation Systems: Aimed at improving plant performance by designing fault-tolerant and intelligent control systems that incorporate inherent plant dynamics and process prediction capabilities. Automation systems are designed to improve overall productivity of plant and engineering personnel through use of CASE and CAD/CAE tools, integration with AGC and dispatch centers and well-designed advanced control rooms.

Monitoring and Diagnostic Systems: Performance monitoring systems are aimed at improving the unit heat rates throughout the load range and dispatch of units based on incremental heat rates. Diagnostic systems are designed to demonstrate improvement in plant availability through selection and use of equipment monitoring systems to improve the unit availability.

Emission Control Systems: Aimed at meeting the acid rain legislation at the lowest cost and to integrate traditional emission monitoring systems with distributed plant controls and to move towards automation of emission control systems.

Decision Support Methods for Assets Management: Help utilities develop an integrated approach to allocating resources to meet corporate objectives across the company. These will include tools to make decisions on run/repair/replace based on the availability equipment condition, as well as plant managers work bench to analyze maintenance planning (Ref. 21).

Personnel Training and Advanced Technology Transfer: Assist utilities in making a smooth transition to use computer technology most efficiently. A client/server based information network will allow utilities to access the information readily from various sources: vendors, EPRI and demonstrating host sites to evaluate developing technologies expeditiously and from remote sites.

Among the five major areas that contribute towards integrated fossil plant operation, only the controls and automation and the monitoring and diagnostic systems are discussed somewhat in detail in the following paragraphs.

CONTROL AND AUTOMATION SYSTEMS

Utilities spend about $400 million/year in control retrofits and about $2-$5 million/year on computer system and information systems. The primary focus of EPRI R&D program in control systems is to enhance the state-of-the-art of the algorithms being used by the utilities in changing to digital distributed systems, as shown in Figure 11. To utilize the power offered by these microprocessors, many new adaptive and multi-variable control options as well as supervisory control methods are being evaluated as a part of the controls R&D program. As a preliminary step, EPRI produced a three volume control retrofit guidelines to assist utilities in planning their upgrades. These three volumes address: 1) cost justification of control retrofits, 2) technical assessment for control upgrades and 3) case studies of successful utility upgrades.

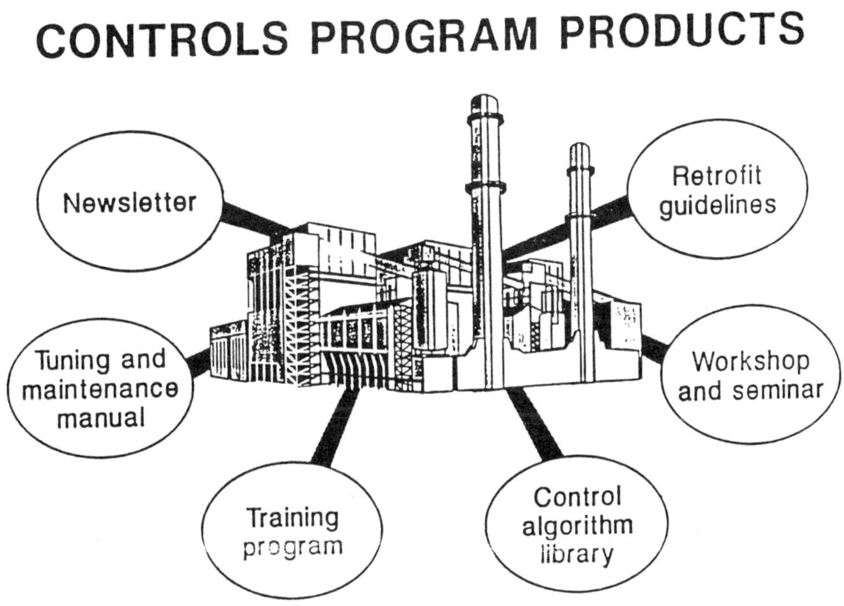

Figure 11: EPRI's short-term R&D focus is to develop products to assist utility control retrofit programs.

EPRI is also addressing the needs of the utilities who want to maintain and enhance the analog systems modifications. Innovative tuning techniques utilizing a model-based approach are also planned to assist utility control engineers in improving control gains for the control loops.

The major emphasis of EPRI's control program is on the algorithm development. Four new algorithms have been investigated to date:

- State Variable Estimation Algorithm - Uses a model of the superheater to better control the outlet temperature of the heat exchanger. The state variable estimator has a self-correcting action that allow latitude in tuning constants without seriously affecting the system stability or performance.

- Nonlinear Process Model-Based Algorithm - Uses a nonlinear approximate model directly for control purposes. This method provides nonlinear feed-forward performance for changes in process throughout, nonlinear decoupling of manipulated variables and nonlinear feedback control.

- Optimal regulator for steam temperature control - Predicts future behavior of the process variables using the state equation for the control purposes. Using the plant response data, an auto regressive (AR) model is generated and validated. An optimal feedback gain matrix is determined using dynamic programming techniques. This regulator design has been demonstrated to have improved plant performance in five Japanese power plants.

- Concurrent front-end controller - Performs a supervisory control function in a multi-mode operation. Performance optimization as well as control objectives are accomplished during startup/shutdown, normal operation and transients using the rate optimal controllers (ROCs). This uses the cycling controllers outlined earlier in the paper.

  EPRI has embarked on many integration projects to demonstrate the benefits of automation technologies:

- The El Segundo Demonstration - In this project with Southern California Edison, diagnostic and performance monitoring, on-line advisory systems and advanced controls are integrated at El Segundo power plant units 3 & 4. The project is unique in that it has required an industry control system supplier to integrate distinct diagnostic monitoring systems from various vendors with the control systems. Guidelines from the El Segundo project will facilitate integration in subsequent power plant control retrofit projects (Ref. 22).

- The Roxboro Demonstration - This project with Carolina Power & Light will move integration to the next step through utilization of on-line testing, 'what-if' type expert systems, imported European proven advanced control algorithms and information management systems into a plant wide automation system based on EPRI's utility communications architecture (UCA) and ABB's plant network schemes (Ref. 19).

- East Lake Units Demonstration - Centerior Energy's East Lake units site will be the first host utility to demonstrate EPRI's control retrofit and man-machine interface guidelines. Interfaces to a training simulator and continuous emission monitoring systems will be other aspects of this control retrofit and integration project.

Future Challenges in Controls & Automation

EPRI's efforts in controls and automation are in their infancy and many new technical challenges are ahead. In addition to fully automated plants, new technologies that include fuzzy-logic, neural networking schemes and closed-loop intelligent controls are already finding their way into Japanese and European power plants (Ref. 23-26). EPRI will be working with the U.S. utilities and the control suppliers to adapt selected Japanese and European proven methods in the U.S. operating fossil power plants. The Japanese progress in developing these advanced technologies and implementing them in operating power plants has been impressive.

Most newly built plants in Japan have introduced totally computerized and distributed control systems, and they generally include the typical functions of plant automation, data acquisition, boiler control, burner management, turbine control, and automated environmental control systems. Table 2 lists seven such units in operation since 1983. Each of these was integrated into a single plant-wide system and has contributed to the plant automation, swing load operation, heat rate and availability improvements (Ref. 23). Table 3 lists major needs seen by plant personnel and how they are achieved in the paper by Hitachi Ltd.

In another application by Hitachi (Ref. 24), an expert system for optimum startup scheduling is developed and is modified on-line using fuzzy reasoning to reduce the operator workload. The functional structure of the system is shown in Figure 12. Through appropriate man-machine interface and interactive graphic displays, fuzzy reasoning is introduced to represent the way experts think, and to utilize the knowledge-base to improve schedule optimization.

A control system developed by IHI can automatically and optimally modify its own control parameters according to the used coal characteristics when the coal is changed. As shown in Figure 13, the control system modifies its own control parameters based on the input information of coal characteristics and feedback information of the plant state variables (Ref. 25).

EPRI is working with Leeds and Northrup Company to use a neural network based boiler learning model to optimize boiler performance. In another effort under EPRI funding, a software package for coal quality evaluation system (C-QEUL) which is highly modular is being developed, promising potential integration in the future with the control hardware (Ref. 8).

TABLE 2: Total computerized, distributed control systems in service in Japan (Ref 23).

Utility	Station	Type	In Service
Electric power Development Co.	Takehara #3, 700 MW	Once-through Coal	Mar-83
Hokkaido Electric Power Co.	Tomato-Atsuma #2, 600 MW	Once-through (Sliding press) Coal	Oct-85
Tohoku Electric Power Co.	Higashi-Niigata #3, 1090 MW	Combined cycle LNG	Dec-85
Chuugoku Electric Power Co.	Shin-Onoda #1, #2, 500 MW x 2	Once-through Coal	Apr-86 Feb-87
Electric Power Development Co.	Ishikawa #1, #2, 156 MW x 2	Once-through (Sliding press) Coal	Nov-86 Feb-87

Table 3: Distributed Digital Control Application Details in Japanese Fossil Power Plants (Ref. 23)

Plant Needs	Related Technologies	System Configuration
Plant automation	• Automatic plant startup and shutdown control	• Totally computerized monitoring and control system
Cycling operation	• Steam temp. prediction control	• A microprocessor; stand-alone or connected on μΣ NETWORK
Plant life extension	• Thermal stress monitoring and control for turbine and steam pipes • Thermal stress monitoring for boiler	• Added in a turbine controller or process computer • A microcomputer; stand-alone or connected on μΣ NETWORK
Low NO_x operation	• Fuel/air ratio control at microscopic level	• Coordinate modulating with logic control in the subsystem microcomputer
Heat rate improvement	• Above control and operation • Sliding pressure control and major auxiliaries variable speed control	• Above configurations • Added in each subsystem control microcomputer
Availability improvement	• Diagnostics of control system • Non-stop control	• Added in each subsystem control microcomputer • Redundancy of controller/and or CPU

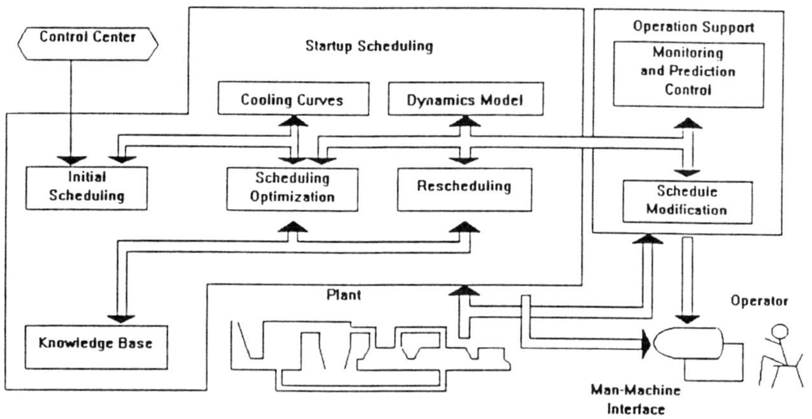

Figure 12: Functional Structure of a fuzzy - logic based Startup system developed by Hitachi Ltd (Ref 24)

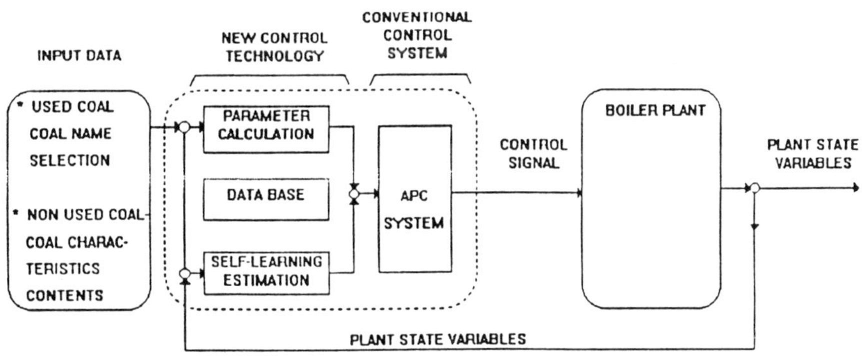

Figure 13: An Adaptive control system developed by IHI for Coal Switching (Ref 25.)

MONITORING AND DIAGNOSTIC SYSTEMS

On-line diagnostic monitoring helps utilities reduce outage time with an estimated 10-30% reduction in the typical multimillion-dollar annual plant maintenance budget. Savings in outage time and repair costs frequently amount to millions of dollars, as reported by the utilities at the Incipient Failure Detection Conference. EPRI conservatively estimates that the across-the-board application of monitoring technology could enhance plant availability at least 2%, bringing annual savings of more than $400 million to the industry (Ref. 27). Diagnostic monitoring and inspection equipment are very diverse and use a broad spectrum of technologies. Tables 4 and 5 give a synopsis of benefits accrued by various utilities in using diagnostic monitoring systems (Ref. 27, 28).

Table 4: Utility Experience in Incipient Failure Detection (Ref. 27) *

Vibration Monitoring and Vibration Signature Analysis

Central Electricity Generating Board (Great Britain) (3)
Electricite de France (36)
Detroit Edison (4),(11),(35)
New England Power (5)
United Illuminating (6),(10)
Southern California Edison (7)
Ontario Hydro (8)
Pacific Gas & Electric (9)
Ente Nazionale per l'Energia Elettricia (Italy) (12)
Comision Federal de Electricidad (Mexico) (28)

Acoustic Crack Detection

Philadelphia Electric (19)
New England Power (5)
Ohio Edison/Pennsylvania Power (20)

Boiler Creep Monitoring

Central Electricity Generating Board (3)

Utility Monitoring Programs

Middle South Utilities (24)
Georgia Power (25)
San Diego Gas & Electric (26),(27)
Electricite de France (29)
Philadelphia Electric (30)
Detroit Edison (11)

Acoustic Leak Detection

Central Electricity Generating Board (3),(33)
New England Power (5)
Ente Nazionale per l'Energia Elettricia (13),(17),(34)
Philadelphia Electric (14),(16),(18)
Public Service Electric & Gas (15)
Ohio Edison (16)
Cleveland Electric Illuminating (16)
Salt River Project (16)
Texas Utililities Electric (16)
Baltimore Gas & Electric (16)

Rotor Crack Detection

New England Power (5)
Tennessee Valley Authority (21)
Florida Power and Light (22)

Fatigue Damage Monitoring

Southern California Edison (23)

Generator and Electrical Equipment Monitoring

Electricite de France (29)
Philadelphia Electric (30)
Alleghany Power System (31)
Arizona Public Service (31)
Central Hudson Gas & Electric (31)
Intermountain Power Agency (31)
Ontario Hydro (31)
Salt River Project (31)
Southern California Edison (31)
Tennessee Valley Authority (31)
Texas Utilities Electric (31),(32)
Indianapolis Power & Light (37)

* The numbers in paranthesis are the references given in the paper by M.A. Blanco et.al (Ref. 27).

Table 5: EPRI First Uses: Documented Savings by Utilities Through Use of On-line Diagnostic Monitoring. (Ref 27,28)

DATE	TECHNOLOGY	UTILITY	ANNUAL SAVING
12/80	Acoustic Monitoring of Nuclear Plant Relief Valve Position	Philadelphia Electric	$651,000
12/80	On-Line Monitors for Electric Generators	TU Electric	$287,700
12/83	Turbine Generator Shaft Torsional Monitoring	Southern California Edison	$746,000
12/85	On-Line Turbine Vibration Monitoring System	United Illuminating	$1,016,000
10/86	On-Line Monitoring of Solid Particle Erosion	Dayton Power & Light	$137,800
4/87	Laser Doppler Vibrometer	United Illuminating	$42,000
12/87	Acoustic Boiler Tube Leak Detection System	Philadelphia Electric	$480,000
4/88	New Crack Detection Technique for Pump Shafts	General Public Utilities Nuclear Corporation	$748,000
8/88	Boiler Stress and Condition Analyzer	Consolidated Edison	$659,000
11/88	High Frequency Vibration Monitoring for Fans	Pennsylvania Electric	$245,000
12/88	Automated Vibration Monitoring of Rotating Machinery	Pacific Gas & Electric	$1,438,000
12/88	On-Line Transient Monitor for Tracking Fatigue	Southern California Edison	$165,000 *
10/89	Thermal Optical Scanner to Detect and Diagnose Generator Hot Spots	Seattle City Light	$4,738,000 **
11/89	Continuous Valve-Leak Detection System	Philadelphia Electric	$393,000
3/90	Hydrostatic Acoustic Emission Technique for Inspection of Reheat Piping	Pennsylvania Power & Light	$500,000-900,000 *

* First Year Savings
** Total Savings

Diagnostic Monitoring: Promise vs. Reality

If utilities are to reap the potential benefit of diagnostic monitoring, systems must be available that perform reliably and well at sensor data acquisition, fault detection, and diagnosis levels. For their part, utilities must integrate diagnostic monitoring into predictive maintenance programs. Many practical considerations, such as sensor calibration and signal validation, software integrity, personnel training, and maintenance management structure must be taken into account in setting up a predictive maintenance program. In nuclear plants, we have an abundance of system and equipment data, but often lack the knowledge to tie what we measure to equipment condition in an unambiguous way. Vibration monitoring of reactor coolant pumps is a case in point. EPRI believes that the following concerns at each level of diagnostic monitoring systems must be addressed before the widespread, successful application of predictive maintenance becomes a reality:

Level 1: Sensor Data Acquisition

Monitoring system reliability is a major concern here. Sensor maintenance, calibration, and signal validation are a practical necessity requiring continual attention. Widespread use of diagnostic monitoring systems will only occur if technical developments can be translated into reliable, engineered monitoring systems. To ensure high reliability in many systems, sensor fault tolerance and self-checking electronics will be required.

Level 2: Fault Detection and Data Management

As more machines are monitored, increasing quantities of data must be screened rapidly to focus on the critical symptoms of incipient failure. In addition, increasing quantities of historical monitoring data must be stored for subsequent use in trending and severity evaluations. In other words, effective data management is mandatory. Eventually, the required computing and display devices will have to be integrated into a unified diagnostic system designed to avoid overburdening operating staff with data.

Level 3: Diagnosis

To improve availability, accurate timely diagnoses must affect maintenance and operations decision making. Therefore, as utility predictive maintenance programs grow in importance, EPRI expects diagnostic engineering to evolve into a full-time plant maintenance function, warranting a centralized area and staffing to coordinate the review of detailed diagnostic data and equipment-failure histories with the scheduling of predictive maintenance.

Full-Scale Predictive Maintenance: Monitoring & Diagnostic Center

To demonstrate the full potential of the available and emerging monitoring and diagnostic technologies, EPRI and Philadelphia Electric Company (PECo) are integrating diagnostic monitoring systems into a predictive maintenance program for boilers, turbine generators, environmental controls, and balance-of-plant equipment at PECo's Eddystone station.

The goal of advanced plant instrumentation is to integrate machinery control, protection, performance and condition functions into a single coordinated network. With virtually every major plant component permanently wired with sensors and feeding continuous data through a fiber-optic highway, operators and maintenance engineers at Eddystone will be able, for the first time and in real time, to observe the actual effects of operation on the performance and wear of critical parts and subsystems. The signals from several computer monitoring systems are integrated for display and analysis in a predictive maintenance diagnostic center that addresses operator needs. Special hardware and software interfaces have been designed to bring the data from incompatible computer operating systems under a common set of diagnostic display terminals.

The Monitoring & Diagnostic (M&D) Center at Philadelphia Electric is an industry resource for training in monitoring technology and for guidance on implementing predictive maintenance programs. The M&D Center will continue to assist utilities in developing the multi-disciplined diagnostic engineering talent essential for an effective predictive maintenance program. The Center demonstrates and validates diagnostic sensors, instrumentation, and diagnostic techniques.

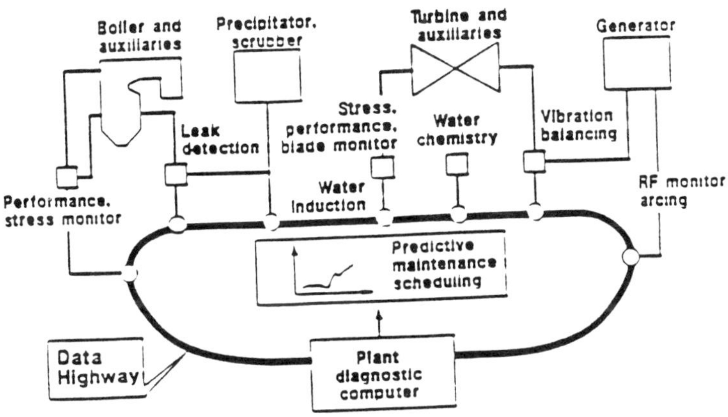

Figure 14: **EPRI and PECo are integrating various diagnostic monitoring systems into a predictive maintenance program for the plant equipment.**

Based on experience at Eddystone and other utility plants, EPRI will prepare guidelines covering the selection and installation of diagnostic monitoring systems, the interpretation of data supplied by those systems, and the incorporation of diagnostic monitoring systems in a plant maintenance strategy.

Future Trends in Monitoring and Diagnostics

EPRI has recently identified a number of plant failure mechanisms (Ref. 27) which still await suitable on-line monitors (see Table 6). In considering what approaches might be taken for new systems, it is helpful to note that:

(1) Sensors that can operate in high-temperature locations are available. These include thermocouples, strain gauges, dynamic pressure sensors and accelerometers.

(2) Extremely powerful sensors, able to detect minute magnetic fields, can be formed from superconductors. Called SQUIDS, these sensors may permit detailed study of damage mechanisms in power plant materials.

(3) Non-contacting measuring instruments — including lasers, infrared devices, non-intrusive and gamma ray detectors — have been used successfully in hostile environments.

(4) There are no significant limitations on the computing ability or the degree of data manipulation that can be built into systems.

(5) Optical sensing techniques which are now under development can provide measurements not currently possible and on a distributed basis.

Table 6: Oppurtunities for new on-line monitors (Ref. 27)

- Direct monitoring and measurement of creep damage in thick wall boiler and turbine components
- Detection of coal ash corrosion of superheater and reheater tubing
- On-line monitoring of acid deposition and corrosion of air heaters and other back-end equipment
- Detection of punching shorts and circulating currents in stator cores
- Monitoring of negative sequence currents and overheating of generator rotor iron
- Tracking of the onset and growth of "alligator" cracking in supercritical boiler water walls
- Detection of the growth and release of exfoliants on the steamside of boiler tubing
- Monitoring of hard particle erosion of turbine blades, nozzles and diaphragms
- Detection of tube leaks in fluidized bed boiler tubing
- Monitoring of creep-fatigue crack growth in steam lines
- Diagnostic monitoring of burner imbalances in boilers
- On-line detection of fouling of cooling tower fills and cooling tower performance deterioration
- On-line monitoring of fouling in condenser tubing

In addition to work on monitoring systems, EPRI will continue to support work on expert systems designed (1) to receive data from sensor-based monitoring and data-collection systems and respond in real time to incipient problems, and (2) to improve incipient-failure forecasting and follow-up action by channeling expertise to power plant operating and maintenance personnel.

CONCLUDING REMARKS

Driven by competitive and regulatory pressures, utilities will seek new methods and technologies to maintain or increase profitability in the 1990s. Utility managers will increasingly turn to computerized techniques and information technology to enhance plant productivity. The top priorities of the U.S. utilities are to improve the operating heat rate and availability of their existing power generation base, and to be prepared with the best available technologies for new capacity additions.

EPRI is responding to utility needs for better information-based technology in the following ways:

- Development of predictive control and automation technologies, working with the major U.S. vendors,
- Demonstration of benefits from integrated monitoring and diagnostic systems,
- Demonstration of comprehensive training programs,
- Integration and improvement of environmental control and monitoring systems,
- Development of decision-making tools to assist utilities with assets management decision.

These strategies will permit the U.S. utilities to be better prepared for competitive operating climate and to maximize company profits in the 1990s. The successful utility of the future will be the one that embraces techniques which are at the cutting edge of this "information" age. Controls and automation engineers will play a significant role in bringing about the necessary cultural change in the utility industry, as they themselves will be taking on the broader challenge of being systems engineers, instead of traditional instrumentation and logic specialists.

ACKNOWLEDGEMENTS

The authors would like to acknowledge the contributions to this paper from M.A.Blanco, J.R.Scheibel, G.Poe, A.F.Armor and S.M.Gehl of EPRI and Ravi Shanker of Indus Technology Inc.

REFERENCES

1. National Energy Strategy, First Edition 1991/1992, February, 1991, Washington, D.C.

3. A. F. Armor, "The Growing Need for Better Fossil Plant controls," EPRI Conference on Fossil Plant Controls, February 24-26, 1987.

4. S. M. Dalton, "The 1990 U.S. Clean Air Act - The Effect of FGD Market and Design," Second International Conference on Desulphurization, Sheffield, England, March 20-21, 1991.

5. A. F. Armor and G. L. Touchton, "The Changing Utility Business Environment," EPRI Conference on Power Plant and Power System Training Simulators and Modeling," April 17-19, 1991, Miami Beach, FL.

6. S. M. Divakaruni, R. Ghosh, "The Role of Training and Simulators in Improving Fossil Plant Productivity in the 1990's," EPRI Power Plant and Power System Simulator Conference, April 17-19, 1991, Miami Beach, FL.

7. Jason Makansi, Move Toward Process Control For CEM – Natural, But Slow Power, August 1989.

8. Dan Mitus, Anderson-Higgs, Dave O'Connor, "Coal Quality Analysis Linked to Power Plant Control," Power Engineering, May 1991.

9. G. G. Poe, et al., "Thermal Efficiency Improvement of Existing and New Fossil Plant – The EPRI Program," 1989 EPRI Heat Rate Improvement Conference, Richland, WA.

10. K. Muroki et al, "Operating Experience of the Control System on the 700 MW Supercritical Spirally Wound Sliding Pressure Operating Plant," ISA Power Conference, 1982.

11. Jan Jelinek, "Cycling Controllers for Electric Utilities," EPRI Cycling Conference, October 1987, Princeton, N.J.

12. EPRI Internal R&D Planning Document.

13. D. Smith, Computers Maximize Productivity of Power Engineers, Power Engineering, December 1988.

14. S. M. Divakaruni, et al., Control System Retrofit Guidelines, American Power Conference, April 29 - May 1, 1991, Chicago, IL.

15. Man and Machine in Control: Assessing Their Performance, Process Engineering, December 1984.

16. Application for Fiber Optics Show Diversity, Cost-Effectiveness, Power, February 1988.

17. N. Y. Michael, "Analysis of Control Systems for Life Extension," Sargent & Lundy Engineering Conference, Spring 1986, Chicago, IL.

18. Thomas Elliott, "modernization Controls to Promote Plant Life Extension," Power, February 1986.

19. G. L. Touchton et al., "Future Power Plant Networks for Integrated Control and Diagnostic Monitoring," EEI Prime Movers Presentation Subcommittee Meeting, February 4, 1991, Long Beach, CA.

20. S. M. Divakaruni, A. K. Agarwal, "Plant Control System Retrofits and Improvement Project," American Power Conference, April 1990, Chicago, IL.

21. A. F. Armor, H. A. Mueller and G. L. Touchton, "Managing Plant Assets for Profitability," American Power Conference, April 29 - May 1, 1991, Chicago, IL.

22. M. Blanco and J. McKinley, "Integrated Controls and Diagnostics Retrofit Project at El Segundo Power Plant," Fourth Incipient Failure Detection Conference, October 17, 1990, Philadelphia, PA.

23. M. Iioka et al, Hierarchial Function-Structured and Autonomous Control Systems for Fossil Power Plants, IEEE/PES Joint Power Generation Conference, Florida, October 4-8, 1987.

24. H. Matsumoto, et al, "An Expert System Based on Fuzzy Reasoning for Start Up Scheduling and Operation Support in Fossil Power Plant," 91 WM 127-1 EC, IEEE/PES 1991 Winter Meeting, February 3-7, 1991, New York, NY.

25. M. Abe, "IHI New Control Technology," 1988 FWEC International Engineering Conference.

26. K. Yahagi, "Development of Boiler Operation Supporting System," IHI.

27. M. A. Blanco, et al., "Diagnostic Monitoring of Power Plants: The Key to Predictive Maintenance," Fourth Incipient Failure Detection Conference, October 17-19, 1990, Philadelphia, PA.

28. R. J. Colsher, et al., "A Method of Combining Many Diagnostic Monitors into One Integrated System," Fourth Incipient Failure Detection Conference, October, 1990, Philadelphia, PA.

©1991 - Paper #91-701
0074-056X/91/029-036/$0 + .50pp

THE FUTURE OF NUCLEAR I&C APPLICATIONS

V. K. Chexal J. F. Lang T. U. Marston K. E. Stahlkopf

Electric Power Research Institute
3412 Hillview Avenue, Palo Alto, CA 94303

ABSTRACT

In response to industry's recognition that a substantial portion of the nuclear plant I&C capability in the U.S. is outdated and without vendor support, EPRI has set a strategic goal to complete the modernization of I&C systems in at least 10 nuclear plants by the year 2000. Cooperation is being sought from suppliers, utilities, and government agencies. The draft I&C plan is now out for industry review and feedback. Based on this input the plan will be revised to take advantage of industry insights. The next step will be to select contractors and alliance partners, and get the work underway.

INTRODUCTION

Utility industry health over the next few years will be influenced by its ability to achieve and maintain high availability and reasonable operation and maintenance (O&M) costs. Any improvements that can contribute significantly to these attributes deserve the closest attention. The nuclear power industry is continually searching for, investigating and implementing such improvements. EPRI has maintained major programs to support industry efforts. Most improvements that are identified are piece-meal but together they make a major difference.

There is one potential area for improving reliability and reducing O&M costs at existing nuclear power plants that only recently has come to be appreciated. This is the adoption of modern instrument and control (I&C) technology. The I&C area may present the single largest opportunity to improve reliability and control O&M costs. This is because no other class of equipment at nuclear plants has become more outdated and surrounded with problems. At the same time, there have been dramatic advancements in the I&C area outside of the nuclear industry. Of all of the plant systems, I&C is the most amenable to physical change. Modern I&C systems can outperform their predecessors of only a few years ago in many ways. They are smarter, can provide highly analyzed status information, can control complex functions in a more refined way, can assist in diagnostic efforts, are more reliable and economical, and make possible a higher level of automation.

In spite of the advances in I & C technology, the nuclear power industry has been slow to use this new technology primarily because of the stalled growth of nuclear power in the United States. However, the growing unavailability of replacement parts and other service problems with present equipment, along with the much-superior monitoring and control capability of the modern technology, have prompted a new look at I&C. The Electric Power Research Institute (EPRI) is leading an effort to bridge the gap between existing and future I&C systems.

EPRI's comprehensive I&C initiative is directed at formulating a plan and at integrating the major technical elements. Nuclear industry suppliers, I&C vendors, national laboratories, and other development agencies will be actively involved. The initiative will give each existing nuclear plant the capability to upgrade its I&C systems to the extent that best serves its particular situation and needs. At the same time, the initiative will provide state-of-the-art I&C capabilities to meet requirements for future nuclear plants.

BACKGROUND

When the nation's first generation of commercial nuclear plants was designed in the 1960's, the I&C systems were a combination of old and new art. They employed the traditional balance-of-plant systems that had served well in the petrochemical, pulp, and paper industries, as well as in fossil fuel power plants. For the control and monitoring of the nuclear part of the plants – the reactor and its component systems – industry turned to technology that had been developed primarily in the national laboratories, where nuclear energy was harnessed.

In the first nuclear plants, the I&C systems for both the nuclear island and the balance of plant were analog; the age of digital technology was not yet at hand. These original systems have done what was expected of them. They have provided comprehensive pictures of plant conditions and have allowed adequate plant control. These systems have brought nuclear power a long way, but modern technology provides the opportunity to:

- address obsolescence,
- handle more complex problems,
- increase measurement accuracy,

- improve system reliability,
- keep the operator better informed,
- perform more effective testing and surveillance, and
- facilitate the development of improved equipment and architecture.

I&C technology has come a long way since the original, analog systems were designed [References 1-9]. Modern systems now perform functions that were unimaginable when the existing plants were conceived. Advanced instrumentation can project the status of a process in more ways than current instrumentation. Moreover, modern equipment can do this with outstanding reliability. Modern systems can be designed to be more fault-tolerant and redundant. Utilizing digital control flexibility, control functions that previously required difficult manual action can now be performed automatically. Although the older systems do their job, they contribute to a significant number of trips in both nuclear and fossil plants. Advanced equipment has the potential to eliminate many I&C-caused outages and to minimize current maintenance requirements.

The plants that can benefit are those that are now operating, those that are partially completed and will someday be operating, and the next generation that is currently being designed.

THE CHALLENGE AHEAD

Utilities have much to gain from focussing attention on improving the reliability and capability of their I&C systems. As used here, I&C has a broader scope than in the past. It includes not only the traditional I&C functions but a wide range of other functions that support power plant operation, maintenance and engineering (Figure 1 and Table 1). Considering the broad scope, utilities have much to gain from modernizing their I&C systems. Modern I&C technology has the potential to improve reliability, reduce O&M costs and improve overall plant performance.

To meet the I&C challenge the following are needed:

- active participation by utilities to define I&C requirements,
- a coordinated strategy to pool resources among EPRI, utilities, vendors, government and standards groups by forming alliances because none can do it alone,
- alliances with manufacturers, utilities, and government organizations to pool resources, and
- an approach that provides vendors a larger potential market with a consistent set of products.

The benefits from this initiative will be greater standardization, lower costs to utilities, generic regulatory review to minimize licensing costs and schedules, and a reliable vendor pool for technology support.

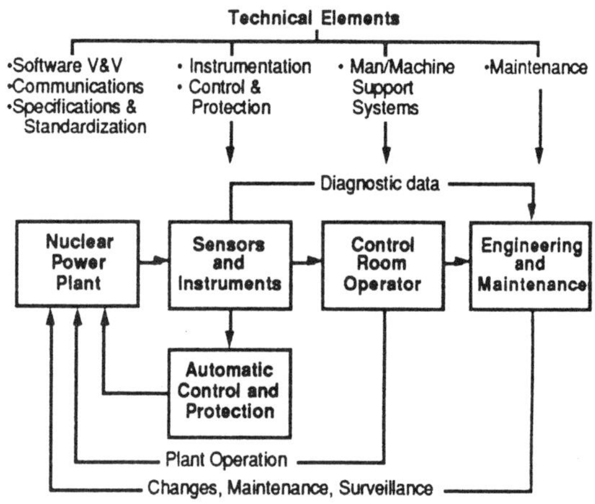

Figure 1. Integration of I&C with other functions.

Table 1. I&C in action for plant operation.

Functions	*Systems*	*Instrumentation*
• Reactivity Control • Reactor Core Heat Removal • Secondary Heat Removal • Primary Coolant Inventory • Containment Integrity • Balance of Plant Performance	• Reactor Protection • Engineered Safety Features • Reactivity Control – Primary and Secondary • Steam and Power Conversion • Auxiliary	• Pressure Sensors • Flux Sensors • Temperature Sensors • Flow Sensors • Breaker Contacts • Interlocks, etc. • Radiation Sensors • Electrical Sensors • Valve Position Sensors • Chemistry Sensors • Level Sensors • Position/Vibration Sensors

EPRI'S ROLE

EPRI will play a major role in this initiative by being the focal point for coordinating and integrating research and development, making sure the technology is transferred, and arranging for adequate funding. An integrated I&c upgrade planning matrix has been prepared (Table 2). The work and funding categories are shown in Fig. 2.

Table 2. Integrated I&C Upgrade Planning Matrix

	R&D	*I&C Upgrade Demonstration Plants*	*I&C User Group*
Goals	Develop new technology for current and future plants	Ten plants complete upgrade by year 2000	Continued information interchange among utilities
Emphasis	New technology. Solve problems/add functions by developing and applying new technology	Upgrade using current technology. Resolve near-term I&C problems.	Promote information interchange to reduce O&M costs and handle current obsolete equipment.
Vehicle	R&D - EPRI plan	I&C Upgrade demonstration plants	I&C User Group/Ad Hoc Working Groups
Initial Projects	EPRI R&D tasks linked to plan	4 Demo Plants - BG&E (CC) - TVA (BF) - Other 1 - Other 2	Working Groups, eg: - Setpoints - Obsolescence - Cali. Reduction - Etc.
EPRI Products	New components, systems, guides, specs, etc.	Planning, and Upgrade Guidelines	Organization, Administration and Info. Base
Funding Sources	Internal Funds DOE CRADA Cost sharing Tailored Collab. Strategic Alliances Co-funding	Tailored Collaboration	Utility members of User Group

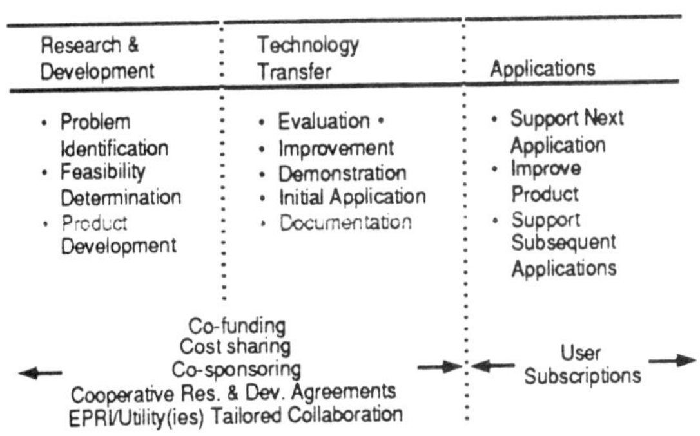

Figure 2. Work/funding categories. (See Appendix A for definition of funding categories.)

Research and Development

I&C technology has made dramatic advances in recent years, and EPRI has been in the forefront of this activity. The Institute has developed expertise in I&C and has provided I&C assistance to utilities, utility owners groups, and standards organizations. Its involvement includes R&D projects in the following areas:

- Safety parameter display systems
- Power shape monitoring systems
- Digital control system development
- Signal validation techniques
- Sensor calibration reduction methodology
- Surveillance and testing optimization
- Cost-benefit methodology
- Control room enhancements
- Sensor development
- Process instrument accuracy
- Alarm system improvements
- Software verification and validation techniques
- Development of radiation-hardened components
- Simulator qualification
- Computerized O&M aids
- Instrumentation systems

- Design change guidelines
- Set-point methodology
- Human factors guidelines

The advanced I&C equipment now available can perform with outstanding accuracy and reliability. Fault-tolerant, redundant digital systems have the potential to greatly improve overall I&C system reliability. The low-drift and self-checking capabilities of modern systems significantly reduce surveillance and testing requirements. Moreover, modern control equipment makes possible new control algorithms, data fusion, and enhanced analysis and display. It can perform many useful functions that are impossible or very difficult to perform with existing equipment.

The digital electronic systems now available make it possible to use distributed control techniques. There is a growing use of such techniques in fossil power plants and petro-chemical plants. Microprocessors are located close to the points of control to store readings, process data, and, with the operators' oversight, actually control the plant. A major benefit of distributed control is that if one aspect of a control system fails, other aspects can still carry on the affected operation because of the distributed functionality.

Both U.S. and foreign utilities have reported excellent results with new maintenance techniques that use the predictive capability available from diagnostic and surveillance information to optimize equipment replacement and overhaul. A significant improvement in plant availability has been attained through enhanced ability to schedule repair and replacement work. Unnecessary maintenance, which imposes costs in outage time, man-hours, and man-rems, can be reduced.

EPRI's plan is to address the following broad based I&C technical elements:

- Instrumentation
- Control and protection systems
- Man-machine support systems
- Software verification and validation techniques
- Communication protocols
- Maintenance procedures
- Specifications and standardized components

The future plan activities will be carried out with R&D guidance from the newly formed I&C Advisory Committee of utility representatives.

Technology Transfer

In support of the capability to upgrade existing plants, cooperative programs are planned at four utilities. The goal is to demonstrate upgrades at plants of the four major U.S. vendor designs. Discussions with several utilities are currently underway to seek their agreement for developing overall plant specific architecture needs, and management plans for implementation. The cooperative programs will give EPRI a chance to perfect and demonstrate generic upgrade guidelines. It will also show how the generic guidance can be tailored to make it most valuable and most easily used by individual utilities.

There are at least two utilities that are planning to finish construction on partially finished nuclear plants. A unique opportunity is being explored for the application of advanced I&C technology for such plants. The scope includes the concept, architecture, requirements, relative costs, control room layout, etc. The wide-scale upgrades possible in partially constructed plants can serve as a demonstration vehicle to help the industry develop an overall system approach to upgrading I&C systems and assess the cost/benefit of advanced technology in existing plants.

In addition to conducting cooperative programs, EPRI will disseminate information via I&C workshops, newsletters and expert advisors.

I&C PLAN

At EPRI, technical work in the I&C area is being carried out in several divisions. The emphasis is to build on the existing I&C infrastructure as much as possible. A draft integrated I&C plan prepared by the Nuclear Power Division at EPRI provides the details of activities and resource requirements for the various technical elements. An EPRI R&D plan to address the needs in fossil plants [Reference 10] is available separately.

The I&C plan is designed to serve as: (a) a guide for the direction of EPRI R&D programs in terms of project identification, anticipated scheduling, and funding requirements; (b) an integrated baseline document that enables monitoring and evaluating EPRI responsiveness to utility industry needs; and (c) a source document for utility industry use in identifying, planning, and implementing projects for successful technology transfer.

The plan for each element (Fig. 1) covers the objectives, the significance, problems experienced or expected, what has been done, what needs to be done, the schedule, cost/benefit, resources, implementation issues, and any ALWR considerations.

The information for prioritization of the tasks in the plan was obtained from 1987-88 Licensee Event Reports data, 1987-88 Lost Production data, 1987-88 Maintenance Reports, Nuclear Maintenance Assistance Center Maintenance Survey information, I&C Advisory Committee survey information, and I&C Workshop information and direct utility communications.

Based on input from the large number of utilities, the tasks for each technical element have been prioritized in Table 3:

Table 3. Prioritization of technical element tasks.

Instrumentation	*Control and Protection*
High Priority: • Radiation hardened electronics • FW flow measurement accuracy • Sensor health monitoring • I&C surveillance and test reduction • Radiation hardened fiber optic comm cables • High temp, In-Situ chemistry sensors (pH) • H_2/O_2 sensor assessment • Area and process radiation monitors	*High priority* • Network based Safety Grade Digital Controller (PLC) - Fault-tolerant, modular design with self-testing and ease of maintenance • EH turbine control system • Steam dump digital control system • Reactor rod control system • Overall guidelines for digital control retrofitting, replacement, and maintenance

Man-Machine Support Systems	*Maintenance*
High Priority: • Human Factors/Engineering Guidelines - Control room display - Human reliability /error in O&M - CRT based controls - Modular and piecewise control room modification/upgrade • Man-Machine Systems - EOP systems to support operator training - EAL systems to support emergency exercise and drill - Workstation design to support operation, maintenance, and engineering	*High Priority Tasks:* • RCM Workstation • I&C Maintenance Guide • Optimization of RCM for I&C • Reliability Study of NIS and RPS Systems • Circuit Breaker-Relay Health Monitoring • Wireless Automatic Sensor Monitoring • Instrument Calibration Reduction Program • Sensor Calibration Advisors *Medium Priority Tasks:* • Valve Diagnostic Advisor • Pump Diagnostic Advisor

Communications	*Verification and Validation*
High Priority: • Demonstration plants - select architecture using todays technologies . communications (data acquisition and control) . information (acquisition storage, display) - incorporate UCA as applicable - adopt planning documents for entire upgrade • Research and Development - standards and protocols (e.g., field bus) - open architectures - identify special requirements for nuclear plants, e.g., reactor protection systems - cabling applications, e.g., fiber optic	*High Priority:* • Demonstration Plants - Evaluate previous use of software and protocols - Adapt to NPP's . non-safety . safety - NRC licensing basis . open architecture - testing • Research and Development - Methodology for protection system V&V - Methodology for neural networks - Study requirements for system software tools: . control, expert, protection, neural networks . pre-qualifying reusable software

Specifications and Standardization
High Priority: • User group basis, as required • Demonstration plant - Applicability - as built vs current - NRC requirements and involvement • Future Specs, Standards - Develop specifically for modern I&C design specifications - Work with NRC on future requirements • Promote standardization by vendors for future upgrading

STRATEGY

Introducing modern I&C technology on a plantwide scale requires a coordinated effort by several areas of the industry. This effort can benefit from strategic alliances among these various areas that ensure a user base of developmental support and future sales. Such a base will encourage industrial suppliers to participate. Ongoing technology exchanges and cooperative efforts between EPRI and other domestic and foreign organizations including power plant designers and architects, I&C system vendors, and the U.S. Department of Energy will be expanded.

The level of research conducted by manufacturers, the government, and others in support of electric utilities has been decreasing. This is of concern because technology has been a primary source of competitive advantage for regulated electric utilities. Even though great strides have been made in I&C technology, the new concepts have not been applied widely, particularly in the nuclear power industry. Part of the reluctance to utilize new I&C technology stems from regulatory considerations but there is also the fear that markets for new technology are not sufficient to justify the necessary R&D investment. In order to counter this, EPRI, as an established collaborative R&D entity, can provide a vehicle for pooling the research resources of many in the industry in a way that:

- benefits EPRI's member utilities,
- focuses on raising the state of the art in installed I&C systems,
- does not infringe on antitrust laws,
- provides open system results that can be applied by participating suppliers at reduced costs for all,
- provides a technical forum where world specialists can interact, share knowledge, solve state-of-the-art problems, stimulate progress, and foster innovation but still remain competitive,
- steers and integrates the research of others to meet the present and future needs of the utilities, and
- integrates the EPRI research program more closely with that of commercial organizations.

The concept of the strategic technical alliance reflects that EPRI produces and is a repository of information, know-how, and technologies for solving a wide variety of I&C problems. Through the process of technology transfer, the information and results become part of the design basis for new products and operating practices. This process can be improved through multipartite international strategic technology alliances among EPRI, utilities, manufacturers, and other research organizations. Building on the results to date and on information that may be contributed from participating organizations, multipartite alliance members will be able to identify and complete the remaining needed technical work, organize the results in a way that facilitates use, and provide for the implementation of these results by the participating members.

The objectives of I&C strategic technical alliances will be to:

- make modern cost effective I&C systems available to the electric power industry,
- enhance the competitive position of alliance members through the development of commercially attractive products,
- produce techniques for utilities to use in addressing plant specific I&C issues,
- provide comprehensive support packages through user's groups that include technical workshops, training, hotline support, and plant visits, and
- reduce the growth of operation and maintenance costs through the use of improved I&C technology.

CONCLUSIONS

Modern I&C technology can help to improve plant availability and reduce O&M costs. In addition to traditional I&C functions, it can support other plant operating activities such as maintenance and engineering. EPRI has put in place an initiative to form *strategic technical alliances* to address the need for improved I&C capabilities. The alliances will be led by EPRI and will include the technical service industry, foreign utilities, and pertinent government agencies as partners.

As is true with any new technology, solving old I&C problems and providing broader I&C capabilities will uncover new problems and challenges. The initiative will require careful management, from development to testing to application. It will be necessary to assure that regulators are kept fully apprised of the status and goals of the program to expedite final approval if necessary. This initiative has the attributes that are needed to complete the task and to incorporate the lessons learned from the problems encountered by other users of advanced I&C technology.

A plan to make the best of the microprocessor evolution in the '90's and beyond is now in place. The vision of the I&C initiative is that I&C systems will be designed to make effective use of modern technology, and standardization in a way that will help utilities control costs and improve plant performance. The improved systems will be available both for elective upgrades at existing plants and for new plants.

APPENDIX A: DEFINITION OF FUNDING CATEGORIES

Co-funding: Funds received from a non-EPRI source to support research activities approved as a part of the base EPRI research plan. It is a bilateral contractual arrangement between EPRI and another organization (separate from the agreement between EPRI and the research contractor) which sets forth the extent of the other organization's financial participation and the terms of that participation. EPRI is responsible for the direction of the work.

Cost-Sharing: Research activities funded by an EPRI contractor as part of a contribution to an EPRI research program. It represents the portion of the full costs to perform the research that is absorbed (discounted) by the contractor. EPRI is responsible for the direction of the work.

Co-Sponsoring: A multi-lateral contractual arrangement between EPRI, other sponsoring organizations and contractors to accomplish a research objective wherein the financial participation of all parties is clearly identified. Responsibility for the work direction is defined in the agreements. The co-sponsor funds do not flow through EPRI to the contractor.

Cooperative Research and Development Agreement (CRADA): This is an agreement between one or more federal laboratories and one or more non-federal parties. Resources such as funds, personnel, services, facilities, equipment, etc., are shared. No funds flow back and forth between the parties. Participants may be industry, universities, non-profit institutes, and state and local governments.

Tailored Collaboration: Gives EPRI members the ability to direct where a portion of their R&D money is spent. In this program a member may target a specified amount of its EPRI dues to an eligible R&D project. (The amount is limited to 10% of the member's dues in 1991 but will increase to 25% in following years.) Some restrictions apply to the program.

1. The total amount targeted must be $200,000 or more over the lifetime of the project.
2. The projects must be in the scope of the overall EPRI program plan approved by the advisory committee.
3. Only R&D projects are eligible.

ACKNOWLEDGEMENTS

The work reported here is a synopsis of the effort being carried on by several individuals of the EPRI staff. The authors gratefully acknowledge the contributions of Lance Agee, Sid Bhatt, Rob Colley, Mike Downs, Joe Naser, Les Oakes, John O'Brien, Seung Oh, Bill Reuland, Ed Rumble, Avtar Singh, George Sliter, Bill Sugnet, Bill Sun, Joe Weiss, and Dan Wilkinson.

REFERENCES

[1] L. Oakes, et al, *Instrumentation and Control Strategies for Power Plants, Volume 1: Notes on the Instrumentation and Control Workshop,* Electric Power Research Institute, Palo Alto, CA, NSAC-153 Vol. 1, August 1990.

[2] L. Oakes, et al, *Instrumentation and Control Strategies for Power Plants, Volume 2: Problem Definition and Recommendations,* Electric Power Research Institute, Palo Alto, CA, NSAC-153 Vol. 2, December 1990.

[3] J. A. Naser, Ed., *Proceedings: Conference on Expert Systems Applications for the Electric Power Industry.* Hemisphere Publishing Company (1990).

[4] J. J. Taylor and B. K. H. Sun, "Applications of Computers to Nuclear Power Plant Operations," *Nuclear News,* October 1990, pp. 38-40.

[5] B. Chexal and M. Divakaruni, *Control, Instrumentation, and Automation News,* Quarterly Newsletter, Spring 1991.

[6] *Proceedings: 1986 Seminar on Emergency Response Facilities and Implementation of Safety Parameter Display System,* EPRI Report NP-5510-SR. Electric Power Research Institute (Nov. 1987).

[7] *Proceedings: 1986 Integrated Power Plant Computer Communications Seminar,* EPRI Report NP-5641-SR. Electric Power Research Institute (Feb. 1988).

[8] *Proceedings: 1985 Seminar on Power Plant Digital Control and Fault-Tolerance Microcomputers,* EPRI Report NP-4769-SR. Electric Power Research Institute (Sept. 1986).

[9] *Integrated I&C Upgrade Plan,* Rev. 0, EPRI Nuclear Power Division, December 1990.

[10] *R&D Plans for Advanced Computer and Control Technologies,* Volumes 1-4, EPRI GS-6673, Final Report. Electric Power Research Institute (Feb. 1990).

V. K. (Bindi) Chexal is the Director of the Nuclear Safety Analysis Center (NSAC), Nuclear Power Division at the Electric Power Research Institute (EPRI) in Palo Alto, California.

He received his B.S. in Mechanical Engineering from Thapar Institute, India in 1970 and his M.S. in Mechanical Engineering from Georgia Tech, Atlanta in 1972.

From 1974 to 1977 he was a design engineer with Atomic Energy of Canada Ltd, Mississauga, Ontario working on the system design and safety analysis of CANDU heavy water reactors. From 1972 to 1973 and again from 1977 to 1979 he was with Quadrex Corporation, Campbell, CA as a design engineer performing research in thermal hydraulics, structural integrity and heat balance analyses of light water reactors.

In 1979 he joined NSAC at EPRI and in 1989 became the director. NSAC directs industry research to evaluate power plant response during low probability high consequence accidents. Mr. Chexal has successfully applied matrix management techniques in developing solutions to a host of generic technical issues. Mr. Chexal is active in research and development in corrosion, two-phase flow, and power plant safety.

Mr. Chexal is a fellow of the American Society of Mechanical Engineers.

J. F. (Jim) Lang is Director of the Engineering and Operations Department at the Electric Power Research Institute (EPRI) in Palo Alto, CA. He joined the Institute in 1979 as Manager of the Steam Generator Thermal Hydraulic, Structural and Safety Programs in the EPRI Steam Generator Project Office. He subsequently managed Steam Generator Integrity and Containment Integrity issues in the Nuclear Safety Analysis Center and the Operations and Maintenance program in the Engineering and Operations Department.

Previously, Mr. Lang spent nine years with the U.S. Navy's Nuclear Power Directorate. From 1977 to 1979, he served as Director of the Reactor Plant Valve Division. Prior to that, he held various positions involving the design and maintenance of steam generators and other primary components. Mr. Lang began his work in the nuclear power field in 1968 as a Physicist with the Phillips Petroleum Company and Idaho Nuclear Corporation.

Mr. Lang received a B.S. degree in Physics from Michigan State University in 1966 and a Certificate in Engineering, Bettis Reactor Engineering School, in 1971.

Dr. T. U. (Ted) Marston is the Director of the Advanced Reactors Department, Nuclear Power Division at the Electric Power Research Institute (EPRI) in Palo Alto, California. He joined the Institute in 1976 as a Project Manager in the Pressure Boundary Technology Program, becoming Manager of the Structural Mechanics Program in 1979. He was assigned to Electricite de France with the Service D'Etude et Projets Thermiques at Nucleaires (SEPTEN) in Paris from July 1983 to July 1984. He became manager of the Operations and Maintenance Program in October 1984 and Director in October 1986 of the Engineering and Operations Department. He assumed his present position in November, 1990.

Before joining EPRI, Dr. Marston spent three years with Combustion Engineering, Inc., Chattanooga, Tennessee. Starting as a Research and Product Development Engineer, he moved to Supervisor of Materials Properties in the Metallurgical and Materials Laboratory.

Dr. Marston received a B.S. degree in 1969 and an M.S. degree in 1970 in Mechanical Engineering from the University of Michigan. He was awarded the Ph.D. degree from the University in 1972.

Dr. K. E. (Karl) Stahlkopf is the Director of the Safety and Reliability Department, Nuclear Power Division, for the Electric Power Research Institute (EPRI) in Palo Alto, California. Dr. Stahlkopf joined the Institute in 1973 as a Project Manager, Pressure Boundary Technology. He advanced to Program Manager in 1974, moving to Director, Nuclear Systems and Materials Department in 1980, and to his present position in 1989.

Before moving to EPRI, Dr. Stahlkopf was a Research Fellow in the Department of Engineering at the University of California, Berkeley. In 1969 he was a member of the Chief of Naval Operations Staff, Pentagon, Washington, D.C., as a technical consultant on nuclear submarine propulsion. During the previous four years, Dr. Stahlkopf was a Division Officer assigned to construction and operation of nuclear submarines.

Dr Stahlkopf received a B.S. degree in Electrical Engineering and one in Naval Science from the University of Wisconsin. He was awarded the M.S. and Ph.D. degrees in Nuclear Engineering from the University of California, Berkeley.

The author of numerous publications, Dr. Stahlkopf is a member of AAAS, the American Nuclear Society, and the Welding Research Council. He is a Registered Professional Engineer in the State of California.

©1991 - Paper #91-702
0074-056X/91/037-042/$0 + .50pp

First Annual ISA/EPRI Joint Controls and Automation Conference, June 3-5, 1991, St. Petersburg, FL

Future of Intelligent Control Systems

John R. James
United States Army
Training and Doctrine Command
Fort Monroe, VA 23651

Abstract - The future of intelligent control systems depends upon the extent to which Artificial Intelligence (AI) technologies can help control engineers deliver practical solutions to hard control engineering problems. Conventional control design approaches have achieved notable successes in the design and implementation of robust, adaptive controllers for systems with well-defined mathematical models. However, conventional approaches have had difficulty supporting engineers in the design and implementation of control systems when an accurate mathematical model is not available. Also, verification that computer-controlled systems perform to specifications, validation of the specifications, higher-level control, operator decision aids, system diagnosis, operator alerting, and reconfiguration of systems which experience large changes over time or potentially catastrophic failures are significant challenges to control science and engineering. It is in these hard areas where the AI technologies of knowledge representation, learning, search, diagnosis, planning, and decision are being used to aid control engineers. Algorithms for computer-controlled systems and software tools to help implement these algorithms have been a subject of research and commercialization for decades. Computer-Aided Control Engineering (CACE) tools have achieved a degree of success in the past decade based on their ability to assist in the control system design and implementation process. Specialized tools have been made available for system identification, system simulation, controller design and controller implementation. Recently, efforts have been made to build integrated CACE environments. Also, some current research is aimed at increasing the utility of available systems by creating a mathematical basis and a software architecture for efficiently describing complex systems and using these as a means of achieving a higher level of integration of the diverse tools already available. A recent *Workshop on Software Tools for Distributed Intelligent Control Systems* was sponsored by the U.S. Army and The Defense Advanced Research Projects Agency (DARPA). This paper will describe the results of the workshop and subsequent efforts to use these results to shape a DARPA software development project. The first section of the paper provides a brief review of the current applications of AI in the design and implementation of control systems. The second section discusses areas where AI can be applied in the near term to help solve challenges in the implementation of computer-controlled systems. The third section gives an overview of the development of CACE tools. The fourth section provides a review of the Army/DARPA workshop and the last section discusses the use of the results of the workshop.

Key Words - intelligent control, expert systems, knowledge representation, distributed processing, learning, propagation of uncertainty, possibility theory, diagnosis, reconfiguration, decision aids.

1. Intelligent Control Today

Use of AI technology in the design and implementation of control systems grew rapidly in the 1980s. Today there are many examples of expert systems being used to assist in system identification (Nagy and Ljung, 1989), diagnose failure or degradation of performance of complex systems (Enand et al., 1991), alert operators to possibly catastrophic failures (Barry, 1989), design control systems (Birdwell et al., 1985; James 1985; Trankel, 1985), and implement control systems (Salle and Arzen, 1989). Fuzzy controllers have been used commercially since the late 1970s and are currently being applied to a wide range of consumer products (Zadeh, 1990). Expert systems have been used from low-level tuning of Proportional-Integral-Derivative (PID) control loops (Astrom et al., 1986) to advisory systems for control of dispersed military forces (SDIO, 1989). Early efforts in the mid-1980s to build real-time expert control systems have resulted in a broad range of technology demonstrations and industrial implementations (James and Suski, 1988). Several of the companies which specialize in the process control industry are now offering artificial intelligence options for their control products (Honeywell and AEG among them). Also, the theme of the Advanced Control Conference for 1991 sponsored by Control Engineering and the Purdue Laboratory for Applied Industrial Control will be "Expert Systems Applications in Advanced Control (Successes, Techniques, Requirements, and Limitations," (Control Engineering, 1991). Given the range of applications and the sustained active interest in the area, it may be useful to consider what has not yet been achieved.

2. Near-Term Applications of AI to Control Engineering

While a number of projects have established the technical feasibility of building expert advisory systems for control analysis and design and some have also demonstrated on-line redesign of controllers, the difficulties in coupling the numeric and symbolic computations have slowed widespread use of these techniques. However, there is now available a tool which provides close coupling of symbolic and numeric computations. This tool, Stable Factorization Package (SFPAC), is available at no charge for academic and research work (Pang et al., 1990). Commercially available control design tools currently support either numeric or symbolic simulation of system performance. It is reasonable to expect that commercial tools will soon follow this research tool in supporting both.

Neural networks have been remarkably successful in learning the control necessary to smoothly perform motion control for robotic arms for fixed paths and loads as well as in performing other difficult pattern recognition tasks. However, one shortfall of neural networks has been that they must be treated as "black boxes" which can rapidly provide outputs for inputs but whose operations are not subject to analysis. However, a recent work has indicated some success in using neural networks to perform system identification (Bialasiewicz and Soloway, 1990). In addition, recent work has been done in the use of neural nets to generate fuzzy rules (Zadeh, 1990). These efforts indicate that it is possible to construct learning systems which can assist in the on-line recognition of both minor and major changes in system dynamics and use this knowledge in the application of conventional control analysis and design methods as well as in the application of expert systems technology. A near-term realization of this capability would be a significant advancement in the realization of adaptive control systems.

Architectures for complex computer-controlled systems have been repeatedly proposed (Albus and Quintero, 1990) but it has been difficult to establish a national standard for the representation of the elements of an Architecture for Real-Time Intelligent Control Systems (ARTICS). This has been proposed by Albus and Quintero but there is still no general agreement on the composition of such a reference architecture for systems such as flexible manufacturing systems, power generation and distribution systems, military command and control systems, and other man-in-the-loop, complex machines. Agreement on a reference architecture would ease the development of more capable software packages for building intelligent control systems.

The underlying mathematical representations of a complex computer-controlled system is still insufficient to create a model which accurately captures the dynamics of the system over the entire range of system operation. We remain in a situation where we must tradeoff the accuracy of our models with the manageability of the models. Closed-form solutions of mathematical models are almost exclusively limited to linear system models. Computer simulations of nonlinear and discrete-event models provide a means for off-line design of control systems through iterative search but such simulations cannot perform exhaustive search due to the complexity of the problem. Guarantees of system performance are limited to those regions where the robustness conditions apply. These conditions may not apply during startup and shutdown or during periods of anomalous operation. Excellent results are available for cases where adequate mathematical models are known and the system is operating "close enough" to a linear region.

Also, effective tools are available to model high-level system changes as a finite state machine. Several attempts to improve our modeling capabilities are focused on mapping the continuous world into a discrete one (Ramadge and Wonham, 1987; Ho, 1987; Benveniste and Le Guernic, 1990; Inan and Varaiya, 1988). However, repeated results are available which indicate that large interactive systems evolve into states where minor events can lead to a catastrophe (Bak and Chen, 1991). We are left with the result that there is a pressing need for a more adequate theory and mathematical basis for representing and predicting the performance of hybrid (continuous and discrete) dynamical systems. In the near term we will probably be able to mathematically prove (verify) that the implementation of a subset of software for computer-controlled systems performs to specifications but will have to use conventional metrics for verification of the majority of the software being used.

3. Computer-Aided Control Engineering

The availability of increasingly more economical mainframe computer systems in the 1970s led to the proliferation of first-generation software tools to aid in the analysis and design of control systems (Grabow et al., 1977; Smith et al., 1976; Dongerra et al., 1979; Edmonds, 1979) as well as to improvements of the algorithms and numerical analysis routines available. The MATLAB program became widely used in the early eighties (Moler, 1982) and has been since used as the basis for a number of commercial and research tools (Integrated Systems, 1985; Pang et al., 1990). The IEEE Control System Society Technical Committee on Computer-Aided Control System Design (CACSD) maintains the Extended List of Control Software which contains one-page abstracts of a large number of commercial, proprietary, and public domain software. By the early-1980s, efforts had begun to try to provide integrated environments (Spang, 1984; Taylor et al., 1989) and knowledge-based tools to assist control engineers in the application of the powerful analysis, design and implementation software already available. The rapid increase in the number of engineering workstations and the growth of national computer communication networks in the last half of the 1980s helped to fuel a corresponding increase in the number and diversity of the tools available. Also, in recent years, research in the use of knowledge-based systems in Computer-Aided Control System Design (CACSD) has increased significantly (James,1988; MacFarlane et al., 1987, Lewin and Morari, 1988; Pang et al., 1990). The triennial IFAC Conference on Computer-Aided Design of Control Systems and the series of IEEE workshops on CACSD have documented these progressive changes.

During the same period, examples of building knowledge-based control systems have increased dramatically, especially in the process control industry (Astrom et al., 1986; Basila et al., 1990; James and Suski, 1988; Karsai et al., 1987; Liu and Gertler, 1987; Moore et al., 1987; Le Clair and Abrams, 1988). As previously stated, the SFPAC software combines the ability to perform symbolic and mathematical computations in a single software tool. Also, there are now several commercially available programs which will help build and implement knowledge-based control systems. It should be noted that there is a growing appreciation that the fields of off-line CACSD and on-line adaptive control tend to merge in the implementation of supervisory intelligent controllers (Sanoff and Wellstead, 1984; James and Rapisarda, 1988; Astrom, 1991, Basila et al., 1990). The knowledge-based controllers implemented thus far are all *ad hoc* systems which require a substantial investment in construction and maintenance. As noted in Section 2 above, advances in the mathematical foundations of intelligent control are needed to lower the cost of building hybrid systems.

4.Workshop on Software Tools for Distributed Intelligent Control

However, there remains a very large gap between those tools which a controls engineer can use today to assist in the design and implementation of control systems and those which will be needed to achieve the next level of system complexity and integration at an affordable price. Current digital control practice is largely focused on single processor implementation of fixed controllers. While significant results are available concerning multiprocessor control systems design and implementation for communicating sequential processes (*e.g.* the SIGNAL system of INRIA; Benveniste and Le Guernic, 1990), there remains a dearth of theory or tools to assist in the design and implementation of distributed intelligent control systems. In July of 1990 the U. S. Army and DARPA sponsored a *Workshop on Software Tools for Distributed Intelligent Control Systems* conducted by the Lawrence Livermore National Laboratory at Pacifica, California (Herget, 1990). Forty-eight attendees met for three days to (1) identify the current state of the art in tools which support control systems engineering design and implementation, (2) identify research issues associated with writing software tools which would provide a design environment to assist engineers in multidisciplinary control design and implementation, (3) formulate a potential investment strategy to resolve the research issues and develop public domain code which can form the core of more powerful engineering design tools, and (4) recommend test cases to focus the software development process and test associated performance metrics. In September of 1990, DARPA released a Broad Agency Announcement requesting interested activities to submit proposals to conduct research in a number of areas. The proceedings of the Pacifica workshop were made available to all attendees of a DARPA bidders workshop conducted in Pittsburgh, Pennsylvania in October of 1990. Five-year projects in the Domain-Specific Software Architectures (DSSA) area will start in the Summer of 1991 with the goal of reducing life cycle software costs for computer-controlled systems by providing a set of tools to support component-based programming..

5. Applying the Results of the Workshop

The workshop resulted in a number of lengthy discussions, the most notable being a general agreement about the breadth and quality of the tools currently available and an exchange of ideas about the fact that distributed intelligent control is ill-defined. The major recommendations were: (1) perform a review of the current state of the art, (2) Develop a taxonomy of currently available tools, (3) develop a high-level integration tool, (4) develop a technology transfer plan, and (5) Establish a repository for the software developed.

The United States Army has a number of technology insertion programs either currently underway or scheduled to start which may be able to apply the software tools being developed by the DARPA DSSA projects. These technology insertion programs include the Rotocraft Pilot's Associate Advanced Technology Transition Demonstration (ATTD), the AirLand Battle Management ATTD, the Lower Echelon Command, Control, Communications and Intelligence ATTD and the Advanced Field Artillery System ATTD. Each of these technology insertion efforts have specific goals associated with increasing the flexibility and functionality of the systems through the use of computers. Hard problems include those associated with embedded training aids, smart sensors, crew station decision aids, automatic reporting of location and status, driver vision enhancement, navigational aids, interface with current and future command and control systems, electronic technical documentation, and self diagnostics and prognostics. It is possible that near-term advances in intelligent control can be applied to achieve these requirements more cheaply.

6. References

Albus, J. S. and R. Quintero, Toward a Reference Model Architecture for Real-Time Intelligent Control Systems (ARTICS), *Proceedings of the Workshop on Software Tools for Distributed Intelligent Control Systems*. Pacifica, CA, 1-10, (1990).

Astrom K. J., J. J. Anton and K. E. Arzen, Expert Control, *Automatica* **22**, 277-286 (1986).

Bak, P. and K. Chen, Self-Organized Critically, *Scientific American*, pp. 46-53, January, 1991.

Barry, M. R. PX1: A Space Shuttle Mission Operations Knowledge-Based Systems Project, *Proceedings of the AI Systems in Government Conference*, Washington, D.C. (1989).

Bialasiewicz, J. T. and D. Soloway, Neural Network Modeling of Dynamical Systems, *Proceedings of the 5th IEEE International Symposium on Intelligent Control*, Philadelphia, PA (1990).

Basila M. R. Jr., G. Stefanek, and A. Cinar, A Model-Object Based Supervisory Expert System For Fault Tolerant Chemical Reactor Control, *Comput. chem. Engn,* **14**, 551-560 (1990).

Benveniste, A. and P. Le Guernic, Hybrid Dynamical Systems Theory and the SIGNAL Language, *IEEE Transactions on Automatic Control*, **35**, 535-546 (1990).

Birdwell, J. B., J. R. B. Cockett, R. Heller, R. W. Rochelle, A. J. Laub, M. Athens, and L. Hatfield, Expert Systems Techniques in a Computer-Based Control System Analysis and Design Environment, *Proceedings of the 3rd IFAC Symposium on Computer Aided Design in Control and Engineering Systems*, Lyngby, Denmark (1985).

Birdwell, J. D. and S. Liang, Concurrent Processing Environments for Distributed Intelligent Control Systems, *Proceedings of the Workshop on Software Tools for Distributed Intelligent Control Systems.* Pacifica, CA 21-29, (1990).

Boeing Commercial Airplane Company, Integrated Application of Active Controls (IAAC) to an Advanced Subsonic Transport Project, *NASA Contractor Report 3880,* (1986).

Control Engineering, Papers Sought for the 17th Purdue Conference, *News of Control*, p. 23 (January, 1991).

Dongerra, J. J, C. B. Moler, J. R. Bunch, and G. W. Stewart, *LINPAK User's Guide*, SIAM Press, Philadelphia, PA (1979).

Edmonds, J. M., Cambridge Linear Analysis and Design Program, *IFAC Symposium on Computer-Aided Design of Control Systems*, Zurich, (1979).

Enand, R., K. Cowan, and T. Knutilla, Taking Expert Systems to the Field: Lessons Learned from the Deployment of PRIDE, *Proceedings of the Symposium on Artificial Intelligence Applications for Military Logistics.* Williamsburg, VA (1991).

Gavel, D. T., Control Systems Engineering in an Open-Architectured Object-Oriented Software Environment, *Proceedings of the Workshop on Software Tools for Distributed Intelligent Control Systems.* Pacifica, CA 73-82, (1990).

Glass, B. J., Hierarchical Heterogeneous Symbolic Control: Lessons Learned from TEXSYS, *Proceedings of the Workshop on Software Tools for Distributed Intelligent Control Systems.* Pacifica, CA, 83-87, (1990).

Garbow, B. S., S. Nash, and C. Van Loan, *Matrix Eigensystem Routines - EISPAK Guide Extension,* Lecture Notes in Computer Science, **51**, Springer-Verlag, New York, (1977).

Herget, C. J. (ed.), Proceedings of the Workshop on Software Tools for Distributed Intelligent Control Systems. Pacifica, CA (1990).

Ho, Y. C., Performance Evaluation and Perturbation Analysis of Discrete Event Dynamic Systems, *IEEE Transaction on Automatic Control*, **32**, 563-572 (1988).

Inan, K and P. Varaiya, Finitely Recursive Process Models for Discrete Event Systems, *IEEE Transactions on Automatic Control,* **33**, 626-639 (1988).

Integrated Systems Inc., *MATRIX$_X$ User's Guide*, Palo Alto, CA 1985.

James J. R., Expert System Shells for Combining Symbolic and Numeric Processing in CADCS, *Proceedings of the 4th IFAC Symposium on Computer-Aided Design in Control Systems,* Beijing, PRC (1988).

James, J. R. and L. Rapisarda, "Digital Control Workstation," *Proceedings of the IEEE/ASEE Frontiers in Education Conference* (1988).

James J. R. and G. J. Suski, A Survey of Some Implementations of Knowledge-Based Systems for Real-Time Control, *Proceedings of the 27th IEEE Conference on Decision and Control,* Austin, TX. (1988).

James, J.R., J. H. Taylor and D. K. Frederick, An Expert System Architecture for Coping With Complexity in Computer-Aided Control Engineering, *Proceedings of the 3rd IFAC Symposium on*

Computer Aided Design in Control and Engineering Systems, Lyngby, Denmark (1985)

Karsai G., E. Blokland, C. Biegl, J. Sztipanovits, K. Kawamura, N. Miyasaka and M. Inui, Intelligent Supervisory Controller for Gas Distribution System, *Proceedings of the 1987 American Control Conference.* pp. 1353-1358 (1987).

Klose, D.R., L.U. Dworkin and L. Gorr, Software Tools for Lower Echelon Systems Development, *Proceedings of the Workshop on Software Tools for Distributed Intelligent Control Systems.* Pacifica, CA, 131-140, (1990).

LeClair, S., and F. L. Abrams, "Qualitative Process Automation," *Proceedings of the 27th IEEE Conference on Decision and Control*, Austin, TX (1988)

Liu K. and J. Gertler, A Supervisory (Expert) Adaptive Control Scheme. *Proceedings of the 10th IFAC World Congress on Automatic Control,* Munich FRG (1987).

MacFarlane A. G., L. G. Gruebel and J. Ackerman, Future Design Environments for Control Engineering, *Proceedings of the 10th IFAC World Congress on Automatic Control,* Munich FRG (1987).

Mann, R. C., A Hardware/Software Environment to Support R&D in Intelligent Machines and Mobile Robotic Systems, *Proceedings of the Workshop on Software Tools for Distributed Intelligent Control Systems.* Pacifica, CA 203-207, (1990).

Moler, C., MATLAB User's Guide, *Department of Computer Science,* University of New Mexico, Albuquerque, NM, (1982).

Moore R. L., L. B. Hawkinson, M. Lewin, A. G. Hoffman, B. L. Matthews and M. H. David, Expert System Methodology for Real-Time Process Control. *Proceedings of the 10th IFAC World Congress on Automatic Control,* Munich FRG (1987).

Nagy, P. A. J., and L. Ljung, An Intelligent Tool for System Identification, *Proceedings of the IEEE Control Systems Society Workshop on Computer-Aided Control System Design (CACSD), Tampa*, FL (1989).

Nerode, A., Modeling Intelligent Control, *Proceedings of the Workshop on Software Tools for Distributed Intelligent Control Systems.* Pacifica, CA 219-226, (1990).

Pang, G. K. H., M. Vidyasagar, A. Heunis, Development of a New Generation of Interactive CACSD Environments, IEEE Control Systems Magazine, **Vol. 10, No. 5**, pp 40-44 (1990).

Rimvall, M. Integrating Controls System Design with Systems and Software Engineering, *Proceedings of the Workshop on Software Tools for Distributed Intelligent Control Systems.* Pacifica, CA 233-243, (1990).

Salle, S. E. and K-E Arzen, A Comparison Between Three Development Tools For Real-Time Expert Systems: Chronos, G2 and Muse, *Proceedings of the 1989 IEEE Control Systems Society Workshop on Computer-Aided Control System Design (CACSD),* Tampa, FL (1989).

Sanoff, S. P. and P. E. Wellstead, "Expert Identification and Control," *Control Systems Centre Report No. 618*, Manchester, UK (1984).

SDIO Technology Transfer Workshop, AI in Command and Control, *Proceedings of the AI Systems in Government Conference*, Washington, D.C. (1989).

Smith, B. T., J. M. Boyle, et al., *Matrix Eigensystem Routines - EISPAK Guide*, Springer-Verlag, Berlin, 1976.

Spang, H. A. , The Federated Computer-Aided Control Design System, *IEEE Proceedings,* **72**, 1724-1731, (1984).

Symon, J. R. and H. Watanabe, Single Board System for Fuzzy Inference, *Proceedings of the Workshop on Software Tools for Distributed Intelligent Control Systems.* Pacifica, CA 253-261, (1990).

Taylor, J. H. , D. K. Frederick, C. M. Rimvall and H. Sutherland, A Computer-Aided Control Engineering Environment with Expert Aiding and Data-Base Management, *Proceedings of the IEEE Control Systems Society Workshop on Computer-Aided Control System Design (CACSD), Tampa*, FL (1989).

Trankle, T. L. and L. Z. Markosian, An Expert System for Control System Design, *Proceedings of Control '85*, Cambridge, UK (1985).

Zadeh, L. ,personal discussion, NSF/EPRI Workshop on Intelligent Control, Palo Alto, CA (1990).

© 1991 - Paper #91-703
0074-056X/91/043-054/$0 + .50pp

RATE-OPTIMAL MULTI-VARIABLE CONTROL FOR FOSSIL-FIRED POWER PLANTS

Dr. Jan Jelinek Ken Anderson

ABSTRACT

Application of new technology for optimally controlling the response and operation of fossil-fired power plants is described. The technology is based on a new structured multi-input/multi-output controller. Discussed in detail are the technology, methodology, and response results of load and setpoint changes at optimal rates with minimal process variations. Comparisons of the new application with conventional feed-forward PID techniques are described with emphasis on the robustness of the new technology. Included is a description of system's ability to use stress, environmental, and performance information to optimize their operational activities at the controller level without the need for continuous control system tuning. Some simulation results are provided along with a discussion of enhanced features available with such technology. The first use of this control technology was made possible through close sponsorships of the Pennsylvania Electric Company, Honeywell, and the Electric Power Research Institute.

INTRODUCTION

This paper describes the application of a proprietary multiple-input/single output controllers as the primary technology to control fossil-fired power plants.

A power generation unit represents a tightly coupled, non-linear time-varying, multi-input/multi-output (MIMO) system [1][2][3]. Most variables within the process change as a function of unit load (operating point), fuel supply, or process disturbances (trips). The conventional solution used from the days of pneumatics to today's state-of-the-art solutions has been to use single-loop proportional, integral, derivative (PID) controllers with some form of feedforward and *ad hoc* logic to assure adequate performance throughout the load range[4]. The penalty for such interaction can be seen in the normal tradeoffs involved in tuning conventional coordinated control schemes, as shown in Figure 1. These tradeoffs are necessary to maintain both stability and responsiveness with conventional linear control algorithms such as PID.

Self-tuning or optimal regulators, gain scheduling, dynamic matrix compensation, and Inverse Nyquist Array decouplers have been all used in an attempt to solve the interaction and response problems associated with fossil-fueled plants. The primary drawback with these solutions, beyond their expense to implement and maintain, is that they rely on implementation strategies developed for continuous control, PID. The digital implementation of PID is only a good approximation of its pneumatic or analog equivalent, sometimes enhanced with some nice features. The PID implemented as a discrete control algorithm is still a linear controller solution: it takes little advantage of the capabilities and knowledge that can be incorporated into a modern digital control system.

The development of the non-linear finite-horizon Rate Optimal Controller (R.O.C.) in 1988 by Dr. Jan Jelinek[5] has provided an opportunity to replace conventional PID control schemes and finally reap some of the benefit of advanced control for power generation. A complete discussion of the structure of the R.O.C. and its capabilities can be found in Anderson[6].

The first part of this paper is a technical description of the implementation of multi-input /single-output (MISO) ROC's into a "structured" MIMO technology. For continuity a brief description of the implementation of this technology is included. The description outlines the application of R.O.C. technology as a structured MIMO solution for coordination of the boiler and turbine control in modern fossil-fueled power plants. In the final sections, the current results of the application and its future are discussed. This development which holds great promise, is being made possible through the close sponsorships of the Pennsylvania Electric Company, Honeywell, and the Electric Power Research Institute.

MIMO SYSTEMS WITH SPARSE TRANSFER-FUNCTION MATRICES

A Typical Power Plant

The controller presented in this paper applies model-based predictive control concepts on a plant-wide scale. It promises a unified and systematic approach to building robust, high-performance control systems that could involve hundreds of control loops exhibiting mutual coupling (Patent application: A Structured Controller for Multiple-Inputs/Multiple-Outputs (MIMO) Plants).

In the conventional approach, the control system is built incrementally from the bottom up, starting at the level of individual loops. Each process variable is assigned a closed loop driven by a single-input/single-output (SISO) controller, usually a PID. This controllers setpoint represents the desired value of the variable. Because of physical interactions in the controlled process, however, the loops are not independent. Their mutual coupling not only hurts overall performance but also makes the system difficult to operate and maintain. The effects of coupling can be reduced by feed-forward links and cascaded loops, which are usually added *ad hoc* to fix problems encountered in the first design pass. The procedure must often be repeated several times to achieve satisfactory system behavior. Implementing and maintaining these solutions requires considerable control expertise and extensive knowledge of the controlled process.

Modern control theory seemingly offers the exact solution to this problem: the plant is treated as a MIMO system for which the optimal controller can be designed using appropriate methods. This by-the-textbook approach, as yet, is not commercially viable, because it does not meet some important criteria. First of all, for plants with many process variables, the design algorithms are computationally intractable on feasible hardware. In addition, the optimal controller is not structured, cannot be developed incrementally, and does not support modular design, distributed architectures, or multiple sampling rates. Jelinek discusses criteria at length[7].

The structured MIMO control described below combines the benefits of the modern and conventional approaches. It offers control performance and robustness approaching true MIMO optimal control. The internal structure of the described controller gives the user the security of knowing what is going on inside the controller. In addition, it can be developed incrementally; is modular, distributable and computationally acceptable; and allows for multiple sampling rates.

From the system point of view, any power plant is a MIMO system. For the sake of simplicity, consider a linear plant with equal numbers of inputs (i.e., manipulated variables) and outputs (i.e., process variables). If u(s) and y(s) are the input and output vectors, then

$$y(s) = G(s) \cdot u(s) \qquad (1)$$

where the transfer function matrix, G(s), is compiled from the transfer functions $G_{ij}(s)$, ij = 1,...,N, relating the jth input to the ith output of the system.

Generally speaking, most, if not all, the entries for any plant will be different from zero, suggesting that any input has some impact on every output. The key observation is that, although non-zero, many entries will have very small gains, so their contribution to an output will be small. If we neglect them, a controller designed using the simplified matrix will not be optimal; nevertheless, its performance will be close to optimum. We estimate that a typical process variable does not strongly depend on more than five manipulated variables. Most of these variables will be directly manipulating the boiler-turbine unit at which the process variable is measured. Most fossil units are naturally distributed allowing for the strongly interacting variables to be handled by a single controller node on the unit where the variables originate. Only distant interactions such as a plant coordinator or system dispatcher need be communicated through a supervisor connecting individual controllers. This situation greatly reduces the volume of data to be exchanged among controllers at every sampling period.

In neglecting the weak interactions we turn G(s) into sparse matrix. Its ith row lists up to, say, five non-zero transfer functions in the columns corresponding to the plant inputs that have a strong influence on the ith plant output. (Hereafter, we will always assume five important inputs.) For example, in an extreme case, such as a 100 by 100 plant, out of 10,000 possible I\O interactions, at least 9,500 will be neglected. If we find a way to identify the important inputs for each output and apply a controller design algorithm that takes advantage of the transfer matrix sparseness, we have, in principle, a MIMO control system that is about as good as the optimal system. However, the computational costs associated with this type of system are only a fraction of similar costs for the optimal system. This design philosophy has motivated the structured MIMO control system.

THE STRUCTURED MIMO CONTROL ALGORITHM

Consider a MIMO plant with N inputs $u_1(k), u_2(k) \ldots u_N(k)$ and N outputs $y1(k), y2(k), \ldots y_N(k)$ sampled at time kT, T being the sampling period. The performance criteria to be minimized is

$(u_1,u_2,.....u_N)$ = \I\SU(m=1,N, $v_m(y_m - r_m)^T(y_m - r_m)$) + \I\SU(m=1,N, $w_n(q \cdot u_n - u_n)^T(q \cdot u_n - u_n)$) (2)

where q is the forward shift operator.

The optimal control strategy is defined as a set of sequences $\ddot{u}_1, \ldots, \ddot{u}_N$ (¨ denotes a sequence) which minimizes (2):

$$J(\ddot{u}_1,\ddot{u}_2,\ldots\ddot{u}_N) = \min_{u_1,u_2,\; u_N} J(u_1,u_2,\ldots u_N) \tag{3}$$

The structured MIMO controller uses a variant of the gradient descent algorithm approach that has been simplified and termed "steepest partial gradient descent algorithm". A very complete theoretical base for this subject has been explored by Stengal[8]. For criterion (2), the partial gradients are computed for all control sequences, $\ddot{u}_1,\ddot{u}_2,\ldots\ddot{u}_N$, and applied to individual plant inputs within the computational window. Because no plant output is assumed to depend on more than, say, five plant inputs, y_n is function of at most five sequences $\ddot{u}_{n1},\ddot{u}_{n2},\ldots\ddot{u}_{n5}$ out of N, making the partial gradient evaluation a simple task.

The simplification is important for two reasons. First, it cuts down the computational complexity and makes the algorithm more balanced so every controller will be computing the same code. Second, it gives every MISO the appearance of a stand-alone controller, so the user can view the structured MIMO controller more as a multitude of single SISO controllers than as black box with N inputs and N outputs. Of course, the constituent controllers internally cannot be SISO, because in addition to the process variable input they also need other inputs for exchanging the information necessary for their mutual coordination. The idea is to make the coordination inputs and the associated controller interconnection supervisor as user-transparent as possible, so the user is not necessarily distracted and confused by the variable interactions. This is accomplished by pairing the plant inputs and outputs.

In a MIMO plant, the nth input is not related to the nth output more than to any other output, because it affects all outputs to some extent. Nevertheless, from the practical point of view, such as a power plant, an input can often be perceived as primarily responsible for the manipulation of a particular output (i.e., process variable) and thus associated with it. Such input will be called dominant for the output. All other inputs whose contributions to the output were not neglected up front, are called interfering. In the above example, there might be up to four interfering inputs for each output since the fifth input is deemed dominant. Usually, the dominant input-to-output transfer function has a larger gain than any of the interfering ones for any particular output of the plant.

Let us now return to the algorithm. The partial gradient of (3) by $\ddot{u}_n$ is

$$\begin{aligned}\mathbf{grad}_{\ddot{u}_n} J(\ddot{u}_1,\ddot{u}_2,\ldots\ddot{u}_N) = & \sum_{m=1}^{N} v_m \mathbf{grad}_{\ddot{u}_n}(y_m - r_m)^T(y_m - r_m) + \\ & \sum_{m=1}^{N} w_n \mathbf{grad}_{\ddot{u}_n}(q \bullet u_n - u_n)^T(q \bullet u_n - u_n)\end{aligned} \tag{4}$$

Even though the sum appears to imply N terms, actually most of them will be zero. Only the terms in which one of the five control sequences affecting the output $y_m, m = 1,\ldots,N$ happens to be u_n will be different from zero. The number of non-zero terms will vary for individual partial gradients, because u_n may be present in an unspecified number of y_m's. In one and only one of the terms u_n will represent the dominant input; in all others, the same u_n will play the role of an interfering input. If we drop all the interfering terms, the partial gradient will be replaced by its estimate

$$\begin{aligned}\mathbf{grad}_{\ddot{u}_n} J(\ddot{u}_1,\ddot{u}_2,\ldots\ddot{u}_N) = & v_m \mathbf{grad}_{\ddot{u}_n}(y_m - r_m)^T(y_m - r_m) \\ & + w_n \mathbf{grad}_{\ddot{u}_n}(q \bullet u_n - u_n)^T(q \bullet u_n - u_n)\end{aligned} \tag{5}$$

where we now assume y_m to be the dominant output for u_n. It should be noted here that, unlike the interference concept, the dominance concept works in both directions. Every output has just one dominant input, and every input is dominant for just one output. Obviously, in introducing dominance we impose a somewhat arbitrary one-to-one pairing between inputs and outputs of the plant. To simplify the subscripts in (5) we may renumber the plant outputs so that m = n:

$$\begin{aligned}\mathbf{grad}_{\ddot{u}_n} J(\ddot{u}_1,\ddot{u}_2,\ldots\ddot{u}_N) = & v_n \mathbf{grad}_{\ddot{u}_n}(y_n - r_n)^T(y_n - r_n) \\ & + w_n \mathbf{grad}_{\ddot{u}_n}(q \bullet u_n - u_n)^T(q \bullet u_n - u_n)\end{aligned} \tag{6}$$

where the output y_n depends on the dominant input u_n and up to four interfering inputs $u_{n1},\ldots,u_{n4}$:

$$y_n = f(u_n, u_{n1}, \ldots, u_{n4}) \tag{7}$$

This formal change will also rearrange the plant matrix so that the dominant transfer functions will always be on its main diagonal.

IMPLEMENTATION OF THE STRUCTURED MIMO CONTROLLER ALGORITHM

The structured MIMO controller uses the minimization algorithm described in (3) with the partial gradients replaced by their estimates (6). More formally, the algorithm is

$$|u_n|_{i+1} = |u_n|_i - \partial_{n,i} \bullet \mathbf{grad}_{\ddot{u}_n} J'(\ddot{u}_1,\ddot{u}_2,\ldots\ddot{u}_N)|_i \quad \text{for } n = 1,\ldots,N \tag{8}$$

with ∂_n minimizing $J'(\ddot{u}_1,\ddot{u}_2,\ldots\ddot{u}_N)|_{i+1}$ in every iteration. As earlier, each of the N recurrent procedures in (8) updates the control sequence on one plant input, and may be pictured as running in a stand-alone controller. In (8) we explicitly emphasize that $\mathbf{grad}_{\ddot{u}_n} J'(\ddot{u}_1,\ddot{u}_2,\ldots\ddot{u}_N)$ actually depends only on the dominant input and up to four interfering inputs for each plant input:

$$\mathbf{grad}_{\ddot{u}_n} J'(\ddot{u}_1,\ddot{u}_2,\ldots\ddot{u}_N) = \mathbf{grad}_{\ddot{u}_n} J(\ddot{u}_1,\ddot{u}_2,\ldots\ddot{u}_N) \tag{9}$$

We have not yet discussed how the step lengths in (8) are computed. In fact, determining the optimal length minimizing the criterion in every iteration is by far the most computationally demanding task of the algorithm.

In the structured MIMO controller, the algorithm functions according to (8), but the actual implementation is not a coded version of the procedures of (8). If we take another look at (8), we may also interpret it as a set of procedures for minimizing the criteria

$$J(u_1,u_2,\ldots u_N) = v_n(y_n - r_n)^T(y_n - r_n) + w_n(q\bullet u_n - u_n)^T(q\bullet u_n - u_n) \quad \text{for } n = 1,\ldots,N \tag{10}$$

at every iteration. Note that each of the N criteria above is to be minimized with respect to the dominant input, only even though y_n in its body also depends on the interfering inputs, which, in turn, are dominant inputs of some other criterion in the set (10). Though the criteria are mutually coupled through their variables, the coupling does not show up directly in the computations. In the course of computing (8), the interfering variables are thought of as fixed parameters and are updated only at the end of every iteration. This is a result of the very definition of the partial gradient (or derivative).

The control sequence that minimizes $J(u_n)$ for fixed $u_{n1}\ldots,u_{n4}$ and any given (dominant) transfer function can be found by control theory methods. In our implementation, the sequence is also required to be deadbeat, leaving the output at a steady state if the sequence was ever fully executed. The algorithm generating the control sequence on-line is called the multiple-input/single-output rate-optimal controller (MISO R.O.C.) (Patent application Docket 1185(T1013563/US O Controller for Electric Utilities). Using the terminology of the previous sections, the MISO R.O.C. is the controller. Each plant input is served by one MISO R.O.C. whose setpoint refers to the input's dominant output. The MISO R.O.C. 's are interconnected by a supervisor. However, now only the control sequences need to be transmitted, never the output and setpoint sequences, as was the case for the partial gradient descent algorithm. Also, the MISO R.O.C. 's are identical and therefore can be thought of as generic building blocks which are hooked to the plant inputs and interconnected by the supervisor as in Figure 2.

When a MISO R.O.C. completes the computation of the updated control sequence, it must make this information available to any MISO R.O.C. 's that may need to incorporate its consequences into their own control strategies. One can imagine each MISO R.O.C. as posting its sequence on a public bulletin board. Among all the posted sequences, each MISO R.O.C. is interested only in those few that interfere with its control objectives through the process variable interactions in the plant. The control engineer setting up the controller must specify for each MISO R.O.C. which four (in case there are four interfering inputs) out of the posted sequences are of interest to it. Technically, this is implemented by feeding the sequences into the MISO R.O.C. through a properly laid out supervisor. The engineer sets up the supervisor connections by means of the configuration table. As described above, in every sampling period as specified by the designer the controller executes one or more iterations of the algorithm. After the last one is completed, the first element of the control sequence that represents the current control signal value is sent to the corresponding actuator. The remaining ones are planned actions that are likely to be modified in the next sampling period as a result of unanticipated disturbances acting on the plant, as depicted in Figure 3 for a 3X3 plant.

Coping with process variable interactions is the most difficult aspect of designing controllers for MIMO plants. At the outset, the designer may lack knowledge concerning the interactions. With the structured MIMO controller, he may begin by assuming the most optimistic scenario, namely that there are no interactions at all. That is, the MIMO plant, in fact, is just a collection of N mutually independent SISO plants. Once the controller is in place, a short attempt at running the system usually proves the assumption wrong. The important advantage of the presented approach is that the configuration table can be built up incrementally and in a systematic fashion, without turning the controller off. It is impossible to uncover all interactions at once, because stronger ones tend to bury weaker ones. However, once the strongest one is compensated for and becomes invisible, the next strongest one can be identified, and so on. In this way, the designer can proceed in suppressing the interactions as long as he considers necessary.

CONCURRENT FRONT END
The First Applied Structured MIMO

Penelec, EPRI, and Honeywell in a joint effort have developed an advanced unit master controller under the project name "Concurrent Front End" (CFE). The new application is based on the enabling technology of a proprietary rate-optimal algorithm and its structured MIMO implementation. This new algorithm can accommodate process variations and make changes to control constraints based on information developed by other programs running anywhere and communicated to the distributed control system. The application of this concept will be the first modern control offering that solves the problems associated with the dynamic characteristics of the generating unit. Using true modern digital control technology the control system will be integrated to coordinate turbine and steam generator responses. The new front end is the enabling technology that links the control system to process conditions and constraints being generated by process-optimization packages running in modern power plants.

The position of the "Concurrent Front End" (CFE) in the control system is similar to the old unit master controllers for the appropriate sub-loops. As depicted in Figure 4, the CFE acts as an interface between the operator or load dispatch on one side and the boiler, turbine, and other sub-loop components on the other. The CFE includes interfaces that give the dispatcher accurate information on energy available to respond to generation demand change and system frequency stability. The CFE does not replace the existing sub-loop control clusters around the fuel feed, attemperation, and some other system components. The CFE strategy has been developed to balance process stress, environmental constraints, energy storage changes, firing rate, and megawatt production under varying generation demands.

The totally independent handling of reference and disturbance changes provided by the R.O.C. algorithm allows for practical implementation of three new modes of control, shown in Figure 5. Application of the CFE concept solves the problems described in the introduction in a systematic way. Through different operating ranges, each plant will have varying process dynamics related to both the equipment in service and the length of time at that load range. As illustrated by the Figure 2, different "knowledge packets" of the current state of operations can be developed for varying conditions. These packets are switched into use by the supervisor instantaneously on any given sampling period. In this way the CFE can be constructed to compensate its operating rates, constraints, and process interpretations based on process conditions such as run backs, rate limits, or response requirements.

The first implementation of the CFE using the structured MIMO is depicted in Figure 6 for the supercritical boiler-turbine power plant. A complete analytical discussion of the use of this methodology is described in detail by Anderson [9]. As indicated by this structure, the turbine, fuel, and feed-water masters are still available as independent controllers if boiler- follow or turbine-follow modes are required. But the response-based mode of the system tends to make these modes generally unnecessary, even under run-back conditions. In the application shown in Figure 6, R.O.C.#1, R.O.C.#2, and R.O.C.#3 are independent controllers running sequentially in the same sampling period. The input and output sequences $\ddot{u}_m$ of each of the R.O.C. 's will be updated once each time step. At a given processing instant, if R.O.C.#1 is currently being evaluated, its control sequence, $\ddot{u}_1$, will be computed using the previously computed sequences for R.O.C.#2 $\ddot{u}_2$ and R.O.C.#3 $\ddot{u}_3$. As previously described, the controls $\ddot{u}_2$ and $\ddot{u}_3$ are seen as known disturbances in the process. In each step n, after $\ddot{u}_m(k+n-1)$ is computed, new sequences, y_{M1}, y_{M2}, and y_{M3}, can be updated for each controller.

Operation of a structured MIMO (SMIMO) for power plant control can best be described by example. Initially, the form of the SMIMO for a supercritical boiler turbine construction is shown in Figure 7. At one load point the input-output transfer functions for the MIMO plant are identified, as shown in Figure 8. The resulting gains show some significant coupling. Figure 9 shows a step response for which both the setpoint change for plant (controller 3) of 10 percent. The perturbations to be eliminated at a rate of .5 units/sample. As the results show, even when taking advantage of the MISO R.O.C.'s robustness, the coupling effects of the load change cannot be eliminated by controller 2. But given the information of the coupling sequences, Figure 10 clearly shows almost all signs of the interference between plant variables are overcome. Yet, as shown in Figures 11 & 12 disturbances on the other plant outputs will be eliminated at the proper rates with no change in the tuning of the system. This again indicates the ability to separate, in the discrete domain, the control response desired and the tuning of the controller[5].

A significant amount of energy must be absorbed or extracted by the process as a result of control actions. This is often undesirable if the process rates of change cannot properly be controlled. Current digital technology allows power plants to be moved at rates of change that can create both thermal fatigue and emissions excursions that can heavily penalize current and future production. The use of *ad hoc* logic combined with classical control schemes to automatically limit such behavior is inefficient and can unnecessarily reduce unit response and heat rate. Improved station heat rate through variable pressure operation has been well documented [10]. The problem associated with this form of operation is related to decreased response when in variable pressure modes. The CFE, using true predictive coordination between turbine and boiler, can provide a very stable response by moving in the fastest rate-optimal manner, not in the current rate limited fashion. It is expected that the power generation industry will see significant benefits from this development. The successful completion of the first implementation of the CFE will quantify the following expected benefits:

◊ Increased plant life and reduced component failure through fixed rates of change based on critical stress levels.

◊ Improved system frequency response though rapid control mode switching for fixed-duration, fixed-rate.
◊ Optimized response to dispatched load changes through RBM feedback to the dispatch center.
◊ Improved heat rate through improved process stability, through fixed-rate control rather than through plant response feedback.
◊ More avenues for sub-loop improvements becoming available with the application of an advanced control technology as the foundation for plant-wide control.

The joint EPRI - PENELEC - Honeywell development is scheduled to complete demonstration by June 1991. The complete results will then be made available to the utility industry. The CFE is the first use of the R.O.C. and SMIMO in a power plant application.

SUMMARY

The CFE as implemented will be targeted to provide or improve plant stability, quality of steam conditions, thermal storage, sliding pressure operation for drum boilers, turbine valve management, and plant response to dispatched load.

In any fossil-fueled plant, the need for coordination of the boiler energies and turbine energy is the fundamental design requirement of the plant control system. In the case of the supercritical once-through boilers, the process interactions among feedwater-rate, firing- rate, and turbine valve movements are more closely coupled than is the case of drum-type boilers. In drum boilers, with their large saturated liquid storage and the insulating effect of having saturation linked to the drum pressure. For drum boilers there is a large pressure stored energy compared to the changes in feed-water flow [11]. In fact, an important difference in the characteristic and, hence, control requirements of once-through versus drum-type boilers is the steady state furnace tube flow rate being proportional to load in a once-through boiler. This results in the thermal storage playing a large role in solving optimum rates of change in both feed-water flow rate and firing rate for supercritical units. The CFE therefore has only two basic constructions, one for drum-type boilers and the other for once-through construction. The first application of the CFE is a 850 megawatt supercritical boiler-turbine unit.

The CFE provides the control mechanism for taking advantage of the information being acquired by other sub-loops and optimization routines in the power plant. This information is not easily integrated into classical linear control algorithms or compensators that are tied to more orthodox strategies. Access to this information allows the CFE to provide maximum unit response under constrained plant operation while maintaining pressure and thermal balance. The priority of a particular parameter or rate configuration in the CFE should attempt to optimally control is operator- or program-selectable. No *ad hoc* logic, feed-forwards, tuning, or algorithm modifications are required if these priorities or rates are changed. Plant parameters and constraints are effectively utilized in the CFE structure to provide a robust controller for the process, not another advisor to the operator. To maximize the benefits of the CFE, an output of available, predicted, and scheduled load rate can be integrated with the utility's economic dispatch system. The CFE is expected to be a leading technology for integration of the roles of the turbine, boiler, and dispatcher to respond to load demand changes efficiently and reliably.

ACKNOWLEDGMENTS

This work could not have proceeded without the support of a utility participant. I would like to thank Pennsylvania Electric, and especially Marty Heeter and Rick Seese, for their support and insight.

Conventional Control

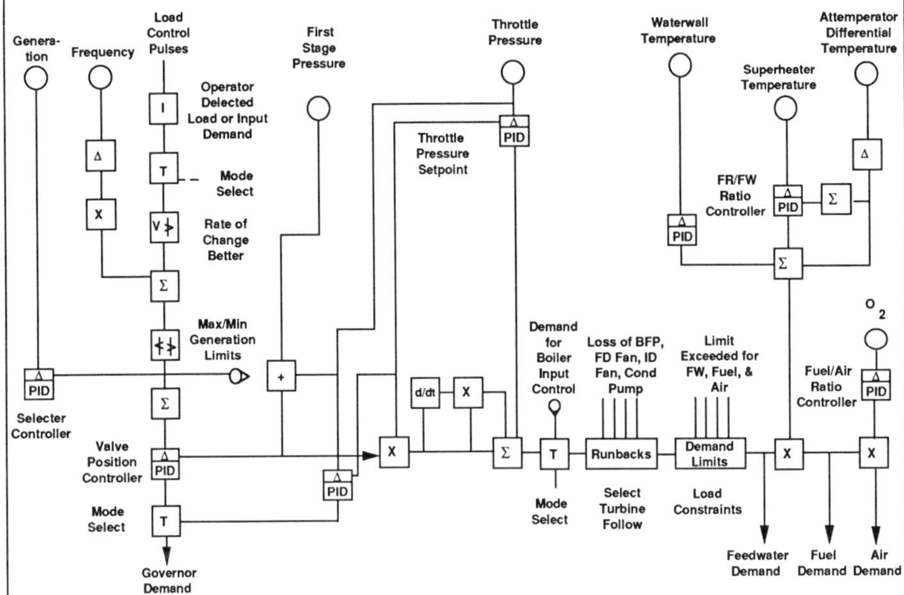

Figure 1

Concurrent Front End Implementation

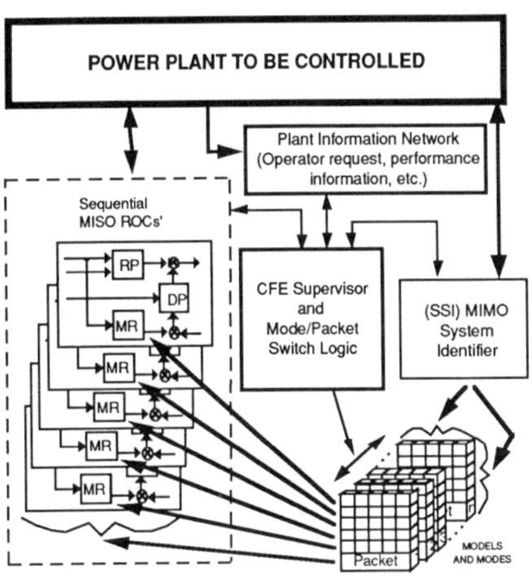

Figure 2

Control Front End with MISO Controller

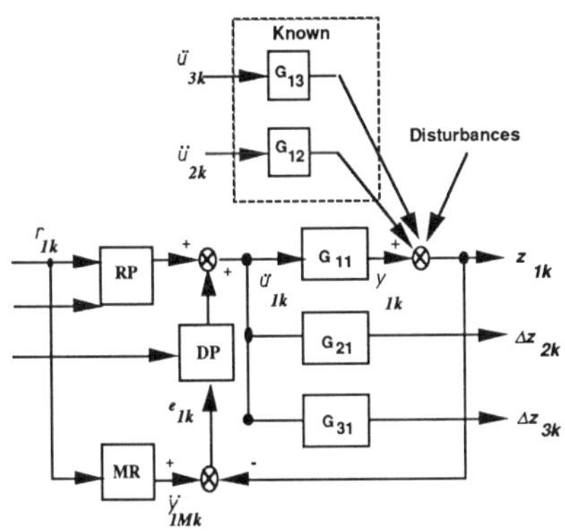

Figure 3

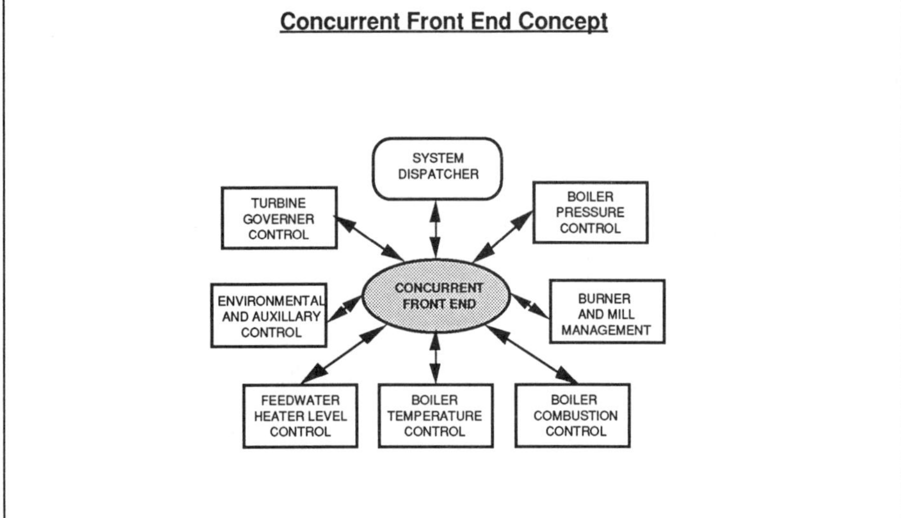

Figure 4

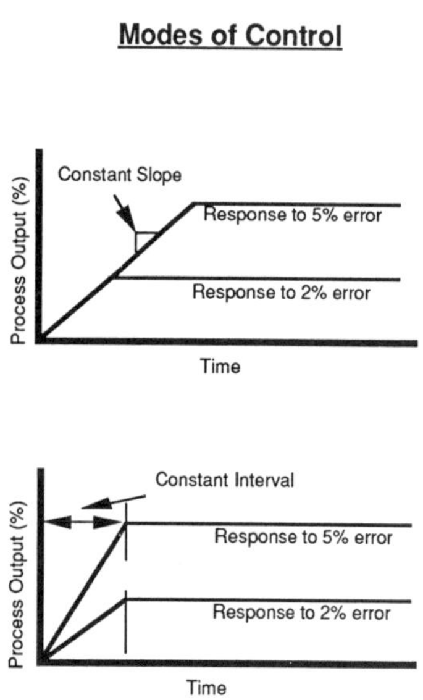

Figure 5

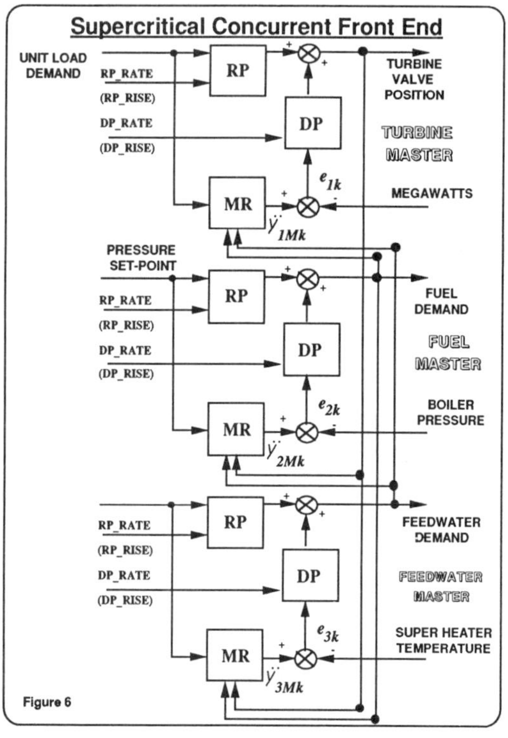

Figure 6

Conemaugh Concurrent Front End Implementation

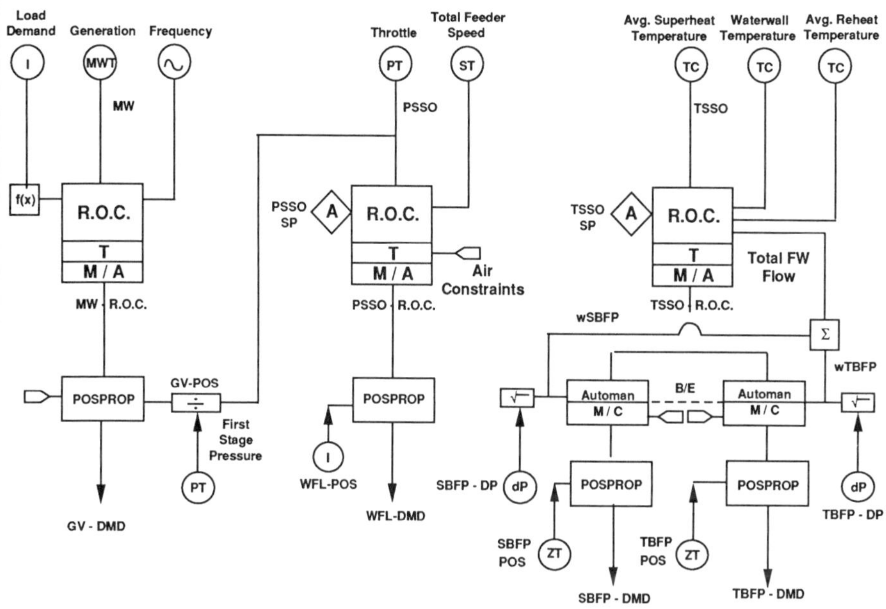

Figure 7

3 x 3 MIMO PLANT

$$
\begin{bmatrix}
\dfrac{20}{S+2} & \dfrac{-1.594}{S+10} & \dfrac{0.5}{S+0.8196} \\[2ex]
\dfrac{-0.015s}{(S+2)(S+0.216)^3} & \dfrac{-6.11\,e^{-3S}}{(S+0.8333)(S+0.216)^3} & \dfrac{1.0797\;\;e^{-4S}}{(S+0.416)(S+0.8333)} \\[2ex]
\dfrac{-0.1866}{S+0.5465} & \dfrac{0.5787}{(S+0.8333)(S+0.3125)} & \dfrac{1.42\,(S+0.6)\,e^{-12S}}{(S+0.8196)(S+0.416)(S+0.8333)}
\end{bmatrix}
$$

GAINS

10.0	-0.1594	0.61
-0.744	-727.576	3.1146
-0.158	2.22	3.0

NORMALIZED GAINS (Columns Divided by Dominant Gain)

1.0	2.191×10^{-4}	0.203
-0.0744	1.0	1.038
-0.0158	3.05×10^{-3}	1.0

Figure 8

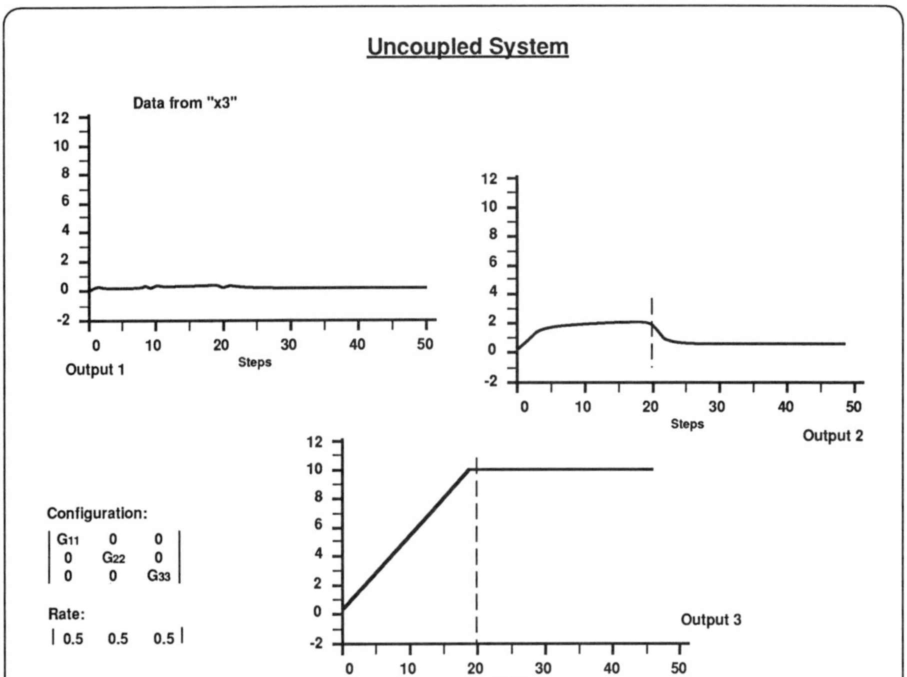

Figure 9

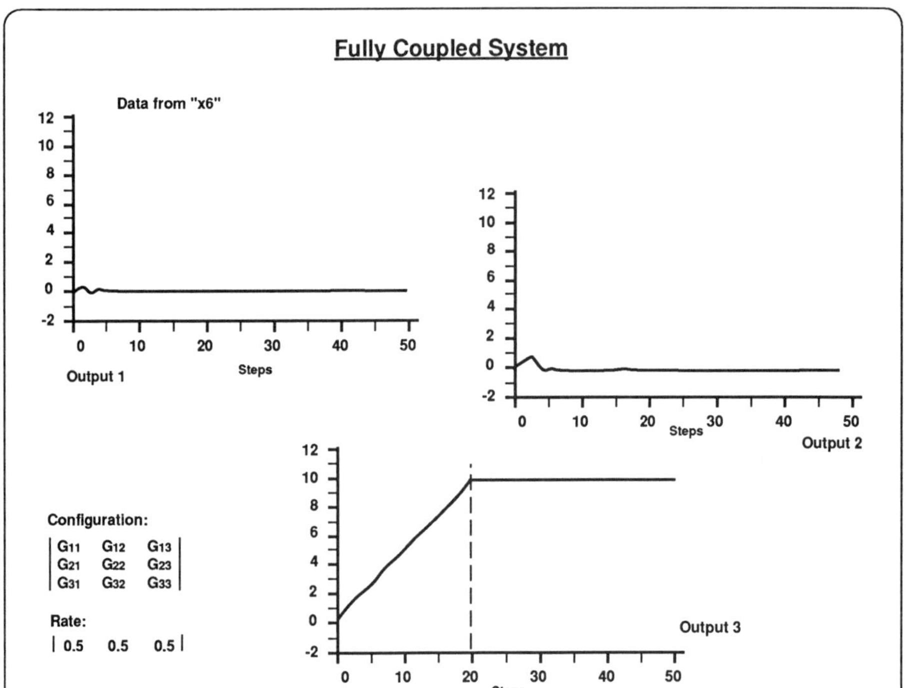

Figure 10

REFERENCE

[1] Amin M.,et al., "Application of a Multivariable Self-Tuning Controller to a Power Plant Boiler", ACC, Proceedings, San Diego, CA, June 1984.

[2] Ohki T., Inoue T., "Dynamic Characteristics and Control System for Variable Pressure Operation Supercritical Once-Through Steam Generator",

[3] Hope J., "Multivariable Control System CAD: Power Plant Application and Experience", EPRI Conference on Power Plant Controls and Automation, Miami, Fl, 1989.

[4] Gery H.C., "The Evolution of Coordinated Control", ISA, St Petersburg, FL, May,1988.

[5] Jelinek J., "Cycling Controllers for Electric Utilities" EPRI Conference on Power Plant Controls and Automation, Miami, Fl, 1989.

[6] Anderson K., "Power Generation Control Strategies for the Decade of the 90's", EEI 1990 Engineering and Operating Computer Forum, St. Louis, MO, Sept. 1990.

[7] Jelinek J.: Towards Plantwide Model-based Control, Honeywell SSDC Report (in preparation).

[8] Stengal R.F., Stochastic Optimal Control Theory and Application, Wiley, New York, N.Y., 1986, pp. 259-270.

[9] Anderson K.J., "Advanced Control Technology to Reduce Power Plant Stress", ASME Publication, Paper #90-JPGC/Pwr-2, Boston, MA, Oct. 1990.

[10] Lansing E.G., "Variable Pressure Peaking Boiler, Operation, Testing, and Control", ASME Publication, Paper #74-WA/Pwr-2, NY, 1975.

[11] Jenkens T.W. and Littman B., "Response Capability in the Control of Large Generating Units," IEEE Winter Power Meeting, NY, 1971. 71 CP 73 - PWR

APPENDIX 1

Definitions, Acronyms, Abbreviations

"^" - Denotes a sequence of discrete values.

CFE - Concurrent Front End - Program name of an advanced control application for power utilities based on the R.O.C. algorithm.

Coordinated Action - Occurs when the disturbance processor appears to share its setting with the reference processor. Actually, the disturbance processor only provides active robustness and does not directly participate in control.

DP - Disturbance Processor - The closed-loop response procedure internal to the R.O.C..

e(k)- Processor Error - The difference between Y_Mk and Yk at sampling instant (k).

EPRI - Electric Power Research Institute, Palo Alto, CA.

V(k) - Data Validation Indicator - An output produced by the R.O.C. which indicates the health of the Uk to Yk relationship which can be corrupted by load disturbances.

Fixed Duration - A control sequence that forces the process to respond at a constant desired interval with zero overshoot.

Fixed Rate - A control sequence that forces the process to respond at desired constant slope which shall never exceeds the specified rate.

LMSE - Least Mean Square Error

MIMO - Multiple-Input/Multiple-Output - A controller structure based on many process variables and many manipulated variables.

MISO - Multiple Input/Single Output - A controller structure based on many process variables and one manipulated variable.

MR - Model Reference Processor - The procedure that determines the expected plant response to the scheduled control action.

PENELEC - Pennsylvania Electric Company, Johnstown, PA.

PID - The classic proportional, integral and derivative linear control algorithm used in conventional control schemes.

R.O.C. - Rate Optimal Controller - An advanced control algorithm developed at Honeywell SSDC by Dr. Jan Jelinek.

RP - Reference Processor - The open-loop response procedure internal to the R.O.C..

S - Supervisor - Coordinates procedures for determining the dominant control action, reference model obsolescence, decision and transient behavior synthesis, etc.

SISO - Single Input/Single Output - A controller structure based on one process variable and one manipulated variable.

SMIMO - Structured MIMO - A control methodology in which discrete MISO controllers are synthetically coupled to create a MIMO structure.

SSI - Stochastic System Identifier - The system identifier algorithm that produces the model coefficients and process delay values to be used by the model reference processor in the R.O.C..

U(k) - Control Amplitude/Action - The output of the R.O.C. at a given sampling instant (k).

Ur(k) - Reference Processor Control Amplitude/Action - The output of the R.O.C. RP at a given sampling instant (k).

Ud(k) - Disturbance Processor Control Amplitude/Action - The output of the R.O.C. DP at a given sampling instant (k).

Uncoordinated Action - Occurs when the reference and disturbance processors are set up independently to operate either in the fixed-duration or fixed-rate fashion.

Y(k) - Process Amplitude/Output - The feedback variable to the R.O.C. at a given sampling instant (k).

YM(k) - Model Reference Processor - The output of the R.O.C. MR at a given sampling instant (k).

©1991 - Paper #91-704
074-056x/91/055-068/$0 + .50pp

GUIDE TO BOILER CONTROL SYSTEM CHECKOUT AND START-UP

C. W. Taft G. R. McFarland

ABSTRACT

A thorough and proper boiler control system checkout and start-up is crucial to obtain optimum performance from a new control system. A systematic approach with extensive documentation is required to achieve a successful and rapid start-up. The paper presents such an approach based on the authors' personal experience with many start-ups. Procedures for checking power, grounding, field wiring, final drive units, sensors and transmitters are described. Initial operation guidelines and initial and final calibration and tuning of the system are also discussed. The paper will serve as a useful reference for vendor start-up engineers and utility engineers responsible for the maintenance of a system. Keywords: Boiler control, control system, start-up, checkout.

INTRODUCTION

A thorough and proper boiler control system checkout and start-up is crucial to obtain optimum performance from a new control system. By the time a new boiler control system is ready for checkout and start-up, many manhours have been expended on the important tasks of engineering studies for project justification, definition of project scope, writing the technical system specification and procuring the boiler control system. Once the system is purchased, either the vendor or the user develops the control system philosophy and logic and implements the design into the hardware.

The final phases of the project, system checkout and start-up, are just as important, as they can mean the difference between a successful and unsuccessful project. Because the checkout and start-up of a new system is a long complicated process, it is important to have a systematic approach for the entire process. This paper describes one approach which has been used successfully by the authors. The major steps in the process are as follows:

1. Review installation,
2. Verify proper power and grounding,
3. Calibrate and check input signals,
4. Stroke final drive units,
5. Perform initial system calibration and scaling,
6. Initial tuning,
7. Final tuning,
8. Acceptance testing.

Following the steps listed above will ensure that the start-up progresses in an orderly fashion and can save the start-up engineer many hours of unnecessary backtracking and rechecking of earlier work.

The checkout and start-up of a new boiler control system is a joint venture between the end user and the boiler control system vendor, and the better the working relationship, the better the odds for a successful project.

CONFIGURATION

Although configuration is not technically part of system checkout and start-up it is vitally important to the start-up and therefore warrants discussion in this paper. With today's distributed digital control systems, the system configuration is typically done by one of three parties: the manufacturer of the system, the architect/engineer for the project or the final user of the system. It is very desirable that the engineer who will perform the final system start-up is involved somehow in the configuration of the system. The more involvement there is the better. As a minimum the start-up engineer should be involved in the factory checkout of the configuration software. The best situation is one in which the start-up engineer is part of the configuration team and heavily involved throughout the configuration work. This is rare but ensures that the start-up engineer is intimately familiar with the control system design and implementation long before he has to make the system perform on a real boiler.

It is not uncommon for a start-up engineer to modify a control strategy in the field simply because he is not familiar with the concept behind the strategy and has trouble making it work. If he is involved in the configuration effort he will understand the concept behind the strategy.

FACTORY CHECKOUT

The checkout of the boiler control system starts at the vendor's facilities. Once the vendor has completed manufacturing and configuration of the control system, a thorough checkout of the system's hardware and software are performed. After this checkout is completed, the end user (client) and the project architect/engineer (A/E) will visit the vendor's facilities to witness a demonstration of the proper operation of the boiler control system. This demonstration is called the factory acceptance test (FAT). The type of test performed will vary depending on the vendor and the customer's requirements. The type of test should be defined in advance and agreed to by both parties. A detailed test procedure should be submitted by the vendor in advance for approval by the end user. Once the test procedure is agreed upon, it will be used during the test as the check-off document.

Except for the simplest control systems it is very desirable to use closed loop simulation to verify proper operation of the system during the factory checkout. There are currently three different methods of doing closed loop simulation checkout. The first is to close the loop in software within the existing control system hardware as shown in Figure 1. This is done by modifying the configuration software so the points which are normally read from or written to real input/output hardware are connected together. This way the output to a control drive for instance is connected to the input which is affected by the drive. Simple dynamic models using lead/lag blocks and integrators can be placed between the software output and the software input to represent the process being controlled. This type of test can verify most of the control logic and strategies and requires no additional checkout hardware. Its main limitation is that it provides no check of the real I/O hardware and whether the I/O hardware is properly addressed. Another drawback is that the configuration software must be modified to perform the checkout and then returned to its original state when shipped to the site. There is always the possibility of introducing errors in the software during the change from the checkout version back to the original version. Another point is that the dynamic modeling capability is very limited using this type of FAT and therefore it is not possible to do any realistic tuning during the FAT.

Another approach to a closed loop checkout is to add another drop on the distributed control system (DCS) highway strictly for doing the dynamic modeling of the plant. This approach, shown if Figure 2, is similar to the one discussed previously except that much more sophisticated models of the process can be constructed and therefore it may be possible to pre-tune some of the loops during the factory checkout. It still does not verify the I/O hardware and the software together in one integrated test.

The third method, shown in Figure 3, is to use a completely separate real time checkout simulator and connect it to the control system exactly as the real plant equipment is connected. In all cases today, the separate simulator is a computer with I/O equipment which emulates the real transmitters, thermocouples and final control drives used in the plant. Since the simulator is a general purpose computer it is practical to construct sophisticated non-linear dynamic models of the process and use them for a detailed evaluation of the control system. This system will do the best job of checking out the system but is also more expensive and sophisticated. It may be overkill to use such a simulator for checking out a very simple control system, but for complex boiler control systems such a simulator can easily pay for itself with reduced start-up costs. Start-up costs are reduced because the control system is fairly well tuned when it reaches the plant and therefore the start-up time is shortened.

After completion of the FAT and sign-off by the customer, the system is shipped to the plant site for installation and the next phase of system checkout.

INSTALLATION

The installation includes the physical placement of the system components, the connection of proper grounding and power systems and the connection of the field wiring to the system termination panels.

The key to a successful installation is an accurate and detailed set of drawings from the vendor and the project's engineers. Drawings for the physical location of equipment, the power and grounding cabling, termination drawings or lists and instrument installation drawings are typically provided. The start-up engineer generally will not be involved in the installation but will begin work when the installation is complete.

LOGBOOK

The first thing the start-up engineer should do upon arrival at the jobsite is prepare a start-up logbook using a permanently bound notebook with numbered pages. It is very important that the start-up engineer keep a detailed record of all significant activities in this book. Circuit card failures, calibration activities, milestones and tuning progress are some of the items to be recorded in the logbook. As a minimum the start-up engineer should make entries in the logbook at the end of each day although it is better to make entries throughout the day when events happen. This leads to more accurate and complete entries and less forgotten items at day's end. The logbook will contain a record of the entire job and will be a valuable aid to the start-up engineer in remembering activities throughout the start-up. It is particularly useful when there is a change in start-up engineers during the project.

SYSTEM CHECKOUT

The first major checkout activity for the start-up engineer is the verification of the power and grounding installation.

POWER

Generally verifying the power connections is a straight forward process of comparing actual connections with those shown on the design drawings. The power requirements of the system must be defined in advance to ensure that the correct size power main and size and number of feeder circuits are provided. Power should never be applied to the system until power supply connections have been checked against the drawings. This includes external power and internal power distribution. Hopefully, the internal power wiring was fully checked at the Factory Acceptance Test but it is worth rechecking to be sure. Some things to check on the external power connections are the wire size, feeder breaker size, and that the supply is from the correct feeder panel. If the system includes printers, CRT consoles or other remote devices those should all be powered from the same feeder circuit. They should not be plugged into the nearest convenience outlet.

GROUNDING

Unfortunately, checking the grounding system is not nearly as straight forward as checking the power supply system. Before powering up any distributed control system, the start up engineer should completely check and verify the installation for proper grounding. The term grounding refers to two completely different electrical techniques and therefore is frequently misunderstood. One aspect of grounding is for the safety of personnel and equipment and requires a direct connection of one of the AC power conductors of the control system to the ground mat of the plant. Also, all the non-current carrying metal parts of the control system must be electrically bonded together and connected to the ground mat. The other aspect of grounding is providing a zero voltage reference for the measurement signals in the system.

The manufacturer's documentation should be used to design the grounding system but there are frequently special situations which are not described in the manufacturer's instructions. It is not the start-up engineer's job to design the ground system but only to check that it is installed correctly. Because each distributed control system vendor has unique recommended grounding procedures there are no universally accepted rules and almost every installation is a little different. The start up engineer should not only verify that the vendor's procedures have been followed for the individual pieces of equipment, but that basic grounding procedures have been followed for the complete control system. This includes control cabinets, power supplies, control consoles, computer peripherals, transmitters, final drive devices, and other manufacturers' equipment integrated into the control system.

Different manufacturers use different terminology for the various grounds within a systems. Below are some of the more common terms.

Protective Grounds	Circuit Reference Grounds
Power Ground	Logic Ground
Chassis Ground	Power Supply Ground
Shield Ground	Digital Ground
Earth Ground	Analog Ground
Equipment Ground	System Ground
Plant Ground	Signal Ground
Safety Ground	Circuit Ground
	Bus Ground

Many of the protective and circuit reference grounds within the system will be pre-wired to designated terminals. The start-up engineer must verify correct ground wiring and procedures from these designated terminals to the plant's ground grid. Protective grounds are for the safety of both personnel and equipment. The non-current carrying metal parts of the system are grounded to prevent these parts from becoming energized if they should come in contact with dangerous potentials. All control systems have designated cabinet or equipment ground terminations for protective grounding. The start-up engineer must verify that the cabinet or equipment ground has not been connected to a ground used by any high voltage equipment, has not been grounded through other equipment grounds, or vice versa, or has not been grounded to structural members such as I-beams or columns. The protective ground must be connected directly to the plant grounding grid. Other problems to look for include multiple connections between the different ground systems, use of inadequate wire size, loose connection bolts and unintentional connections between the copper ground bus and the cabinet frame. The type and size wire used will depend upon the system's size and the vendor's requirements. If the proper wire size is in doubt, it is better to have it too big than too small.

Circuit referencing or signal grounding is the grounding of current carrying return conductors in AC or DC power distribution schemes. This includes the external signals to the control system as well as internal signals. Each vendor's procedures for the proper grounding of power grounds, shields, signal grounds, etc. should be followed and verified.

It is necessary to keep the different ground systems separate from each other except at one designated point. This is called a single point ground system. The common ground point is connected directly to the plant's ground grid.

Many signals in today's distributed control systems are high frequency, low-level signals which are susceptible to electrical noise. Ground wiring must present a low impedance, as well as low resistance, since even minimal reactance in ground wiring can result in electrical noise being induced into the ground system, and in turn, cause ground currents to flow. For these reasons, ground leads should be kept as short as possible.

Each vendor will include ground wiring diagrams specific to the project. Figure 4 shows a typical arrangement for single point grounding, which gives independent circuits for each load and eliminates the additive voltage drops of daisy-chain ground currents. Control system vendors will supply more realistic single point diagrams, including both digital and analog circuits,and their power distribution schemes and ground requirements. Single point grounding also applies to the grounding of multiple control system cabinets and equipment.

An important ground to check on today's distributed control systems is that of the communication wiring used between system components (e.g., nodes, drops, etc.). The communication network or data highway may be susceptible to excessive circulating currents if it is grounded at different points in the plant where the ground potential is not equal. The vendor's recommended grounding of the communication network must be checked and verified to avoid degrading the total system performance. If there are large potential differences between nodes of the system it may be necessary to use a fiber optic communication link or other form of electrically isolated system.

In checking the grounding system, the start-up engineer should verify that all grounding cables are installed according to the design drawings paying particular attention to the field installed ground cables between system cabinets or components.

One factor which makes checking out grounding systems difficult is that frequently problems will not show up during normal operation of the system, but will show up during unusual events such as nearby lightning strikes or power system transients. Because of this it is difficult to determine whether the grounding system is really satisfactory during normal operation.

Additional information and guidance on plant grounding is available from Article 250 of the National Electric Code [5] and IEEE Standard 518-1982 [6].

INITIAL SYSTEM POWERUP

Once the proper power and grounding wiring has been assured, the boiler control system is ready to be commissioned by the vendor's service personnel. It is important that the commissioning of the system be performed by an authorized vendor employee/representative, in order to ensure correct equipment power-up and to prevent voiding the vendor's warranty. This is especially true with today's advanced microprocessor based distributed control systems.

At this stage, the commissioning service includes:

1. applying power to the control system,
2. verifying proper operation of all the elements of the system, such as, processors, operator's consoles, manual/auto stations and peripheral devices like printers and recorders,
3. installing all of the system's configuration software for the controllers and consoles,
4. checking the bill of material to insure that all system equipment was received and documenting any equipment that was not.

FIELD INPUT OUTPUT CHECKS

After the system has been powered up, the software loaded and the initial operation checked, the field input and output wiring should be verified. A good time to check sensor input signals is when the sensor is being calibrated in the field. As the sensor is checked for zero and span, the signal received at the control system can also be checked. If the system is digital, the software scaling should also be verified at this time.

Verifying field wiring is pretty routine usually but there are a couple of areas which can be a little tricky. The first is grounding the shield on analog input signals. Many books and manufacturers recommend grounding the shield at the source end (i.e., at the transmitter or thermocouple) but in practice it is usually better to ground all shields at the control system terminals. It makes it easy to verify when all shields are grounded in one place and the quality of the ground is generally better at the cabinets than out at the instrument. In all cases shields should not be grounded at both ends because if the two grounds are not at exactly the same potential, current will flow in the shield which can induce noise in the signal wires. If shielded wires go through intermediate junction boxes, the continuity of the shield through the junction boxes should be assured.

When checking thermocouple wiring it is normal to ensure that the extension wire polarity matches the thermocouple and the terminal block. Occasionally there will be a double reversal of polarities in intermediate junction boxes which causes slight errors (less than 20 degrees F) in thermocouple readings.

Another thing to look out for when checking thermocouples is that the thermocouple is definitely bottomed in its well. If it is not fully bottomed it will read lower than the actual fluid temperature but will otherwise appear OK.

When checking pressure or differential pressure transmitters, the routing of the tubing from the pressure tap to the instrument should be carefully examined. In general for a water or steam connection the transmitter should be below the pressure tap. For an air or gas connection the transmitter should be above the tap.

Also, care should be used in accommodating the water leg on transmitters measuring steam or water pressure. The water leg can be handled in the calibration of the transmitter or in the scaling of the transmitter in the control system configuration. Either way will work, just make sure one and only one of the ways is used.

DRIVE UNIT CHECKOUT

After the sensors and their wiring have been verified the checkout of the final drive units should begin. There are many objectives of checking the drive units. First is to ensure that the operator can manually position the drive from the control room. Second is to ensure that the linkage arrangement allows full travel of the drive unit and thereby ensures maximum torque is delivered to the driven arm. Third is to establish a linear relationship between the control system output signal and the parameter being regulated by the drive (e.g., air flow).

The first thing to do is to physically inspect the drive units and their linkages. Drive units must be securely bolted in place before stroking. Linkage should be examined for reasonable arrangement and should be disconnected from the drive unit before energizing the drive. The pins which connect the linkage together must be carefully checked to ensure there is no play or lost motion in the joints. The drive unit housings should be checked for proper lubrication.

Linkage design is a more complex than most people think and as a result is not well understood. It is not the intent of this paper to cover linkage design, but a few key points will be discussed.

1. When rotary actuators are connected to rotary driven arms there are several possible ways to arrange the linkage, as shown in Figure 5. Figure 5a shows the most common arrangement, parallel and equal length arms. Figure 5b is similar but the arms are facing in opposite directions. Figure 5c is an

arrangement in which the torque delivered to the driven arm can be greater or less than the torque produced by the drive arm. This can be very useful on applications which require a lot of breakaway or seating force. Figure 5d shows an arrangement where the arc of travel of the driven arm is less than the arc available from the drive arm. Torque multiplication is available here also. Figure 6 shows a plot of torque supplied to the driven arm as a function of the angle of the drive arm for the four cases shown in Figure 5.

2. The linkage designer can select the length of each arm (within limits), the starting angle of each arm and the length of the connecting link. The given quantities are the maximum stroke angle of the drive unit and the required stroke angle of the driven arm. The design objective should be to use as much of the stroke angle of the drive arm as possible to maximize the torque transferred to the driven arm.

3. When long connecting links are necessary, the weight of the link should be factored into the torque calculation.

Once the linkage arrangement has been checked it is time to stroke the actuator with the linkage disconnected. On electric actuators this verifies that the power switch can start and stop the actuator and can move it in both directions. It also allows the end of the travel limit switches to be roughly set. On pneumatic actuators this checks the I/P converter and the pneumatic positioner. After the drive has been successfully stroked with the linkage off, the linkage should be re-connected and the drive carefully stroked again using a jogging technique at first. After ensuring that the drive unit can fully stroke its load, the drive should be calibrated for 0% and 100% position and the end of travel limits switches set.

On electric drive units it may be necessary to tune the position control loop for the drive unit. All electric drives have a position control loop but it may not be tunable by the user. Tuning the position control loop involves adjusting the deadband and adjusting the raise and lower pulse widths as a function of the position error. The first adjustment to make is the minimum raise and lower pulse widths. These should be set as short as possible but long enough to always move the drive unit under its worst case load. In other words the pulse must have enough energy to overcome the static friction within the drive and actually change its position. You don't want to be sending pulses to the drive and not moving it.

After the minimum pulse widths have been set, the deadband should be adjusted. Again the deadband should be as small as possible but large enough that the minimum pulse width set previously cannot move the drive unit position all the way across the deadband. If the position does move all the way through the dead zone, the position controller will continuously oscillate about the deadband.. Actually the total deadband (plus and minus) should be about twice the amount of travel that results from the minimum pulse width. This way when the error increases to the deadband, a minimum pulse will bring it back to zero.

Once the minimum pulse width and the deadband have been adjusted all that remains is to set the pulse width versus position error curve. There are two types of systems, continuously variable and step function. Figure 7 shows the relationship between pulse width and position error for both a continuously variable and a step type system. In the continuously variable system the slope of the line is adjusted to alter the relationship. In the step system, the error magnitude at which the steps occur and the height of the steps are the typical adjustments. The adjustments are made to get the fastest possible position response without overshoot under all conditions. Since the load on the drive unit can vary tremendously depending on the direction of movement and boiler load, the drive unit's response must be verified at several different conditions. Do not set the response to optimum at one condition and assume it is optimum for all conditions. Use conservative tuning if several conditions cannot be checked.

INITIAL CALIBRATION AND TUNING

The purpose of initial calibration and tuning is to set up the control system so the unit can be operated in automatic mode in a safe and stable fashion while other parts of the plant are operationally checked out. This means that the unit should be stable during steady load operation and should be able to make slow load changes without constant operator assistance.

All control systems contain thousands of adjustable parameters which must be set before the system will perform correctly. In analog control systems, all settings are made by adjusting hardware such as gain resistors or pots. In distributed control systems almost all of the adjustments are made in the software of the system. This greatly simplifies the process but does not eliminate it.

Most adjustments associated with input and output scaling are done during the configuration process. Items such as transmitter scaling and temperature compensations fall into this category. The next major group of adjustments are those that match the values in the control system to the operating values of the plant. This is called calibrating or scaling the control system and will be described in the next section. The final type of adjustment is the tuning of the controllers to optimize the dynamic response of the controlled variables. Tuning will also be discussed in more detail later.

Scaling the control system is very necessary to make it perform optimally. Since a load index feedforward is the main demand signal to the fuel, air and feedwater control loops, it must be calibrated to give the proper amount of action throughout the load range or the feedback controllers on each loop will be forced the compensate for its inaccuracy. If the unit being started up is a new unit which has never operated, then design data must be used for the initial scaling. On a retrofit job you may be fortunate enough to have accurate data from the unit's previous operation. Part of all retrofit jobs should be to gather data throughout the load range before the installation outage begins. Unfortunately it rarely happens.

To begin the scaling process, select a load point to be the full load point. This is usually the rated load point of the turbine or the maximum calculated load at rated pressure of the turbine (100% valves wide open) or the maximum continuous rating (MCR) point on the boiler. Typically the boiler MCR point corresponds closely to the maximum calculated point on the turbine and since both boiler and turbine data is available there, it makes a good full load point. Once the full load point is selected, it is assigned a numerical value within the control system. On digital control systems it makes sense to assign the value 100.0 to this point. On analog systems, the value will be a voltage or a current and so must depend on the design range of the hardware. If the system is a 0 to +10V range, then a value of +8V is good because it allows about 25% over-ranging of the signal. Never scale the full load point to be at the maximum range of the hardware. On digital systems this is not an issue because the range of allowable numbers is essentially unlimited.

Once the value at the full load point has been selected, then the values for the feedforward load index signal, fuel flow, air flow and feedwater flow must be scaled to equal the selected value at the full load point. Consider fuel flow for an example. Assume the design fuel flow at full load is 160,000 lbs/hr. Also assume the unit has four feeders each rated at 50,000 lbs/hr. The analog inputs for the feeders are scaled so that in the control system 50,000 lbs/hr equals 100.0. The four feeder signals are summed to give a total fuel flow signal which must then be scaled to equal 100.0 at full load. There are several different ways to approach this problem. Probably the easiest is to calculate the output of the fuel flow summer at a total flow of 160,000 lbs/hr and then determine the multiplier to reduce it to 100.0. Since 160,000 is 80% of 200,000, the output of the fuel flow summer at 160,000 lbs/hr would be 0.8 X 400.0 = 320.0. This needs to be multiplied by 100.0/320.0 to give the required 100.0. This example shows the general approach for scaling the primary flow loops, fuel, air and feedwater.

After the initial scaling has been performed, the initial tuning settings must be determined. Again if it is a retrofit project, the tuning parameters from the original control system are a good starting point. Be careful when using tuning constants from the original system that other parameters in the loop have not been changed. A change in transmitter scaling means that the original tuning will not preform as before. If the original tuning constants are not available or it is a new installation then other means of determining initial tuning must be used. One way is through modeling and analysis of the control system using a computer aided control system design package. Unfortunately this method is rarely used in the utility industry because of unfamiliarity and a perception that it is expensive. The usual way for the utility industry is to just make a conservative guess. A good starting point is to use a gain of 1.0 and a reset of 0.5.

When the initial tuning is a guess, it is necessary to observe the response of the loop closely when it is first placed in automatic mode to make sure that the loop does not go unstable.

The general procedure for tuning any control loop is to start at the lowest level of control and work your way upward. On a cascade control loop for example, this means tuning the drive unit positioning loop, then the inner or slave control loop and finally the outer or master loop. In addition to this priority within a loop there is also a priority within the overall control systems.

The first priority of the initial scaling and tuning effort is to make the drum level and furnace pressure operate on automatic and be stable. These are the two loops which are not self-regulating and therefore require the most attention from the operator. The next step is to get the fuel and air control on auto so the operator can regulate boiler load more safely. This allows the unit to be operated over its load range to verify all the plant systems are functioning properly. It also provides an opportunity to collect the data necessary for the final scaling of the control system. Steam temperature is the next loop to get on auto although sometimes this is not done in the initial tuning.

The objective of the initial tuning is to have stable automatic control during steady load operation and during slow load changes. The tuning parameters selected at this time should give a well damped response which favors stability over quick response. When tuning a loop it is good to start with the reset and rate settings at zero and tune the proportional gain first. The reset should be done next, followed by the rate. Rate action is generally under-utilized on boiler control systems because people believe it destabilizes the control. In fact it tends to stabilize the control but, it will amplify any noise in the system giving the appearance of instability. To take full advantage of rate action in a control loop, the signals in the loop must be as noise free as possible. Simple resistor-capacitor filters on the inputs can greatly reduce noise and not degrade response if their time constant is about 10 times faster than the dominant time constant of the control loop. For 60 Hertz noise, a simple filter with a time constant of 1 second will remove about 98.4% of the noise.

Initial tuning of the loops is generally done by introducing step changes in the setpoint and watching the response. A good actual closed loop response to shoot for in the initial tuning is somewhere between 5% and 20% overshoot. This is highly damped but still has some overshoot. You really don't know where you stand in tuning a loop until some overshoot is achieved, so never stop increasing the gain until the response overshoots at least a little.

Once this initial operation period is completed the final scaling and tuning can be started.

FINAL TUNING AND SCALING

The final scaling process contains three steps which are similar to those in the initial scaling process. They are:

1. setup the final scaling using actual plant data,
2. tune the PID controllers,
3. setup the feedforward gains for the loops that have them.

The final scaling process is the same as the initial scaling process except that actual plant data is used for the scaling. The scaling also should be set up at several different load points using function generators if necessary. When collecting the data, the unit must be at a steady state condition and in the normal design configuration (e.g., no feedwater heaters out of service, excess oxygen at the design level). It is also nice to collect several sets of data at each point and use the average for scaling.

Determining the final tuning settings is also similar to the initial tuning process. In the final tuning process the objective is the optimum response not just a satisfactory response. Determining the optimum response is significantly more difficult than determining a satisfactory one. For the final tuning the best response may be less damped than for the initial tuning. It is important to follow the correct sequence when doing the tuning. In [2], the authors describe a sequence for a once-through unit which is excellent. For a drum unit a good sequence is listed below:

1. Furnace pressure
2. Drum level and feedwater
3. Condensate flow
4. Pulverizer temperature and primary air flow
5. Secondary air flow
6. Fuel flow
7. Steam temperature
8. Excess oxygen
9. Throttle pressure
10. Megawatts

The order is important because of interactions between control loops. The lower level loops should be tuned first so that when the upper level loops are tuned you can be sure that any instability is due to the tuning of the upper loop.

In addition to the normal tuning of the PID controllers, the final tuning involves the setting of any dynamic feedforward gains on the control loops. A dynamic feedforward is not the same thing as the load index feedforward signal which is really a static feedforward. The dynamic feedforward gains contain dynamic elements such as lags and rate kickers. These are sometimes used on the fuel flow, air flow and steam temperature loops to improve response during load changes.

CONCLUSION

The checkout and start-up of a boiler control system is a complex task that requires a systematic approach if it is to be done well. The approach described in this paper is one that has been used successfully by the authors on several start-ups. In addition many practical suggestions have been made to assist the working start-up engineer in doing a better job. The keys to a successful boiler control system start-up are following a proper plan, having a good working relationship between all the parties involved and having a knowledgeable start-up engineer.

REFERENCES

1. Lewis, Roger A., "Control Drive Linkage Design Guide," Internal Report, Southern Company Services, Inc., Birmingham, Alabama, 1987.
2. Morse, Richard H., Nishikawa, Shoji and Sato, Michinori, "Aspects of Tuning a Boiler Control System - A Strategy for Optimization," Power Industry Symposium, San Fransisco, California, May 10-12, 1976.
3. Martz, Lyle, "Practical Instrument Engineering," Lecture Notes, Hagan Chemical and Controls, Inc.
4. Driskell, Les, Control Valve Selection and Sizing, Instrument Society of America, Research Triangle Park, North Carolina, 1983.
5. Grader, Jerome E., "Basic Process Control," Technical Data Publication, TDS-10A120, Taylor Instrument Co.
6. National Electric Code - 1990, Article 250, National Fire Protection Association, Quincy, MA, 1989.
7. IEEE Standard 518-1982, "IEEE Guide for the Installation of Electrical Equipment to Minimize Electrical Noise Inputs to Controllers from External Sources," Institute of Electrical and Electronics Engineers, New York, New York, 1982.

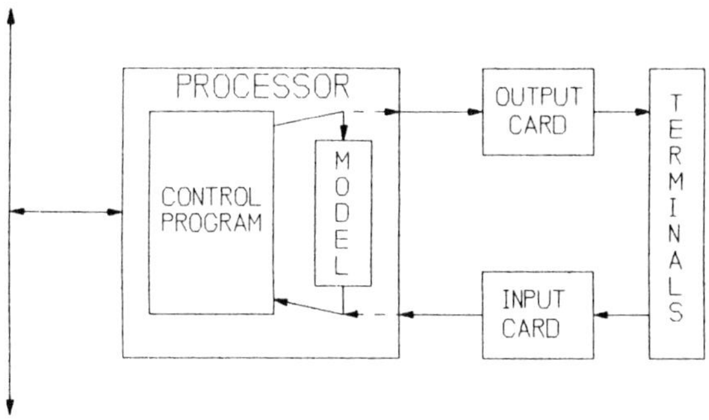

Figure 1 - Model in Same Processor

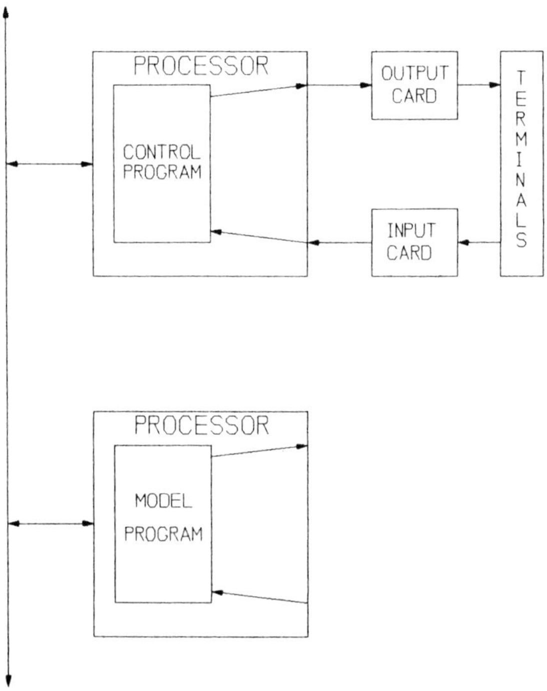

Figure 2 - Model in Separate Processor

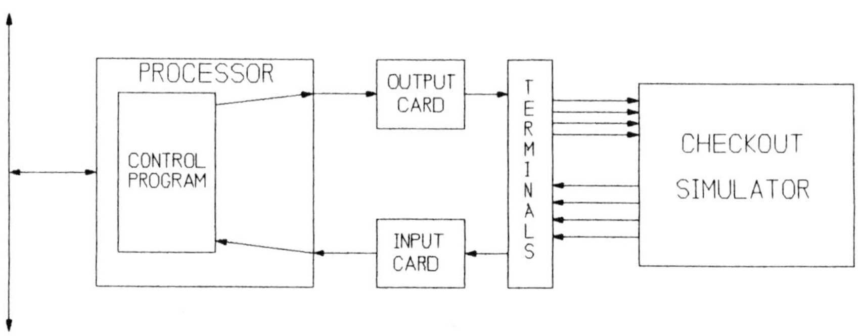

Figure 3 - External Checkout Simulator

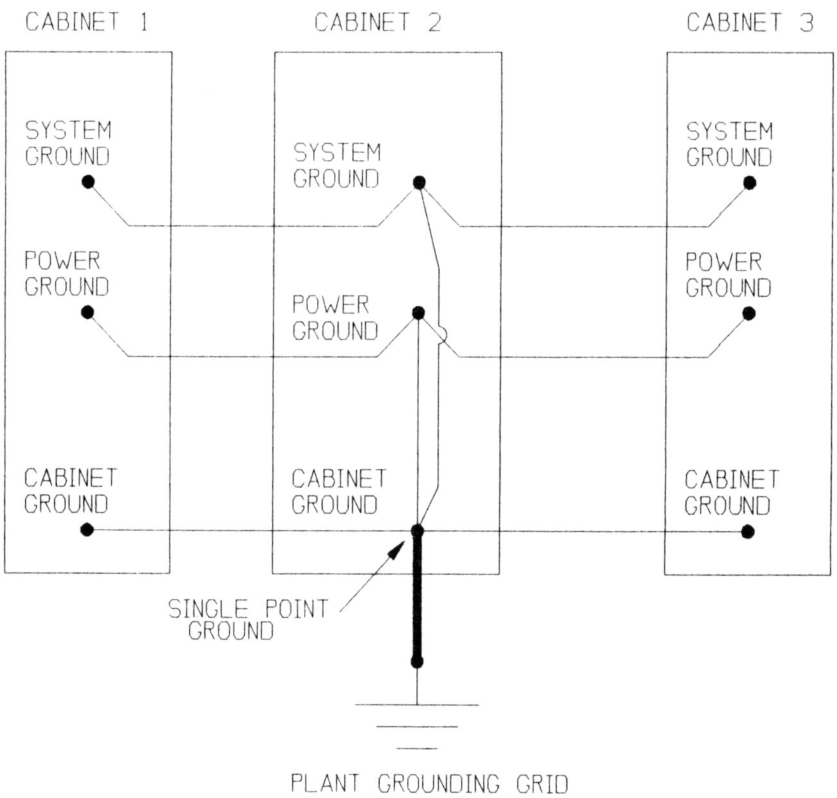

Figure 4 - Single Point Ground System

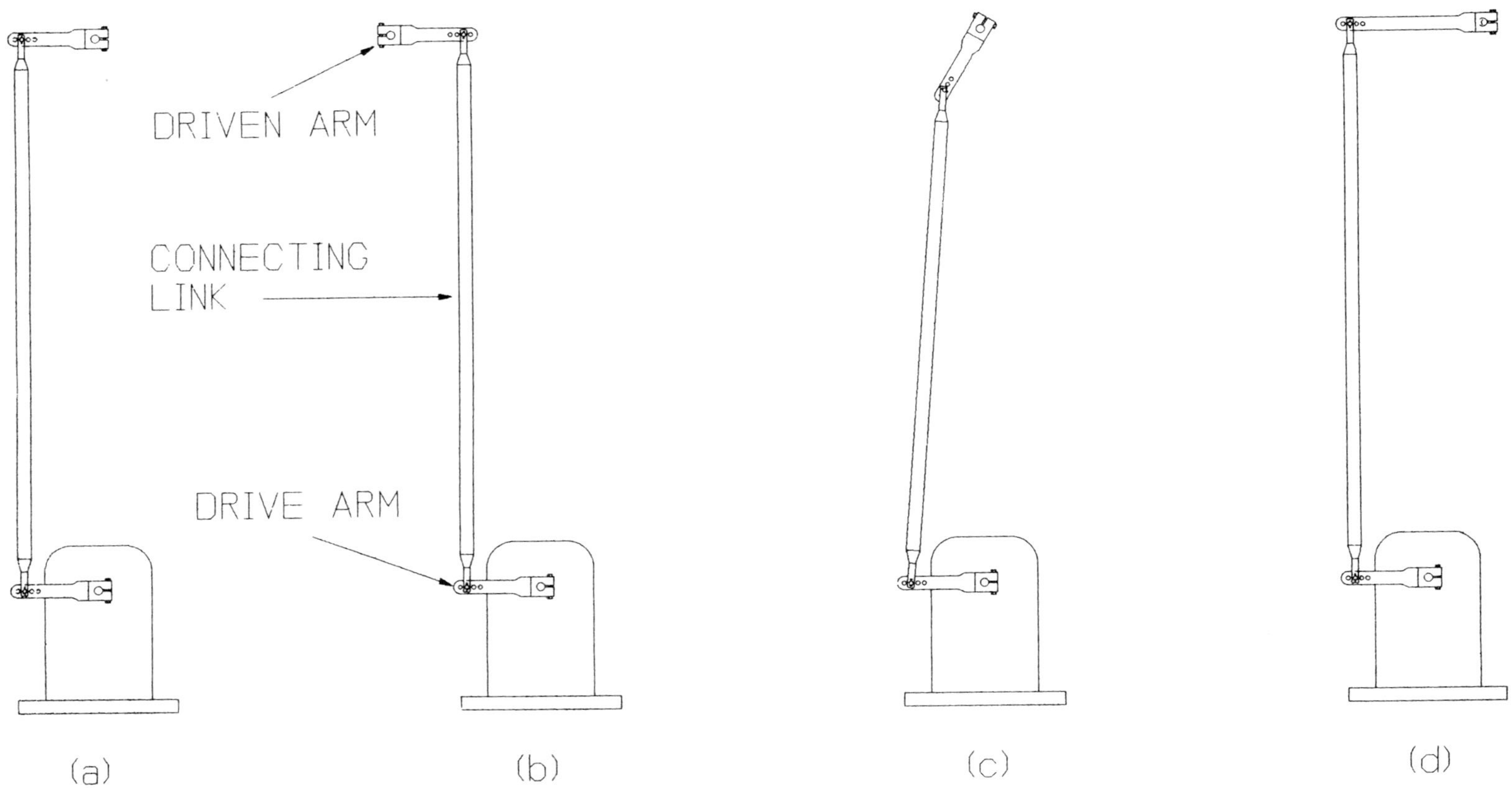

Figure 5 - Drive Unit Linkage Arrangements

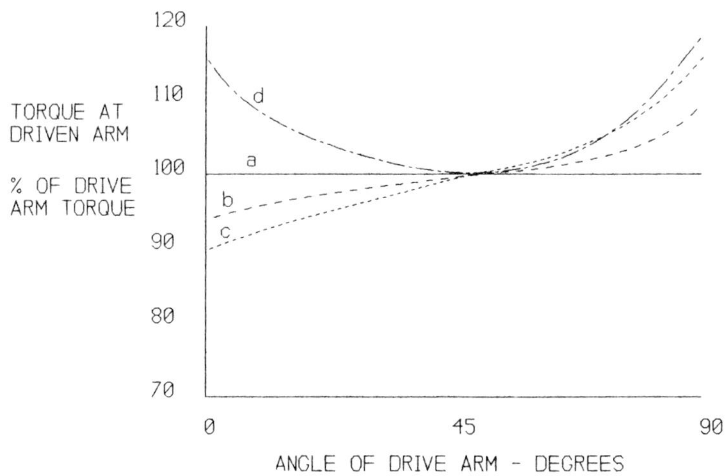

Figure 6 - Torque at Driven Arm

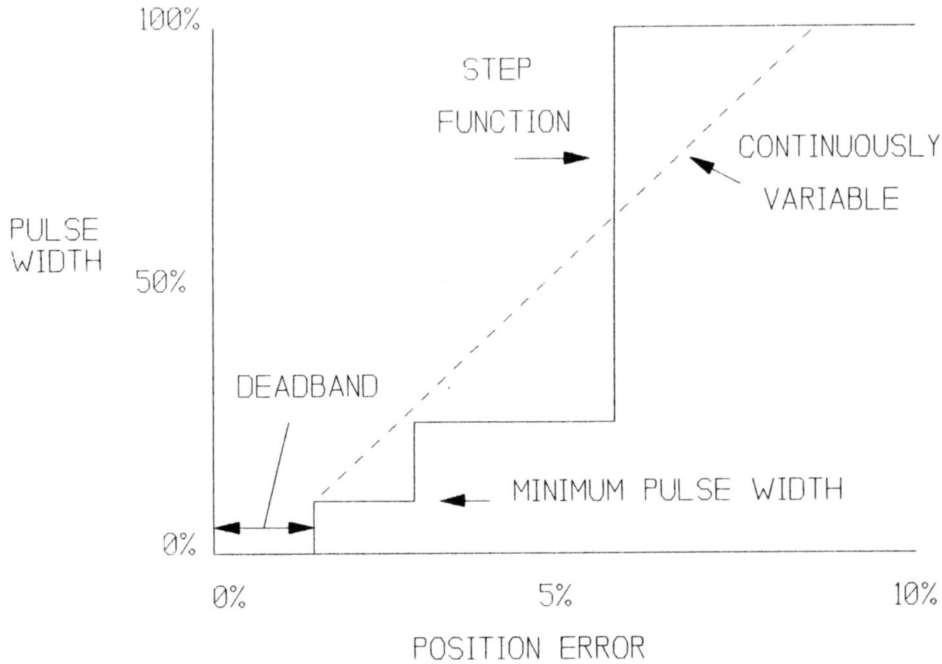

Figure 7 - Pulse Width vs. Position Error Curve

©1991 - Paper #91-705
0074-056X/91/069-095/$0 + .50pp

A BTU COMPENSATED COAL POWDER FLOW MEASUREMENT
An "Intelligent" Sensor

Antonio De Marco
ENEL-DSR-CRA
Research & Development Dept.
Automation and Computing
20093 Cologno Monzese (MI) Italy
Via Volta, I

Attilio Radice
ENEL-DCO
Design & Construction Dept.
Control System Design
20124 Milano Italy
Via Cardano, 10

Robert N. Hubby
Applications Specialist
Leeds & Northrup Company
North Wales, Pennsylvania

ABSTRACT

A major limitation to the effectiveness of traditional combustion control systems in controlling fuel input, and the resultant control of excess air has been the inability to accurately measure the fuel feed to the furnace on the basis of actual heat energy released. While a heat release calculation has been available for drum type unit for many years, the control of once through units has depended on some form of mill model driven by feeder speed, or some other index of mill throughput for its fuel feedback signal. This paper describes a promising new technique to measure mill coal powder flow to the boiler and to compensate this powder flow measurement for actual heat energy released. Discussion of recent field experience with the coal powder flow measurement will also be included in the paper.

1. BACKGROUND

Today, most coal-fired thermal units both in the United States and Italy have to participate in load changes: they are no longer base load units. Their overall design has to take into account load variations in order to respond correctly to frequency control.

Dynamic conditions must therefore be satisfied. This means that in a boiler control system the mills mass flow rate inner control loops must have a rapid, dynamic response and ensure mass flow rate values rather close to the desired values. This can be achieved only if a suitable measurement of coal mass flow rate of each mill is available.

Furthermore during a mill start-up or shut-down an accurate and fast measurement of the coal mass flow rate of each mill would allow the maneuver to be executed with small deviations in total coal flow because the other mills can counterbalance the flow variation of the mill being started up or shut down.

The ideal boiler control system matches the output energy of the boiler to the energy demand of its load (the turbine) under all plant operating conditions. Modern control strategies regulate the boiler inputs in parallel to accomplish this energy match. The key to any parallel regulation strategy is the accurate measurement of each boiler input.

Providing proper input regulation assures stable control of the controlled variables of megawatts, throttle pressure, steam temperatures, drum level and excess air. Accurate boiler input measurements also allow the monitoring of critical relationships between inputs such as fuel, air, and feedwater as is needed for boiler protective systems.

This brings us to the fuel measurement problem. One of the most difficult problems for the designer of control systems for fossil fired boilers is the development of a fuel measurement that is adequate for a meaningful fuel/air ratio control. For the once-through unit, the ramifications of an adequate fuel measurement are even more serious since the control of final steam temperature is wholly dependent on accurate control of firing rate/feedwater ratio.

The major problems associated with obtaining an adequate fuel measurement are:

- Varying heating value content
- Measuring fuel delivery
- Accounting for unmetered fuel

Varying heating value, while a problem for all fuels, is particularly pronounced with coal. As noted Reference 10, the BTU value for coal depends on weather, coal storage conditions and the consistency of the fired coal. Purchase of coal from different sources further aggravates the problem of determining the BTU value.

Measurement of actual coal delivery presents an almost impossible task due to two significant problems: first, mill storage capacity; and second, mill output response. For example, coal flow signal from the mill outlet to the boiler was until now estimated using the feeder speed signal as if it were pulverized-fuel flow. While it is correct from a steady state point of view, dynamically it is not as true as real-time response of the mill because grinding deadtime and storage lag are not taken in account even if some open loop artifices are applied in order to reproduce an estimated response.

The problems of measuring coal delivery are discussed in detail in Reference 9 which is included as Appendix A of this paper for convenience. Let it suffice to say that as the mill storage capacity increases, so does the complication of modeling the fuel delivery system response including elements of the storage mechanism. In addition, conventional methods of measuring fuel delivery such as feeder speed applied to a mill model or mill primary airflow have no way of compensating for mill start-up or shutdown or for changing physical characteristics of mills.

Finally, unmetered fuel inputs such as ignitors or warm up guns remain unaccounted for by conventional metering systems.

This brings us to the central issues of providing an accurate measurement of coal delivery that reflects the actual energy released from the fuel. We will consider topics as follows:

1. The method of pulverized coal mass flow rate measurement including process measurements

2. The underlying principles for measuring coal mass flow rate.

3. Field test validation and calibration of the coal mass flow measurement.

4. Compensation of coal powder flow measurement for actual energy released.

Let us begin with the development of the method of the powder coal flow measurement.

2. METHOD OF PULVERIZED COAL MASS FLOW RATE MEASUREMENT

A method was investigated and developed in Ente Nazionale per l'Energia Elettrica (ENEL) to measure pulverized-coal flow rate based on the computing in a microprocessor the friction pressure drop between the pulverized fuel duct inlet and the furnace. The mixture of pulverized fuel and air is considered to be homogeneous and the algorithm for computation of the pulverized fuel flow is derived from consideration of equations of conservation of mass and momentum. Allowance is made for the air mass flow and temperature compensation. The plant measurements required, which are normally present as standard mill instrumentation for control, are:

- primary air differential pressure
- primary air temperature
- mill inlet pressure
- mill outlet pressure
- mill outlet temperature
- furnace pressure

Features of the method are:

- use of normal industrial instruments
- computation of the pulverized mass flow rate through a model of flow, performed by a microprocessor as is available in today's boiler control system
- accuracy comparable with that of industrial instrumentation for other mass flow rate measurements
- dynamic characteristics and noise levels basically equivalent to the ones of air mass flow rate measurements
- calibration in conjunction with another coal flow rate measurement (feeder speed, oxygen concentration, boiler heat release, fuel oil equivalence, etc.).

Advantages of the method are:

- real-time mass flow rate feedback availability
- combustion control improvement leading to better boiler and unit control
- combustion efficiency improvement
- mill safety condition improvement.

Validation of the industrial measurement was carried on by tests based on oxygen concentration in flue gases. A calibration procedure has also been established.

Consideration of such a method involving a computation which would have been practically unfeasible with analog control systems, is practical only in today's microprocessor based technology.

3. PRINCIPLE OF THE METHOD OF PULVERIZED COAL MASS FLOW RATE MEASUREMENT

The relationship of the pressure drop between the duct inlet and the combustion chamber and the relevant process variables, and the pulverized coal mass flow rate is developed as follows. The flow of air and pulverized coal through the ducts can be considered a two-phase flow and can be studied with an homogeneous model, where the solid phase has the same velocity of the gas phase. Furthermore the energy conservation equation need not be considered in the analysis because the heat exchanged between this mixture and the walls of the ducts is negligible and the coal particles are assumed to be in thermal equilibrium with the flowing air. Only the mass conservation and linear momentum conservation equations have, therefore, been considered (hydrodynamic equations). It can be shown that the dynamics associated with the hydrodynamic equations, for the usual dimensions of the ducts carrying the pulverized coal, can be ignored, so that (Reference 12):

i) the mass conservation equation is

$$Wt - Wa - Wp = 0 \tag{3.1}$$

Note all symbols are listed in Appendix B.

This equation, if we assume that there is no slip between air and pulverized coal, defines the mean density of the mixture

$$\mu = \frac{(1 + Wp/Wa)\ \mu a}{1 + (Wp/Wa)\ \mu a/\mu p} \cong (1 + Wp/Wa)\ \mu a \tag{3.2}$$

because μp is much greater than μa

ii) The linear momentum conservation equation assuming of ducts of uniform section is:

$$\frac{\delta (v^2/\mu)}{\delta x} + \mu g \frac{dz}{dx} + \frac{\delta P}{\delta x} + \frac{Cf\ 2\ w\ v[v]}{2\ s\ \mu} = 0 \tag{3.3}$$

integrating (3.3) between 0 and 1 the following relation can be obtained:

$$Wa\ \alpha\ Tp\ \frac{Po\ -\ P1}{Po\ P1} + S^2\ g\ \frac{D}{2}\frac{P1\ +\ Po}{Wa\alpha Ta} + \frac{Cf\ w}{4\ S}\ Wa\ \alpha\ Tp\ \frac{Po\ +\ P1}{Po\ P1} = \frac{S^2}{Wa\ +\ Wp}\ (Po - P1) \tag{3.4}$$

where the density of the air μa is obtained from the perfect gas equation. The two phase friction factor Cf may be defined as a function of some adimensional parameters. However for a defined geometry and given fixed air properties and pulverized coal with a reasonable fixed distribution, this friction coefficient can be considered a function only of the ratio between the pulverized mass flow rate and the air mass flow rate. Then

$$\frac{Cf\ w}{4\ S} = f\ (Wp/Wo) \tag{3.5}$$

and a linear approximation of this relation gives:

$$\frac{Cf\ w}{4\ S} = Kfo + \beta\ (Wp/Wa) \tag{3.6}$$

From equation (3.4) the value of the pulverized coal mass flow rate can be obtained (Reference 13):

$$Wp = Wa\ [\ -Q + \sqrt{(Q\ -\ 1)^2 + R}\ ] \tag{3.7}$$

$$Q = \frac{1}{2}\left[\frac{Z}{\beta} + \frac{S^2 X}{\beta} + \frac{Kfo}{\beta} + 1\right] \tag{3.8}$$

$$R = \frac{S^2}{\beta}\ Z\ \frac{Po\ P1}{Wa^2\ \alpha\ Tp} \tag{3.9}$$

$$Z = \frac{Po\ -\ P1}{Po\ +\ P1} \tag{3.10}$$

$$X = \frac{g\ D\ Po\ P1}{2(Wa\ \alpha\ Tp)^2} \tag{3.11}$$

Note all symbols are listed in Appendix B.

The actual equipment arrangement is shown in fig. 1. Since the pressure transducer P1 is not located at the exit of the ducts but at the top of the furnace, then a correction term P has to be inserted in the computation of R (eq. 3.9):

$$R = \frac{S^2}{\beta} Z \frac{Po\ P1}{Wa\,\alpha\,Tp_a} + X\,P\,S^2/\beta = \frac{R^*}{\alpha} \qquad (3.12)$$

$$P - \frac{Tp}{Ta}\frac{Dc}{D}2\frac{Pa}{Po + P1} - \frac{Tp}{Tc}(\frac{Dc}{D} - 1)\frac{2\ P1}{Po + P1} \qquad (3.13)$$

$$\underline{a}\,\frac{Tp}{Ta}\frac{DC}{D} - \frac{Tp}{Tc}(\frac{Dc}{D} - 1)$$

Then the pulverized coal mass flow rate can be evaluated from the duct inlet pressure measurement Po, the furnace pressure P1, the air mass flow rate, the air-pulverized coal temperature together with an estimation of the ambient and furnace temperature.

The parameters Kfo and β of the formula (3.6) have to be estimated from static tests.

The estimation of Kfo is performed by m static tests (m>1) with no pulverized coal flowing in the ducts and with different air flow rate values. The following quantities

$$(P_o)_i,\ (P1)_i,\ (Wa)_i,\ (Tp)_i;$$

are measured in the i-th test. Then from equations (3.7), (3.8), (3.10), (3.11), (3.12) setting Wp = 0 the value of (Kfo) i of the i-th test can be obtained by:

$$(Kfo)_i = R^* - (Z + S^2\,X)$$

where $R^* = \alpha R$

The estimate of Kfo is given by the following weighted mean:

$$Kfo = \frac{\sum_{1}^{m} i\,(Kfo)i\,(Wa)i^2}{\sum_{1}^{m} i\,(Wa)i^2}$$

The estimation of the coefficient β is performed by n static tests (n>1) with the presence of pulverized coal and different values of the ration Wp/Wa - Φ, once the coefficient Kfo is known. The following values are measured in the i-th test.

$$(Po)_i,\ (Pl)_i,\ (Wa)_i,\ (Tp)_i,\ (Wp)_i$$

Note all symbols are listed in Appendix B.

The evaluation of the pulverized coal mass flow rate can be obtained either from mill feeder speed or oxygen measurements at the economizer. The value of $(\beta)_i$ of the i-th test can be obtained:

$$\beta i = \frac{R^*}{1 + (\Phi)_i} - D^*)\frac{1}{\Phi_i}$$

where $D^* = Z + S^2X + Kfo$
and $\Phi_i = (Wp)_i/(Wa)_i$

The estimate of β is given by the following weighted mean:

$$\beta = \frac{\sum_{i=1}^{m} (\Phi)_i (Wa)_i^2}{\sum_{i=1}^{m} (\Phi)_i}$$

A rough but useful value of mass flow rate of water in the coal can be easily obtained from the measurement of primary air mass flow rate and temperature and from the measurement of the temperature of the mixture of air and pulverized coal at the outlet of the mill.

The mass flow rate of this stream is added to the primary air mass flow rate and the measurement of dried pulverized coal can be obtained.

4. VALIDATION OF THE METHOD

The validation of the proposed method of measurement has been carried out both with static and dynamic plant tests (as described in Section 5).

In steady state conditions, the values of pulverized mass flow rate obtained by this method are in agreement with the values of measurement obtained with the mill feeder speed and the oxygen concentration (see Table 3).

The dynamic conditions have been obtained with the following type of tests:

1) Variation of mill feeder speed
2) Variation of mill primary air
3) Variation of both mill feeder speed and primary air flow rate.

An indirect measurement of the pulverized coal mass flow rate has been obtained by measuring the oxygen concentration at the boiler economizer. This indirect measurement is based on the relation between oxygen concentration, coal mass flow rate, oil and air mass flow rate, the coal and oil composition being known and with the assumption of constant combustion efficiency.

The dynamics of the Zirconium-Oxide instrument used to measure O_2 and the transport delay time of the flue gas into the boiler have been identified:

Note all symbols are listed in Appendix B.

$$G(s) = \frac{1}{1 + st_s} e^{-st}$$

where t_s = 80 seconds and t = 30 seconds

The pulverized coal mass flow rate supplied into the combustion chamber and measured directly with the proposed instrument, has been delayed by the dynamics represented by the transfer function. This delayed measurement has then been compared with the pulverized coal mass flow rate indirectly determined from Oxygen concentration measurement (see fig. 2).

The results of this comparison are shown in fig. 3, fig. 4 and fig. 5, in Appendix B.

5. FIELD TESTS AND CALIBRATION

5.1. Field tests

Before the development of the pulverized coal flow measurement described above a lot of identification tests were carried on to investigate mill dynamics: the scope was to tune related control loops and to provide equivalent open loop calculations to fulfill the lack of measurement. (documentation is available from the Authors)

The tests were based on oxygen response at the boiler outlet and required a long execution times to reach a full steady state condition: control engineers who are acquainted with testing in a thermal power plant do know how time and labor intensive these procedures can be.

To execute these tests, take a step of mill primary air whose contribution in pulverized coal to the furnace has to be correctly detected by oxygen probes. The conditions for this test are as follows: all other mills control loops have to be in manual and also other variables of the boiler which could affect flue gas equilibrium have to be in manual; only turbine and feedwater actuators position can be changed by operators.

Figures 7 and 8 summarize the difference in control strategies when in absence and when in presence of a pulverized coal flow signal.

Recorded variable in the figures are (from top to bottom):

- coal demand
- feeder speed
- coal flow (pulverized coal mass flow rate)
- mill pressure
- primary air damper position
- primary air flow

Figure 7 shows a test related to current control design making use of open loop terms such as derivative of demand feedforward and mill start-up and shut-down compensation (control diagrams are available from the Authors). These terms are incorporated in both the primary air and the feeder control as a result of previous identification tests through oxygen response.

A simplified diagram of mill control is reported in fig. 6.

A fast response of coal flow is obtained through a derivative demand action which reflects itself in feeder speed and primary air flow. Units equipped with such a control scheme can participate in frequency regulating action.

In figure 7 pulverized coal flow is only recorded while acting feedback signal is feeder speed.

Figure 8 shows the results of a step in fuel flow demand related to a future control strategy. The illustrated coal flow is the pulverized coal flow estimate value which is used now as a feedback signal. It can be seen that feeder speed and primary air flow responses, now obtained under the only action of a PI controller to achieve a fast response of pulverized coal flow, are practically the same as the test in figure 7, except for the action of the derivative of demand feedforward.

Furthermore pulverized coal flow which achieves its desired value in about one minute, though with modern micro-processor based control equipment, this time could be reduced. Tests were carried out in Vado Ligure Power Plant, 4x320 MW units, with an analog relay-type control to which a modern computational device was added for coal mass flow rate computation.

The fuel flow estimation technique not only leads to a desirable dynamic response from the mill but also avoids the need for open loop terms. The closed loop tests were limited to a few days but demonstrated, in particular, the ability of the mill to respond to a step change in demand.

This strategy has been included by ENEL in the control design proposed for retrofitting and new plants, in the ENEL system. It is pertinent to note that with 6 months operating experience, there has been no significant change in the constants Kfo and β_i as a function of mill wear.

5.2. Calibration

Calibration of the mass flow measurement can be carried on when the mill itself has been correctly set up in order to obtain the desired mechanical characteristics such as finess, grindability, rejects, etc. and when related control loops have been satisfactorily tuned.

Calibration consists in two steps in order to determine the basic two parameters:

- air friction factor Kfo
- air coal mixture friction factor b

Air friction factor Kfo is an output of the computing program when only primary air is flowing through the mill to the furnace. Normally less than ten different steady state conditions are enough: a weighted mean value computation is also included in the microprocessor based program.

Having now air friction factor Kfo as an input of the program, air-coal mixture friction factor β can be obtained when coal with air is flowing through the mill to the furnace. As for Kfo, also in this case, less than ten different steady state conditions are required.

A reference coal mass flow has to be introduced as an input to have friction factor β as an output. Normally feeder speed value, suitably corrected, can be used.

Introducing as a second parameter the weighted mean value of β computed by the program, coal mass flow is now obtained as the final output.

As an example the evaluation of the parameters Kfo = 1.41 and $\beta = 0.41$ has been performed, as indicated in paragraph 3, with m=8 static tests for Kfo (see Table 1 and Appendix B) and n = 6 static tests for b (see Table 2 and Appendix II). In these tests the pulverized coal mass flow rate through the ducts has been evaluated from mill feeder speed.

6. COMPENSATION OF COAL POWDER FLOW FOR ACTUAL ENERGY RELEASED

With an accurate coal mass flow measurement developed as described above, it is now necessary to calibrate this signal for actual heat energy released in the furnace.

The development of the heat release calculation for drum type boilers as a method for measuring actual heat released in the furnace has been well established (Reference 9, part 3). Since its inception in 1962 the heat release calculation has been used either as the primary fuel feedback signal for control or as a calibrator of the total fuel flow signal for actual fuel energy released (Reference 10).

The heat release calculation is based on two measurements; boiler steam flow or first stage pressure which is an excellent index of boiler output energy; the second measurement is the rate of change of drum pressure which provides a representation of the rate at which the boiler stored energy is changing. These signals are summed to compute the actual fuel energy released.

The initial introduction of the heat release calculation applied only to the drum type unit. For once through boiler applications, development of the fuel flow signal was wholly dependent on measurement techniques using mill feeder speed in conjunction with a mill model. As reported in References 9 and 11, such a signal could be calibrated for actual heat energy released by computing the steady state ratio of boiler output energy to total fuel (P_1/Total Fuel Input). By applying this as a scaler to the total fuel measurement, the resultant signal for total fuel will be equal to boiler output energy under steady state conditions.

An alternate approach for developing an on-line heat release computation is shown in Appendix C. This computation could be used as a back-up to the simplified approach suggested in this section.

The coal powder flow measurement now overcomes the many limitations associated with using either mill feeder speed or mill primary air flow since it reflects actual coal powder flow to the boiler. Furthermore the coal powder flow reflects not only the steady state but also transient coal delivery to the furnace. When we calibrate this signal for actual steady state fuel energy released, we finally have a fuel feedback signal representative of total fuel energy released.

7 CONCLUSION

The benefits of the coal powder flow measurement are substantial:

1. Boiler coordinated control is improved since coal mass flow rate can be maintained close to the required values under all operating conditions.

2. A coal mass flow rate is available for every mill reflecting actual mill performance during mill start-up or shutdown. Thus when mills started or stopped, the remaining mills can compensate automatically for the attendant variations in coal delivery.

3. The total coal mass flow rate can be compensated for actual steady state heat energy released thus providing for a tighter fuel/air ratio control and temperature control (for once through boilers) as well as the close monitoring of critical relationships between boiler inputs.

4. More effective fuel control ultimately means more stable control of generation, throttle pressure, final steam temperatures, oxygen and drum level (as applicable).

5. Since deviation of controlled variables from their respective setpoints ultimately limits a unit's load regulating capabilities, close control of all controlled variables in the last analysis means a unit that is more responsive to load regulation requirements.

6. The overall system, including the computing device and the field transducers, can be considered as an "intelligent" sensor: future field bus techniques associated with microprocessor based control systems could improve the measurement efficiency.

ACKNOWLEDGEMENTS

The Authors thank their colleagues for providing information included in this paper and expecially R. Catananti (ENEL-DCO), A. Zizzo (ENEL-DSR), S. Menini (ENEL-DCO Lab.) and ENEL power plants staff for having improved knowledge on the subject through theory development and field tests.

REFERENCES

1. T. W. Jenkins Jr, I Garber, F.L. Stewart, *Heat Release Computer for Combustion Control Systems* Proceedings of the American Power Conference, Vol. XXVI, pp. 333-343, 1964.

2. James H. Daniels, *Boiler Turbine Control System* Patent 3,247,671, April 1966.

3. T. W. Jenkins Jr, B. Littmann, *Response Capability in the Control of Large Generating Units* IEEE Paper No. 71CP73 PWR presented at the IEEE Winter Power Meeting, New York, NY, January 31-February 5, 1971.

4. F. P. Ryan, *Modernization - A Key to Improved Performance* Instrumentation in the Power Industry, Volume 25, ISA Power Symposium, Phoenix Arizona, May 24-26, 1982.

5. Richard H. Morse, *Boiler Control System* U.S. Patent 4,213,304, July 22, 1980.

6. Richard H. Morse, *Balancing Boiler Inputs to Energy Demand - One of the Basic Considerations in the Design of Controls for Drum Type Boilers*, ISA Paper No. CI.82-745, ISA Conference Exhibit, Philadelphia, PA, October 17-21, 1982.

7. Cyrus W. Taft, *Performance Evaluation of Boiler Control System Strategies* Seminar: Control Systems for Fossil Fuel Power Plants EPRI Research Project 2710 Session 4, Atlanta Georgia, February 24-26, 1987.

8. H. C. Gery, *The Evolution of Coordinated Control Instrumentation in the Power Industry*, Volume 31, ISA Power Symposium, St. Petersburg, FL, May 23-25, 1988.

9. Robert N. Hubby, *Improved Fuel Measurement* presented to EPRI Improved Coal Firing Conference, Palo Alto California, November 2-4, 1988, Proceedings GS6422 Volume 2, pages 62-1 through 62-14.

10. EPRI R&D Plans for Advanced Computer and Control Technologies - EPRI Ref. GS-6673, Volume 2, page 3-7.

11. Robert N. Hubby, *Control System For Once-Through Boilers* U.S. Patent 4,909,037, March 20, 1990.

12. M. Agostinelli, T. Busi, R. Cori, A. Possenti, *Proposta per il miglioramento del controllo automatico dei mulini nei Gruppi Termoelettrici* - ENEL Report June 1975 (A proposal for an improved control of coal mills in Thermal Power Plants).

13. C. Maffezzoni, *Specification For the Realization of a Pulverized Coal Flow Meter* - ENEL Report March 1984.

14. R. Catananti, A. De Marco, A. Zizzo, S. Menini, A. Radice, *Sperimentazione di un metode per la misura della portata di polverino di carbone nelle Centrali Termoelettriche* - ENEL Report November 1987 (Validation tests of a coal mass flow rate measurement in ENEL Thermal Power Plants).

15. L. Corti, A. De Marco, A. Radice, A. Zizzo, *Control and Modelling of Coal Mills*, Second IFAC Workshop on *Modelling and Control of Electric Power Plants*, Philadelphia, USA, September 1986.

APPENDIX A

THE FUEL MEASUREMENT PROBLEM

One of the most difficult problems that face the designer of control systems for fossil fuel fired boilers is the development of a fuel measurement that is adequate for meaningful fuel/air ratio control. And in the case of once-through units, an adequate fuel measurement is critical to control of firing rate/feedwater ratio as well. The problems of measuring fuel input are as follows:

1. Varying Heat Value Content

   It used to be that utilities could depend on single sources of fuel supply for long periods of time. All that changed with the energy crisis of the 70's. High costs for energy as well as stringent emission standards have forced utilities to source fuel supply from various suppliers depending on price and availability. Thus a constant heating value for any given fuel can no longer be assumed. This problem is further complicated by the industry shift back to coal firing which has always been subject to significant heating value variations. The heating value problem exists for all fuels, gas, oil or coal.

2. Measuring Fuel Delivery

   Measuring fuel delivery has been convenient with gas, more of a problem with oil, and nearly impossible with coal as is described below:

   Gas and Oil measurements

   - Conventional orifice measurement techniques have limited turndown capability—typically 5:1. Wider range measurements require dual-range metering with the attendant complications of making a smooth transfer between low range and high range meters.

   - In-line devices can offer wider turndowns — typically 10:1, but with the attendant problem of having to shut down the line should maintenance of the in-line device be required.

   Coal Measurements

   Coal delivery systems don't reflect the actual coal flow to the furnace due to two main considerations: mill storage capacity and mill response capability. Let's examine the effects of mill capacity and mill response as they relate to the method of control of mill output coal flow to the furnace.

   - Control of Mill Output Coal Flow by Feeder Speed

     When mill coal flow is regulated by adjusting feeder speed, totalized tachometer output is used as a measure of fuel input. A typical example would be a CE-Raymond Bowl Mill with gravimetric feeder. The total feeder speed measurement must be compensated for the following factors in addition to variations in heating value.

     - *Mill Capacity* - All mills have significant capacity with respect to both the grinding and also the storage of coal. Thus, during load changes the fuel flow into the mill does not necessarily represent the actual fuel going to the furnace. A typical storage time for a mill is 100-160 seconds. That means that it takes tens of seconds before a change of mill input will appear as a significant useful output. Equating air flow to measured fuel input under these conditions guarantees an erroneous fuel/air ratio. The classical solution to this problem is to emulate the mill dynamics with a "Mill Model" using feeder speed as its input. This approach is limited in that the actual time constant for a mill is quite complex and varies not only as a function of load but also with primary air flow and the way primary air flow is regulated.

- *Loss of Coal* - Coal hangups, a common occurrence, are not reflected as a change in the feeder speed signal even though there is no coal on the feeder belt. "No coal flow" contacts can help to correct this problem by removing the feeder speed measurement from the total. However, loss of coal tends to be a sporadic event which can result in frequent switching of the individual feeder signal out of the total which if not properly accounted for by the control system can cause significant upsets to the unit.

- *Mill start-up/shutdown* procedures can affect the accuracy of the feeder speed based fuel measurement. A feeder may be shut down to relieve a mill overload or to sweep a mill clean—a condition where useful mill output exists even though the feeder speed is zero. Conversely, if the feeder is run to fill a mill, the result is a feeder speed signal that represents no fuel output.

- Control of Mill Output Coal Flow by Primary Air Flow Control

When mill output is controlled by regulating mill primary air flow and the coal feed to the mill is regulated as a function of mill level, the total of the individual primary air flows is used as a measure of total fuel input. A typical example is the B&W type EL pulverizer.

- Limitations

Use of primary air flow as a measure of fuel input has severe limitations in that the amount of coal in a given primary air stream can vary significantly. This is caused by a number of factors.

- The amount of coal flow to the furnace depends on mill level. Some mill level controls are limited to three values of coal feed rate (high, low and off in the case of most feeder tables). This approach causes continuous oscillations in level, hence fuel to the furnace.

- There are inaccuracies associated with making the basic primary air flow measurement such as temperature variations. Even obtaining a reasonable mill air flow differential is difficult due to the short piping runs around the mill.

- Physical differences between mills (differences in grinding surfaces, classifier settings, etc.) create errors in the relationship between primary air flow and coal flow to the furnace.

- Coal flow versus primary air is only roughly linear over the mills operating range. During mill start-up and shut-down, a significant primary air flow is present. This flow can have little or no fuel content as the mill is either filled up or swept empty.

- Ball tube Mills

A special case is the ball tube mill. From the control point of view, a most important characteristic of this mill is its very large capacity. Up to 15-20 tons of coal may be stored in this mill as compared to the 3-5 tons storage in a more conventional pulverizer. Thus it is not unusual for this mill to maintain full output for 10-20 minutes after its feeder has been shut down. Thus feeder speed is not a viable measurement of ball mill output.

The classical approach to measuring fuel output of a ball mill utilizes mill primary air flow as an index of coal that is swept up from the mill and delivered to the furnace. In addition to the limitations of primary air flow as a fuel measurement cited above, ball mills have a number of additional factors that affect the coal flow to mill air flow relationship.

- The status of ball mill charge is a major variant that affects the grinding efficiency of the mill, hence the reservoir of fines that will be transported by a given air flow.

- The measurement itself may be suspect since it may use as its primary element, the pressure drop across the classifier. There has been considerable debate as to whether coal flow is best represented by the raw differential pressure or by the air flow measurement.

- The non-linearities in the relationship of individual mill coal flow to mill air flow result in differences in computed fuel input depending on the number of mills in service and the coal output supported by each mill.

3. Unmetered Fuel

Additional fuel inputs may be unaccounted for since these inputs are not metered. Examples of unmetered fuel inputs include ignitors and/or warm up guns.

APPENDIX B

LIST OF SYMBOLS

Wp	pulverized coal mass flow rate
Wa	air mass flow rate
Wt	total mass flow rate
μp	coal density
μa	air density
μ	average density
v	total mass flow rate per unit section
g	gravity acceleration
p	pressure
Pa	atmospheric pressure
x	curvilinear coordinate along the ducts
w	total perimeter of the ducts
S	total cross section area of the ducts
z	vertical coordinate
Tp	air-pulverized coal temperature
α	air constant
D	discharge ducts height
Dc	height of the pressure transducer in the furnace
Ta	ambient temperature
Tc	furnace temperature
Po	duct inlet pressure
P1	furnace pressure
Cf	friction factor
Kfo	air friction factor
β	air-coal mixture friction factor
Φ	correction term

APPENDIX C

ON-LINE HEAT RELEASE COMPUTATION

An heat release computation of the radiated heat to the boiler can be obtained also for once through boilers. A simplified model of the evaporator of this boiler is described by the following equations:

1) $a_{11}\dot{P} + a_{12}\dot{h}_u = Qw\,(h_i - h_u) + Q_{mf}$

2) $a_{21}\dot{P} + a_{22}\dot{h}_u = q_w - q_u$

3) $C_m M_m \dot{T}_m = Q_{rad} - Q_{mf}$

4) $Q_{mf} = \Gamma\,(T_m - T_u)$

This model can be implemented in real time, and a system like the one represented in the attached figure can be used to have an estimation of Q_{rad}.

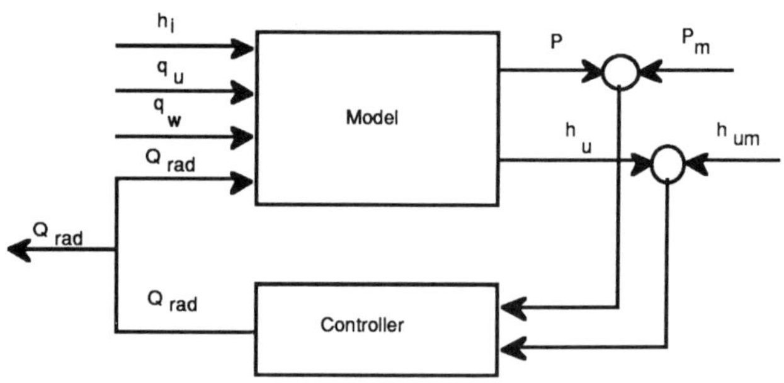

List of symbols

$a_{11}......_{122}$	coefficients (computer off-line with steam tables or derived from field test)
P	pressure
h_u	outlet enthalpy (when superheated steam or supercritical fluid is available)
h_i	inlet enthalpy
q_w	feedwater flow
q_u	steam flow
C_m	metal specific heat
M_m	metal mass
T_m	metal temperature
Q_{mf}	heat exchanged between metal and fluid
Q_{rad}	heat exchanged in the furnace
T_u	fluid mean temperature (temperature at h_u entholpy)
Γ	exchange coefficient
P_m	measured pressure
h_{um}	measured outlet enthalpy

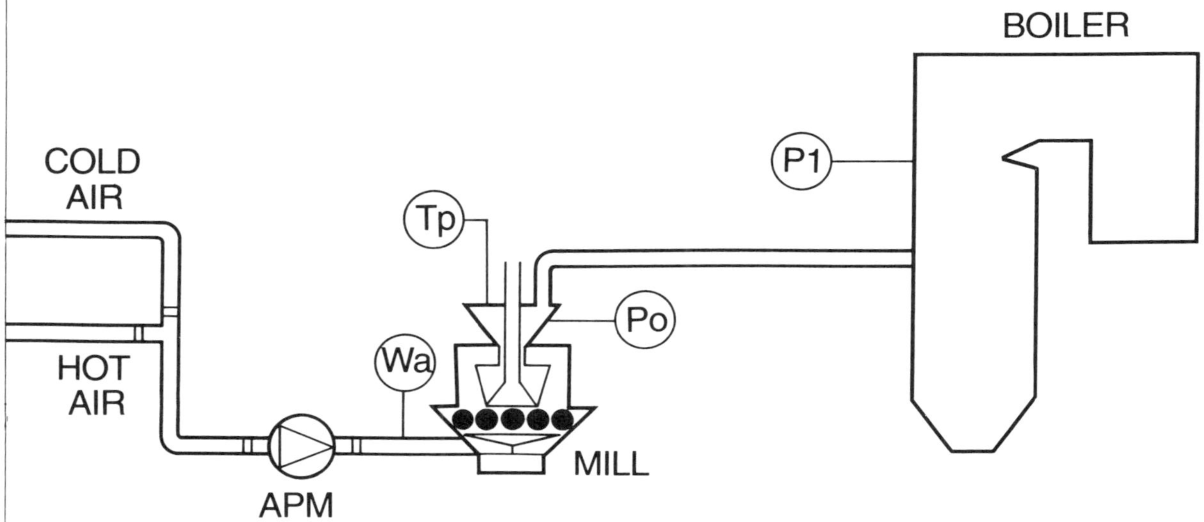

fig. 1 - PLANT LAY-OUT

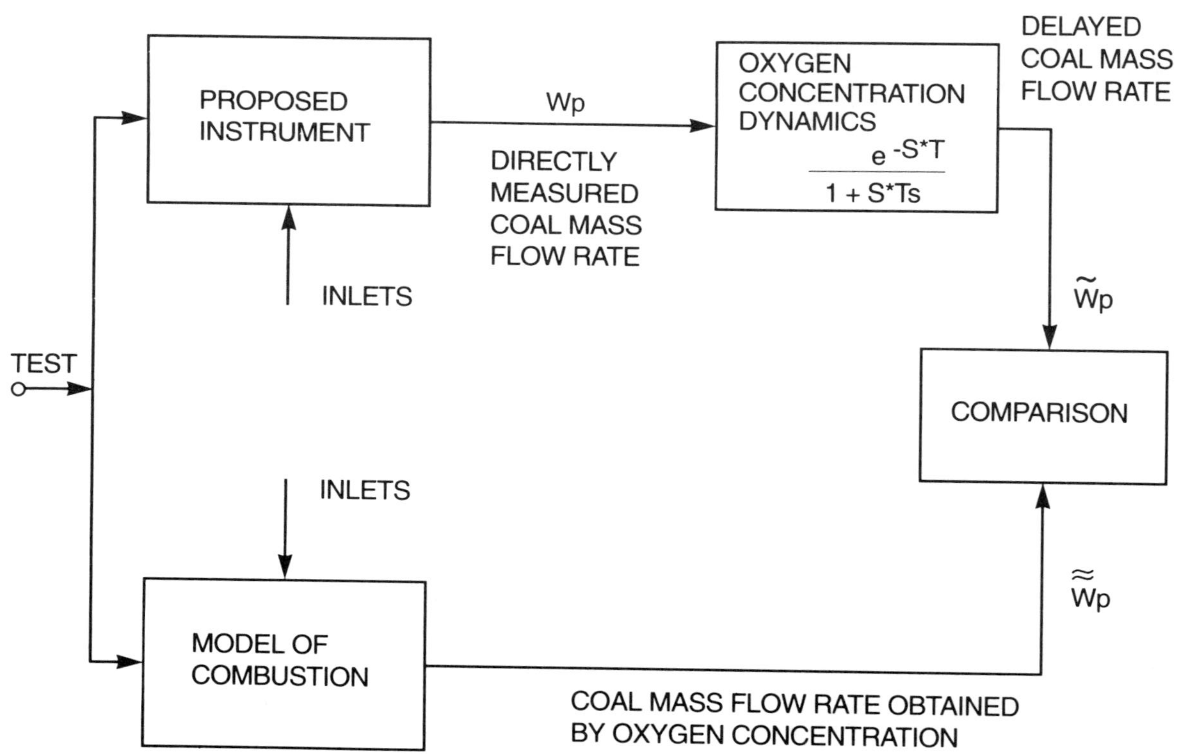

fig. 2 - **VALIDATION PROCESS**

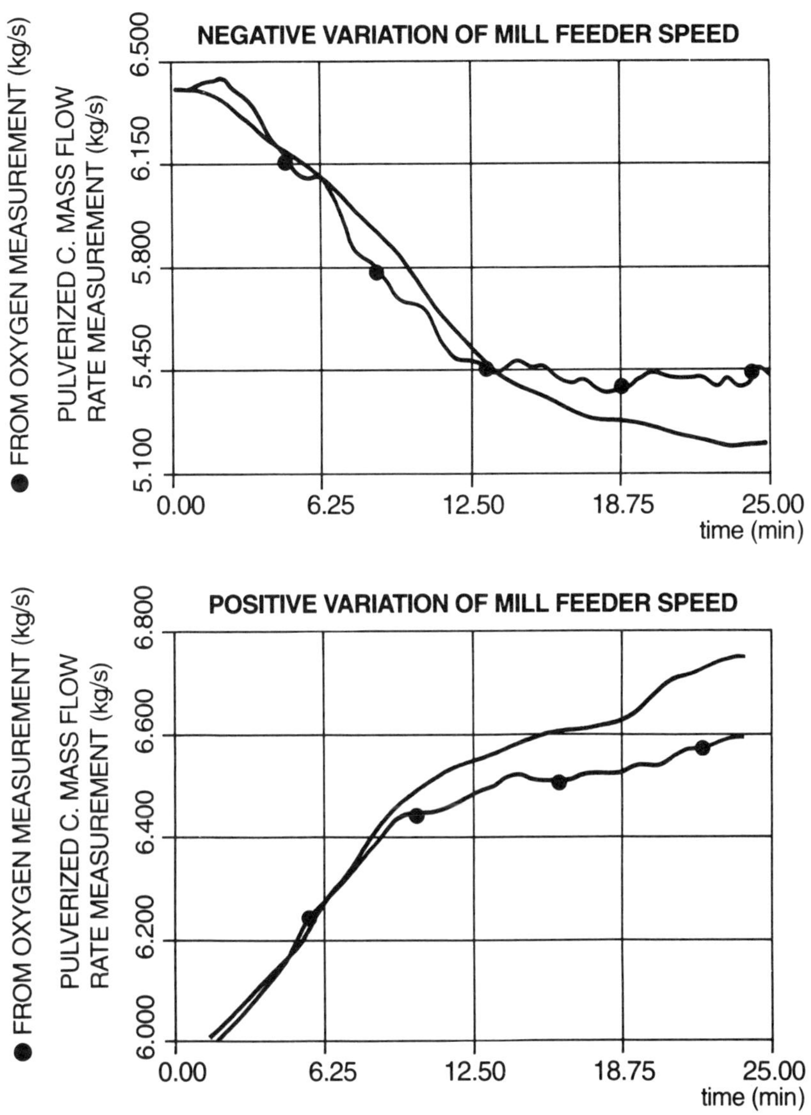

fig. 3 - VARIATION OF MILL FEEDER SPEED

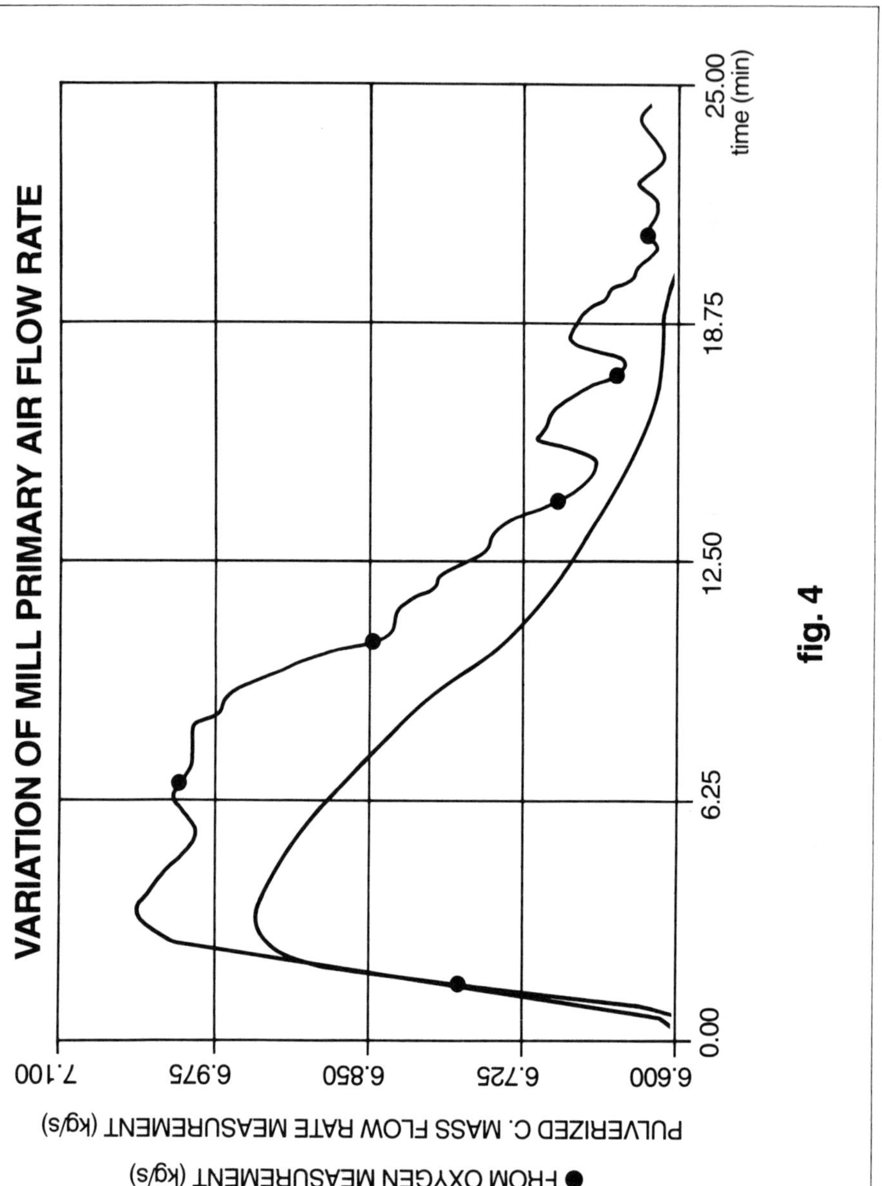
VARIATION OF MILL PRIMARY AIR FLOW RATE
7.100
6.975
6.850
6.725
6.600
0.00
6.25
12.50
18.75
25.00
time (min)
PULVERIZED C. MASS FLOW RATE MEASUREMENT (kg/s)
● FROM OXYGEN MEASUREMENT (kg/s)

fig. 4

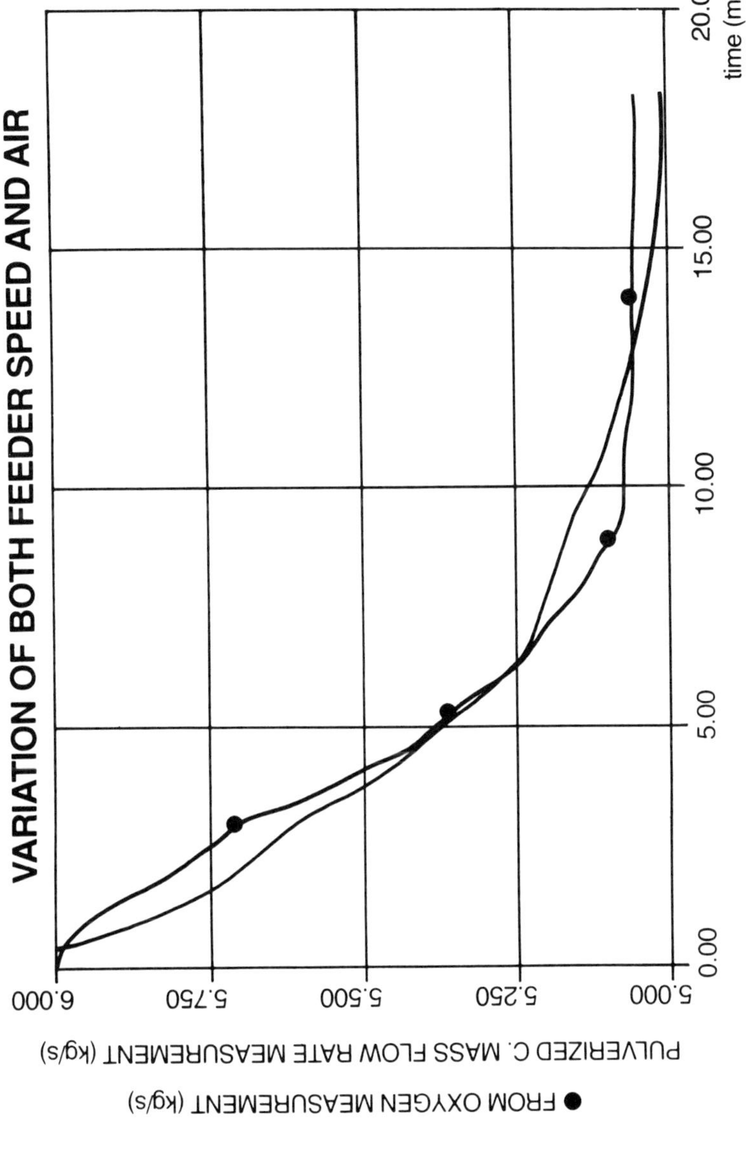
VARIATION OF BOTH FEEDER SPEED AND AIR
6.000
5.750
5.500
5.250
5.000
0.00
5.00
10.00
15.00
20.00
time (min)
PULVERIZED C. MASS FLOW RATE MEASUREMENT (kg/s)
● FROM OXYGEN MEASUREMENT (kg/s)

fig. 5

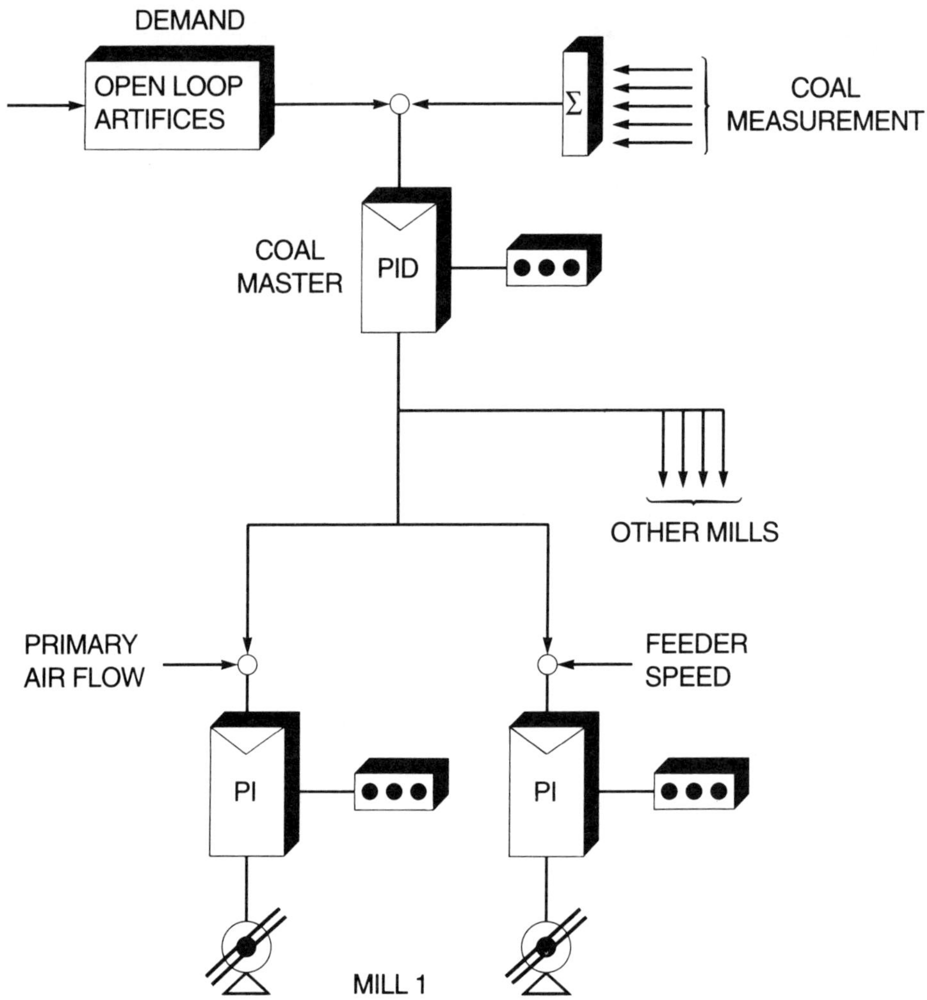

fig. 6

SIMPLIFIED DIAGRAM OF MILL CONTROL

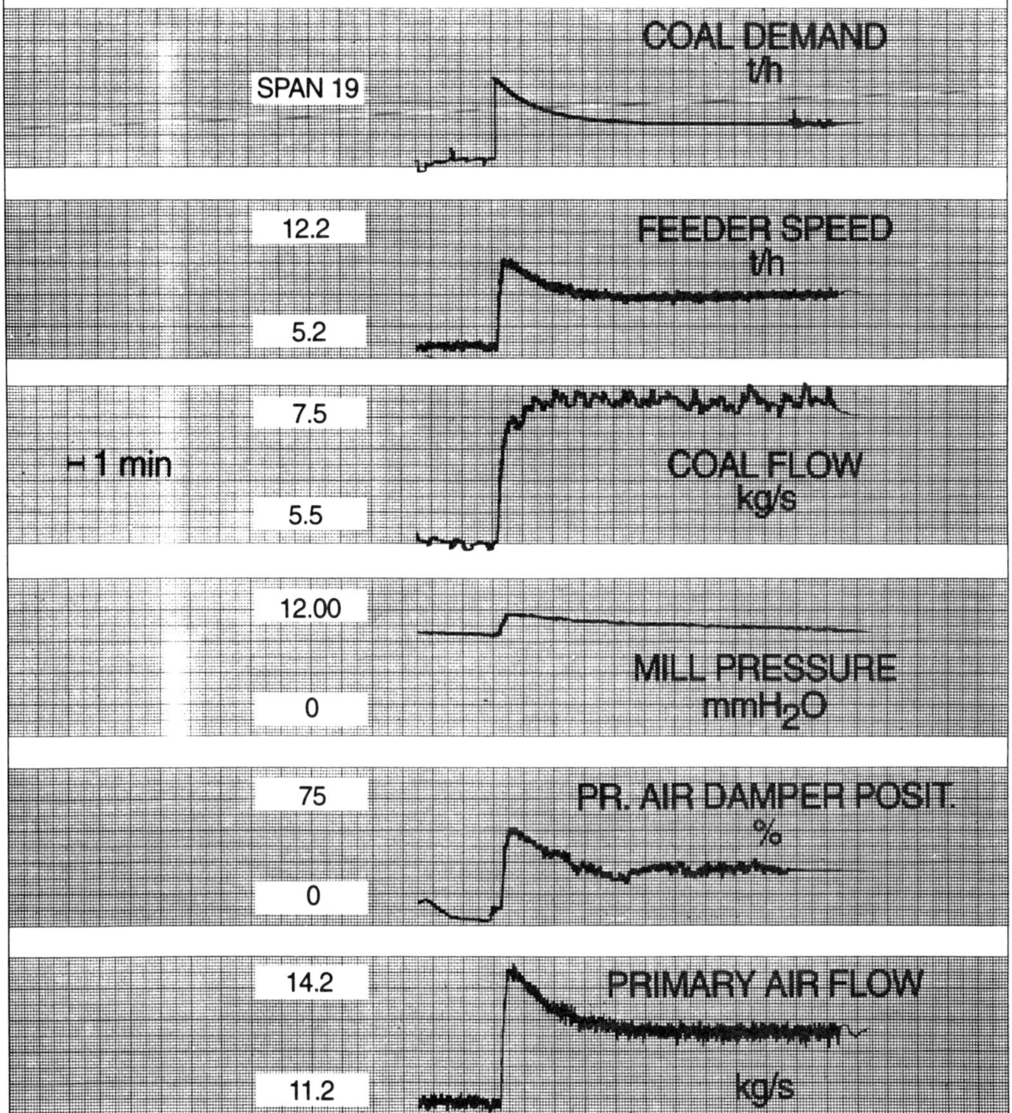

fig. 7 CENTRALE DI VADO LIGURE - SEZ. 2
OPEN LOOP CONTROL

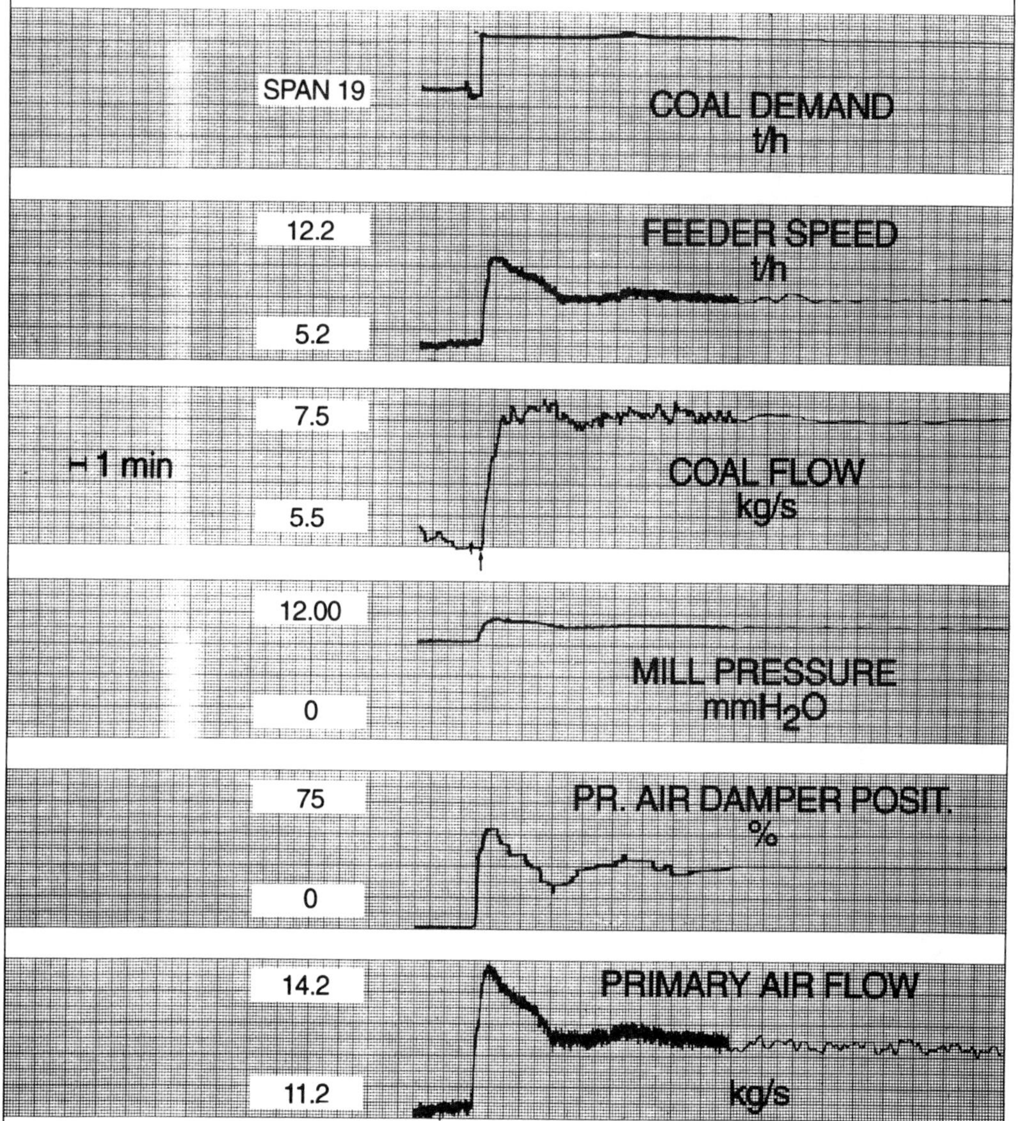

fig. 8 CENTRALE DI VADO LIGURE - SEZ. 2

CLOSED LOOP CONTROL

tab. 1 - Calibration - Friction factor kfo

DATA	N°	WA (kg/sec)	P0 (mm H_2O)	P1 (mm H_2O)	Tp (°C)	Kfo
19/2/87	1	10,7	515	347	87	1,43
19/2/87	2	11,1	535	355	82	1,41
19/2/87	3	11,35	543	357	73,6	1,42
19/2/87	4	12,3	570	360	58	1,41
19/2/87	5	13,3	608	364	58,5	1,39
19/2/87	6	12,3	575	366	58,5	1,40
19/2/87	7	11,7	569	379	58,3	1,41
19/2/87	8	10,9	535	366	73,3	1,42

tab. 2 - Calibration - Friction factor ß

DATA	N°	WA (kg/sec)	P0 (mm H_2O)	P1 (mm H_2O)	Tp (°C)	K fo	ß
20/2/87	1	11,7	728	346	73,2	1,41	0,42
20/2/87	2	12,2	800	401	75,42	1,41	0,40
20/2/87	3	11,2	775	406	71,8	1,41	0,41
20/2/87	4	13,3	880	436	73,9	1,41	0,39
20/2/87	5	12,2	830	419	77	1,41	0,43
20/2/87	6	12,1	743	378	62,9	1,41	0,41

tab. 3 - Validation

N°	TYPE OF TEST	POINT 1				POINT 2			
		Wc kg/s	W kg/s	WD kg/s	Wp kg/s	Wc kg/s	W kg/s	WD kg/s	Wp kg/s
10	FEEDER SPEED STEP	19.89	19.3	6.69	6.6	19.3	18.7	6.2	5.9
11	FEEDER SPEED STEP	19.36	18.6	6.1	6.0	20.2	19.3	6.88	6.7
12	FEEDER SPEED STEP	20.44	19.2	6.74	6.4	19.4	18.0	5.52	5.2
13	PRIMARY AIR MASS FLOW RATE STEP	22.35	20.5	7.2	6.6	22.2	20.5	7.2	6.6
14	STEP OF FEEDER SPEED AND AIR	30.5	28.5	5.2	5.0	31.4	29.3	6.1	5.8
15	STEP OF FEEDER SPEED AND AIR	31.3	30.7	6.1	6.0	29.9	29.6	4.9	5.0

Legenda:
Wc = Total coal mass flow rate from 02 measurement
W = Total coal mass flow rate from feeder speed measurement
WD = Coal mass flow rate of mill D obtained from the total coal mass flow rate Wc
Wp = Pulverized coal mass flow rate from instrument

0074-056X/91/097-105/$0 + .50pp

APPLICATION OF NUCLEAR COAL ANALYZERS IN COAL HANDLING SYSTEMS

AUTHORS
E. A. PFROGNER
Mechanical Engineer
United Engineers & Constructors
600 Penn Center Blvd.
Pittsburgh, PA 15235
and
R. G. SENSENIG
Instrumentation and Controls Engineer

ABSTRACT

Coal analysis has traditionally been performed by a laboratory to determine the sulfur content, ash content, fusion temperature, coal BTU value, moisture content, and other elements of interest. There are now a number of commercially available nuclear coal analyzers that can be applied to coal handling systems to provide a complete elemental analysis of coal in as little as one minute. This data can be used for continuous blending of coals for the content of sulfur, moisture, ash, or lbs. SO_2/MMBTU which can provide significant savings by optimizing the use of less expensive high sulfur coal. Also, the analysis information has many other potential uses within the power plant such as a feed forward signal to boiler controls or plant performance optimization packages.

This paper provides an overview of typical applications of nuclear coal analyzers in coal fired power plants. It also discusses the major issues to be considered by a utility company when preparing a specification for the analyzer and the installation design.

TYPES OF ANALYZERS AND THEORY OF OPERATION

Nuclear-based instruments are not new to coal fired plants. In the past, these nuclear instruments have been used to measure slurry densities or to determine point levels of dry materials in hoppers and tanks. Today, nuclear type coal analyzers can provide much more diverse information on a real-time basis. Data on ash content, sulfur content, and ash constituency may be determined, and, with an interfaced moisture meter, the BTU content, emissions potential, proximate analysis, ultimate analysis, and an estimate of ash fusion temperature may also be obtained.

There are three types of analyzers that generate information about coal. They are:

1. Prompt Gamma Neutron Activation Analysis (PGNAA) utilizing a Californium source. (See Figure 1)

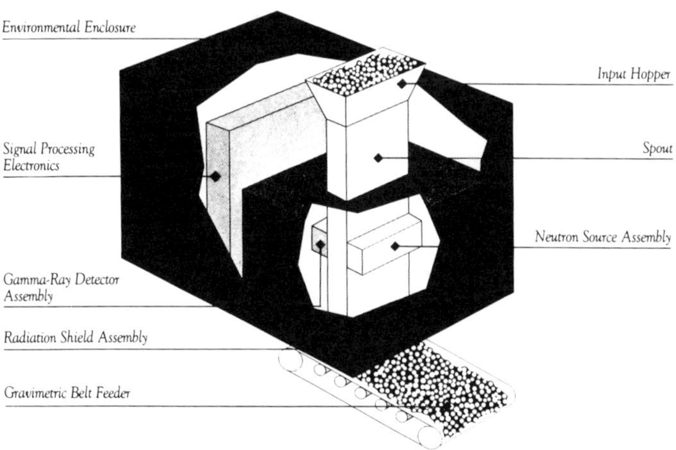

Figure 1

2. Elemental analyzer utilizing a Radium source. (See Figure 2).
3 Elemental analyzer utilizing a Cesium source and an Americium source. (See Figure 3).

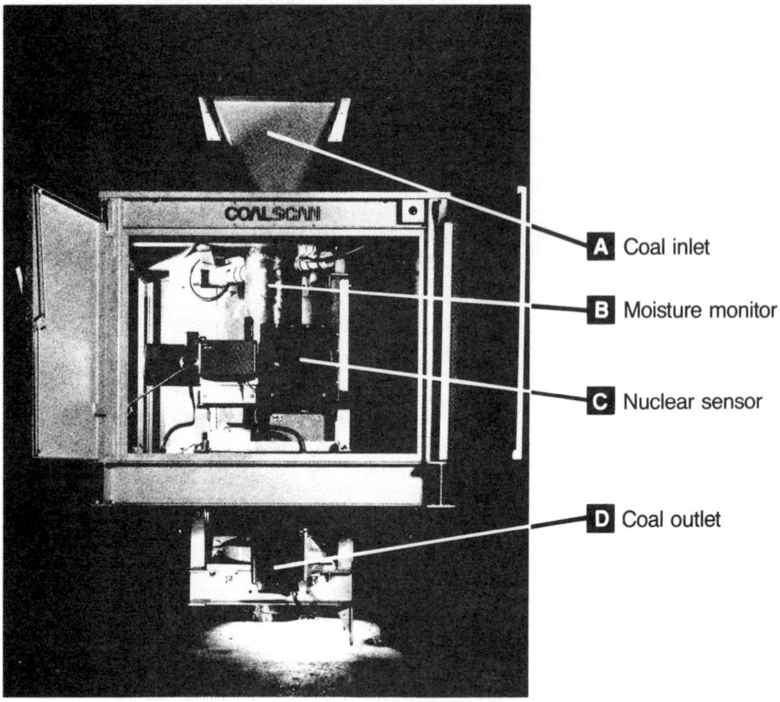

Figure 2

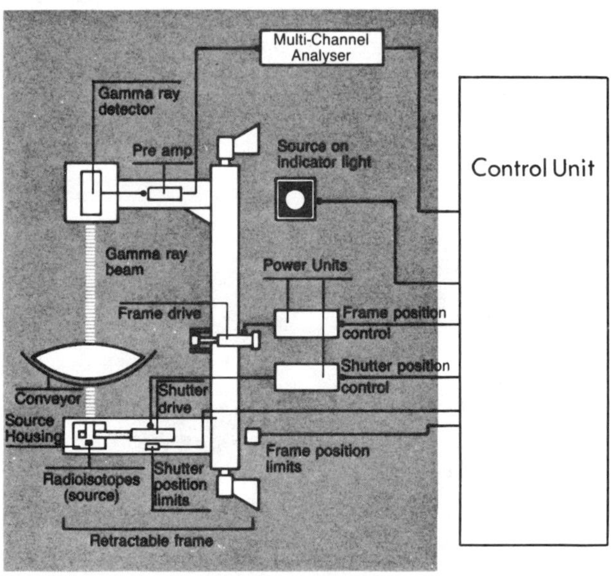

Figure 3

The first two types can be combined with a moisture meter which is installed in or near the analyzers inlet hopper. The meter is a non contacting device which normally employs microwave technology to analyze the calorific value and moisture content of the coal. This information is then fed into the analyzer control algorithms.

The Prompt Gamma Neutron Activation Analysis

The PGNAA analyzer, with a moisture meter, produces a full ultimate and proximate coal analysis. This analyzer requires a stream of coal to fall vertically through its internal chute. Once the coal is in the detection zone of the chute, it is bombarded with neutrons produced from the Cf-252 nuclear source (see Figure 4). The nuclei of the elements composing the coal capture the neutrons. In the excited state, the nuclei emit energy in the form of gamma rays. Each element in the coal produces a gamma ray of unique energy. A detection element on the wall opposite the source counts each gamma ray and measures the energy level. This information is then processed by the analyzer's computer which converts all the different gamma ray pulses into the elements found in the coal (See Figure 5). With this accomplished, the computer then displays this information. The whole process takes as little as one minute, depending upon the manufacturer. The PGNAA analyzer is the type most widely selected for coal fired power plants probably since it can produce both an ultimate and proximate analysis. When packaged with appropriate software and additional I/O, it can provide for boiler performance calculations and coal blending.

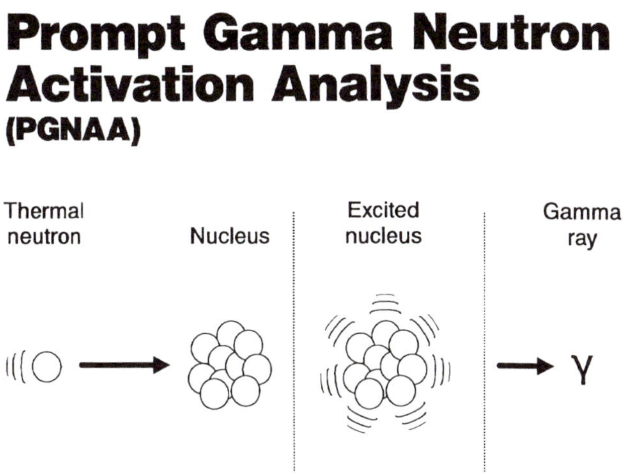

Figure 4. Prompt Gamma Neutron Activation Analysis (PGNAA)

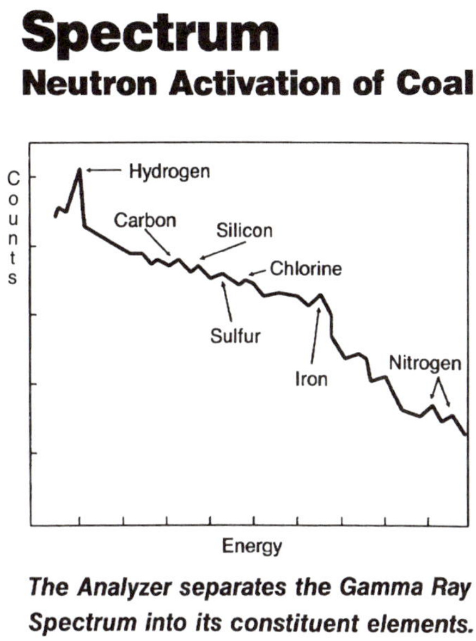

Figure 5. Spectrum (neutron activation of coal)

The following list of elements is typical of the information given by the PGNAA analyzer which may be provided for output to an operator.

Ultimate Analysis	Ash Analysis
Sulfur	Silicon Dioxide (SiO_2)
Ash	Aluminum Oxide (Al_2O_3)
Carbon	Calcium Oxide (CaO)
Hydrogen	Ferric Oxide (Fe_2O_3)
Moisture	Titanium Oxide (TiO_2)
Nitrogen	Sodium Oxide (Na_2O)
Chlorine	Potassium Oxide (K_2O)
Oxygen	Trace Elements

The PGNAA analyzer is supplied with its own protective housing to ensure the safety of plant personnel. The equipment is provided with a locked box that shields the outside from the sources. Only licensed personnel from the manufacturer are permitted to access the sources directly. The radiation levels outside of the housing are low enough that posting signs are not necessary, although many stations still place signs as a precautionary measure. Licensing and permitting are necessary for this apparatus. The manufacturer is available to assist with this phase as he will maintain ownership of the nuclear sources. The manufacturer will also provide training to those plant personnel who will be operating and interfacing with the equipment.

Elemental Analysis Utilizing a Radium Source

The radium source analyzer, will give the power station information on the coal's ash content and moisture content (if a moisture meter similar to the one mentioned is implemented). This analyzer gives a high precision reading (±2%) for the ash content of the coal irrespective of the composition of the coal. The technology employed for this analyzer is pair production.

For pair production, a gamma ray is emitted by the nuclear source Ra-226. When the ray approaches the coal constituents nuclei, the absorption of its energy creates an electron-positron pair. The production of the pairing is measured; the amount of pairing is proportional to the ash content of the coal.

This type of analyzer gives a reliable reading of ash content and has installation costs similar to the PGNAA analyzer since this analyzer has almost identical size and flow requirements as the PGNAA analyzer and both must have coal placed inside its internal chute during coal analysis. If blending of coal on ash content only with a high degree of accuracy is the purpose of the analyzer, this second type of analyzer would be a good selection.

Elemental Analysis Utilizing Cesium and Americium Sources

The analyzer utilizing Cesium and Americium source measures only the ash content to an accuracy of ±10%, much less than the pair production analyzer. This lower accuracy results from the dependence of this model on the iron content of the coal. This analyzer uses Cs-137 and Am-241 nuclear sources to generate gamma rays. Two sources are utilized so that the change in ash or a change in coal depth on the belt can be differentiated. The number of gamma rays absorbed by the coal determines the ash content.

The advantage of this unit is that it is a continuous on-line through belt analyzer. It is mounted above and below the conveyor belt with no direct contact with the coal. The through belt analyzer has lower installation costs and provides reasonable ash content results.

TYPICAL APPLICATIONS OF ANALYZERS

Coal Quality Monitoring

The nuclear coal analyzer has been used to perform numerous coal quality monitoring and data logging tasks in coal handling systems. Coal analyzers have been applied to perform analysis of coal truck samples. This analysis could be relied upon to reject coal shipments if the supplier has not met coal specifications agreed upon within a supply contract. The coal analysis can be performed within one minute of the sample bombardment. The data generated is then displayed on a CRT in the operator's station. This truck sample application will allow for an immediate rejection of the coal being delivered to the plant which exceeds the specification for ash, sulfur, or any other element. The data supplied by the analyzer may be transmitted to a computer for historical collection of data in an analysis database. This database will associate consignments of coal with analysis data. Each analysis could then be used to compile a complete database of as-received coal analysis from each coal supplier. The compiled information could also be used to formulate future coal supplier specifications or assess penalties to off-specification coal suppliers.

Coal fired power plants could be provided significant data necessary to monitor and control their SO_2 flue gas emissions by the PGNAA nuclear coal analyzer. A nuclear coal analyzer equipped with full elemental analysis capabilities can provide sulfur content analysis, and, combined with a moisture analyzer, can provide a lbs. SO_2/MMBTU value for the coal. This analysis can be directly compared with the emissions standards imposed by the EPA for SO_2/MMBTU. Performing this comparison of coal fired to the target EPA regulations can assure that emissions regulations are not exceeded since any coal not conforming to the standards can be diverted from the bunkers. There has been some interest by power plants in utilizing a nuclear coal analyzer to provide the official reports to the EPA on the SO_2/MMBTU quantities rather than using a flue gas SO_2 emissions monitors to perform this task. However this concept has yet to be agreed to by the EPA.

The PGNAA nuclear coal analyzer is claimed to be suitable for monitoring ash fusion temperature of coal. (Coal testing laboratories claim to be able to estimate ash fusion temperatures within 50 °F based on testing. They expect calculation of ash fusion temperatures based solely on analyses of constituents to be accurate to no better than 100 °F since exact formulas do not exist.) The ash fusion data can be used to reject coal suppliers and also be used by operators to optimize boiler tube cleaning. Ash fusion temperature can be used to determine the slagging characteristics of the coal being fired in the boiler. By monitoring this data and providing the knowledge of the slagging index of coal, the operator can be forewarned of increased requirements for boiler soot-blowing.

The moisture analysis of the coal can be used to adjust fuel/air ratios required in the pulverizers to dry and convey the coal to the burners. During high moisture periods, the coal pulverization and conveying system suffers and this in turn reduces the net generation during that period. Therefore, by using the moisture analysis signal of coal being loaded into the unit's bunkers and feed forwarding this signal to a distributed control system, the pulverizer's operation could be optimized during varying moisture periods.

The BTU value of coal can be determined using the nuclear coal analyzer. This signal can be used as a feed forward signal to the boiler control to provide adjustment in the BTU compensation logic. The practical problems with performing this type of compensation are bunker storage capacity and lag time between the analysis and actual coal burning time. If a coal analysis is performed on the coal belt loading into a bunker area, a data acquisition and allocation system may be designed to compensate for the coal destination, the bunker profile, and the capacity of the bunker. Ideally, this modeling of the bunker will allow a computer to calculate the actual BTU value for coal being fed to the pulverizer on a real time basis. The major obstacle is the unpredicable pattern of coal removal from the bunker due to bridging.

Coal Blending Control

Utility companies are attempting to reduce their operating costs on a continuous basis because of increased competition, public pressure and to attract new growth to their service territories. One major cost of operating a coal fired plant is fuel, which amounts to approximately 30% of the operating expenses. Therefore, by installing a coal blending system, a utility could save on operating costs by blending less expensive coals and still meet EPA emissions standards. A nuclear coal analyzer can be used as a feedback device to blend coal for content of sulfur, ash, moisture, or lbs. SO_2/MMBTU blending parameters. Most stations perform coal blending utilizing laboratory analysis results. This method has a feedback time of two hours or more before blend changes can be made. Also when performing blending using a manual sample, a large margin from setpoint must be maintained so that the statistical average of sulfur, for example, does not exceed the mandated EPA limits. The continuous feedback control of a nuclear coal analyzer blending system will reduce blend fluctuations and increase the target blend closer to the EPA limits. As a result, the proper amount of cheaper high sulfur coal can be blended to achieve a similar BTU value as before.

With the PGNAA analyzer, many of the component results given may be used as the target setpoint. Such things as ash, BTU, moisture, sulfur, lbs. SO_2/MMBTU, etc. could be selected as the control element. Most often sulfur, lbs. SO_2/MMBTU or ash content is used. This is because of either contract regulations between the buyer and seller of the coal, or performance emission regulations mandated by the state.

For blending controls, more periphery equipment is required. After the basic analyzer and analyzer controller, an additional blending computer is necessary. The accuracy of the belt scales, reclaim coal feeder, and the moisture analyzer are important in acheiving the maximum cost savings.

Any nuclear coal analyzer could be used to blend for ash content. This blending would enable the station to predict slagging characteristics and sootblowing requirements.

SPECIFICATION OF ANALYZER

One key issue to specifying a nuclear analyzer is the accuracy necessary for the application. A system providing a signal output for reviewing ash content or recording sulfur content could use an analyzer with 1% constituent accuracy. An analyzer to measure the process variable for a coal blending system will require a minimum constituent accuracy of 0.5%. The accuracy of the instrument used in blending coal is also very critical to the economic payback of a blending system. However, if the ash accuracy required is only 3% there would be no value to purchase equipment that provided much greater accuracy (at a higher cost).

A moisture analyzer is required if a measurement of SO_2 or BTU is required. Some moisture analyzers require a crushed size coal sample (maximum 2 inch lump size) to obtain an accurate reading from their instrument. For blending systems utilizing a target variable which is developed from the moisture analysis, it is very important to have a very high precision moisture instrument (0.5% or better).

Another specification item that is very important to the success of a blending control system is the performance warranties requested of the vendors. It is desirable that the vendor supplying the nuclear coal analyzer and blending control algorithm take the responsibility to meet cost savings requirements that have been used to justify the project. The specification should include a blend accuracy target, because, if achieved, it could save hundreds of thousands of dollars in fuel costs.

Other issues to be concerned with when specifying a nuclear coal analyzer are:

- Nuclear element half-life and periodic replacement.
- Element temperature control unit.
- Hopper level controller.
- Microprocessor for nuclear analyzer.
- Blending computer functions and capability reports, long term data storage, and data links to power station DCS systems.
- Delay time from analyzer to coal feeders.
- Upgrade of belt coal feeders for response and linear control in blending.
- Power supplies and lightning protection for computers and microprocessors in coal handling area.

- Reliability and accuracy of existing belt scales or the purchase of new scales.
- Conveying equipment necessary to provide coal to and from the analyzer.

Many decisions must be made to determine what system to purchase and the purpose of the system. The results of these decisions will determine the type of analyzer to be purchased (whether it is ash only, or a full analysis system) and its application. Investigation of the sample system capacity will be necessary. PGNAA analyzer can utilize up to 500 tons per hour (however, most manufacturers process less) at a maximum coal size of two inches. Other influencing factors such as power requirements, backup systems and material handling transfer equipment must also be considered so that a complete and accurate system is provided.

INSTALLATION DESIGN

Analyzer Space Requirements

Nuclear analyzers come in many different sizes depending upon the operation desired. If the ash measuring analyzer is desired, the space limitations are minimal size. These systems are installed directly above and below the coal belt. The installation usually only requires a platform to be constructed and space to be provided for the placement of the analyzer computer.

PGNAA nuclear analyzers and the more accurate ash analyzers (±2%) require much more space. Different manufacturers have different sized "boxes"; so it is important to consider available space for each specific application. Additional space is required for the conveyors, inlet and outlet chutes, and a moisture monitor. In many, if not most instances, an area will need to be reserved, or designed to house the analyzer and its control components. While some analyzers are constructed to withstand the outside elements, the control components must be placed in a controlled environment to protect the electronics.

BLENDING CONTROL SYSTEM

A coal blending control system has three major elements: the nuclear coal analyzer, the reclaim coal feeders, and the belt weight scales (see Figure 6). The microprocessor or PLC of the nuclear coal analyzer which is used to interpret the ele-

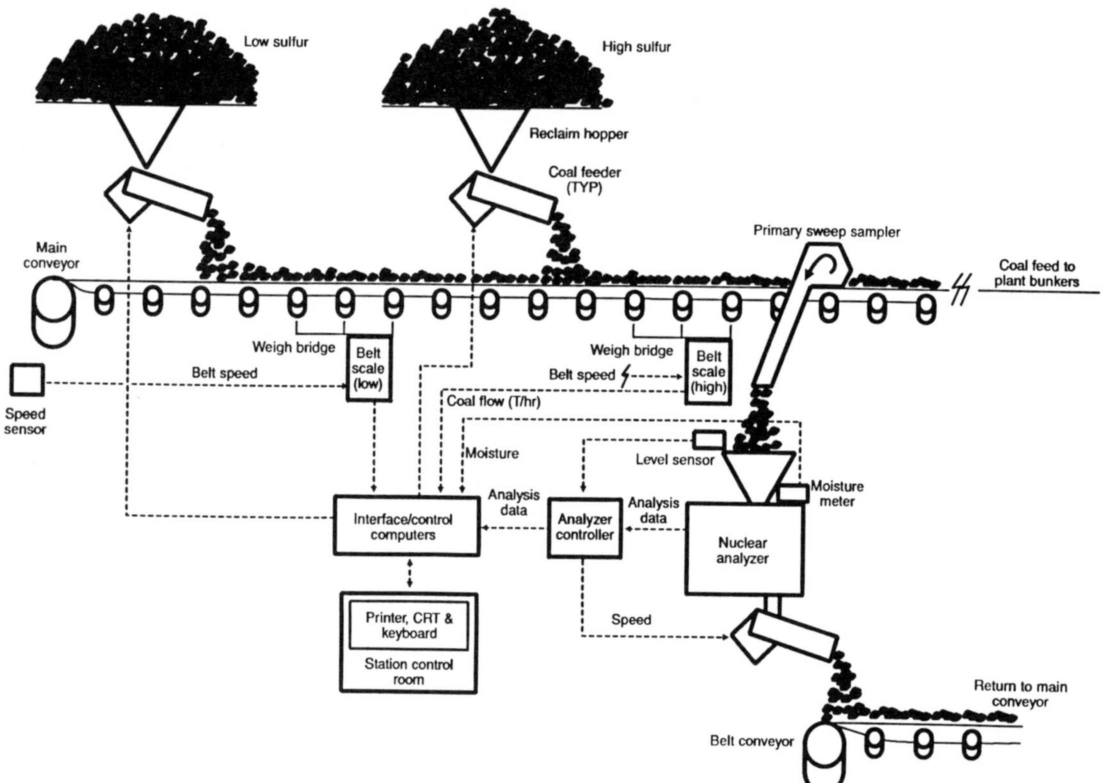

Figure 6. Coal blending system

ment spectrum is also used to maintain the detector element at a constant temperature either by heating or cooling the system. Density measurement of the coal may be performed, along with a moisture analysis of coal (by a separate analyzer), if required. Most nuclear analyzers use a variable speed coal feeder or conveyor belt on the outlet to maintain coal in the analyzer chute at all times. The analyzer computer controls the feeder to maintain a level setpoint of coal in the hopper above the analyzer. This hopper level is determined by a ultrasonic level sensor.

The analyzer is fed coal at a constant rate every time the main belt is running. The blending computer utilizes the analysis signals from the analyzer controller and the weight signals from the belt scales located at the load out points of the reclaim coal feeders to perform computations using statistical algorithms to determine the output signal. This output signal generated by the analyzer sets the reclaim feeder output rates needed to achieve the target element - usually sulfur or SO_2/MMBTU. The reclaim coal feeders for the blending systems are located under the respective coal piles containing high, medium and low sulfur coals. Usually, the coal feeders are vibrating type feeders with variable speed motor drives or clutches to control coal flow on the conveyor belt. There could be numerous feeders for each grade of coal and the control system has to take into account the number of feeders in service when sending demand signal.

The blending control system operator interface is usually installed in either the coal handling control room or the main control room. The operator interface usually consists of a personal computer that has a CRT display which allows for changing the blend setpoint and for viewing coal analysis data. Manual/automatic control stations should be utilized on the reclaim coal feeders. If the blending computer fails or the analyzer is out of service, the coal blending can be performed manually. The blending computer also performs the historical collection of coal analyses data which is stored on hard disk or optical disk every few minutes for future retrieval. In addition, the blending computer calculates cumulative values and weighted averages of coal analysis data for display and use as feedforward to other control systems. Eventually many of these data acquisition and retrieval functions can be incorporated into a plantwide distributed control system for universal use of data.

Power Supply and Power Conditioning Requirements

The power supply and power conditioning for the microcomputer based equipment is very critical in power plant coal handling areas. The coal handling electrical systems are usually very remote from their primary bus and have numerous large motor loads that can cause regular large fluctuations in voltage. The nuclear coal analyzer requires a power source with a constant voltage transformer to regulate voltage during voltage fluctuations and isolation to prevent electrical noise problems. In addition, the personal computer used for the analyzer program must be equipped to reboot the program after a power outage. The analysis data should be stored on fixed media (hard disk) after every analysis or the data will be lost on a power outage.

Special attention should be given to the grounding system installation to limit the effect of lightning and noise on the analyzer components. Surge protection should be provided on all electronic power supplies. The data highway links to any remote blending computer should provide adequate electrical isolation to prevent electrical noise pickup. The belt scale electronics also require the same electrical power supply considerations to provide an accurate and noise-free signal to the blending computer.

Coal Handling Equipment Requirements

(If only a through-belt on-line ash monitoring system is required, additional coal handling equipment is not necessary since the ash monitor mounts directly above the belt line and requires no external equipment to feed any coal through the analysis zone).

The reliability of the transfer equipment is crucial to the operation of the system. This equipment includes: the reclaim vibrating feeders, belt scales, the outlet feeder of the analyzer, any transfer equipment (usually some type of primary sampler equipment and conveyors), and all chutes and hoppers within the system.

It is preferred that upgrades to existing equipment be accomplished by plant personnel due to cost and downtime. As long as the existing sample system is in reasonable condition, a retrofit can be accomplished.

The reclaim vibrating feeders on the coal loadout need to be designed with the ability for remote signal interface to control the output of the feeders. Feeders loading the same coal (for instance low sulfur coal) will be clustered together and receive the same signal for their output demand from the analyzer. The vibrating feeders must be capable of providing a repeatable and linear flow control. Replacement of the feeder pans with liners is usually recommended to reduce the friction of the coal flow.

With any blending system, the quantities of coal to be blended must be known. To gain this information, the system must be interfaced with signals from all belt scales. For each designated blending source, an accurate coal flow signal should be sent to the blending computer. The accuracy requirements of the belt scale varies depending on the analyzer manufacturer; as a rule, the more accurate, the better. More accurate scales cost more to purchase and have higher installation costs. As the accuracy increases, the idler quality must get better; therefore, the cost to replace and install new idlers and the number of idlers needed is directly affected by the scale chosen. The accuracy may be selected based on a scale signal needed for other plant systems. It is recommended that scales with at least 1% accuracy be used. Quite often the belt scale already exist on the conveyor line at the location required for the analyzer so they may be used for the analyzer. The location of any belt scale, existing or new should conform to the scale manufacturer's installation recommendations.

In the projects that we have encountered, there has been little room to install the analyzer directly in the stream of coal. This stems from the fact that many analyzers are not yet capable of evaluating a full belt stream of coal. As mentioned previously, maximum volume that any analyzer today can handle is 500 tons per hour (tph), but many analyzers can only operate with about 25 tph. Therefore, most installations are utilizing the coal from a primary sampler which reduces the volume of coal.

In some instances, the flow from a reject chute for the sampler can be intercepted and diverted to the analyzer for processing. This can be accomplished using a diverter/flop gate and conveyors. Another method of accomplishing this task is to install a new sampler on the coal belt. If space permits at the head pulley, a conventional cutter type sampler may be used; however, in most cases, available space is at a minimum. In these cases, a sweep-type sample cutter may be used. This type of cutter installs directly above the belt and takes a cross-stream cut of coal at timed intervals. The cut then will be transferred to the analyzer through chutes, hoppers, and conveyors.

The cutter must be accurate in its removal of the coal from the belt. Since the nuclear analyzer system's purpose is to duplicate the ASTM laboratory testing, all of the sampler equipment (or diversion equipment) must meet ASTM standards for removal of coal from a belt. Any coal left on the belt will result in an inaccurate sample that will cause a bias reading from the analyzer. If an existing sampler is used, be sure there is no bias (or minimal bias) present in the system. If purchasing a new cutter for this application, be sure to specify all bias and precision requirements.

For all chutes and hoppers, the flow must be kept steady. The sides of this equipment should be lined with a low coefficient of friction material that can withstand the impact of coal (such as stainless steel at the impact points and a PVC liner elsewhere). All hoppers and chutes should be installed with a minimum slope angle of 65°, preferably 70°. For some analyzers, the ability to gain access inside the analyzer chute is necessary for calibration. In these cases, a removable cover should be used over the inlet hopper. Be sure to check the calibration requirements of the manufacturer.

Conveyors are required to move the coal from transfer points. Conveyors should be installed such that the angle of inclination is no greater than 15°. If a conveyor is necessary to transfer the coal from the sampler to the inlet hopper of the analyzer, the installation of the head pulley must allow for the placement of a calibration block; therefore, the clearance requirements must be considered. All conveyors must have belt scrapers installed on them to insure complete removal of all the sample coal so that a representative sample is given to the analyzer.

The analyzer operates with the use of either an outlet vibrating feeder or a variable speed belt feeder. In either case, the purpose of the equipment is to regulate the flow of the coal through the analyzer. The analyzer controls the feeder drive so that the chute remains full during the session of analysis. Another concern with this equipment is the space requirements. The belt feeder should be as horizontal as possible and its length should be 10 feet or less so that the system control is maximized. The vibrating feeder must be adaptable to slow flow restrictions.

After the analysis process is complete, the coal should be returned to the original source flow. This may be accomplished using belt conveyors, bucket elevators, or other similar equipment.

COST IMPLICATIONS

Analyzer manufacturer's equipment vary significantly in cost, depending upon the application of the analyzer, the type of equipment, the accuracy desired, and installation requirements.

The PGNAA Analyzer varies in cost from $300,000 to $500,000. The upper range is due to the precision that the manufacturer is willing to guarantee for each individual element analyzed. Depending upon the requirements of the EPA, the state, and the actual application the most precise model may be necessary.

All PGNAA analyzers require the additional equipment already mentioned. Upgrades of the reclaim vibrating feeders can cost up to $42,000 per feeder (if new electric feeders are needed). Upgrading the existing feeder control can run from $24,000 for pneumatic control retrofit to $60,000 for new hydraulic control on existing pneumatic feeders (cost quoted include installation costs). The belt scale for the system can range in cost from $6,000 each for a 1% accuracy scale to $15,000 each for an 1/8% accuracy certifiable scale. The primary sweep sampler, if required, costs approximately $20,000 each to purchase and install. In some cases, this equipment may be replaced by a flop gate that would be only about $6,000. Other equipment that is necessary for the operation of the system (conveyors, chutes, hoppers, etc.) can increase the cost of the system. Conveyors cost between $8,000 to $10,000 for up to 30 feet. Chutes with internal liners are $6,000. Therefore, final purchase and installation costs for a PGNAA analyzer coal blending control system, excluding any structural modifications or housing enclosures for the analyzer, are approximately $1,000,000 for each installation.

The second type of ash analyzer using a radium source costs around $220,000. This system can be used for blending, but only on ash content. To install this analyzer, all of the earlier mentioned material handling equipment must be provided. Installation costs are similar to the PGNAA Analyzer since this system requires flow through an internal chute (minus the amount of the analyzer proper).

The most basic ash analyzer's utilizing the Cesuim and Americanium source is approximately $75,000 to $110,000. As mentioned earlier, this analyzer is only able to provide ash content of the coal. However, this system requires the least installation costs due to its direct mounting above the coal conveyor belt.

CONCLUSIONS AND RECOMMENDATIONS

Nuclear coal analyzers can be applied in numerous monitoring and control functions in coal fired power plants. The following are five (5) major applications where nuclear coal analyzers can provide very beneficial results.

1. Coal quality monitoring
2. Coal blending control
3. Emissions control
4. Boiler performance monitoring
5. Ash content monitoring

A typical nuclear coal analyzer implemented in a continuous coal blending control application can provide a significant savings in low sulfur coal usage. By applying a continuous feedback blending control using a nuclear analyzer, there is potential to save in excess of 200,000 tons of low sulfur coal per year at a typical large coal fired power plant.

The monitoring applications of the nuclear coal analyzer can improve boiler availability and efficiency through diverting coals that provide high slagging characteristics. When an ash fusion temperature target value has been reached, a signal could be sent to change boiler sootblowing frequency. These are some of the benefits of monitoring coal quality.

The following conclusions and recommendations have been formulated as a result of experience in specifying and procuring nuclear coal analyzers:

1. The nuclear analyzer has potential to be applied in numerous monitoring and control applications within coal fired power plants that will provide cost benefits including elimination of some periodic laboratory testing.
2. Coal suppliers have been using nuclear coal analyzers for sulfur and ash content blending of coals extensively. Utilities specifying coal qualities in their coal supply contracts receive blended coals to meet those specified parameters. The nuclear coal analyzer can be used to verify these coal quality values.
3. The new EPA emissions regulations require a reduction in SO_2 emissions at many coal fired power plants. Coal blending controls are beneficial if blending setpoint tolerances can be reduced and large quantities of lower priced coal can be utilized. Therefore, coal blending benefits have to be evaluated based on coal suppliers, sulfur content, and the ability to use less expensive higher sulfur coal. It is recommended that a study of coal supply cost benefits be conducted prior to budgeting funds for a coal blending system.
4 The current cost for an elemental nuclear analyzer is high. Installation costs for the high accuracy analyzers are high and in some cases do not provide adequate cost/benefits.
5. It would be beneficial to have a "low cost" coal analyzer that could be used to monitor BTU values of coal entering pulverizers from the coal bunkers.

REFERENCES

1. D. Devines, M. Samples, M. Hescox, K. Jarrell (Coalscan Inc., York, PA) Coalscan Application & Benefit Seminar, August 1989.
2. Gamma-Metrics, "On-line Elemental Analyzer for Bulk Materials", Sales literature, 1987.
3. MDH Motherwell, "On-line Rapid Analysis of Bulk Coal, Summary of Technology and Application for the Coal-Fired Utility".

©1991 - Paper #91-707
0074-056X/91/107-116/$0 + .50pp

ENHANCED PLANT PERFORMANCE THROUGH OPTIMIZED SOOTBLOWING

L. J. Line
Virginia Power Company
Mount Storm, West Virginia 26739

D. E. Labbe
C. T. Geary
Stone & Webster Engineering Corporation
245 Summer St.
Boston, Massachusetts 02107

ABSTRACT

The combustion of coal in a furnace produces ash and soot deposit on the heat transfer surfaces and gradually reduces the heat exchange from the hot gas to the steam side. These surfaces are cleaned using sootblowers installed throughout the furnace. When a section of the furnace is cleaned, the energy distribution to the steam side of the boiler is disrupted. This may result in an unfavorable energy distribution, which can lead to a loss of thermal performance. Also, since sootblowers use pressurized air or steam, excessive sootblowing wastes energy.

A sootblow operator advisor was installed at Virginia Power's Mount Storm Unit 1 in late 1989 as part of a control upgrade project to improve unit performance. Unit 1 is a 565 MWe Combustion Engineering pulverized coal-fired drum-type unit that became operational in 1964. The sootblow advisor provides operator guidance in sootblow selection based on continuous plant monitoring and optimization calculations. This information is presented graphically on the operator console.

The sootblow operator advisor was integrated into the plant distributed control system (DCS) and required no custom hardware. The calculations and logic are performed in the plant computer attached to the data highway. A model based optimization program predicts the effects of sootblowing on unit performance. The program uses as inputs the conventional measurements of flow, pressure, and temperature in addition to sootblower status data. No additional sensors were required.

The sootblow operator advisor has assisted the operator in enhancing unit performance by optimizing boiler efficiency and maintaining steam temperatures within the established control range. Major achievements include heat rate improvements and an increase in unit capacity factor. The success of the sootblow operator advisor is the basis for Virginia Power's decision to fully automate the sootblow system in 1991.

INTRODUCTION

The combustion of coal in a boiler furnace generates large quantities of ash and soot, a portion of which deposits on the heat transfer surfaces of the furnace. As the layer of soot builds up, the heat transfer rate from the hot gas to the steam side is reduced. To remove the soot a jet of air or steam is directed by a sootblower to clean a local area. Many sootblowers are installed throughout the furnace. While the fouling of the furnace is a gradual process, sootblowing results in an abrupt change in local heat transfer and in the distribution of energy to the steam side components. The former can cause transients in steam temperature as the control system responds to the disturbance. The latter can cause a loss of performance by driving steam temperatures below the range of control and reducing the achievable steam temperatures. Also, the activation of certain sootblowers that have little effect on performance results in ineffective use of energy.

Virginia Power's Mount Storm Unit 1 contains 95 furnace sootblowers. The large number of sootblowers and the continuously changing unit conditions have made it difficult for the operator to determine which sootblower to activate and when. The operators typically have resorted to a time basis for the selection of sootblower operation. As a result, unfavorable furnace energy distributions frequently would develop, leading to a loss of performance. One indication of a poor distribution was the inability to achieve temperature set point with reheat steam. The reheat steam temperature controls would simply not be able to raise steam temperatures to set point.

Additionally, the abrupt change in heat transfer due to sootblowing often resulted in large transients in steam temperatures that challenged the steam temperature control system. The conventional steam temperature control system did not anticipate the impact of sootblowing on steam temperature. Therefore, the control system relied solely on feedback action to respond to the effect of sootblowing. This resulted in significant deviations from set point followed by several temperature cycles as steam temperatures gradually returned to set point or to the limit of the control range.

To quantify the effects of sootblowing on unit performance, the gradual fouling of the furnace and sootblower cleaning effects were incorporated into the plant simulation model. This simulation model was originally developed to investigate the steam temperature control system and to design the Model Reference Feedforward (MRFF) control (1).

The simulation model predicted the characteristics of sootblowing observed at the station including the inability to achieve reheat steam temperature set point and the large oscillations in steam temperature following certain sootblowing actions. The model quantified the heat rate penalty caused by low reheat steam temperatures. The model also predicted the effects of other fouling related conditions. For example, unit heat rate was influenced by air preheater fouling and a significant improvement in heat rate could be achieved by timely sootblowing. Additionally, sootblowing of certain furnace sections was predicted to improve boiler efficiency while adversely affecting plant heat rate slightly.

The success of the model in quantifying the effects of sootblowing prompted the development of the sootblow optimizer program. The designated function of this program was to predict the optimum sootblower sequence based on the criteria of heat rate and steam temperature control.

Virginia Power was in the process of specifying a DCS for the unit. They had decided that a subset of the DCS would be installed as a pilot project and the control system would include Stone & Webster's MRFF steam temperature control (2). Based on the potential improvements in plant performance that could be achieved through sootblow optimization, Virginia Power requested that the sootblow optimizer be incorporated into the DCS as an operator advisor. The control upgrade project became known as the Advanced Steam Temperature Control (ASTC) project.

The design features of the sootblow operator advisor are described below including the optimization basis, plant input data requirements, DCS integration, and the operator interface.

SOOTBLOW OPTIMIZER DESIGN

Mount Storm Unit 1 is a Combustion Engineering 565 MWe coal-fired drum-type unit that operates at 1000/1000 F and 2400 psig. The twin furnace design has two equally divided sides with waterwall, superheat, and reheat heat transfer surfaces. Superheat steam temperatures are controlled by attemperation spray and reheat steam temperatures by burner tilt position.

Each boiler has unique features that distinguish it from other types of boilers. The sootblow optimizer was custom designed for the Unit 1 boiler by assembling standard component models according to the configuration of the unit. The optimizer was specifically designed to determine the optimal furnace section to clean based on control and performance criteria.

Unit 1 has 95 sootblowers including 12 lance type blowers in the convective passes, several air preheater blowers, and numerous wall type blowers. The sootblowers have been geometrically grouped into 25 sections according to furnace location. The optimizer predicts the effects of cleaning each of the sections on plant performance. Based on the criteria of heat rate and steam temperature control, the section that best achieves the desired effect is recommended for sootblow activation.

A design feature of the sootblow optimizer is that the effects on the overall cycle, illustrated in figure 1, are predicted and used as the basis for decision. The optimizer does not limit itself to boiler considerations only. In the early stages of product development it was determined that certain sootblower activations which may improve boiler efficiency actually reduce cycle efficiency. For example, excessive cleaning of the water walls increases steam production, which reduces gas temperatures to the superheat and reheat sections while increasing steam flows. This may result in the loss of capability to hold steam temperatures at set point. If steam temperatures drop below set point, the increase in boiler efficiency is more than offset by the loss of turbine cycle efficiency. According to the plant model each degree drop of reheat steam temperature corresponds to a 1 Btu/kWhr increase in plant heat rate.

As another example, consider the effects of cleaning a superheat section assuming that final steam temperature is regulated with a spray water type temperature control. For certain conditions the increase in spray flow resulting from higher heat transfer to the superheat section can cause a reduction in overall cycle efficiency. A penalty of 4 Btu/kWhr is predicted for an increase in spray flow of 1% based on steam flow at the throttle valve. This effect is a function of the cycle design and whether the spray water bypasses the economizer and one or more high pressure feedwater heaters.

By including the cycle effects the optimizer can predict the effect of sootblower activation on steam temperatures, control parameters, and cycle efficiency. Criteria have been established and prioritized to achieve the desired control range and heat rate effects.

DCS HARDWARE/SOFTWARE

The sootblow advisor was implemented on a Fisher Provox distributed control system with a DEC Microvax-3300 plant computer as illustrated in figure 2.

Plant signals were terminated at the remote I/O cabinets and brought into a controller. The controllers, operator console and plant computer are tied together by a data highway.

The sootblow optimizer programs which perform the logic and calculations are loaded onto the plant computer. Access to data on the data highway is provided by Fisher's CHIP software package. Plant data is automatically updated in the plant computer using an unsolicited refresh technique to minimize highway traffic. Using a combination threshold criteria of time and deviation, the data is automatically transmitted from the source to the plant computer. Upon receipt of the raw data, the database values for the raw signals and the engineering values are updated. Additionally, data from the local data base within the computer are targeted to other components on the data highway by the same technique. A set of subroutine calls are provided to allow user programs to access the data and write to the data base.

By means of these hardware and software components, current plant data is made available to the sootblow optimizer. At the completion of the calculation cycle the optimizer writes a number of outputs to the data base which are targeted to the operator console.

OPERATOR INTERFACE

To be effective the information developed by the sootblow optimizer must be presented clearly and concisely to the operator. To meet this objective several graphic screens were developed and arranged in hierarchical fashion as shown in figure 3. The graphic facilities of the operator console were utilized including dynamic bar graphs, color coding and messages.

The highest level screen, the overview, states the governing criteria and the recommended section to sootblow, if any. If the operator desires support information, a single key stroke will page down through the screens displaying more detailed information. The support information not only defines the reasons why the particular sootblower section has been selected, but also quantifies the effectiveness of other sootblow sections. This approach has proven very effective in establishing operator confidence in the advisor.

The operator interface made use of the standard features available to the operator console. No special hardware or software was required. Custom screens were developed from the general base designs to meet the specific requirements of the unit. The sootblow advisor interface screens have become one of many sets of screens available to the operator at a touch of a key.

DATA REQUIREMENTS AND VALIDATION

A primary objective of the sootblow optimizer was to limit the required data to the existing available plant data base. The addition of a large number of tube metal temperature sensors and the maintenance of those sensors was not considered to be a viable option. The existing data base included all of the typical primary flow, pressure, temperature and level measurements used in an automatic control system.

To predict the effects of fouling without the availability of detailed temperature measurements a model based approach was utilized. The model predicts the rate and effect of fouling on energy distribution in the boiler. The model then predicts the effect of sootblowing each section on unit performance. A comparison of the predicted and observed effects are used to update the fouling model and maintain consistency with the actual unit performance.

A number of valid measurements are required as input to the sootblow optimizer. To ensure data validity a custom data validation routine was prepared to process the available plant data into the required measurements. The validation routine used a number of techniques based on analytical and sensor redundancy. A typical validation approach is illustrated in figure 4. A local database was established which included the raw measurements, validated measurements and sensor status values. As an added option these values were made available to the data highway for engineering and maintenance applications.

PROGRAM STRUCTURE

The sootblow optimizer is divided into three programs operating on the plant computer; a data retrieval and validation program, an update program and the optimization program. The data retrieval and validation program accesses the plant data from the CHIP data base and processes the data validation routines. The validated data is then output to the two other programs.

The update program ensures that the model used in the sootblow optimizer remains consistent with the plant. The update program calculates the model coefficients required to match the current plant conditions. These are then passed to the optimizer. With this approach the model automatically adapts to changing plant conditions.

The optimizer program processes the plant data and model coefficients to predict the effects of sootblowing. These results are then input to a decision logic section which rates the sections according to established criteria. The predicted effects and logic results are output to the local CHIP data base and targeted to the operator console.

RESULTS OF SOOTBLOW ADVISOR IMPLEMENTATION AT MOUNT STORM

The Mount Storm sootblow advisor has been well received by the station operators and engineers. The system has provided reliable indication of the appropriate sootblow action on a continuous basis. The system has been operational for approximately one year and has been on-line throughout that period.

The improvement in station operation with the sootblow advisor has been impressive. Superheat and reheat steam temperatures have been maintained within their control range on a near continuous basis. Operators have been effectively guided to take action to minimize plant heat rate. Also, superheater spray flow has been substantially reduced.

At the Mount Storm Station dissolved solids in spray water has been a major factor in the gradual reduction of turbine capacity. The turbine deposits that gradually develop from the dissolved solids in the spray reduce turbine efficiency and restrict steam flow. The reduction in spray flows made possible by the sootblow advisor translates into improved unit capacity and unit heat rate.

The sootblow advisor has provided a consistent means of operating the large number of sootblowers at Mount Storm, and based on the success demonstrated to date, Virginia Power plans to fully automate the sootblow system in 1991 by extending the advisor to directly activate the sootblow sequencer.

The reduction in average unit heat rate due solely to the sootblow advisor is projected at 40 BTU/kWhr. An increase of 7 MWe of unit capacity over a one year period is expected from the reduction of spray flow. These benefits translate to a rapid payback period.

CONCLUSIONS

The process of sootblowing can have serious consequences on plant performance, if not properly carried out. The large number of sootblowers, typically around 100, and the complex interactions, makes it a very difficult task for the plant operator to achieve optimum results. A sootblow advisor was developed to undertake the difficult task of determining the most appropriate sootblow action.

This sootblow advisor has been directly installed into a plant distributed control system. The existing control hardware and interface software was used. No special or custom hardware was required. The existing plant data base was sufficient to drive the optimization algorithms and no additional plant sensors were required.

The sootblow advisor was customized to represent the specific features of the unit by assembling standard component models. The advisor includes an optimization based on the overall unit cycle and is not limited to the boiler system. Using this approach it predicts the overall effects of sootblowing action on the control and efficiency of the cycle and has assisted the operators in achieving optimum performance.

The output from the sootblow advisor has been readily adapted to the graphic operator interface available on today's DCS. A hierarchical approach has proven effective in providing information to the operator concisely and with sufficient back-up data to build confidence in the system.

The system has been instrumental in improving plant heat rate and maintaining plant capacity factor at higher levels. A payback period of one year has been projected.

REFERENCES

1. L. Line, D. Labbe, and J. Donohue, "Advanced Controls Help Cycling Plants Cope with Increased Stress", presented at EPRI Fossil Plant Cycling Conference, Washington, DC, Dec 4-6, 1990.

2. M. Leigh and D.E. Labbe, "A Plant Controls Retrofit Using a DCS", presented at the Edison Electric Institute's Engineering and Operating Computer Forum, "Innovations for the 90's", St. Louis, MO, Sept. 9-12, 1990.

Fig. 1
Mt. Storm Unit 1 Plant Model

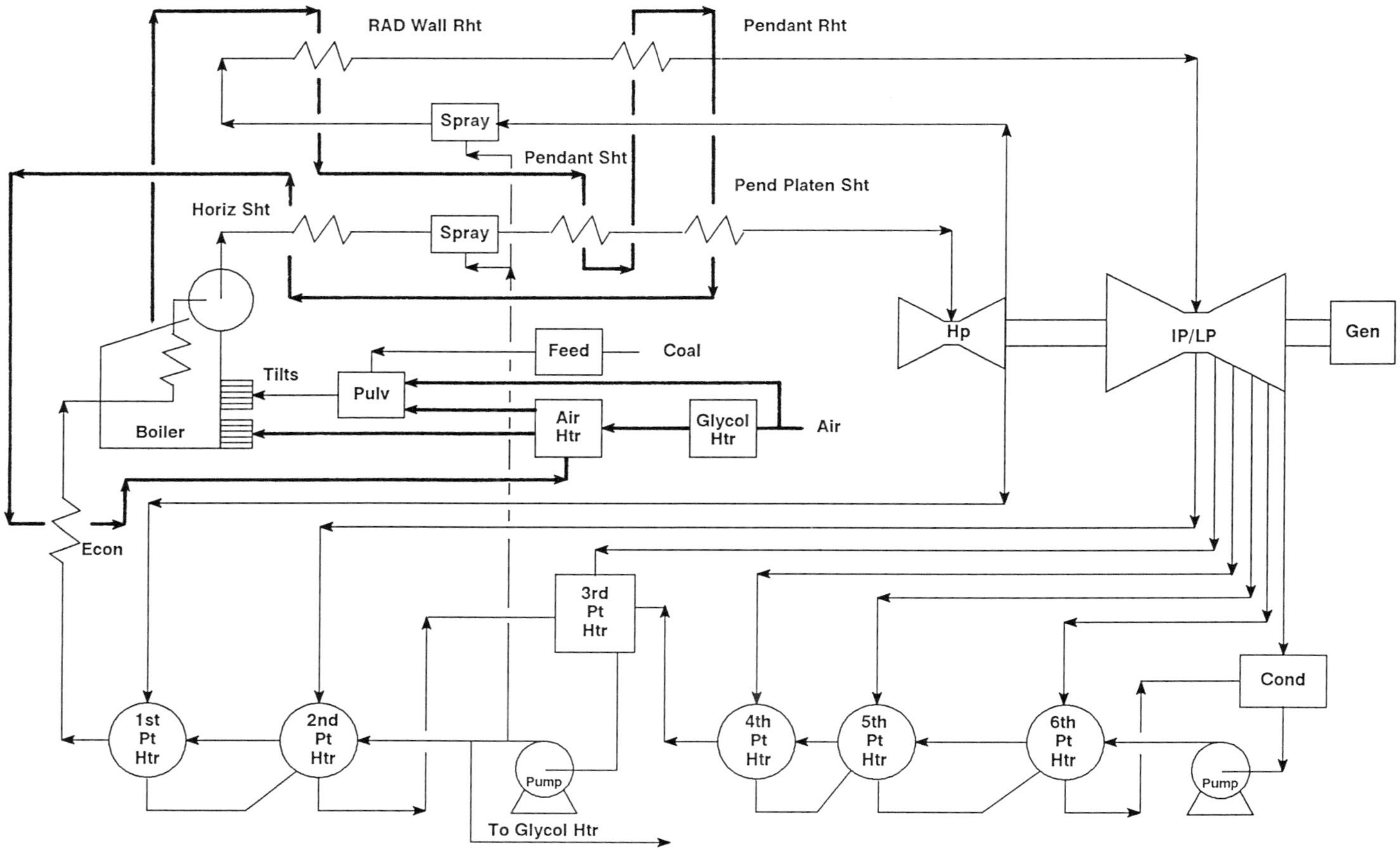

Fig. 2
DCS Hardware Configuration

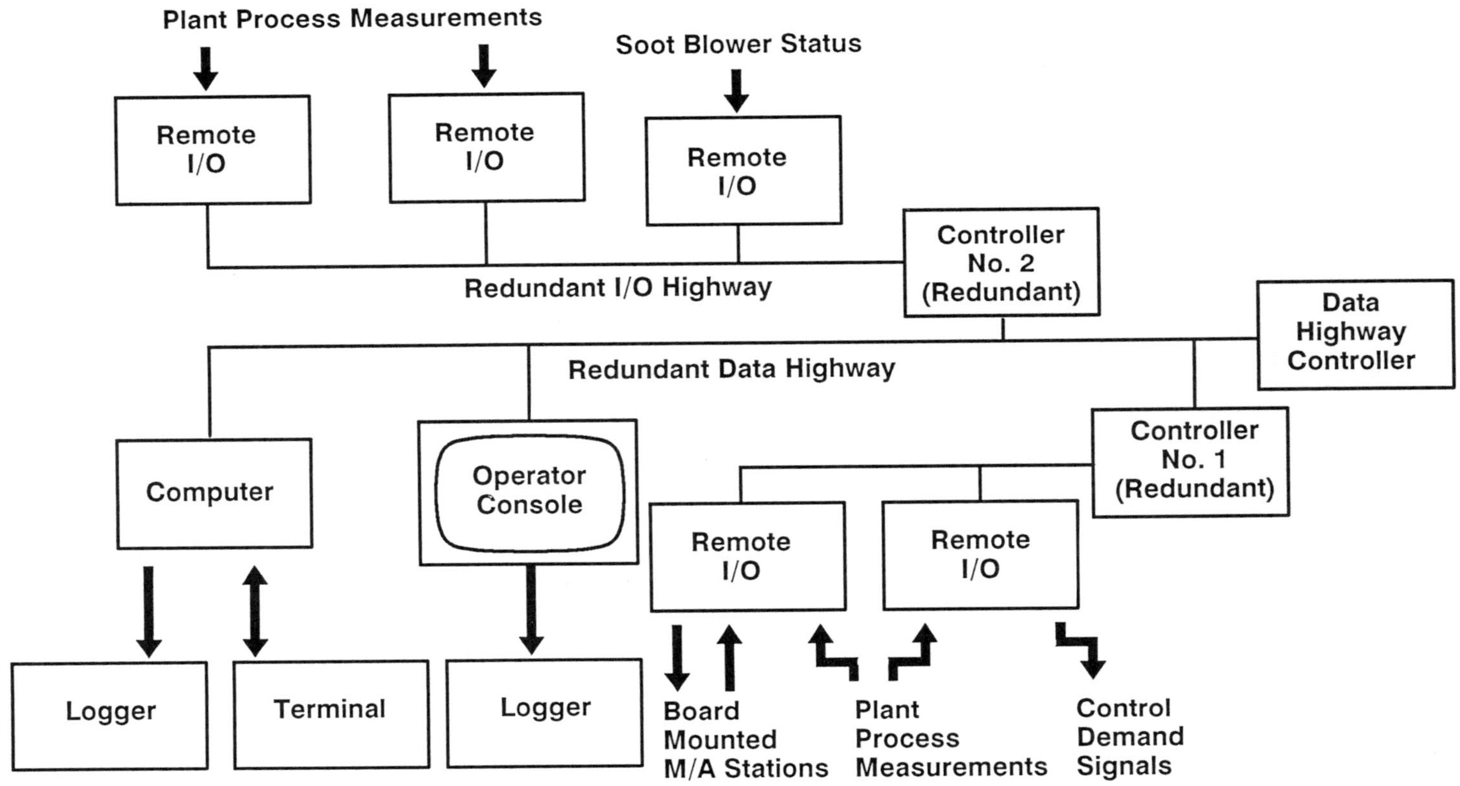

Fig. 3
Sootblow Advisor Display Hierarchy

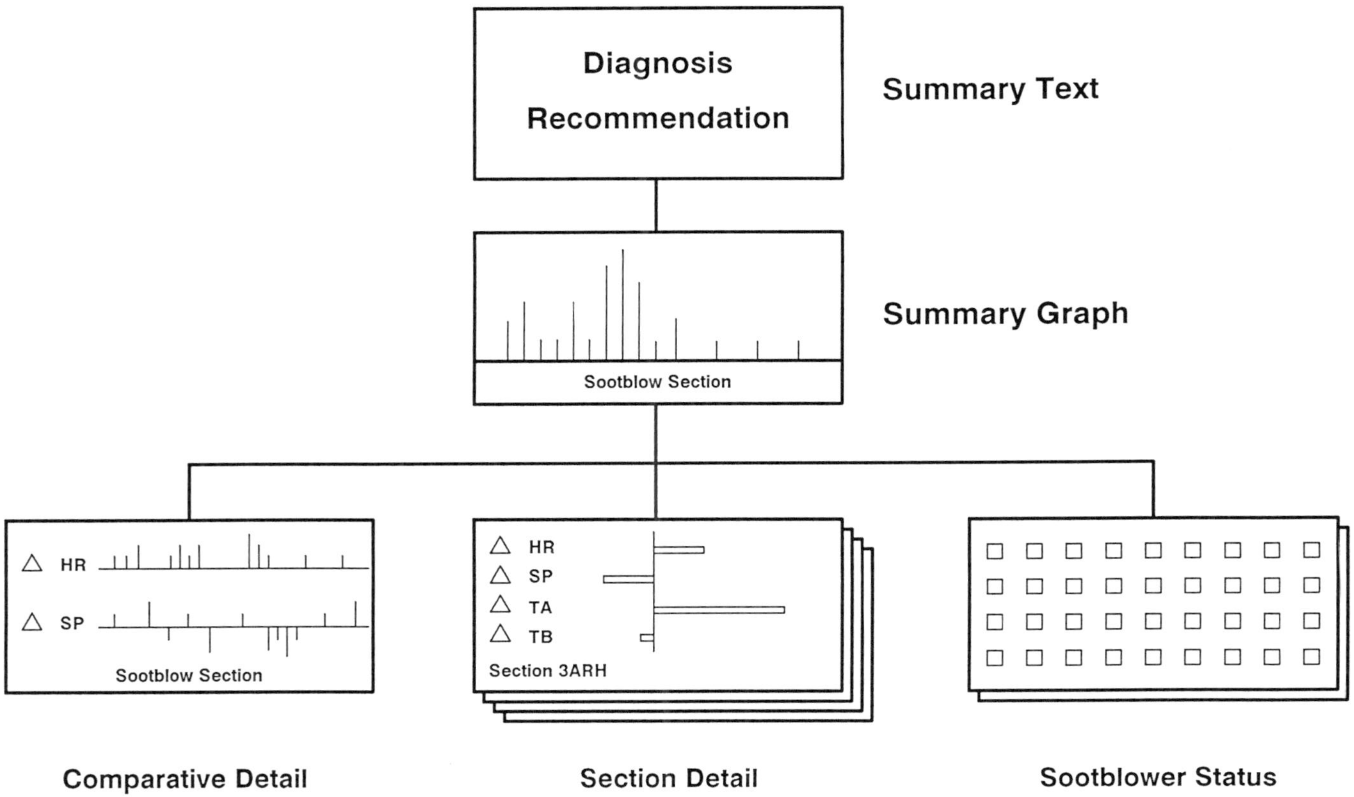

Fig. 4

Sensor Validation Checks System Robustness

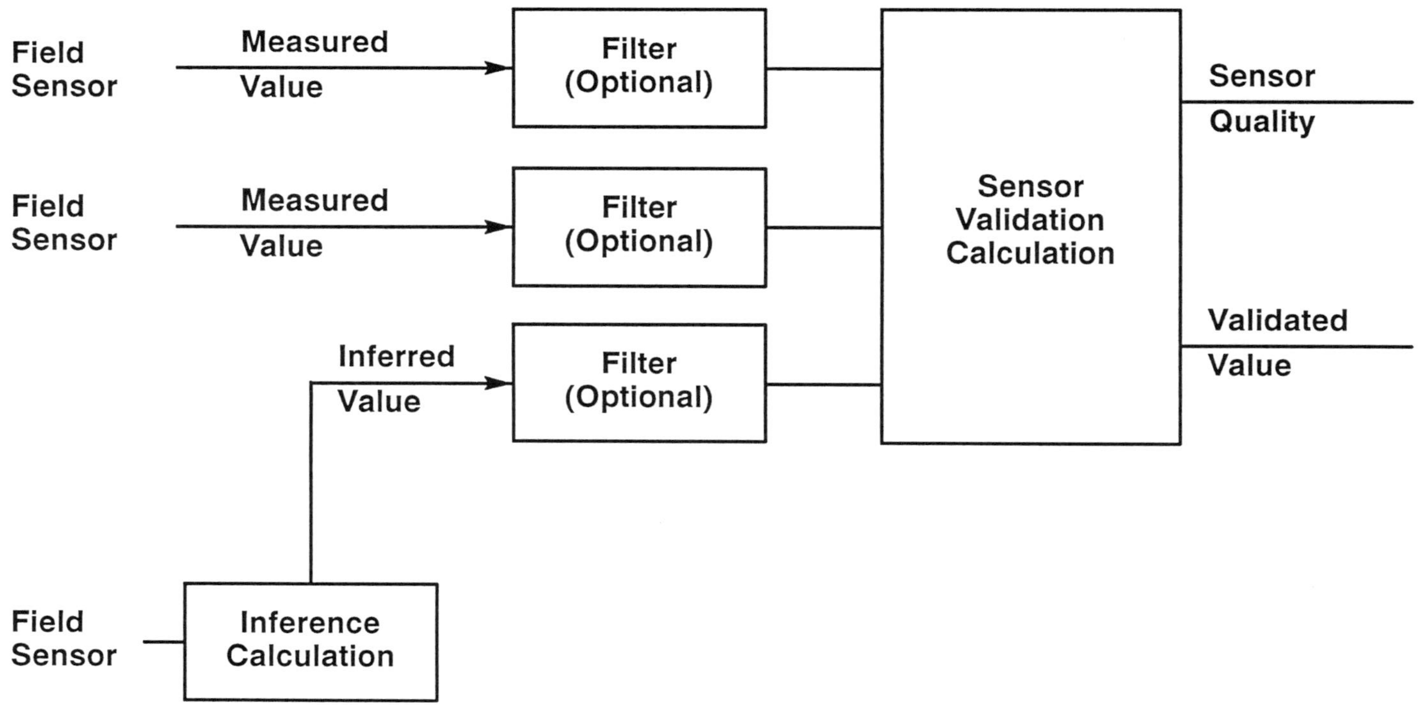

0074-056X/91/117-131/$0 + .50pp

AN ECONOMICAL PERFORMANCE MONITOR CUSTOMIZED TO YOUR PLANT

G. R. Jones

ABSTRACT

Many utilities are now installing on-line continuous Performance Monitors. The cost however can be well over $200,000 for such systems from outside vendors. At Texas Municipal Power Agency, I have built a customized Performance Monitor from scratch that not only is less than one quarter the cost, but also provides an exact match to the "as built" thermodynamics of the plant.

This paper describes the system architecture and development requirements to build this economic customized system.

Key Words: (1) Performance Monitor
(2) Customized
(3) Economical
(4) Continuous On-Line Analyzer

INTRODUCTION

Having installed other outside vendor Performance Monitoring systems in the past, two basic deficiencies always showed up, the cost and the lack of adaptability to your plant. Performance Monitors involve computer hardware, I/O systems and specific software for the system. Although the computer hardware and I/O requirements have a relatively fixed costs, the software to run a Performance Monitor as supplied by outside vendors is generally very expensive and usually constitutes the bulk of the total cost. The software supplied by the outside vendor is not only expensive but it usually has many limitations in terms of adaptability. In other words, once you have your system software which includes a model of your plant, it cannot be easily modified or customized to changes in your plant.

Because of these problems a decision was made at TMPA to build our own Performance Monitor and to do so at the least cost. This meant that a personal computer was used for economical reasons. The I/O system had to communicate with the data collection plant computer already in place. There also had to be a user friendly interface in the PC that displays data in easily referenced formats.

The primary purpose of the Performance Monitor is to provide information to the user on the performance of the unit on a continuous real-time basis. The information is made up of key raw data variables such as feedwater flow, fuel flow, main steam conditions, etc. and calculated data such as heat rate, turbine efficiencies, boiler efficiency, feedwater heater performance, etc. The information is made available on user friendly graphic screens that places the data next to graphic pictures of the major plant components. Thus, it only takes a glance at any screen to get all pertinent data, both raw and calculated, around any plant component.

The system operates on a sixty second cycle, updating all data and performing calculations each minute. Other capabilities such as real time trending, alarm flags and historical trends are included to make a very powerful tool for the end user.

The basic elements of the performance monitoring function as was finally developed are listed below in the sequence that they occur.

Performance Monitoring Elements

I. Receive ASCII data from Plant Computer
II. Transfer data to compiled Performance Program
III. Performance Program runs
    A. Scale and initialize data
    B. Run Performance Calculations
    C. Calculate and assign alarms
    D. Calculate Controllable Losses
    E. Write all raw and calculated data to database
IV. Display System
    A. Read data from database
    B. Display data on user friendly graphic screens
    C. Update data every 60 seconds

The basic elements above all take place in the personal computer. To make the system appear seamless, items I, II and III operate as one task in a multi-tasking environment. Item IV acts independently as separate task. Therefore, there are two tasks operating at the same time.

The complete system description follows the outline below. Each item will be discussed with recommendations on software development. Options will also be presented that users can provide as needed. The system will be described as generically as possible allowing any computer hardware or program language that the reader is most comfortable with.

Performance Monitor System

I. Hardware Requirements
II. System Block Diagram
III. Communications
IV. The Multi-Tasking Environment
V. The Performance Program
VI. The Database
VII. The Display System
VIII. Memory Considerations - System Speed
IX. System Costs
X. Conclusions

HARDWARE REQUIREMENTS

This application assumes that a data acquisition system is already in place in the plant. In TMPA's case, a Honeywell 45000 Process Plant Computer is the primary data source for all computer alarms and CRT displays in the Control Room. About 10,000 points in the plant are monitored in this computer. The Performance Monitor communicates with the Honeywell Plant Computer to receive raw data. In the event that no data acquisition system exists, more effort and resources are required to design and install such a system and is beyond the scope of this paper.

The Performance Monitor itself is a personal computer. Below are minimum specifications for the base computer.

Performance Monitor PC

- Processor - 80386 25 MHz
- Coprocessor - 80387 25 MHz
- Hard disc - 40 Mbyte minimum
- 3.5" floppy - high density
- Random Access Memory - 8 Mbytes minimum
- Monitor - VGA color - 16 bit
- Serial COM Port
- Mouse
- Short Haul Modems - one for each port

At TMPA, two computers were purchased, one for engineering use with a 14" monitor and one for the Main Control Room using a 20" monitor. The communications are paralleled from the Plant Computer to each computer using short haul modems. Shielded twisted pairs are used for interconnecting cables. In very noisy environments or when there are very long cable runs it may be beneficial to use coax cable or even fiber optic cables. These cables however will also increase the cost of the system. The method of paralleling communications gives the system redundancy by using two independent computer systems. If one computer goes down, the other can still function independently. It was decided to use this approach over networking (LAN) from one system to add reliability to the overall system.

SYSTEM BLOCK DIAGRAM

The basic hardware layout is shown in figure 1.

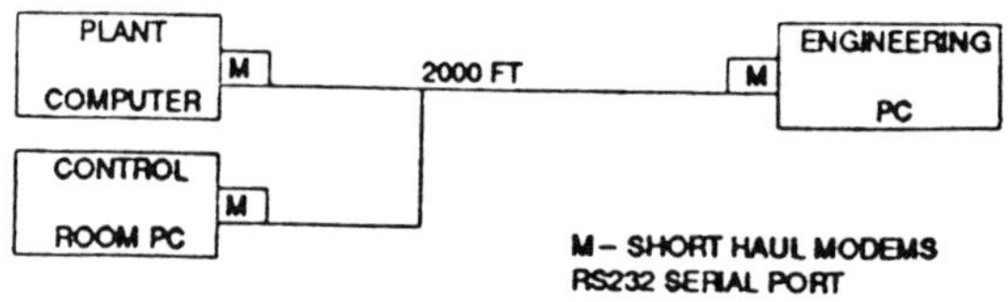

Using short haul modems boosts signal strength and helps in removing noisy data. We were able to use twisted shielded pairs for the 2,000 foot run using these modems with a baud rate of 9600. Faster baud rates can be attained by upgrading the type of transmission cable but again will add extra costs.

COMMUNICATIONS

Establishing communications for raw data from the Process Plant Computer to the PC type of computer presents quite a challenge. Almost every application will be different depending on the type of plant computer used. The easiest approach is to transmit ASCII data strings in a serial line. Most plant computers are capable of down loading serial ASCII files in some format. In our case, the serial ASCII format is designed for a printer device. We used this port to transport data to the PC's.

The first step in development is to start the selected data transmission from the plant computer to the PC. A simple communication program such as PROCOMM can be used in the PC to monitor the data coming in. The data string sent to the PC then needs to be analyzed for clean up. There are usually unwanted characters such as "carriage return" or "line advance" that must be removed. The objective is to remove all unwanted characters and end up with a serial data file that only has the desired data. This clean up activity must be performed by a compiled program written by the system developer and resides in resident memory all the time, waiting for each data transmission.

The communication program residing in active memory in the PC is the trigger to the Performance Monitor. When a data string is detected each minute, the communication program receives the data, cleans up the unwanted characters and then passes the data to the calculation programs. The communication program is the only continuously active program constantly monitoring the PC communication port. When data is received, other programs are called to do the performance calculations and write routines to the database. This approach works well and conserves memory in the PC. The communication program itself is not that large and will "shell" other programs as needed.

A word about program languages:

Computer programming is a very personal thing. Everyone has their favorite programming language to use and most of them will lend themselves to this application. The only recommended consideration is that the language used be a compiled language. Using interpreted basic is possible but would be extremely slow. Languages such as any compiled basic,"C" or Fortran are recommended for speed and enhanced capabilities. For our application the communication and calculation programs were written in Power Basic. The compiled executables run almost as fast as the "C" or Fortran programs.

There was a temptation to include actual programming code in this paper as examples. But due to length constraints it was not practical to do so. Also, the source code written for our application is specific to the Honeywell 45000 Plant Computer. Appendix A however shows a flow description of the source code for this program.

THE MULTI-TASKING ENVIRONMENT

The Performance Monitor requires two programs going on at the same time as described earlier. Task one is the Communication Program receiving data and calling calculation programs. This task is active all the time in RAM but does not interface directly with he user. Therefore the end user does not even need to know that this process is occurring. Task two is the Display System and is responsible for the user screen interface. This task is displaying and updating data on the screen using high resolution graphics. There are two approaches to provide a multi-tasking environment. One is to load an operating system that is designed for multi-tasking. There are several now on the market that will load on a PC. The problem is that these operating systems will not integrate with the vast amount of software available under the old single tasking DOS systems. The second approach is to use the standard DOS and load a special program that will fool the system into multi-tasking. Thus the needed two tasks can run and the benefit of available software is not lost. Programs such as Windows or Desqview are examples of multi-tasking programs designed to run on DOS. Our application uses Desqview on DOS 4.01.

As an added benefit to our approach, another task can be started at any time to run other DOS applications as desired. This means the PC need not be dedicated to the Performance Monitor. Other computer functions can be accessed while the background is still running.

THE PERFORMANCE PROGRAM

The Performance Program is the meat of the Performance Monitor. All of the desired performance calculations are done here each time the Communications Program receives raw data. This program is called by the communication program, runs the calculations and then writes the results to the database. When the program is done, it is erased from resident memory until the next time it is called. This program is again written by the system developer in his favorite compiled language.

The program is basically an assimilation of the various ASME test codes for power plants. Writing this program and verifying its accuracy is a large portion of the system development. It cannot be stressed enough the need to verify your plant model (Performance Program) for "as built" accuracy. The Performance Monitor will have no real value if this program does not represent the actual thermodynamics of your plant. The best person for technical input into this program is someone who has actually run manual performance tests of the plant using the ASME test codes. There is no better way to understand the thermodynamics of the plant than to spend several weeks running the tests and doing the manual calculations. The two primary test codes used are the ASME PTC 4.1 for boiler efficiency and the ASME PTC 6 for turbine performance. Other codes for feedwater heaters, air preheaters, pumps and condensers are also used for heat balance calculations. The key to this program however is that all of the pertinent test codes are run sequentially to evaluate the entire plant. There are also some specially designed calculations to measure the performance of limestone scrubbers used at TMPA.

Again to actually present program source code for this program would take an additional 36 pages of text and is therefore length restrictive. The following program flow description is sufficient for the program developer to write his own program customized to his plant.

PERFORMANCE PROGRAM
Flow Description

I. Read raw data from communications program.

  A. The raw data is read into the program from a data file created by the communications program called INPUT.DAT. This is a sequential file containing only the floating point values of the data.

  B. Assign raw data values to program variables. For our application there are 151 raw data variables read into the program.

II. Perform data error check and set alarms.

Due to the noisy environment of a power plant there are occasionally small errors in the data transmission. Sometimes these errors can throw the calculations wildly off or even cause computer hand ups. Therefore raw data is checked against bogey curve algorithms and values substituted when required. When bogey curve values are substituted, alarm integers are assigned high or low that alert the end user of bad data. This is especially important when a critical transmitter in the field has failed. A consistent alarm on a piece of raw data will flag the end user to have the transmitter repaired. But the failed data will not destroy the calculation routines.

III. Scale data for calculations.

A. Scale raw data for proper units for ASME test code calculations.

B. Create variables using bogey curve algorithms that are not in raw data input. These values are items such as turbine seal leakages or packing leak off that are usually available in the turbine vendors thermal kit.

IV. Turbine calculations per ASME PTC 6.

A. High pressure feedwater heaters extraction flows.

B. Boiler feedpump turbine flow.

C. High pressure and intermediate pressure turbine flows.

D. HP and IP turbine efficiencies.

E. Low pressure turbine swing calculation.

1. Initiate first LP expansion line equation.
2. Feedwater heaters derived extraction enthalpys.
3. Feedwater heaters extraction flows.
4. Machine energy balance calculations to determine LP exhaust enthalpy.
5. Annulus exhaust losses.
6. LP expansion line end point (ELEP).
7. Compare initial ELEP with calculated ELEP.
8. Adjust for errors and iterate until converged.

F. LP turbine efficiencies.

V. Boiler efficiency calculations per ASME PTC 4.1.

Boiler efficiency is calculated using the heat loss procedure to eliminate the errors created by inaccurate fuel flow.

A. Air preheater calculations.

1. APH leakage.
2. APH efficiencies.
3. Corrected exit temperatures.

B. Boiler efficiency calculations.

1. The standard losses are handled normally. Special considerations are added based on empirical data for sootblowing losses, pulverizer rejects and bottom ash losses. This results in an actual boiler efficiency using the long form of the test code.

VI. Heat rate calculations.

A. Turbine heat rate.
B. Net plant heat rate.

VII. Feedwater heater performance.

A. Terminal temperature difference.
B. Drain cooler approach.

VIII. Controllable Losses

These are calculations used to generate the bar graph Controllable Losses screen for the use of the Control Room Operator. Actual values are compared to expected values that are generated with bogey curves based on the best heat rate or the last tested heat rate curve. These values present data to the Operator that he has some control over. The parameters used in our system are:

Throttle Pressure
Throttle Temperature
Reheat Temperature
Condenser Back Pressure
Station Aux Power
Excess 02
Exit Temperatures
Superheater Sprays
Reheat Sprays

IX. Smoothing routine.

Data is received and calculations are made every 60 seconds. Many of these values are very active and fluctuate considerably from minute to minute. Therefore key parameters such as heat rate, feedwater flow, fuel flow, excess 02, etc. are smoothed for display purposes using a rolling average algorithm. Thus the end user can see a trended value to evaluate plant performance easily. The time required for the smoothing will vary with each application. Our system looks at 10 minutes history for good smoothing.

X. Scrubber efficiency calculations

These calculations do S02 removal and efficiency calculations based on mass flow balances. Inputs use fuel analysis (sulphur) and theoretical gas flows.

XI. Rescale raw and calculated values for display purposes in preparation for data output.

XII. Write data output to database to appropriate minute.

This routine writes the data to the database used by the display system and in our case is a random access file using IEEE floating point standards. This routine must be customized to the display system used and will therefore vary for each application.

The Performance Program also requires the use of steam tables for enthalpy and entropy look ups. There are several steam table programs available and the chosen one must be integrated into the Performance Program. One note of caution is that the chosen steam table must be able to run in the wet regions below the saturation line since the LP turbine calculations will dip below the saturation line.

THE DATABASE

The database is a separate random access file that contains 60 minutes worth of raw and calculated data as supplied by the Performance Program. Each minute the Performance Program writes to this file and writes to the minute based on the PC system clock. Due to database requirements of the display system special identifying headers are included in this file. These may or may not be required depending on the type of display system used. Typically the display system will create these headers automatically for its own use. Care must be taken to not overwrite these headers with the Performance Program or the file will be contaminated.

The primary database is a 60 minute cyclical file (the Performance Monitor writes over the one minute records contained in the database as it updates the database). This allows the display system to view 60 minutes of current data at any time. However longer time frames such as 24 hour and 31 days are desired and so a history database exists that holds the same data but is for 24 hour averages and 31 day averages. Thus the display system can trend historical data for 60 minutes, 24 hours or 31 days.

The averaging routines to create the history database are separate executable programs that are called by the continuously resident communications program once each hour and once each day. The 24 hour program is called each hour, averages the data in the 60 minute file and writes to the appropriate hour. The 31 day program reads the 24 hours portion of the history database, averages the data and writes to the appropriate day. Each month the entire history database is automatically copied to an archival directory for historical analysis.

THE DISPLAY SYSTEM

The display system is the only major software purchase that was made for the system. There are several systems available that will take data from a predefined database and, on a real time basis, display it integrated with high resolution graphical background screens. The choice of which system to purchase is dependent on the system developer's preference. Some systems have wider versatility but are harder to set up. As a point of interest, many vendors supplying turn-key Performance Monitoring systems use these same software display systems for their applications. After an extensive bid evaluation, the system of choice for TMPA was EASE+ by Expert-Ease Systems, Inc. This was one of the systems that offered wider versatility, but was more difficult to set up. It was felt that even with more initial effort, the final product would better follow the primary goal of adaptability. The primary development steps following are specific to the EASE+ system and will vary with other systems used.

The Database

It is within the EASE+ system that the databases referenced previously are created. The structure of the database is defined by assigning variables, arrays and memory block sizes. These assignments are used to create the database headers and are used by the display system to find data. The database is created after the developer has defined what raw and calculated variables he wants to work with. In this database system, blocks of memory are predefined for a fixed number of variables. It is therefore recommended that extra memory blocks be allocated in the database for easy future expansions. This will eliminate the need to redefine the database every time extra data points are added.

Graphic Background Screens

These background screens are the graphic background on which the real-time data is superimposed. They are built using graphic elements and icons available in the system. Elements are simply graphic lines, circles, boxes, etc. Icons are active predefined figures that can be placed on the screens. These might be a pump, turbine, feedwater heater, etc. Each screen is built to display specific components in the plant. As an example, each feedwater heater has a separate screen to display all data around that heater. In our system there are about 40 screens available. Other screens were also developed purely as menu screens for branching purposes.

The Real-Time Data

Data is displayed on the graphic background screens by placing data icons on the screen. These icons are created in another section of the system and are linked by the database definitions to the actual data. Thus when a data icon is placed on a screen, it will look up the data in the database through its link and display that data on the screen. The data icons can be placed anywhere on the screen and are placed strategically in their appropriate location on the graphic background.

The Procedure Code

With 40 screens available, a program director is needed to allow all screens to be linked together. This is accomplished with a procedure code that is created in the EASE+ system. This procedure directs the user to each appropriate screen as he selects them. The menu screens are used to facilitate his screen selection. Also, all trending activities are defined in the procedure to look at historical trends in the system.

Graphic Trends

The trends defined in the procedure code can be defined in any format by the developer. In our case four specific trends are defined to match the database definitions. The first three are historical trends to display data for 60 minute, 24 hour or 31 day history. These trends overwrite the graphic screen in a window and when concluded are erased leaving the graphic background intact underneath. The fourth trend is a real-time strip chart that will write data to the line graph each minute as it is happening. Appendix B are photos of some of the screens available in our system. The artistic appearance of these screens is only limited by the talent of the developer. The system as finalized can be modified in any way to accommodate any new data, screens or procedures.

MEMORY CONSIDERATIONS

As can be seen by the previous text, there is a lot going on in the system continuously. This presents another problem when using a PC based system, namely speed. This entire system can be loaded onto a hard disc and run there but because of the constant disc accessing will be quiet slow. One of our considerations was that everything including the data transfer, performance calculations and data display updates had to occur within a 60 second time frame. This is not possible on a hard disc. Therefore the entire run-time system was loaded onto a RAM disc. This is the same as a hard disc but resides in active memory. Our system requires 4 mbytes of disc space and so 8 mbytes of RAM is necessary for the complete system. 4 mbytes is used for the RAM disc and the other 4 mbytes is used for memory managers, multi-tasking and other large DOS programs desired. Using a RAM disc is accomplished through standard DOS software. On computer boot up, a batch file creates the RAM disc and copies the entire system to RAM. This made our system very fast and easily fit into the 60 second time frame.

SYSTEM COSTS

Since one of our key objectives to this system was low cost, this section shows some approximate costs of building such a system. Included are approximate manhours for development since this is one of the primary costs for building your own Performance Monitor.

Hardware:	
Engineering Computer (14" monitor)	$ 2,500
Control Room Computer (20" monitor)	5,000
Modems	300
Software:	
Compiler	$ 100
Multi-tasking	200
Display System	16,900
TOTAL	**$ 25,000**
Development Manhours	800

The 800 manhours included a great deal of learning time for the display system due to its complexity. This time could probably be cut by 300 hours if the developer is familiar with such systems. Using an outside consultant would alleviate much of the development problems but would increase cost. The objective of complete adaptability would still be retained since the system would still be custom designed to the user.

CONCLUSIONS

The Performance Monitor has proven to be a very powerful tool for Operations, Engineering and Management due to the ease of data displays. In our system, extra monitors are being installed with management staff so that unit performance can be monitored easily in their own office.

One further development is needed after the system is installed. The system is so vast in its complete view of the plant that extensive training is required for end users. Training not only includes the use of the system, but also includes guidelines on how to interpret data and trends to improve plant performance.

The Performance Monitor has put the plant in a glass bowl, exposing every aspect of unit performance continuously. This has been a significant advance towards continuous efficient operation of the plant.

Appendix A
Communication Program
Flow Description

I. Open "COM" port and wait for data input

II. As data is received
   A. Place serial data into buffer file.
   B. Count data string as received.
   C. When data count is reached, close buffer file

III. Clean data, remove unwanted characters
   A. Input data from buffer file.
   B. By using count routine
      1. Ignore unwanted characters
      2. Input data to INPUT.DAT file

IV. Call Performance Program using DOS "Shell" command

V. Check PC system clock
   A. If minute is one, call hourly average program to average 60 minute data base file.
   B. If hour is 23 and minute is 55, call daily average program to average 24 hour database file.
   C. If date is one, send copy of history database to archival storage.

VI. Return to open "COM' port and wait for next data string.

Communication program remains active in memory all the time and is the program director for all data input and calculation routines.

Appendix B
Screen Examples

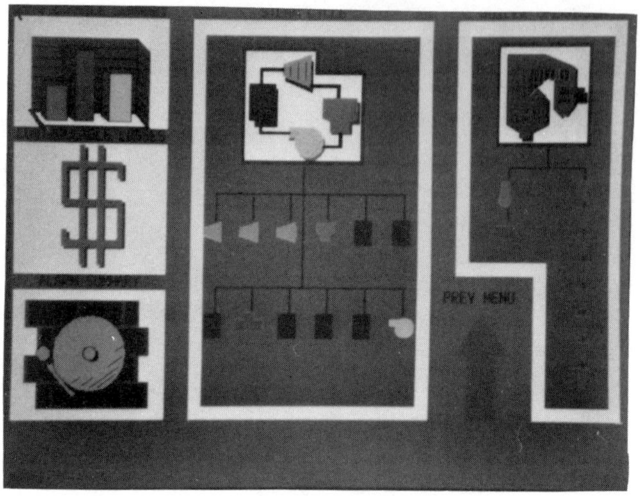

Menu Screen

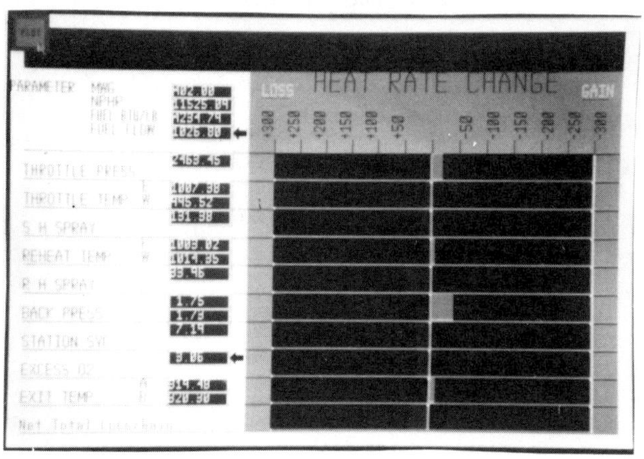

Controllable Losses Screen

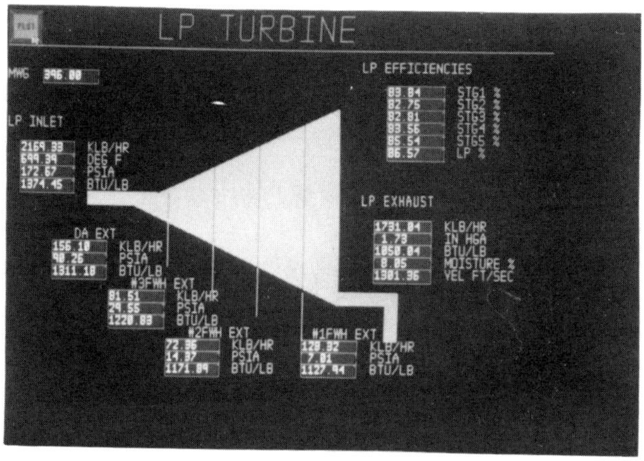

L P Turbine Screen

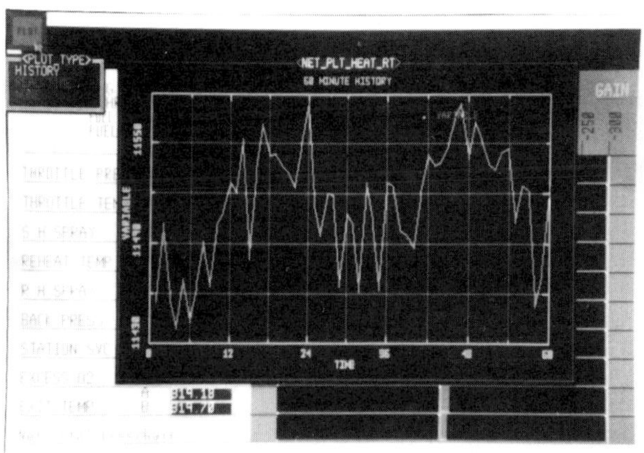

History Graph Screen

0074-056X/91/133-140/$0 + .50pp

THE SIGNIFICANCE OF THE NOMINAL TRIP SETPOINT IN THE WESTINGHOUSE SETPOINT METHODOLOGY

C. R. TULEY
R. B. MILLER

KEYWORDS: Setpoint Methodology
Nominal Trip Setpoints
Technical Specifications

ABSTRACT

The NRC draft "Standard Technical Specifications, Westinghouse Plants" has proposed to reduce the requirements in the Technical Specifications from Trip Setpoint and Allowable Value to Allowable Value only. This results in control of the "as found" condition of the instrument channel and floating of the "as left" condition. Westinghouse calculations use a conservative SRSS technique and are more dependent on the "as left" condition for the continued validity of the uncertainty calculation and thus the determination of an operable channel. This paper briefly defines the methodology and notes the areas of sensitivity where control of the "as left" condition is more important than control of the "as found" condition. It is proposed that the appropriate value in the "Standard Technical Specifications, Westinghouse Plants" be the Nominal Trip Setpoint with allowance for calibration tolerances.

INTRODUCTION

In June of 1978, Westinghouse utilized a square root of the sum of the squares (SRSS) approach for the determination of instrument channel uncertainties. Previous to that, an arithmetic sum was used to make this determination. Use of the SRSS approach resulted in; 1) a considerable gain in margin or 2) allowed for increased error accounting without a significant impact on the Nominal Trip Setpoint or the Safety Analysis Limit. However an additional, but little recognized, side affect of this approach was the requirement of returning an instrument channel to within the calibration tolerance any time it was found outside those tolerances, even though the channel was within the Technical Specification Allowable Value. This requirement was formally recognized in early 1982 with the licensing of the more sophisticated Westinghouse approach. In August 1982, the NRC placed that requirement in the V. C. Summer Technical Specifications, the first plant to utilize this approach[1]. This was the first public indication that the Westinghouse Setpoint Methodology was sensitive to the instrument channel's "as left" condition for continued validity of the uncertainty calculations during plant operation.

BASIC EQUATIONS

As noted, Westinghouse uses SRSS combination after a conservative accounting of dependent effects. The Total Channel Allowance (TCA) is determined by (Eq. 1), with parameters as defined in the Definitions at the close of this paper;

$$TCA=\{(PMA)^2+(PEA)^2+(SCA+SMTE+SD)^2+(SPE)^2+(STE)^2 +(RCA+RMTE+RCSA+RD)^2+(RTE)^2\}^{1/2}+EA \quad \text{(Eq. 1)}$$

This value must be less than or equal to the Total Allowance (TA) which is the Safety Analysis Limit (SAL) minus the Nominal Trip Setpoint.

For current plant Technical Specifications, Westinghouse determines the Allowable Value based on the use of two principle equations. The first (Eq. 2) is a simple arithmetic summation of the process rack uncertainties for which surveillance is performed on a monthly or quarterly basis, i.e.,

$$T_1=RCA+RMTE+RCSA+RD \quad \text{(Eq. 2)}$$

This reflects a conservative interaction assumption for calibration and drift of the process rack modules. The second (Eq. 3) reflects the assumption that the transmitter drifts in a random manner, for both magnitude and direction, and that the parameters that are not checked are also randomly distributed with normal or near normal probability distributions. This condition is conservatively approximated by,

$$T_2=TA-\{(PMA)^2+(PEA)^2+(SCA+SMTE+SD)^2+(SPE)^2+(STE)^2+(RTE)^2\}^{1/2}-EA \quad \text{(Eq. 3)}$$

For protection functions that have more than one input channel to the bistable, e.g., Steamflow/Feedflow Mismatch, the current Allowable Value is determined by the use of either (Eq. 3) or the following,

$$T_3=\{(RCA_1+RMTE_1+RCSA_1+RD_1)^2+(RCA_2+RMTE_2+RCSA_2+RD_2)^2\}^{1/2} \quad \text{(Eq. 4)}$$

where the subscripts 1 and 2 indicate the two different input channels. Currently licensed methodology requires that the smaller magnitude of (Eq. 2) or (Eq. 3) be used for single channel protection functions, and the smaller magnitude of (Eq. 3) or (Eq. 4) be used for multiple channel protection functions. The result is then added (or subtracted) to the Nominal Trip Setpoint to arrive at the Allowable Value.

OPERABLE CHANNEL

From a Westinghouse point of view, channel operability can be determined from two positions, depending on what is determined to be the purpose of surveillance testing. In one instance, operability can be determined by simply verifying that the bistable trips at a value less than the limit. In the second, operability can be defined as consistency with the setpoint methodology and the uncertainty calculation assumptions. Utilization of a simple limit ignores the aspect that a plant may set a bistable in the very conservative direction, e.g., several % span below (above) the Nominal Trip Setpoint and thus allow for more drift before exceeding the limit. This approach does not reflect the uncertainty calculation assumptions, nor does it reflect the hardware design attributes. However, this is the standard approach for many plants that experience significant drift. Rather than investigate what is assumed in the setpoint calculations or determine the cause or implications of such drift, a "conservative" approach is taken to reduce operational margin and allow for increased drift. This reduces the probability of exceeding the limit, but increases the probability of spurious actuation or trip and makes the determination of true hardware operability more difficult. A limited viewpoint may be created, i.e., a legalistic definition of channel operability and the reluctance to investigate significant drift. This approach is even encouraged in the Technical Specifications issued prior to 1975, the "Custom" or vendor Technical Specifications, by not taking calibration or drift into account. Plants are forced to set the Nominal Trip Setpoints in the field more conservative than that noted in the Technical Specifications to ensure not exceeding them. This approach is not discouraged in the final draft version of the "Standard Technical Specifications, Westinghouse Plants"[2] (MERITS for Westinghouse Owners Group) because satisfaction of the Allowable Value or the Total Allowance in the setpoint study does not by definition imply assurance of future operability. Both are necessary, but not sufficient, conditions.

The second approach to operability relies on the determination of consistency with the uncertainty calculations. This approach utilizes the "as left" data to verify that the channel is properly calibrated and the difference between "as found" and "as left" to determine the acceptability of drift over the surveillance interval. With this approach, the "as left" condition is a critical parameter and should be controlled. The point of trip is not as important as the magnitudes of the calibration error and drift. If the magnitudes of the uncertainty calculations are satisfied, the uncertainty calculations and thus the verification of trip is satisfied. This is a more conservative approach to the determination of operability, but proper utilization of "as left" and "as found" data, in conjunction with control of the "as left" condition, results in a necessary and sufficient condition that assures future operability.

The current Westinghouse Technical Specifications determine an acceptable channel condition using both of the above noted approaches. In one instance the channel is determined to be operable by the simple determination that the bistable trip setpoint is measured to be less than or equal to the Allowable Value. This is based on the presumption that the transmitter is behaving in an acceptable manner (no effort over and above a simple channel check is made to make this determination) and that the Allowable Value reflects a maximum of T_1 (T_3 for a multiple channel function). Thus the channel is determined to have an "as found" condition no worse than the arithmetic sum of the process rack calibration and drift terms. For this to be a valid assumption, limitations must be placed on the "as left" condition. If the calibration of the channel is extremely conservative, i.e., significantly more than the calibration tolerances, then an inordinately large drift could still result in the determination that the channel's past (and projected) operation has been (or would be) acceptable. This conclusion with respect to projected operability would be false, based on the use of only one decision input parameter, the current trip setpoint. To avoid this, any time a channel is found to be acceptable with respect to the Allowable Value, but unacceptable with respect to the calibration tolerance, the channel must be recalibrated to within tolerance of the controlled Nominal Trip Setpoint for an "as left" condition for the start of the next surveillance interval.

On the other side, in the event a channel exceeded the Allowable Value, determination of unacceptable operation may be premature. For a protection function with TA close in magnitude to TCA (within approximately one % span), and with the conservatism introduced by the requirement to use the smaller of T_1 or T_2, then T_2 would be the more limiting. This results in the potential for the Allowable Value to be set more conservative than the design tolerances, i.e., the maximum allowed RD will be less than design, or that used in the uncertainty calculation. The current Westinghouse Technical Specifications take both sides into account by (1) the definition in the Bases noting that a channel is considered to be in an acceptable "as left" condition when it is within the calibration tolerance of the Nominal Trip Setpoint (not just set conservatively), and (2) by allowing the plant to demonstrate channel operability by solving an equation using measurement data, i.e., the actual "as left" and "as found" values are used to determine RCA, RCSA, and RD for the process rack modules and the opportunity is presented to use the same for SCA and SD. This allows for conservatism in the transmitter "as left" and "as found" values to counter the nonconservatism in the process racks. Use of the measurement data creates a past history, from which future operation can be extrapolated. Any time a statistical approach is used to perform the uncertainty calculations, periodic evaluation of the calibration and drift data is necessary to assure that the probability and confidence levels are maintained. As noted previously, when this flexibility was introduced, the requirement for the "as left" condition to be within the calibration tolerance of the Nominal Trip Setpoint was also introduced. Thus, even though a channel may be found within the Allowable Value, the validity of the uncertainty calculations (and extrapolation of the future operability of the channel) can only be assured if the starting point for the next surveillance interval is within the calibration tolerance of the Nominal Trip Setpoint. This implies that the most important of the controlled parameters in the Technical Specifications for verification of Westinghouse setpoint uncertainty calculations is the Nominal Trip Setpoint.

NEW STANDARD TECHNICAL SPECIFICATIONS

In the final draft version of the "Standard Technical Specifications, Westinghouse Plants" the NRC provides the following definition for Allowable Value, which is the only control point with regard to channel operability determination;

> "The least conservative value of the process variable at which trip actions must occur. The ALLOWABLE VALUE is measured at the point in the channel defined by the plant-specific setpoint calculations. Surveillance test procedures define specific acceptance criteria that relate to the ALLOWABLE VALUE.
>
> Operation with actual trip values less conservative than nominal trip setpoints is acceptable since an allowance has been made in the setpoint analysis to accommodate this error. Determination of channel inoperability is not the simple exceeding of the ALLOWABLE VALUE, but rather it is the verification that the setpoint calculation's total allowance for instrument and process measurement uncertainties is not exceeded."[3]

Westinghouse has two areas of concern with the use of this definition. The first is that verification of not exceeding the Allowable Value and verification that the Total Allowance value is satisfied are necessary, but not sufficient, conditions for the determination of continued channel operability, as noted previously. The second is that the NRC is allowing the use of the least conservative value as the definition in the plant Technical Specifications. Using the variables of (Eq. 1), the equation for determining the Allowable Value for a Westinghouse plant would be,

$$AV = SAL - \{(PMA)^2 + (PEA)^2 + (SPE)^2 + (STE)^2 + (RTE)^2\}^{1/2} - EA \quad \text{(Eq. 5)}$$

This would appear to provide a considerable margin between the Allowable Value and the Safety Analysis Limit, but appearances can be deceiving. For example, a setpoint based on a simple protection function, like pressure, with no EA term would result in a difference between the Safety Analysis Limit and the Allowable Value of as little as 0.7 % span. The maximum difference would be noted for those functions with EA terms and the value would basically be the magnitude of that term. An investigation of typical instrument errors and uncertainty calculations for recently licensed Westinghouse plants determined that five protection functions would have differences of less than one % span, three functions would have the differences determined by the magnitude of the EA terms, and eight functions would have the differences determined by the magnitude of the PMA terms. From this investigation it can be concluded that none of the Allowable Values would be determined by the magnitude of transmitter or process rack uncertainty values assumed in the setpoint calculations. It appears somewhat incongruous that the NRC mandated limit the plant is supposed to be using to determine channel operability is unrelated to the measurable parameters.

WESTINGHOUSE PROPOSAL

Westinghouse would propose that the appropriate value for inclusion in the Technical Specifications is the Nominal Trip Setpoint. This is based on; 1) the limitation in the final draft Technical Specifications noted above, i.e., only one control point for the determination of channel operability, and 2) the sensitivity of the Westinghouse Setpoint Methodology to the determination that a channel starts each surveillance interval within the calibration tolerance of the Nominal Trip Setpoint. It should be made clear in the document that an acceptable "as left" condition is plus or minus the process rack calibration tolerance about the Nominal Trip Setpoint. Even with calibration of a process rack with a more conservative Nominal Trip Setpoint, verification that the process rack drift is less than that assumed in the uncertainty calculation is still necessary. It should be the verification that the instrument loop is operating within design that provides the confidence that the channel will continue to function in an acceptable manner, not just the absolute value of the "as found" trip setpoint, nor the simple verification that the Total Allowance is satisfied. Drift in excess of design may be acceptable based on margin and compensating effects from the transmitter, but these only provide an indication of past operation and do not provide significant assurance for the future. Extrapolation of channel operability for future operation can only be based on past performance. When a channel demonstrates drift in excess of design, reasonable effort should be made to determine the cause and the prevention of future occurrences. In the event drift in excess of design becomes frequent, every effort should be made to determine the modules experiencing the drift and replacement made prior to failure.

CONCLUSION

An evaluation of the use of only the Allowable Value, as defined in the NRC proposed final draft version of the "Standard Technical Specifications, Westinghouse Plants", NUREG-1431, concludes that use of that parameter to assure instrument channel operability is insufficient. Nor is it sufficient to include as a second parameter the use of the Total Allowance. The Westinghouse Setpoint Methodology is not sensitive to the Allowable Value, but rather is particularly sensitive to the initial condition ("as left") at the start of each surveillance interval. It is proposed that the Technical Specifications be modified to reflect the use of the Nominal Trip Setpoint with inclusion of an acceptable calibration tolerance about that setpoint. Channel operability would then be assured for future operation based on an evaluation of instrument drift relative to the design values assumed in the uncertainty calculations. Surveillance should also include periodic statistical evaluation of the past history of drift for that function to provide confidence in the extrapolation for future operation.

DEFINITIONS

The parameters paraphrased below are the principle uncertainties included in Westinghouse uncertainty calculations[4].

PROCESS MEASUREMENT ACCURACY (PMA) - allowance for noninstrument related effects which have a direct bearing on the accuracy of an instrument channel's reading, e.g., temperature stratification in a large diameter pipe.

PRIMARY ELEMENT ACCURACY (PEA) - error due to the use of a metering device, e.g., venturi.

SENSOR CALIBRATION ACCURACY (SCA) - the reference (calibration) accuracy for a sensor or transmitter as defined by SAMA Standard PMC 20.1-1973[5]. Equivalent definitions for PMC 20.1-1973 may be found in ANSI/ISA-S51.1-1979[6].

SENSOR MEASUREMENT & TEST EQUIPMENT ACCURACY (SMTE) - the accuracy of test equipment used to calibrate a sensor or transmitter in the field or in the laboratory. Westinghouse recommends the use of SAMA Standard PMC 20.1-1973[7] for placing limits on the magnitude of the maximum error, i.e., preferably less than one tenth of the Sensor Calibration Accuracy, but no more than one third, if at all possible.

SENSOR PRESSURE EFFECTS (SPE) - the change in input-output relationship due to a change in the static head pressure from the calibration conditions (if calibration is performed at line pressure) or the accuracy to which a correction factor is introduced for the difference between calibration and operating conditions (principally for Δp transmitters).

SENSOR TEMPERATURE EFFECTS (STE) - the change in input-output relationship due to a change in the ambient temperature (for expected normal operating conditions) from the reference calibration conditions about a transmitter.

SENSOR DRIFT (SD) - the change in input-output relationship over a period of time at reference conditions.

ENVIRONMENTAL ALLOWANCE (EA) - the change in sensor/transmitter output due to adverse radiation and temperature effects from a limiting accident condition.

RACK CALIBRATION ACCURACY (RCA) - the reference (calibration) accuracy, as defined by SAMA Standard PMC 20.1-1973[5], for the string calibration of the process rack modules in an instrument loop.

RACK MEASUREMENT & TEST EQUIPMENT ACCURACY (RMTE) - the accuracy of test equipment used to calibrate a process loop in the racks. Westinghouse recommends the use of SAMA Standard PMC 20.1-1973[7] for placing limits on the magnitude of the maximum error, i.e., preferably less than one tenth of the Rack Calibration Accuracy, but no more than one third, if at all possible.

RACK COMPARATOR SETTING ACCURACY (RCSA) - the reference (calibration) accuracy, as defined by SAMA Standard PMC 20.1-1973[5], of the instrument loop comparator (bistable).

RACK TEMPERATURE EFFECTS (RTE) - change in input-output relationship for the process rack module string due to a change in the ambient temperature from the reference calibration conditions.

RACK DRIFT (RD) - the change in input-output relationship over a period of time at reference conditions.

TOTAL CHANNEL ALLOWANCE (TCA) - the SRSS of the applicable instrument uncertainties for a channel with appropriate accounting for interactive effects.

TOTAL ALLOWANCE (TA) - the difference between the Safety Analysis Limit and the Nominal Trip Setpoint.

SAFETY ANALYSIS LIMIT (SAL) - the instrument trip setpoint for a protection function assumed in the Safety or Transient Analyses.

NOMINAL TRIP SETPOINT (NTS) - the preferred setpoint in the field for a protection function.

ALLOWABLE VALUE (AV) - as used by Westinghouse, it is a control point for process rack drift. The point at which the operability of a process string should be evaluated because of significant drift. In the Westinghouse setpoint methodology, this point is determined by the smaller of two calculated values, T_1 or T_2, which are described by (Eq. 2) or (Eq. 3) for single input protection functions.

REFERENCES

[1] NUREG-0717 Supplement No. 4, "Safety Evaluation Report related to the operation of Virgil C. Summer Nuclear Station, Unit No. 1, Docket No. 50-395, South Carolina Electric and Gas Company," U.S. Nuclear Regulatory Commission, August, 1982, pp 7-1 - 7-3.

[2] NUREG-1431, "Standard Technical Specifications, Westinghouse Plants" Final Draft transmitted via letter from W. T. Russell (NRC) to T. E. Tipton (Nuclear Management and Resources Council), January, 1991.

[3] NUREG-1431, "Standard Technical Specifications, Westinghouse Plants" Final Draft transmitted via letter from W. T. Russell (NRC) to T. E. Tipton (Nuclear Management and Resources Council), January, 1991, p 1.1-1.

[4] Tuley, C. R., Miller, R. B., "Westinghouse Setpoint Methodology for Control and Protection Systems," IEEE Transactions on Nuclear Science, Vol. NS-33 No. 1, San Francisco, CA., February, 1986, pp. 684 - 687.

[5] Scientific Apparatus Makers Association Standard PMC 20.1-1973, "Process Measurement & Control Terminology," New York, NY, 1973, p 4.

[6] Instrument Society of America Standard 51.1-1979, "Process Instrumentation Terminology," Research Triangle Park, NC, 1979, p 6.

[7] Scientific Apparatus Makers Association Standard PMC 20.1-1973, "Process Measurement & Control Terminology," New York, NY, 1973, p 36.

0074-056X/91/141-145/$0 + .50pp

A VIEWPOINT ON POST-ACCIDENT SENSOR UNCERTAINTIES

R. B. Miller
C. R. Tuley

KEYWORDS: Setpoint Methodology
Environmental Allowances

ABSTRACT

Reducing the impact of environmental uncertainties from instrumentation systems on nuclear power plant protection system setpoints can be achieved with confidence through proper design and qualification programs. Two primary methods are utilized to achieve sensor uncertainties with a 95 % probability at a 95 % confidence level. The first method relies primarily on the design process with confirmation provided by the qualification program. The second approach utilizes a sufficient sample population in the qualification program to establish a high confidence in the calculated result. Developing the ability to predict performance of a sensor exposed to harsh environments of different magnitudes is also extremely important in reducing the necessary allowance for total instrument uncertainties and improving plant availability.

INTRODUCTION

Providing an accurate, defendable representation of the post-accident uncertainties regarding sensors and associated hardware, such as cables and reference legs, can have a major impact on plant operability. Most often these uncertainties are based on conservative values reflecting errors anticipated at the maximum environmental conditions enveloping the worst event at several plants. This results in a setpoint that is too close to the operating point for that parameter, risking plant trips for anticipated transient conditions.

SETPOINT METHODOLOGY

A previous publication, "Westinghouse Setpoint Methodology for Control and Protection Systems,"[1] outlined the approach utilized in establishing setpoints since the late 1970's. The methodology used to combine uncertainties into a Total Channel Allowance (TCA) is a square root of the sum of the squares (SRSS) of steady state random variables, with appropriate treatment of statistically dependent parameters, added to non-random terms (biases). Random terms are due to uncertainties from calibration, drift and normal processes and ambient temperature changes. When combined by SRSS, these errors will usually result in an acceptable allowance to add to (or subtract from) the safety analysis limit to establish the parameter setpoint. However, for those functions required post-accident, the effect of the environment on exposed hardware must be considered. Uncertainties will be produced by these environments which may be predictable in direction and are, therefore, treated as biases to determine the TCA. When the Environmental Allowance (EA) is developed from maximum enveloping accident conditions, both radiation and temperature, it may easily exceed the steady state errors by a factor of three or four. For trip functions, the radiation and temperature errors are arithmetically summed since the time period to trip is very short (maximum of few minutes). For post-accident monitoring functions, the radiation and temperature terms are combined via SRSS before adding to the steady state term calculations to provide a more realistic allowance for changing uncertainties over a long period of time.

The SRSS approach is unquestionably valid for large sample populations ($n > 30$) and is considered reasonably accurate for smaller populations when the various components are from normal probability distributions. A rigorous calculation will result in an uncertainty with a 95 % probability, with a confidence level defined by the proper application of tolerance multipliers. The goal of Westinghouse and the regulatory institution is a 95/95 type calculation. For most of the components, a 95/95 value is not too difficult to justify; however, for the EA term, data may be sparse resulting in either a decreased confidence level or an unacceptably high EA magnitude.

ESTABLISHING CONFIDENCE

In order to gain confidence that the probability of the sensor EA terms may be expressed at a level approaching 95 %, two methods of establishing the uncertainty may be considered. The first approach involves a rigorous assessment of the design of both the mechanical and electrical portions of the sensor. Materials, processes and components must be selected for their response to increasing temperatures and radiation. Additional data is usually obtained by engineering tests at the component or subassembly level. It should be possible through a combination of design, circuit analysis and preliminary environmental testing to establish a mathematical algorithm for the expected performance during elevated temperature and radiation conditions. Radiation uncertainties stem from the mechanical to electrical transformation section and the signal amplification and conditioning circuit. These are typically minimized to a predictable value by selecting radiation hardened components by lot testing or by circuit design where, for example, a large change in the gain of a transistor will have a small effect on the output.

Temperature effects on electronic circuits can be predicted by selection and application through the design process. It is sometimes more difficult to accomplish the same result with the mechanical portion of a sensor. For this reason, temperature compensation circuits are utilized to provide adjustment for the resulting temperature induced uncertainty of the assembled unit. Westinghouse has made it a practice to temperature compensate each individual sensor assembly to established uncertainty limits. These limits are then utilized as the environmental allowance. Through this process of design, preliminary testing and temperature compensation, we feel that both the radiation and temperature environmental allowances have been established with a 95 % confidence. An additional important advantage of this approach is the ability to predict performance at different temperatures which may be useful in minimizing the effect on the ultimate parameter setpoint. Qualification testing should then be conducted on a reasonable sample population to verify the design process and also verify that dynamic effects that may occur during the initial phase of an accident condition were conservatively predicted by the model.

The second approach still relies on selection of materials and components and control of the manufacturing process but places less emphasis on establishing a mathematical model of the sensor performance. The performance would be determined by utilizing a significantly larger sample population for the qualification program. Based on the data from the qualification tests, the post-accident uncertainty would be expressed as the mean ± a tolerance factor times the standard deviation ($\mu \pm k\sigma$). A sample population of thirty would minimize the effect of k on the final result. To maintain a high confidence, a significance test at the 5 % level should be performed to confirm the validity of assuming the data is from a normal population. Where it is only necessary to determine that the uncertainty is less than a certain value (e.g. actuation of a trip function), a one-sided confidence interval for this term may be established. For post-accident monitoring functions, a two-sided confidence interval would be more appropriate. One advantage of this approach would be that the bias and random terms are rigorously developed and may be applied in that manner to the calculation of the TCA, i.e., the mean would be considered, and treated, as a bias, while the $k\sigma$ term would be considered, and treated, as a random variable. This would result in reducing the overall effect of the environmental allowance on the setpoint. Separating these terms would be more difficult if the first approach was implemented. A disadvantage to the second approach may be the ability to predict uncertainties at lower temperatures unless testing at more than one temperature was performed during the qualification program.

Either of these approaches would contribute to a 95/95 calculation result for the TCA when the environmental allowance is included for sensor errors. Both have their advantages, and the selection of the method will sometimes be determined on other factors, such as cost. Westinghouse has primarily utilized the first approach to minimize the qualification sample population. The uncertainty at various temperatures is calculated to minimize the effect on the setpoint where possible, based on mathematical models and engineering tests predicting the expected linearity of the response. The effect of radiation may be determined in a similar manner which may be confirmed during the qualification test. The sample population could get quite large if a variety of applications requiring different turndown ratios are involved in the qualification program. Some grouping may be accomplished, but sensors with a high turndown ratio, which significantly amplifies errors, should be included as a separate sample population. In any case, Westinghouse conservatively utilizes the design values (for maximum temperature application) or values determined from the response algorithm (for lower temperature applications) for trip actuation function uncertainty calculations.

UTILIZING QUALIFICATION TEST DATA

Although the qualification test may be used to determine uncertainties with the appropriate sample, it is primarily a program to verify that the design predictions and bounding uncertainties are accurate. Since most testing is performed sequentially, each part of the sequence must be monitored effectively with calibrations between phases of the test so that deviations due to each event may be determined.

THERMAL AGING

The preconditioning activity typically does not lend itself to establishing performance characteristics because it is accomplished at static elevated temperatures. The initial phase of the test could be used to verify uncertainties at the selected aging temperature, but after an extended period of time the data may not be useful, depending on the aging characteristics of the sensor. Due to the higher temperatures utilized, any changes detected during this test phase may not be indicative of normal environment performance.

RADIATION AGING

Most qualification programs expose the sensor to the total expected integrated dose during this part of the sequence to limit program cost. In order to verify performance for trip and actuation functions, Westinghouse performs a calibration check at a dose that conservatively represents the expected radiation level after the sensors installed life plus time to trip. This provides confirming data to establish setpoints from a somewhat larger sample, since it is typical for a vendor to utilize the same electronics for all transmitters that are manufactured for post-accident service. Completing the radiation test will confirm established bounding uncertainties for post-accident monitoring functions.

During some part of the total process from design through engineering tests and the qualification program, it is necessary to demonstrate the sensitivity of the device to dose rates expected during the initial phase of an accident. As long as material considerations do not limit the maximum dose rates, the concern may be easily verified by performing the radiation test at the maximum dose rate expected.

SEISMIC

Since for most functions it is not necessary to consider seismic events simultaneously with other plant disturbances, sensor offsets or oscillations that occur during the test are of no concern. Uncertainties that must be considered are residual effects from the seismic event that could possibly lead to a permanent shift in the sensor output. For example, the shift of calibration potentiometers may occur and the magnitude may be determined by sample testing. This magnitude could then be translated into a change in the sensor output which is included as an environmental allowance.

HIGH ENERGY LINE BREAK SIMULATION

It is typical, for cost effective qualification programs, to expose the test items to the worst postulated high energy line break environment rather than performing a variety of tests to confirm or determine (by sample test) uncertainties for separate feedline break, steamline break or loss of reactor coolant events. This type of program does provide confirmation that the sensor error will be less than the predicted bounding condition. If a model of the sensor has been developed and confirmed, this worst case test is all that is needed. However, performing a simulated test at one lower or intermediate temperature would certainly provide additional confirmation of the model and, therefore, additional confidence in the setpoint. This additional test could serve as part of the total program which might demonstrate reduced dynamic effects if, for example, the initial rise in temperature due to a feedline break is not as severe as a steamline break. Confirming uncertainties due to feedline breaks could be a distinct advantage in many plants where steam generator water level is the primary actuation function to mitigate the effects of a feedline break. Justifying a reduced environmental allowance and, therefore, a minimal setpoint for this critical function has a very positive effect on plant availability.

DOCUMENTATION

The reporting of the results of the design and qualification testing in a usable format is receiving increased importance. ISA RP67.04 - Part II, Draft 8, "Methodologies for the Determination of Setpoints for Nuclear Safety-Related Instrumentation", notes the need for users to work with equipment suppliers in determining the probability/confidence level of instrument uncertainties prior to use in uncertainty calculations. To minimize the need for frequent consultation, the supplier should provide clear identification of these values and any limitations, e.g., one directional bias values, that apply to their use. Clear identification is necessary since there are several interpretations of the probability/confidence level of performance specifications (both steady state and adverse conditions) reported in the vendor supplied literature, e.g., 95/95 two sided (1.96σ), 95/95 one sided (1.645σ), 95/68 two sided (1σ) and 95/99.9 one sided (3σ). An incorrect interpretation of the confidence level could result in the inappropriate determination of the random $k\sigma$ value or application of the error. In addition, if the information is available, specification of the mean (μ) (if the value is not near zero) and the corresponding random component ($k\sigma$) would allow a more rigorous and potentially favorable combination of errors.

CONCLUSION

Either the design process or the qualification test (large sample population) will yield the necessary information to enable calculating a Total Channel Allowance with a 95 % probability at a 95 % confidence level. Both have their advantages, the primary being the ease of predicting errors at various temperatures when the design method is followed and the ability to separate random and bias errors if a large population is tested. Westinghouse has followed the design method for several reasons, including the preference of understanding sensor performance capabilities and the cost of a large qualification program. Whichever method is selected, it is important to understand the significant impact that the environmental allowance for sensors and related hardware has on the setpoint for a post-accident function.

REFERENCES

[1] Tuley, C. R., Miller, R. B., "Westinghouse Setpoint Methodology for Control and Protection Systems," IEEE Transactions on Nuclear Science, Vol. NS-33 #1, San Francisco, CA., February, 1986, pp. 684 - 687.

0074-056X/91/147-149/$0 + .50pp

NUCLEAR INSTRUMENT TECHNICIAN TRAINING; ISN'T WHAT IT USED TO BE!

E.S. Wollesen
April 5, 1991
Revision 1

ABSTRACT

Nuclear Instrument Technician (NIT) Training has developed at an accelerated rate over the past three decades. During the 1960's commercial nuclear power plants were in their infancy. For that reason, there is little wonder that NIT training had little structure and little creditability. NIT training, in many early plants, was little more than On-The Job Training (OJT). The seventies brought changes in Instrumentation and Controls as well as emphasis on the requirements for more in depth training and documentation. As in the seventies, the eighties saw not only changes in technologies but tighter requirements, standardized training and the development of accredited Nuclear Instrument Training; thus the conclusion: Nuclear Instrument Training Isn't What It Used To Be.

NUCLEAR INSTRUMENT TECHNICIAN TRAINING ISN'T WHAT IT USED TO BE!

This paper overviews NIT training from times of OJT to current NIT Training programs with accreditation. Starting with basics of the sixties, 1960's, we progress to details of current courses and accreditation.

The sixties brought about construction and design of commercial nuclear power plants. Moving over to Nuclear power from coal, oil and gas units marked the beginning of an extensive learning process. Nuclear power presents different challenges than conventional fossil fuel plants, requiring different instrumentation. Not only were the instruments different, but the introduction of nuclear power introduced environmental and regulatory differences as well. New instrumentation, techniques and philosophies prompted the specific need for formal NIT Training. NITs working through the construction phase of these new commercial plants gained OJT superior to simulated classroom training. NIT training faced a new challenge; to create classroom training that could duplicate field experience and conditions.

The need was high for qualified NITs. The lack of criteria and training resulted in NITs untrained to the latest technologies, instruments, and regulations. Details of the required training were barely known, however, the **need** for the training surfaced. NITs had to learn what they could, when they could, how they could, and from any one they could. This approach occurred from the philosophy that nuclear power is just a different power source than fossil, not more complex, impacting or critical. However, nuclear power is different and requires in-depth training. Since the NIT's effect much of the plant, their need for training was critical; but not found in the growing commercial Nuclear Power industry. The seventies were going to help.

NIT training in the early seventies continued with OJT and added vendor training. Supplementing vendor training helped for a lack in training on specific equipment but still the nuclear needs exceeded the NIT training. NITs are responsible for virtually all Instruments And Controls (I&C) used by Operators for determining plant operating parameters to assure compliance with operational and legal requirements. NIT training must now include practice, performance measures, interpersonal communication skills, and be creditable.

The NIT of the seventies did not have the time to develop course lists. They were in the field performing tasks for which they needed training. Their performance, at times, proved the need for in-depth training. To assure competence for the present and future NITs, development of formal training was essential. In addition, during the seventies some plants were going through wage negotiations. Some wage negotiations highlighted the need for trained NITs to assure no inadvertant changes would cause reactivity changes in the nuclear core. Therefore, NITs received additional compensation based on the emphasis for additional training. The NITs training course list was growing by leaps with no apparent bounds. As the seventies progressed, NIT training resembled a fluid process. Instruments, control loops, and systems changed requiring constant training updates and additions.

The seventies also highlighted spontaneous response required by the NIT during plant transients. Since plant designs included known transients, unexpected transients presented a new uncharted responsibility for the NIT. The new challenge presented NIT training problems. To develop hands on training for dynamic processes that are critical to the plant without causing plant transients. Simulators and extensive dynamic theory were the answer. To develop training and train NITs for needs identified in the seventies, the eighties had their task set.

The eighties expanded the identified NIT training needs of years past. During the early eighties, a technical training organization developed classification levels for NITs based on their level of training and competence. (1) The classification structure was a topic presented at a conference for the nuclear power industry. Demonstrated competence levels reflected the NIT's ability and classification to the level of responsibilities. Use of the classification levels clearly identified the training needs for the NIT. Not using the structured levels put responsibility on the supervisor to determine the NIT's qualification when assigning work activities.

Efforts to meet the NIT training needs progressed. The eighties brought a change to formalize NIT training. Documented lesson plans, evaluated course content/delivery, and accreditation were milestones in the new NIT training. To meet NIT training needs of industry (for Instrument Technician skills required by all industries), the Instrument Society of America (ISA), colleges and industry joined forces. (2) Professional Training groups presented NIT classification at conferences; ISA, Tampa Bay Section, worked with the State Of Florida Department Of Education to issue "Career Achievement Plan(s) (Career Maps)." (3) The course plans included: Instrumentation Technology, Instrument Repair, Instrument Technician, Test Technician, Calibration Laboratory Technician, Instrument Mechanic, Inspector, Assembler, Calibrator, Instrument Technician Helper, and Electrical Instrument Repairer. Community Colleges researched the industry to develop curriculum for a two year Associate of Applied Science Degree for the Instrument Technician. (4) These efforts developed courses and curriculum to meet the needs and requirements for the entry level NIT only.

The entry level NIT training includes plant specific systems, diagrams, equipment, plant administration and management policies. (5) The plant conducts plant specific training for the NIT. Because of the plant trained NIT skills, operations and plant management depend heavily on the NIT. Plant operators and management look to and rely on the NIT for solutions to plant abnormal responses and transients. Even with the required procedures for the NIT, NIT training must be thorough enough to detect errors with the procedures; prior to causing transients. To follow a procedure takes a level of competence. To oversee the effects of following the procedure takes the skill of the trained NIT. The increasing need for the trained NIT and the dependance of industry on nuclear power, the need for the specialized training is in place to stay. Training the "hands-on" skills, once done with the plant on line, is now turning to the simulators.

Accreditation, as sighted in INPO 85-002 revision 01, is an involved process. (6) Protions of the process include: establishing industry standards, systematic self-evaluations by the utilities, INPO Team evalution, Accreditation by the National Nuclear Accreditation Board, and maintenance of the accreditation. With these standards, experiance from the OJT, and the time to administer the training; NIT Training Isn't What It Used To Be.

REFERENCES

1) Howarth, Robert J.; Newell, William J.; Dr. Stanley L. Cohen, PhD.; THE USE OF A QUALIFICATION PROGRAM FOR INSTRUMENTATION AND CONTROL TECHNICIANS; The American Nuclear Society, WQashington, DC, November 18, 1982.

2) Instrument Society Of America (Sponsor), QUALIFICATIONS AND CERTIFICATION OF INSTRUMENTATION AND CONTROL TECHNICIANS IN NUCLEAR POWER PLANTS (draft standard), Pittsburg, Pennsylvania, June 18, 1981.

Instrument Society Of America (Sponsor), QUALIFICATIONS AND CERTIFICATION OF INSTRUMENTATION AND CONTROL TECHNICIANS IN NUCLEAR POWER PLANTS, ANSI/ISA-S67.14-1983 (AMERICAN NATIONAL STANDARD), Research Triangle Park, North Carolina.

3) State of Florida, Department Of Education, With assistance by the Tampa Bay Section, Instrument Society Of America, CAREER ACHIEVEMENT PLAN (CAREER MAP), IN 15.040400 Instrumentation Technology and IN 47.040100 Instrument Repair, Tallahassee, Florida, June 1984

4) Monroe Community College, Bulletin 1985-86, Volume XXI No 1, page 110, INDUSTRIAL INSTRUMENTATION TECHNOLOGY -AAS DEGREE- (70 credit hour) and page 111, INDUSTRIAL INSTRUMENTATION TECHNOLOGY CERTIFICATE PROGRAM (24 credit hours), Rochester, New York, 1985.

5) Florida Power Corporation, NUCLEAR OPERATIONS TRAINING, TRAINING PROGRAM DEVELOPMENT MANUAL, revision 4, March 1987.

6) The Institute Of Nuclear Power Operations (INPO), Document Number GPG-08, GUIDELINES FOR INSTRUMENT AND CONTROL TECHNICIAN QUALIFICATION, Revision 0, July 31, 1981.

7) The Institute Of Nuclear Power Operations (INPO), INPO 85-002, Revision 01,

©1991 - Paper #91-712
074-056X/91/151-166/$0 + .50pp

APPLYING ISA RECOMMENDED PRACTICE 67.04 TO SETPOINT CALCULATIONS

Michael J. Blomquist, Bechtel Corp.
Kent R. Bjornn, Bechtel Corp.
Robert E. Green, Supply System

ABSTRACT

The ISA issued draft 7 of Recommended Practice (RP) 67.04 "Methodologies for the Determination of Setpoints for Nuclear Safety-Related Instrumentation" in September of 1990. The draft RP describes methods for calculating and combining the uncertainties associated with instrument loop setpoints (e.g., temperature, pressure, and radiation effects, etc.). However, in some cases, the guidance provided on how to apply these uncertainties and other factors to setpoint calculations is not completely developed. In addition the RP does not address how these uncertainties might affect the setpoints of other instrument loops within a system. This paper addresses these issues. The methods for applying the uncertainties to instrument loop setpoints presented in this paper were developed while performing setpoint calculations for instruments located in harsh environments at WNP-2. Considerations for applying uncertainties when determining setpoints include: the effects of cascading setpoints, the effects of setting tolerances, switch reset, and two sided analytical limits.

1.0 INTRODUCTION

Operation of a nuclear power plant requires continuous monitoring of the power production related systems and the systems required to assure a safe shutdown in the event of a plant accident. Should a system failure occur, action must be taken to ensure that the design safety limits are not exceeded, thereby protecting the public and avoiding potential major equipment damage. Plant instrumentation, with its inherent inaccuracies, is designed to monitor system process variables and provide either automatic initiation of protective equipment or alert plant operators when these variables exceed predetermined setpoints, thereby maintaining process variables below design or safety limits.

Accurate determination of a setpoint used to protect plant systems requires two considerations. First, the limiting conditions (i.e. the point at which the safety of the public is endangered (10 CFR 100 limits), or the point at which equipment damage might occur) should not be exceeded, and second, the inaccuracies of the instruments used to monitor system process variables must be determined at a specified confidence level. The instrument setpoint must be such that when all uncertainties are considered, the process variables will not exceed the limiting conditions, within the confidence level of the inaccuracies.

Instruments which monitor the process variables in the systems required to assure a safe shutdown may be exposed to a harsh environment and must continue to function. The harsh environment can have an effect on the accuracy of the instrumentation. To assure proper operation of safety systems, additional uncertainty due to the effect of the harsh environment must be determined and included in the setpoint calculation for the instrument loop. In general, the effect of the harsh environment will be to increase instrument inaccuracies, and therefore, increase the required separation between the setpoint and the limiting condition.

The NRC issued Regulatory Guide 1.105, "Instrument Setpoints for Safety-Related Systems" to address these issues. In Revision 2 of this Regulatory Guide, the NRC endorsed ISA Standard ISA-S67.04-1983, "Setpoints for Nuclear Power Plants" thereby defining the acceptable method for ensuring that safety related setpoints do not exceed technical specification limits. The ISA is currently developing a Recommended Practice dRP 67.04 "Methodologies for the Determination of Setpoints for Nuclear Safety-Related Instrumentation", to complement S67.04. This recommend practice defines specific methods for determining instrument different uncertainties and combining these uncertainties. The NRC is expected to revise Regulatory Guide 1.105 to endorse this recommended practice once it is published by ISA.

Washington Public Power Supply System recently began a program at its WNP-2 plant to insure that harsh environment effects were properly included in the original safety related instrument setpoint calculations. Concurrent with this effort, the completeness and accuracy of the limiting conditions for the systems are being reviewed.

The effort to review the limiting conditions for the many safety systems at WNP-2 involves a review of design documents, and, in some instances, the development of additional calculations depending on the completeness of the original documentation. Determination of the instrument uncertainties is being performed in accordance with the methods discussed in Recommended Practice dRP 67.04. In the process of performing the work at WNP-2, several situations were encountered which were not thoroughly covered in the recommended practice. In particular, cascading setpoints, the effects of setting tolerances, switch reset, and two sided analytical limits are not thoroughly discussed. As a result, methods had to be developed to properly address these situations.

This paper presents the methods developed at WNP-2 for combining limiting conditions, uncertainties and setting tolerances to determine a nominal setpoint. The methods are demonstrated by example. The examples are based on setpoint calculations performed at WNP-2. The system descriptions in the examples have been simplified for presentation purposes and do not reflect the actual plant design.

2.0 DEFINITION OF TERMS

Allowable Value (AV) - The limiting value that a trip setpoint can have when tested periodically, beyond which the instrument channel is declared inoperable and corrective action must be taken [1, page C-11]. The allowable value includes the effects of drift, measuring and test equipment uncertainties, instrument accuracy and the effects of environmental variation during normal plant operation.

Allowable Value Trip Setpoint Margin (AVTSM) - The margin required between the nominal setpoint (NSp) plus or minus the setting tolerance, and the AV to account for instrument drift, instrument calibration uncertainties, and instrument uncertainties during normal operation [2, section 7.3]. The instrument uncertainties during normal operations are listed in the definition of Uncertainties (U), below.

Analytical Limit (AL) - Limit of a measured or calculated variable established by the safety analysis to ensure that a safety limit is not exceeded [1, page C-11]. The analytical limit is developed by considering such parameters as process delays, instrument response times, rod insertion times, reactivity changes, etc., and determining their effect on the safety limit [2, page 43].

Dead Band (DB) - The range through which an input signal may be varied, upon reversal of direction, without initiating an observable change in output signal[3 page 17]. The dead band is the difference between the switch setpoint and reset. The dead band can be fixed or adjustable depending on the type of switch.

Limiting Safety System Settings (LSSS) - Limiting safety system settings for nuclear reactors are settings for automatic protective devices related to those variables having significant safety functions [4].

Lower (Upper) Setpoint - The nominal setpoint minus (plus) the negative (positive) setting tolerance included to aid calibration.

Maximum (Minimum) Setpoint - The extreme value which a setpoint (upper or lower) may have and still properly account for instrument uncertainties to insure that the analytical limit is not exceeded.

Nominal Setpoint (NSp) - The nominal value at which a bi-stable device changes state to indicate that the quantity under surveillance has reached the selected value [1, page 9]. The NSp is usually the midpoint of the setting range (nominal setpoint $\pm$ setting tolerance).

Non-Limiting Safety System Settings (NLSSS) - This includes setpoints within safety-related systems, but are not considered as LSSS by the definition above.

Process Design Limit (DL) - The limit established upon process variables to prevent equipment damage based on plant operating experience, vendor data, or original design criteria. In actual operations or in system design, a margin would be included between the process design limit and the failure limit, at which point equipment damage would occur. For Non-Limiting Safety System Settings, the failure limit is effectively the same as the safety limit, though without the legal implications since it does not imply release of radioactive material or fuel element damage. The process design limit is effectively the same as the analytical limit, though without implication of release of radioactive material or fuel element damage [2, section 8.2]. For Non-Limiting Safety System Settings, a process design limit applies rather than safety and analytical limits.

Reset - The value at which the bi-stable returns to the state which existed before the NSp is reached.

Safety Limit (SL) - Limits upon important process variables that are found to be necessary to reasonably protect the integrity of certain of the physical barriers that guard against the uncontrolled release of radioactivity [3]. Safety limits apply to Limiting Safety System Settings.

Setting Tolerance (ST) - The variation the instrument setpoint can be from the NSp and still be considered as properly calibrated. Since calibrating an instrument to trip at an exact point is rarely possible, a range of acceptable values is provided. This is included as an aid to the person performing a calibration. ST^- is the setting tolerance below, and ST^+ is the tolerance above, the NSp.

Uncertainties (U) - The expected amount of error in an instrument channel. Uncertainties are defined in ISA dRP67.04 [2] and include such effects as reference accuracy, calibration accuracy, which includes setting accuracy and measuring and test equipment effects, drift, insulation resistance, humidity, power supply, pressure, primary element, radiation, seismic, static pressure (of the process), and temperature effects, etc. The methods for determining and combining different uncertainties are defined in [2]. These uncertainties all have a confidence level associated with them. That is, the confidence that other measurements of the same environmental effect will be bounded by the value used in the calculation. This confidence level should be at least 95%. This paper assumes that all uncertainties have been considered and combined correctly according to [2]. U^- is the uncertainty below the lower setpoint, and U^+ is the uncertainty above the upper setpoint.

3.0 EXAMPLES

The examples provided will be based upon two systems. The first three examples are based on the simplified system shown in figure 1. The last example is based upon the system shown in figure 7, and is discussed further in section 3.4.

The system for the first three examples is the High Pressure Core Spray (HPCS). The HPCS system is a safety system designed to provide cooling water to the reactor at a high pressure if a small break loss of coolant accident occurs. The HPCS system can draw water from the condensate storage tank, and deliver it to the reactor via the HPCS pump. If the water level in the condensate storage tank drops below acceptable limits, the HPCS pump can draw from the suppression pool. Switching from the condensate storage tank to the suppression pool is controlled by a low level switch COND-LS-2 located at the condensate storage tank. When this switch trips, the suppression pool suction valve HPCS-V-1 opens, and the condensate storage tank suction valve HPCS-V-2 closes.

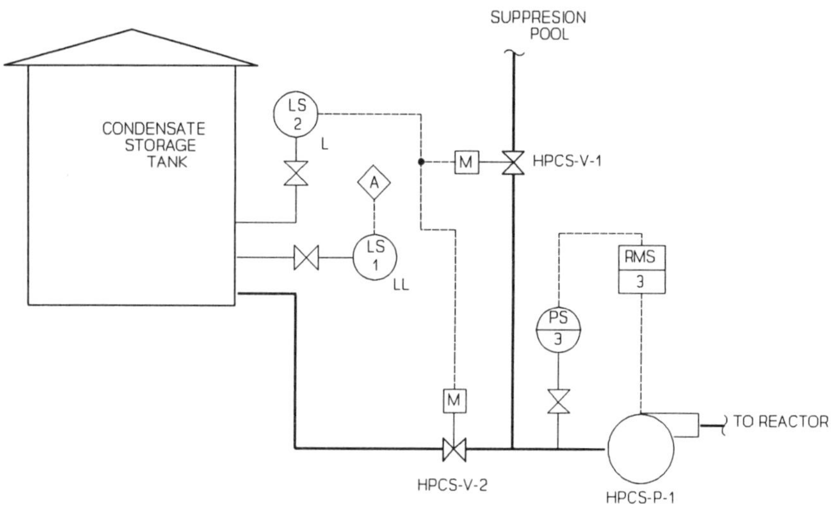

Figure 1
System Sketch
Condensate Storage Tanks & HPCS Pump

In addition to low level switch COND-LS-2, the condensate storage tank has a second low-low level switch COND-LS-1. The purpose of this switch is to alarm in the control room informing the operator that the water level has dropped below normal operating limits thereby indicating a system malfunction.

The HPCS pump HPCS-P-1 has a suction side pressure switch HPCS-PS-3 which is designed to shutdown the pump on low net positive suction head and alarm in the control room.

These three instrument loops: COND-LS-1, COND-LS-2, and HPCS-PS-1 will serve as examples in applying the guidelines of dRP67.04 to determining setpoints.

3.1 EXAMPLE 1 - BASIC SETPOINT RELATIONSHIPS

Instrument loop COND-LS-1 is a simple low-low level alarm. The setpoint calculation for this loop will demonstrate the basic principals discussed in dRP67.04 [2]. To determine the setpoint for low-low level switch COND-LS-1, several different factors must be considered:

- o The Safety Limit (SL) - The SL for COND-LS-1 is the level in the condensate storage tank where vortex formation begins to occur. Based on the geometry of the outlet connection relative to the tank and the velocity of the fluid in the outlet pipe, this point was determined to be 43 inches above the bottom of the tank.

- o The Analytical Limit (AL) - The AL for this switch must include a margin to allow for valves HPCS-V-1 and 2 to change position as required. This margin is required to insure that the SL is not exceeded after the switch has tripped, and the tank is being isolated. Although it does not apply directly to the alarm setpoint, it applies to the COND-LS-2 setpoint. It will be shown later that the COND-LS-2 setpoint is dependent on the COND-LS-1 setpoint. The additional margin is added here for simplicity and conservatism. The required calculations concluded that the required margin was 4.2 inches. Therefore, the AL for COND-LS-1 is 47.2 inches from the bottom of the tank.

- o Uncertainties (U) - As discussed in the introduction, the method for determining the U associated with level switch COND-LS-1 is documented in [2]. It is a combination of different uncertainties that occur during normal and accident conditions. These different uncertainties are listed in the definition of U presented in section 2.0. Based on [2], $U^+ = 1.5$ inches, $U^- = 2.3$ inches.

- o Allowable Value Trip Setpoint Margin (AVTSM) - AVTSM is used in defining the margin between the NSp and the Allowable Value. It is a combination of the uncertainties terms affected by normal plant operation. Based on the direction in [2], $AVTSM^+ = 0.3$ inches, $AVTSM^- = 0.6$ inches

- o Setting Tolerance (ST) - A ST is included to ease the technician's job in calibrating the instrument. The ST for COND-LS-1 was set by plant operations staff at $ST = \pm 1.0$ inches

Using the factors defined above, the minimum setpoint for COND-LS-1 was calculated by adding U^- to the AL as shown in figure 2. The difference between the minimum setpoint and the lower setpoint is margin added by plant operations. For example, if the existing lower setpoint is conservative compared to the calculated minimum setpoint, the existing setpoint would not need to be changed. The margin would be the difference between the existing lower setpoint and the calculated minimum setpoint. In this case the lower setpoint was the same as the minimum setpoint. The NSp was calculated by adding ST^- to the lower setpoint. The upper setpoint was calculated by adding ST^+ to the NSp.

The allowable value was calculated by subtracting the AVTSM from the **lower setpoint**. The allowable value is the limiting value that the setpoint can have when tested periodically. If the setpoint "drifts" beyond the allowable value, the channel is declared inoperable, and corrective action must be taken.

These terms are presented in figure 2. The NSp for low-low alarm COND-LS-1 was calculated to be 50.5 inches above the bottom of the tank. A margin between the minimum setpoint and lower setpoint was not required. The allowable value was calculated to be 48.9 inches.

The basic principals for calculating setpoint described above apply to all instrument setpoints. The different uncertainties, and how they are combined may change, but the relationships between the safety limit, analytical

COND-LS-1	
	inches
Upper Setpoint	51.5
Nominal Setpoint	**50.5**
Lower Setpoint Minimum Setpoint	49.5
Allowable Value	48.9
Analytical Limit	47.2
Safety Limit	43.0

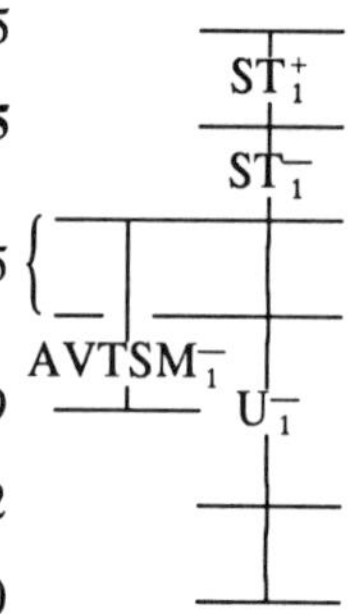

Figure 2
Limits and Setpoints for COND-LS-1

limit (or design process limit), the nominal setpoint and the allowable value does not change. The remainder of the examples will build upon these basic principals.

3.2 EXAMPLE 2 - CASCADING SETPOINTS

The low level switch COND-LS-2 provides the signal to change the HPCS pump suction from the condensate storage tank to the suppression pool. The COND-LS-2 setpoint is dependent on the setpoint of COND-LS-1, in other words, COND-LS-1 and COND-LS-2 are cascading. The COND-LS-2 setpoint must be determined such that:

a. when all COND-LS-1 and COND-LS-2 uncertainties and setting tolerances are considered, the low level setpoint of COND-LS-2 will always (within the confidence level used to determine U^+ and U^-) trip before the low-low alarm setpoint of COND-LS-1, and

b. the usable volume in the condensate storage tank is maximized.

Therefore, the starting point for determining the COND-LS-2 setpoint is the COND-LS-1 NSp calculated in example 1. Using the values for ST_1^+ and U_1^+ given in example 1 and the following equation, the AL for COND-LS-2 was calculated to be 53.0 inches.

$$AL_2 = NSp_1 + ST_1^+ + U_1^+.$$

This is shown on the left side of figure 3.

Once the AL is determined, the method for determining the COND-LS-2 NSp is the same as that used to calculate the COND-LS-1 NSp. Using this method, the following values were calculated

ST_2^+	= 1.0 inch	ST_2^-	= 1.0 inch
U_2^+	= 0.8 inches	U_2^-	= 1.4 inches

$AVTSM_2^+$ = 0.6 inches $AVTSM_2^-$ = 0.3 inches

Using these values, the COND-LS-2 NSp was calculated to be 55.4 inches and the AV was calculated to be 54.1 inches. The relationship between the COND-LS-1 and the COND-LS-2 ALs, AVs and NSps is shown in figure 3. From this figure, it is clear how the COND-LS-2 NSp cascades off the COND-LS-1 setpoint. In calculating the NSp by this method, it is assured that the COND-LS-2 setpoint will always (within the confidence of the uncertainty terms) trip before COND-LS-1. This method also assures that the useable tank volume is maximized.

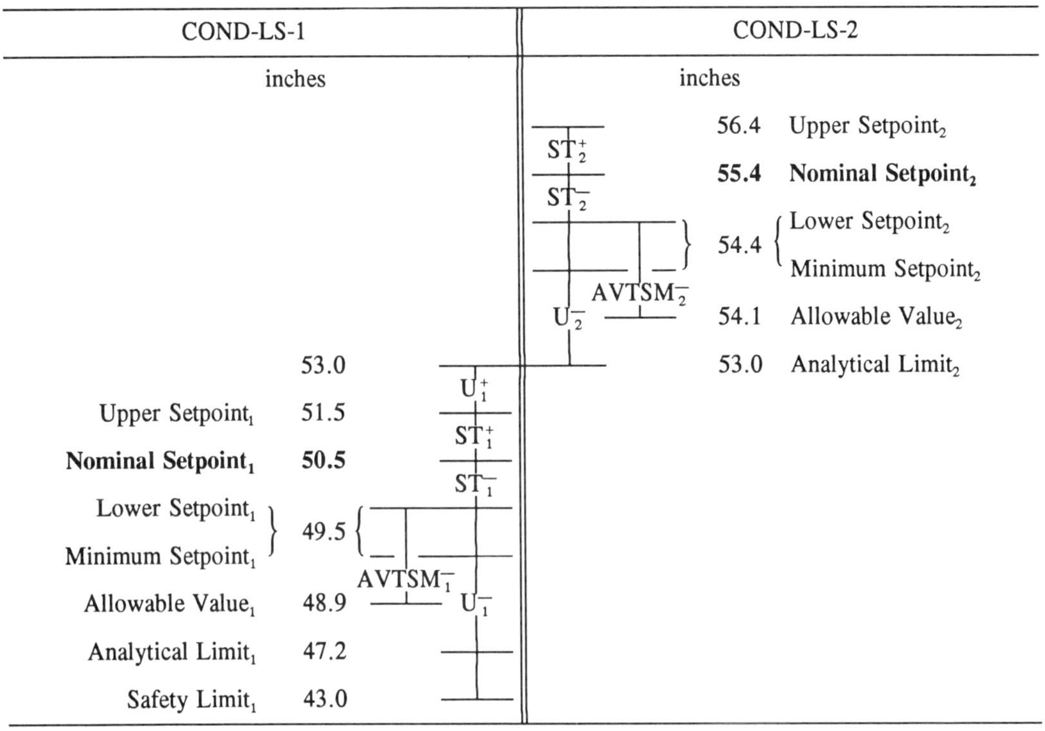

Figure 3
Limits and Setpoints for COND-LS-1 and COND-LS-2

This method applies any time there exists the potential for two NSp to cross over one another due to the setting tolerances and uncertainties.

3.3 EXAMPLE 3 - SETPOINTS WITH TWO SIDED ANALYTICAL LIMITS

Pressure Switch HPCS-PS-1 is designed to protect HPCS Pump HPCS-P-1 from cavitating if the net positive suction head available (NPSHA) falls below the net positive suction head required (NPSHR). This instrument loop does not function to protect a safety system setting. The instrument loop's function is simply to protect the pump from self destructing. Therefore, the NSp is a non-limiting safety system setting (NLSSS). In accordance with section 8.2 of [2], there is no SL for a NLSSS. The NSp for a NLSSS is determined by adding or subtracting instrument uncertainties (and the ST) from the process design limit (DL). Therefore, the DL and the AL are treated as equal within the context of the setpoint calculation. The DL is established based on plant operating experience, vendor data, or original design criteria.

The DL for HPCS-PS-1 is two sided. The lower DL is the maximum NPSHR for the HPCS pump. The upper DL is defined as the minimum NPSHA. If the switch trips below the lower DL, the pump may cavitate before the switch trips and the pump shuts down. If the switch trips above the upper DL, the switch may trip during normal operation. The vendor's pump curve, with the NPSHA and NPSHR curves superimposed, is shown in figure 4. From this curve, the lower DL was determined to be 32.3 feet (14.0 psia). The NPSHA calculation for the HPCS pump concluded that the minimum NPSHA occurs when the HPCS pump draws from the suppression pool. Based on this calculation, the upper DL was determined to be 53.5 feet (23.2 psia).

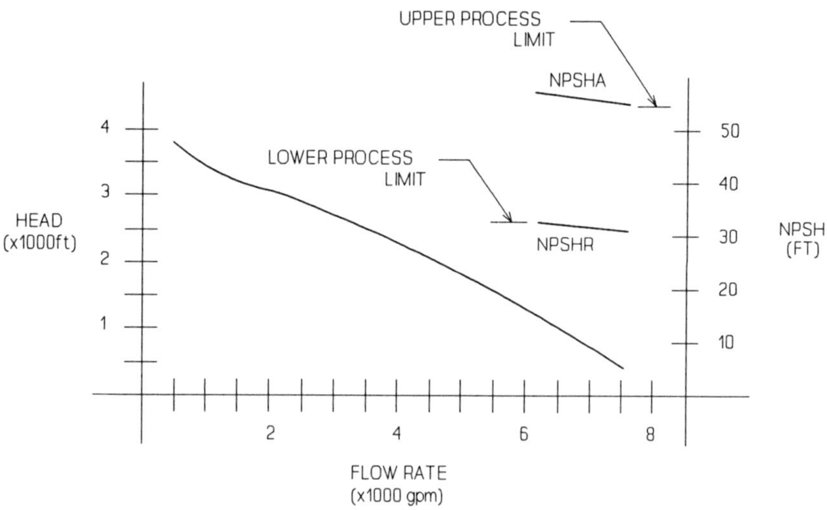

Figure 4
Pump Curve for HPCS-P-1

The NSp should be set as close to the lower DL as possible to minimize the potential for spurious trips. Using the methods described in example 1, the following factors were determined:

ST_3^+ = 1.5 psi
U_3^+ = 2.3 psi
$AVTSM_3^+$ = 0.9 psi

ST_3^- = 1.5 psi
U_3^- = 2.6 psi
$AVTSM_3^-$ = 1.2 psi

These factors were combined as in example 1 and the NSp and AV for HPCS-PS-1 were determined to be 18.1 psia and 15.4 psia respectively. This is shown in figure 5.

Now a check must be made to assure that the upper DL is not exceeded when U_3^+ and the ST_3^+ are added to the NSp calculated above. So long as this value is less than the upper DL, the NSp calculated above is acceptable. Determining acceptability of the NSp can also be expressed in terms of a maximum setpoint which is the upper DL minus U_3^+. So long as the upper setpoint (NSp + ST_3^+) is less than the maximum setpoint, the NSp calculated

HPCS-PS-3	
	psia
Upper Setpoint	19.6
Nominal Setpoint	**18.1**
Lower Setpoint / Minimum Setpoint	16.6
Allowable Value	15.4
Lower Design Limit	14.0

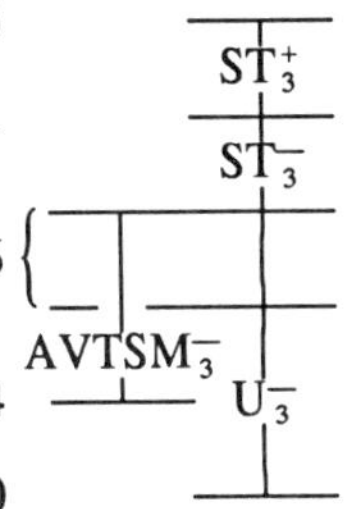

Figure 5
Lower Limits and Setpoints for HPCS-PS-3

above is acceptable. If the upper setpoint is greater than the maximum setpoint, then the uncertainties and/or the setting tolerances associated with the pressure switch are too high, and the pressure switch has the potential to trip during normal operation. In this case, either the ST must be tightened and/or the switch must be replaced with a pressure switch which has lower uncertainties.

For HPCS-PS-1, the upper setpoint (NSp + ST_3^+) equals 19.6 psia, which is less than the maximum setpoint, 20.9 psia (upper DL - U_3^+). Therefore, the calculated NSp is acceptable. The NSp relative to the upper and lower DL and the upper and lower AV is shown in figure 6.

The difference between the upper setpoint (19.6 psia) and the maximum setpoint (20.9 psia), is the margin. The setpoint for HPCS-PS-1 can be set anywhere between the calculated NSp (18.1 psia) and the calculated NSp plus this margin (19.4 psia). As stated above, the NSp for this application should be set as close to the lower DL as possible. Therefore, the calculated NSp is the desired setting.

This method for determining the NSp applies any time an instrument setpoint is bounded by a two sided SL/AL and/or DL. Often, one limit might be the safety/analytical limit and the other limit might be the normal process operating conditions. A case such as this is presented in example 4.

3.4 Example 4 Containment Instrument Air Emergency Supply Control

The Containment Instrument Air (CIA) System is designed to provide an uninterruptible supply of compressed gas to several safety related users within containment. One of these users is the Automatic Depressurization System (ADS) Main Steam Relief Valve (MSRV) accumulators.

The primary source of gas for the CIA system is from the Containment Nitrogen (CN) System. A backup source for the portion of the CIA System serving the ADS MSRV accumulators is provided should the pressure in the primary supply drop below a preset level. This backup consists of a rack of high pressure storage bottles. Switching from the normal supply to the backup supply is controlled by a two out of three logic. The system is shown in figure 7.

HPCS-PS-3	
	psia
Upper Design Limit	23.2
Maximum Setpoint	20.9
Upper Allowable Value	20.5
Upper Setpoint	19.6
Nominal Setpoint	**18.1**
Lower Setpoint / Minimum Setpoint	16.6
Lower Allowable Value	15.4
Lower Design Limit	14.0

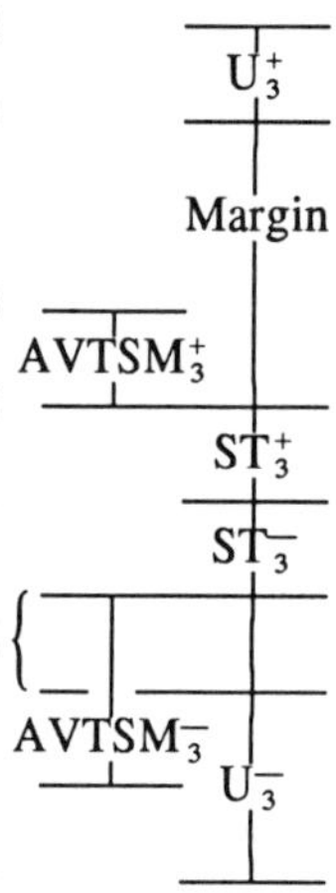

Figure 6
Limits and Setpoints for HPCS-PS-3

The two out of three logic elements are:

1. Valve CIA-V-1 is closed, isolating the ADS supply header from the remainder of the CIA system. This valve closes when CIA system pressure falls below the setpoint for pressure switch CIA-PS-4 for more than a specified time. CIA-PS-4 is located up stream of valve CIA-V-1 and senses the CIA System pressure.

2. Pressure in the ADS supply header is less than the CIA-PS-6 setpoint. This pressure switch is located down stream of valve CIA-V-1, and therefore senses the pressure in the isolated portion of the header. The signal reaches the control logic controller after a time delay. Pressure switches CIA-PS-4 and CIA-PS-6 have the same setpoint.

3. Pressure in the ADS supply header, as sensed by CIA-PS-5, is slightly less than the CIA-PS-6 setpoint. Pressure switch CIA-PS-5 is off the same root valve as CIA-PS-6 and acts as a backup to CIA-PS-5. CIA-PS-5 also activates a low pressure alarm.

The time delays are designed such that the backup system will not initiate due to a pressure transient.

If the pressure in the CIA System header supplying the ADS accumulators falls below the setpoint for CIA-PS-4, valve CIA-V-1 will close, and CIA-PS-6 will also trip. The control logic will then open the first emergency supply high pressure bottle. The pressure in the isolated portion of the CIA header will return to the normal operating pressure, 150 psig, and CIA-PS-6 will reset. So long as the pressure in the remainder of the system stays below CIA-PS-4 setpoint, valve CIA-V-1 will remain closed. If pressure in the isolated portion of the CIA header falls below the CIA-PS-6 setpoint, indicating the bottle is depleted, the next emergency bottle will be open. When the pressure falls to the CIA-PS-5 setpoint, an alarm will be generated in the control room alerting the operators that the emergency supply has been depleted.

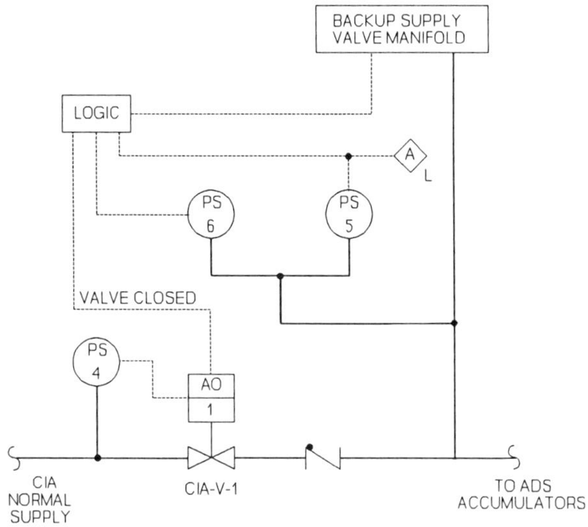

Figure 7
System Sketch
Containment Instrument Air

This example demonstrates the method for determining the setpoints for CIA-PS-5 and CIA-PS-6. The setpoints for CIA-PS-4 and CIA-PS-5 could be calculated using the same method. The CIA-PS-5 and CIA-PS-6 setpoints are cascading like those discussed in example 2. The difference here is that the CIA-PS-5 and CIA-PS-6 setpoints are bounded by a two sided analytical (design) limit (as in example 3), and CIA-PS-6 must reset before the upper AL (DL) reached.

The lower safety limit for CIA-PS-5 and CIA-PS-6 is based on the minimum pressure required to assure proper ADS MSRV operation. Plant documentation defines the minimum pressure to assure ADS actuation as 123 psig. Allowing for the system time delays, the lower analytical limit was calculated to be 125.5 psig. The low pressure alarm, CIA-PS-5, must be set as close to this lower AL as possible to assure that the emergency supply is maximized. Using the basic principals defined in example 1, and the values listed below, the NSp for CIA-PS-5 was calculated. This setpoint calculation is shown in figure 8.

ST_5^+ = 1.0 psi	ST_5^- = 1.0 psi
U_5^+ = 3.5 psi	U_5^- = 2.3 psi
$AVTSM_5^+$ = 1.6 psi	$AVTSM_5^-$ = 1.1 psi

The NSp for CIA-PS-6 cascades off the low pressure alarm, CIA-PS-5, setpoint. Therefore, the lower AL for CIA-PS-6 is the CIA-PS-5 upper setpoint plus the uncertainties and setting tolerance (U_5^+ and ST_5^+) for CIA-PS-5. Using the values shown below, and the method demonstrated in example 2, the NSp for CIA-PS-6 can be calculated. The results of this calculation are shown in figure 9.

ST_6^+ = 1.0 psi	ST_6^- = 1.0 psi
U_6^+ = 4.2 psi	U_6^- = 3.0 psi
$AVTSM_6^+$ = 1.9 psi	$AVTSM_6^-$ = 1.3 psi

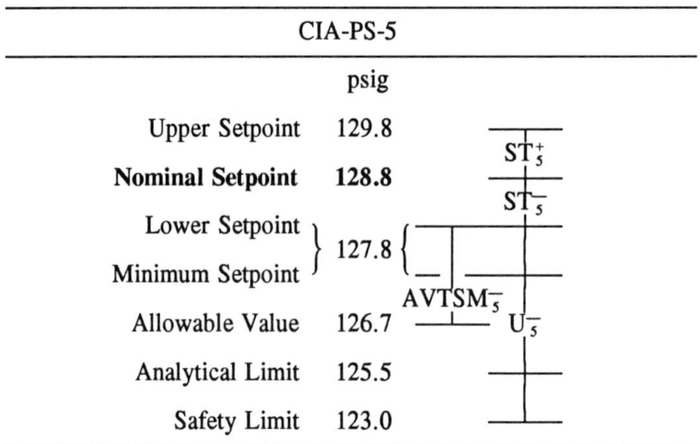

Figure 8
Limits and Setpoints for CIA-PS-5

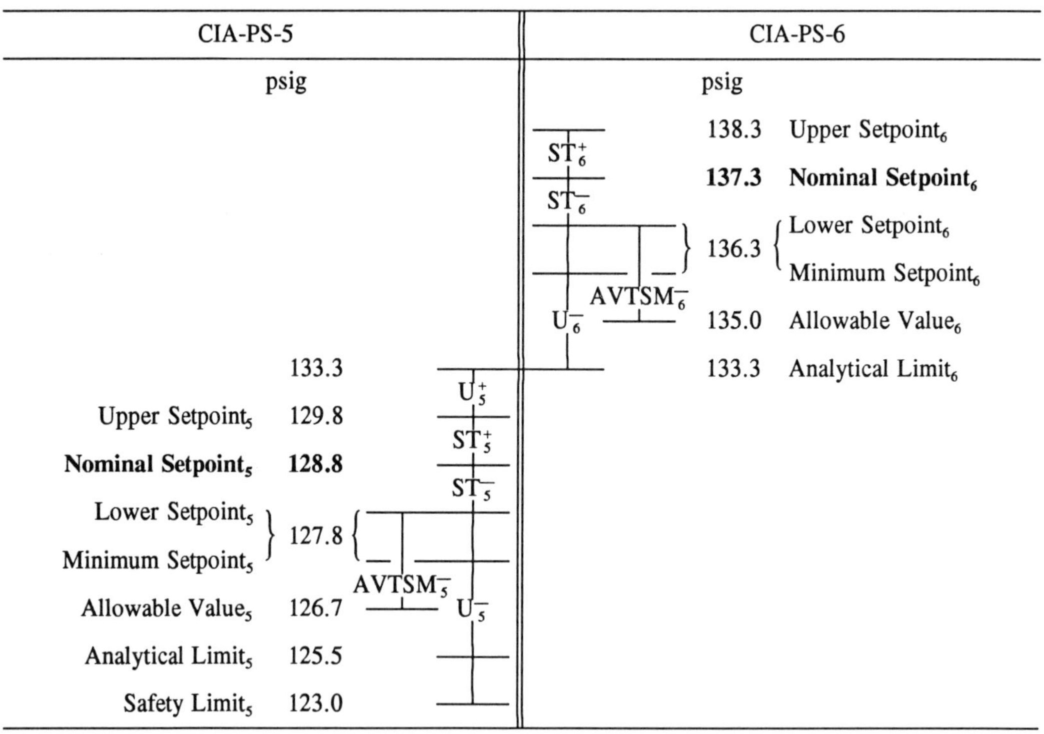

Figure 9
Limits and Setpoints for CIA-PS-5 and CIA-PS-6

In addition to the lower AL defined above, CIA-PS-6 is also bounded by an upper design limit. This DL is the normal system operating pressure, 150 psig. For the emergency backup to function correctly (i.e. reset to the normal supply of compressed gas from the CN System), pressure switch CIA-PS-6 must reset before the system pressure reaches the upper DL. To verify that the upper DL will not be exceeded, the instrument dead band and the uncertainties of the reset must be considered.

The dead band is defined in section 2.0. For CIA-PS-6 the dead band (DB_6) is fixed at 5.0 psi. Although not specifically addressed in [2], there are uncertainties associated with the dead band. These uncertainties include such things as accuracy, temperature, pressure, humidity effects, etc. If the dead band was adjustable, a setting tolerance for the dead band would also need to be considered. For pressure switch CIA-PS-6, the uncertainties for the reset (R) are: U^+_{6R} = 0.5 psi, U^-_{6R} = 0.8 psi. For this switch, there is no calibration (fixed dead band) and no drift, therefore AVTSM is zero and there is no need to include an allowable value for the reset.

The upper reset for CIA-PS-6 was calculated by adding the dead band (DB_6) and the reset uncertainty (U^-_{6R}) to the calculated upper setpoint for CIA-PS-6. Based on the values above and the upper setpoint shown in figure 9, the reset was found to be 148.3 psig. The maximum reset is the DL (150.0 psig) minus the reset positive uncertainties (U^+_{6R} = 0.5 psi). So long as the reset is less than the maximum reset, switch operation is acceptable. This is shown in figure 10.

The last item requiring attention is the reset for the low pressure alarm, CIA-PS-5. The purpose of the alarm is to alert the operator that the pressure in the CIA System header serving the ADS MSRV accumulators has dropped below normal operating pressure, and the emergency supply is either depleted or is not available. So long as the header pressure is less than the pressure at which the emergency supply should be utilized (i.e. the CIA-PS-5 setpoint), the alarm should remain energized. Therefore, the minimum pressure at which the switch should reset is equal to the analytical limit for switch CIA-PS-6.

The calculation to determine the reset point for this alarm is similar to the reset calculation performed for CIA-PS-6 except the CIA-PS-5 dead band is adjustable. The dead band is adjustable and therefore, will have a setting tolerance. The uncertainties and setting tolerances are as below.

ST^+_{5R}	= 0.5 psi	ST^-_{5R}	= 0.5 psi
U^+_{5R}	= 0.4 psi	U^-_{5R}	= 0.5 psi

The dead band must be adjusted to insure that the switch does not reset before the analytical limit of CIA-PS-6. The width can be determined by considering that the instrument's reset will vary with the setpoint. For example, if the deadband is adjusted to an arbitrary 6 psi then the reset will be 6 psi above the point where the instrument trips (plus or minus the uncertainties of the reset itself). Therefore, if the instrument trips just above the lower analytical limit (125.5 psig) then the reset will be 6 psi above that (131.5 psig). In this case, the reset would be below the analytical limit of CIA-PS-6. Therefore, the dead band should be set to be equal to the range from the analytical limit of CIA-PS-5 to the analytical limit of CIA-PS-6. Note that this is the range over which CIA-PS-5 has the specified probability of tripping ($U^-_5 + ST^-_5 + ST^+_5 + U^+_5$; probability here is same as confidence level for the instrument uncertainties). The dead band therefore, should be set at DB_5 = 133.3 - 125.5 = 7.8 psi.

Starting at the CIA-PS-6 analytical limit, the nominal reset was calculated by adding the width of the dead band (DB_5), the negative reset uncertainties (U^-_{5R}), and the negative setting tolerance (ST^-_{5R}). This gives a nominal reset of 142.1 psig. This is shown, along with the other values described above, in figure 11. This figure shows the relationships between the different analytical limits (design limits), allowable values, setpoints and reset points for these two switches.

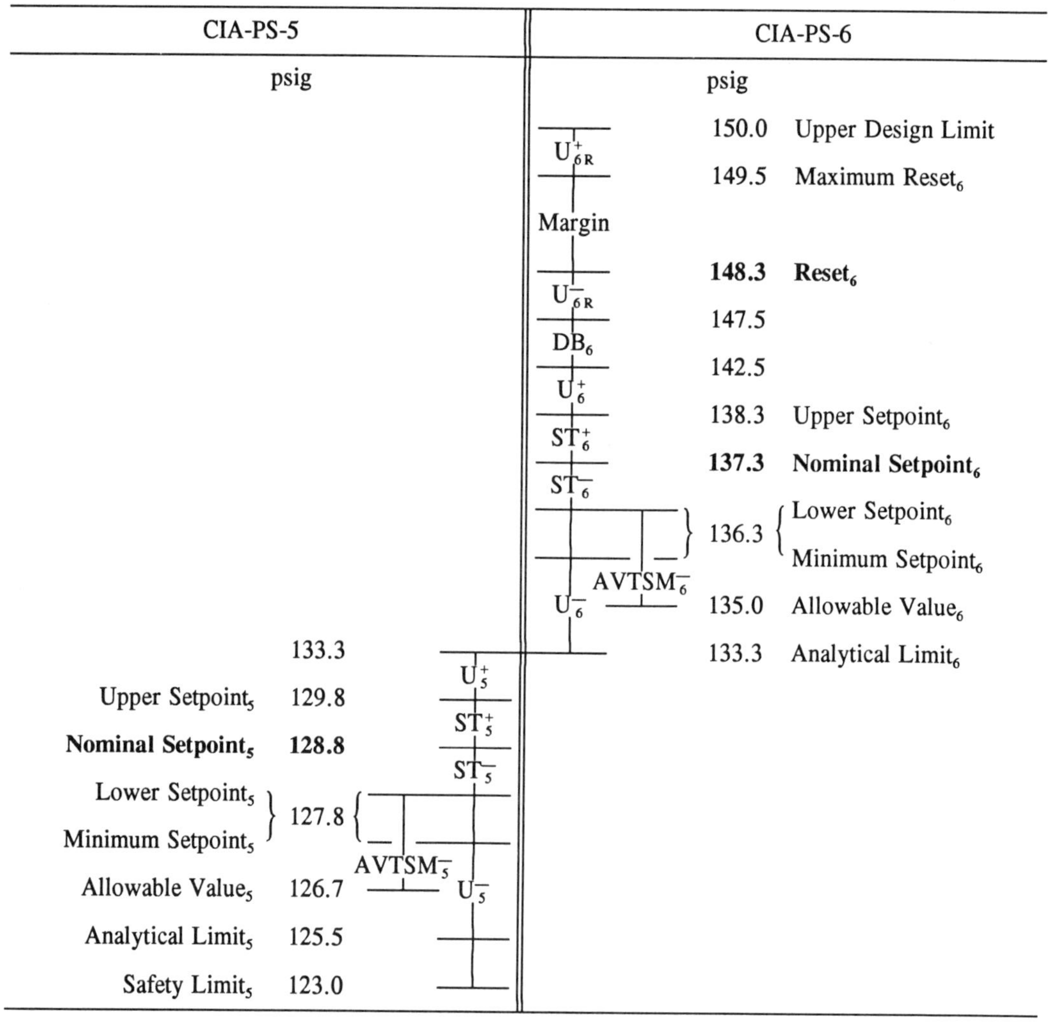

Figure 10
Limits and Setpoints for CIA-PS-5 and CIA-PS-6 and Reset for CIA-PS-6

4.0 CONCLUSION

The preceding examples demonstrate methods developed at WNP-2 for combining the limiting conditions (safety/analytical limits or design limits), the instrument loop uncertainties developed using draft Recommended Practice 67.04 and the required setting tolerances to calculate a setpoint. The methods ensure that sufficient margin exist between the limiting condition and the calculated setpoint (NSp) such that when all uncertainties for the environment in which the instrument is installed are considered, the instrument will always (within the confidence level of the uncertainties) trip before the limiting condition is exceeded. The methods also maximize

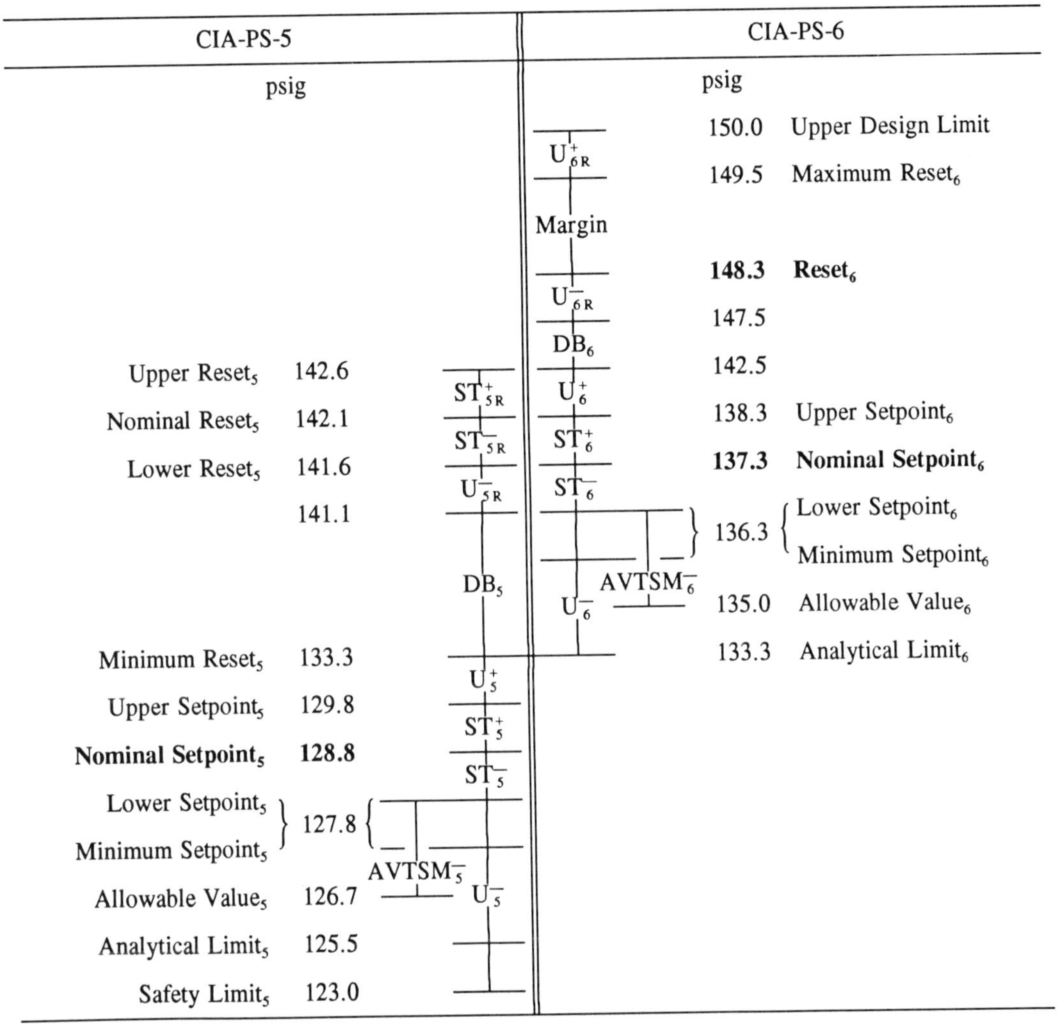

Figure 11
Limits, Setpoints, and Resets for CIA-PS-5 and CIA-PS-6

the margin between the normal system operating conditions and the NSp, thereby providing an additional operating flexibility.

As stated in the introduction, dRP 67.04 does not adequately address applying instrument uncertainties and setting tolerances to setpoint calculations. Nor does it address two sided limiting conditions or cascading instrument setpoints. Once endorsed by the NRC in regulatory guide 1.105, RP67.04 will be the standard for performing setpoint calculations throughout the industry. Therefore, the authors believe that the methods developed here, or similar methods, should be thoroughly covered in RP67.04.

5.0 REFERENCES

[1] ANSI/ISA-S67.04-1988, Setpoints for Nuclear Safety-Related Instrumentation

[2] ISA-dRP67.04-draft 6, Methodologies for the Determination of Setpoints for Nuclear Safety-Related Instrumentation

[3] ANSI/ISA-S51.1-1979, Process Instrumentation Terminology

[4] 10 CFR 50.36, 1-1-90 Edition, Technical Specifications

0074-056X/91/167-187/$0 + .50pp

RP67.04-PART II WHAT THE RECOMMENDED PRACTICE FOR SETPOINT DETERMINATION METHODOLOGIES DOES OR DOES NOT DO

B. E. Beuchel
Chairman SP67.15
&
I&C Engineering Supervisor
Seabrook Station
Seabrook, New Hampshire

M. C. Annon
Secretary SP67.15
&
Owner
I&C Engineering Associates
Waterford, Connecticut

ABSTRACT

The long-awaited Recommended Practice (RP) to complement ISA S67.04-1987 "Setpoints for Nuclear Safety-Related Instrumentation" is very close to publication. The RP, entitled "Methodologies for the Determination of Setpoints for Nuclear Safety-Related Instrumentation," has been prepared through the efforts of over 50 persons representing over 20 utilities, all four LWR NSSS vendors, three instrument manufacturers, several architectural engineering firms, numerous consulting firms and other individuals.

The RP's goals are to aid in promoting consistency of technique and terminology when preparing safety-related setpoint uncertainty calculations and to provide a base of information from which specific setpoint methodology programs can be established. These goals are achieved by describing acceptable methods for determining instrument channel uncertainty and establishing setpoints based on instrument channel uncertainty to supplement the requirements of S67.04.

Throughout the period during which the RP was being prepared, it became apparent to the RP Committee members that the scope and approaches outlined in the RP were not fully understood by the general nuclear community. Therefore, the goals of this paper are to enhance the understanding of the RP and facilitate the utilization of the RP by potential users. The primary objective of the paper is to provide a broad overview of the purpose and scope of the RP. This will be achieved by describing the general organization of the RP as well as a brief explanation of the contents of each section. The paper will also discuss the role of the RP in the development of actual setpoint calculations, i.e., an explanation of the need for a user's specific methodology document to "customize" the RP for a specific user. In addition, the paper will identify, and attempt to clarify, several potential misconceptions by users, as to what the RP does not do.

Key words: Setpoint Calculations, Instrument Setpoints, Uncertainty Determination, Error Identification, Standards

BACKGROUND

The basic underlying requirements for instrument setpoint programs are derived from federal law.

Criterion 13, "Instrumentation and Control," of Appendix A, "General Design Criteria for Nuclear Power Plants," to 10 CFR Part 50, "Domestic Licensing of Production and Utilization Facilities," requires, among other things, that instrumentation be provided to monitor variables and systems and that controls be provided to maintain these variables and systems within prescribed operating ranges.

Criterion 20, "Protection System Functions," of Appendix A to 10 CFR Part 50 requires, among other things, that the protection system be designed to initiate operation of appropriate systems to ensure that specified acceptable fuel design limits are not exceeded.

Paragraph (c)(1)(ii)(A) of 50.36, "Technical Specifications," of 10 CFR Part 50 requires that, where a limiting safety system setting is specified for a variable on which a safety limit has been placed, the setting be so chosen that automatic protective action will correct the most severe abnormal situation anticipated without exceeding a safety limit.

In the past, if a safety related channel was found to exceed its technical specification setpoint it was an event reportable to the NRC. The most common cause for setpoints being exceeded was that setpoints did not account for instrument drift.

The Instrument Society of America sponsored a review of the setpoint drift problem in April 1975 by establishing the SP67.4 committee (now renumbered as SP67.04).

The committee's review indicated that a more thorough consideration of setpoint drift was necessary in the design, test, purchase, installation, and maintenance of nuclear safety-related instrumentation.

Revision 1 to Regulatory Guide 1.105, "Instrument Setpoints," was issued in November 1976 in response to the large number of instances reported in Licensee Event Reports (LERs) of setpoints drifting outside the limits specified in the technical specifications. Revision 1 provided general guidance for (1) specifying setpoints (by considering instrument drift, accuracy, and range) and (2) having a securing device for the setpoint adjustment mechanism.

The method described in Revision 1 to Regulatory Guide 1.105 was incorporated into an Instrument Society of America Standard, ISA-S67.04-1982, "Setpoints for Nuclear Safety-Related Instrumentation Used Nuclear Power Plants." Revision 2 to Regulatory Guide 1.105 was developed to use the guidance of ISA-S67.04-1982 and provides more specific guidance on establishing and maintaining setpoints in response to the needs that were apparent from (1) a continuing large number of reportable occurrences and (2) the licensing review of methodology for specifying allowable values and trip setpoints.

In 1987, a revision to S67.04 was issued for the purpose of clarification and to reflect current industry practice. The term "Trip Setpoint" is consistent with the terminology used in the NRC Standard Technical Specifications and reflects what previously was known as "Upper Setpoint Limit."

In the preparation of ISA-S67.04-1987, it was determined that a need existed to provide additional guidance with regard to methods for implementing the requirements of the standard. In order to address this need, Standard Committee SP67.15 was formed in 1988 and has prepared a Recommended Practice (RP). The RP, entitled "Methodologies for the Determination of Setpoints for Nuclear Safety-Related Instrumentation," has been prepared through the efforts of over 50 persons representing over 20 utilities, all four LWR NSSS vendors, three instrument manufacturers, several architectural engineering firms and numerous consulting firms and other individuals.

It is the intent of the SP67.15 committee that the RP scope be consistent with the scope of S67.04. The RP is to be utilized in conjunction with S67.04, therefore the RP carries the same number, 67.04. The standard is Part I and the RP is Part II. The standard, S67.04, will be identified as Part I on the next printing of the standard.

The RP presents guidelines and examples of methods for the implementation of ISA-S67.04 in order to facilitate the performance of instrument uncertainty calculations and setpoint determination for safety-related instrument setpoints in nuclear power plants.

The RP provides guidance for the implementation of ISA-S67.04 in the following areas:

- Methodologies including sample equations to calculate total channel uncertainty
- Common assumptions and practices in instrument uncertainty calculations
- Equations for estimating uncertainties for commonly used analog and digital modules
- Methods to determine the impact of commonly encountered effects on instrument uncertainty

- Application of instrument channel uncertainty in setpoint determination
- Sources and interpretation of data for uncertainty calculations
- Discussion of the interfaces between setpoint determination and plant operating and calibration procedures, and accident analysis
- Documentation requirements

OVERVIEW

Before utilizing the RP, it is important that the user understand the relevance of instrument channel uncertainty and safety-related setpoint determination for nuclear power plants.

Safety-related instrument setpoints are chosen such that potentially unsafe or damaging process excursions (transients) can be avoided and/or terminated prior to exceeding safety limits (process design limits). The selection of a setpoint requires that consideration be given to much more than just instrumentation.

Experience has shown that an operational limit should be placed on critical process parameters to ensure that, given the most severe operating or accident transient, the safety limits will not be exceeded. Performance of an accident analysis establishes the analytical limits for critical process parameters. Typically, the accident analysis model will include the thermodynamic, hydraulic, and mechanical response of the processes, as well as assumptions regarding the time response of the instrument loop. The analytical limits, as established by an accident analysis, do not normally include considerations for the accuracy (uncertainty) of installed instrumentation.

Instrument channel uncertainty should be determined based on the characteristics of installed instrumentation, the environmental conditions present at the plant locations associated with the instrumentation, as well as process conditions. A properly calculated setpoint will initiate a plant protective action before the process parameter exceeds its analytical limit, which, in turn, ensures that the transient will be avoided and/or terminated before the process parameter exceeds the established safety limit.

The RP does not contain guidance on how to perform this accident analysis nor how to establish the analytical limit. However, the RP does describe acceptable methods for determining instrument channel uncertainty and establishing setpoints based on instrument channel uncertainty to supplement the requirements of S67.04, Part I.

Appendix A provides an outline of the RP's Table of Contents. The Section/Appendix Numbers used in the RP will be referenced throughout this paper. (Note: Any information extracted from the RP, for use within this paper, is based upon resolution of comments on RP Draft 8, as the material will appear in Draft 9.)

The recommended practice is primarily focused on calculating a setpoint for a single safety-related channel, using acceptable statistical methods. These methods represent the most common practices at this time. There are a number of different approaches that are being used in establishing nuclear safety related setpoints. The recommended practice is intended to identify areas that should be evaluated when one does a setpoint calculation, to provide some examples of present thinking in the area of setpoint calculations and to provide some recommendations for a setpoint methodology. The methods discussed in the recommended practice are not intended to be all inclusive.

The recommended practice is structured to mimic the process one would follow to determine an instrument channel setpoint. Sections 5 and 6 of the RP describe a method for calculating channel uncertainties as part of the process of calculating a setpoint. Section 7 of the RP presents methods to determine the instrument channel Allowable Value and Trip Setpoint. Section 8 of the RP discusses selected subjects related to determining setpoints for nuclear plant instrumentation. Section 9 of the RP discusses interfaces. Section 10 of the RP discusses documentation considerations. The appendices provide more in-depth discussions concerning the theory and background for the information presented in Sections 5 through 7 of the RP, as well as extensive and comprehensive examples of setpoint calculations and discussion of unique topics in setpoint determination.

RP CONTENTS

Section 5

This section guides the reader through the basics of the channel layout, functions provided, and sources of uncertainty that may be present. Section 5 also provides references to the appropriate RP section(s) for detailed discussions of particular topics of interest. Figure 1 (RP Figure 5-1) outlines a typical instrument channel layout; while Figure 2 (RP Figure 5-2) outlines a flowchart of the setpoint calculation process.

Section 6

This section is the "heart" of the RP in that the section develops the general uncertainty equation (see Appendix B); classifies the various types of uncertainty (see Figure 3, RP Figure 6-1); outlines the common types of data required to be addressed in the calculation; develops a general equation for calculating a specific module uncertainty (see Appendix B) and calculating total channel uncertainty (see Appendix B).

Section 7

This section outlines the relationships that exist between the various Technical Specifications and S67.04 terms; how to determine a specific trip setpoint; and various ways, currently in use, to determine an allowable value. Figure 4 (RP Figures 7-1 and 7-2) summarizes these concepts.

Section 8

This section discusses selected subjects that may occasionally arise in developing a setpoint/uncertainty calculation. Currently, it addresses only two topics; correction for setpoints with a Single Side of Interest, Non-limiting Safety System Settings (Non-LSSS).

Section 9

This section addresses some of the organizational interfaces that need to be considered in developing a setpoint/ uncertainty calculation. The primary organizations addressed are safety analysis group; design modification group; plant maintenance, instrumentation and operations; licensing and regulatory groups; and vendors.

Section 10

This section outlines the relationship that should exist between the documentation required for the calculation and a facility's quality assurance, design control and configuration management programs. In addition, it provides general guidance as to appropriate documentation to support a setpoint/ uncertainty calculation.

Appendices

The appendices (see Appendix A) include information that is considered directly relevant to the RP but was not included in the main body for one (or more) of the following reasons:

- The volume of the information was considered disruptive to the presentation in the main body
- The information represents detailed development of equations or approaches to technical issues

Sample Calculations

Four sample calculations are provided (in Appendix L) to illustrate selected concepts discussed in the RP and give practical examples of calculations that are consistent with the RP methodology. The samples and the concepts illustrated are outlined in Appendix C.

RP ROLE

The RP is intended for use primarily by the owners/operating companies of nuclear power plant facilities or their agents (nuclear safety steam suppliers, architect/engineers, etc.) in establishing setpoint methodology programs and preparing safety-related instrument setpoint calculations; although the concepts can be used for any setpoint application.

User Application

The RP is intended to serve more as a primary reference document for users rather than a procedure or work instruction that can govern the preparation and approval of a user's specific calculations. For example, it is anticipated that a potential user would need to prepare some sort of administrative guidelines that address, as a minimum, the following:

- Delineation of the RP's applicability for the plant, i.e., Technical Specification parameters, all safety-related setpoints, selected Balance-of-Plant parameters, etc.
- Responsibilities for the preparer, reviewer, verifier (if required), and approver(s) of the calculation
- Format and level of detail to be included in any calculation
- Sources of plant specific data to be used
- Inter-department or Inter-divisional interface requirements
- Plant/System/Application generic or common assumptions
- Definition of what constitutes acceptable documentation of data obtained from communication with manufacturers, vendors, or other utilities
- Plant specific commitments on approaches to certain technical issues, i.e., NSSS unique approaches, Technical Specification commitments, etc.

POTENTIAL MISCONCEPTIONS

"RP is a Cookbook"

A common comment has been, "The RP should be a cookbook" or "I thought that the RP would allow me to just plug-in my values." While on the surface this appears to be a worthwhile goal, the practicality of achieving this goal is not obtainable due to the immense complexity and interactions of the numerous different plant situations and approaches.

"RP has All the Answers"

The RP has not attempted to provide the answers to all of the questions that can arise during the preparation of a specific setpoint calculation. For example, specific guidance is not provided to answer the questions of "What does the user do when actual/specific data are not available?" and "What's enough data?" The RP has focused on outlining issues that may be relevant, as well as, providing a suggested method to properly address these issues, when a consensus could be reached.

"RP Addresses Response Time Testing"

The RP does not address dynamic effects, such as response times on setpoints. If required, the effects are generally included in the analytical limit development and/or specific accident analyses.

"Methods & Formats are Prescriptive"

The method(s) used within the RP to address each issue are not the only way to handle each issue; rather it is a way that has been agreed by the consensus process. It is not intended that the method be used to the exclusion of all others and in fact, in several cases multiple approaches are presented. Each user is responsible for deciding the appropriateness of any approach used.

"The RP is a Rachet for the NRC"

While it is expected that, in the future, the NRC may issue a revision to R.G.1.105 that will address the NRC's position(s) on both ISA-S67.04-1987 and this RP, the RP is not intended to be a tool for regulators. The RP was prepared to help the nuclear industry address some of the issues (related to setpoint calculations) that have been raised by both regulators and preparers of the calculations. Various inspections being conducted by the NRC over the last several years have already raised the issue that the design bases for nuclear power plants need to be better understood including what goes into the determination of a setpoint. This RP is simply a vehicle for the industry's use to enhance the understanding of a complex (and frequently misunderstood) topic, as well as to explain the guidance of S67.04.

CONCLUSION

ISA Recommended Practice 67.04 Part II is a consensus document which will provide additional guidance to the nuclear industry in implementing ISA Standard 67.04. In addition, the RP is a good reference document for use in quantifying the uncertainty associated with an instrument's setpoint. This paper has attempted to provide the background and an overview of the contents of the RP and to introduce it to those that have not participated in its development. We have attempted to define the role that we see for the RP and to address some of the misconceptions associated with it. The RP is not, nor is it intended to be, the definitive word on setpoint calculations or setpoint methodology.

ACKNOWLEDGEMENTS

The concepts outlined in this paper are the results of the efforts by ISA SP67.04 and SP67.15 committees. The authors would like to acknowledge the contributions of all those individuals (and their companies) who have supported the developments of the RP over the last three years. A list of these individuals will be included in the RP.

REFERENCES

(1) ANSI/ISA S67.04-1987, Setpoints for Nuclear Safety-Related Instrumentation

(2) IEEE Standard 603-1980, IEEE Standard Criteria for Safety Systems for Nuclear Power Generating Stations

(3) IEEE Standard 338-1977, IEEE Standard Criteria for the Periodic Testing of Nuclear Power Generating Station Safety Systems

(4) IEEE Standard 498-1980, Standard Requirements for the Calibration and Control of Measuring and Test Equipment Used in the Construction and Maintenance of Nuclear Power Generating Stations

(6) ANSI/ASME PTC 19.1-1985, Measurement Uncertainty, Part I

(7) ANSI N15.15-1974, Assessment of the Assumption of Normality (Employing Individual Observed Values

(8) SAMA: PMC 20.1-1973, Process Measurement and Control Terminology

(9) U.S. Nuclear Regulatory Commission, Regulatory Guide 1.105, Instrument Setpoints For Safety Related Systems, Revision 2, February 1986.

(10) Zalkind, C. S., and F. G. Shinskey (The Foxboro Company), The Application of Statistical Methods in Evaluating the Accuracy of Analog Instruments and Systems.

NOMENCLATURE

The following terms as defined in the RP are used throughout the RP and this paper:

Accuracy Reference. In process instrumentation, a number or quantity that defines a limit that error will not exceed when a device is used under specified operating conditions. Error represents the difference between the measured value and the standard or ideal value.

Allowable Value. The limiting value that the trip setpoint can have when tested periodically, beyond which the instrument channel is evaluated for operability.

Analytical Limit. Limit of a measured or calculated variable established by the safety analysis to ensure that a safety limit is not exceeded.

Arbitrarily Distributed. This term is used in this Recommended Practice to denote uncertainties which do not have a normal distribution. See section 6.2.1.2.2 for further information.

Bias. An uncertainty component which consistently has the same algebraic sign, and is expressed as an estimated limit of error.

Dependent Uncertainty. Uncertainty components are dependent on each other if they possess a significant correlation, for whatever cause, known or unknown. Typically, dependencies form when effects share a common cause.

Drift. An undesired change in output over a period of time, which change is unrelated to the input, environment, or load.

Independent Uncertainty. Uncertainty components are independent of each other if their magnitudes or algebraic signs are not significantly correlated.

Instrument Channel. An arrangement of components and modules as required to generate a single protective action signal when required by a generating station condition. A channel loses its identity where single protective action signals are combined.

Limiting Safety System Setting (LSSS). Limiting Safety System Settings are settings for automatic protective devices related to those variables having significant safety functions.

NOMENCLATURE (Cont'd)

Module. Any assembly of interconnected components which constitutes an identifiable device, instrument, or piece of equipment. A module can be removed as a unit and replaced with a spare. It has definable performance characteristics which permit it to be tested as a unit. A module can be a card, a draw-out circuit breaker, or other subassembly of a larger device, provided it meets the requirements of this definition.

Random. Describing a variable whose value at a particular future instant cannot be predicted exactly, but can only be estimated by a probability distribution function. The algebraic sign of a random uncertainty is equally likely to be positive or negative with respect to some median value. Note: In the context of this document, "random" is an abbreviation for random, approximately normally distributed. Thus random uncertainties are eligible for square root sum of squares combination.

Safety Limit. A limit on an important process variable that is necessary to reasonably protect the integrity of physical barriers that guard against uncontrolled release of radioactivity.

Trip Setpoint. A predetermined value at which a bistable device changes state to indicate that the quantity under surveillance has reached the selected value.

Uncertainty. A value which represents the distribution of possible errors within a probability and confidence level.

APPENDIX A RECOMMENDED PRACTICE TABLE OF CONTENTS

SECTION

APPENDICES

APPENDIX A RECOMMENDED PRACTICE TABLE OF CONTENTS (Cont'd)

APPENDIX B - RP EQUATIONS

Basic Uncertainty Equation

The basic formula for uncertainty calculation takes the form of:

$$Z = \pm \left[(A^2 + B^2 + C^2)^{1/2} + |F|\right] + L - M \qquad \text{(Eq. 6.1)}$$

where: A, B, C = random and independent uncertainty terms. The terms are zero-centered, approximately normally-distributed, and are indicated by a ± sign;

F = arbitrarily-distributed uncertainties and/or biases with unknown direction. The term is used to represent limits of error associated with uncertainties which are not normally distributed and do not have known direction. The magnitude of this term (absolute value) is assumed to contribute to the total uncertainty in a worst-case direction and is also indicated by a ± sign;

L and M = biases with known direction. The terms can impart an uncertainty in a specific direction and, therefore, have a specific + or- contribution to the total uncertainty;

Z = resultant uncertainty. The resultant uncertainty combines the random uncertainty with the positive and negative components of the non-random uncertainty terms separately to give a final uncertainty. The positive and negative non-random terms are <u>not</u> algebraically combined before combination with the random component.

Module Uncertainty Equation

The individual module random uncertainties are in themselves a statistical combination of uncertainties. Depending on the type of module, its location, and the specific factors that can affect its accuracy, the determination of the module uncertainty will vary. For example, the module uncertainty may be calculated as:

$$e^{+} = +[RA^2 + DR^2 + TE^2 + RE^2 + SE^2 + HE^2 + \ldots SP^2 + MTE^2]^{\frac{1}{2}} + B^{+} \qquad \text{(Eq. 6.27a)}$$

$$e^{-} = -[RA^2 + DR^2 + TE^2 + RE^2 + SE^2 + HE^2 + \ldots SP^2 + MTE^2]^{\frac{1}{2}} - B^{-} \qquad \text{(Eq. 6.27b)}$$

APPENDIX B - RP EQUATIONS (Cont'd)

where: e = Uncertainty of module,

RA = Module Reference Accuracy (usually specified by the manufacturer),

DR = Drift of the module over a specific period,

TE = Temperature Effect for the module; the effect of ambient temperature variations on module accuracy; the TE may be a normal operating TE or an accident TE, as required,

RE = Radiation Effect for the module; the effect of radiation exposure on module accuracy; the RE may be normal operating RE, an accident RE, or time of trip RE as required,

SE = Seismic Effect or vibration effect for the module; the effect of seismic or operational vibration on the module accuracy,

HE = Humidity Effect for the module; the effect of changes in ambient humidity on module accuracy, if any,

SP = Static Pressure effects for the module; the effect of changes in process static pressure on module accuracy,

MTE = Maintenance and Test Equipment effect for the module; this accounts for the uncertainties in the equipment utilized for calibration of the module,

B = Biases associated with the module, if any.

Additional terms may have to be included for a particular application. The terms shown are the most common ones encountered for a module.

Total Channel Uncertainty Equation

Using the basic formulas described in Section 6.1, the typical linear channel calculation would take the form:

$$CU^{+} = +[PM^2 + PE^2 + Module_1^{\,2} + Module_2^{\,2} + \ldots Module_n^{\,2}]^{1/2} + B^{+}_{T} \qquad \text{(Eq. 6.26a)}$$

$$CU^{-} = -[PM^2 + PE^2 + Module_1^{\,2} + Module_2^{\,2} + \ldots Module_n^{\,2}]^{1/2} - B^{-}_{T} \qquad \text{(Eq. 6.26b)}$$

APPENDIX B - RP EQUATIONS (Cont'd)

where: CU = Channel Uncertainty (CU) at a specific point in the channel; the CU can be calculated for any point in channel from Module 1 to Module n, as needed.

PM = Random uncertainties that exist in the channel's basic Process Measurement (PM).

PE = Random uncertainties that exist in a channel's Primary Element (PE), if it has one, such as the accuracy of a flowmeter table,

$MODULE_{1,\ 2,\ n}$ = Total random uncertainty of each module that makes up the loop from Module 1 through Module n,

B^{+}_{T} = The total of all positive biases associated with a channel; this would include any uncertainties from PM, PE, or the Modules that could not be combined as a random term (biases, and arbitrarily-distributed uncertainties, discussed in section 6 of the RP),

B^{-}_{T} = The total of all negative biases associated with a channel.

APPENDIX C EXAMPLE UNCERTAINTY/SETPOINT CALCULATIONS

Included in these examples are certain unique aspects that should not be construed as required for all similar calculations. Conversely, the terms deemed applicable for these examples should not be considered to be the only terms applicable to any similar calculation. The preparer of each calculation is responsible for including all terms necessary and appropriate to comply with the guidelines outlined in the RP.

Different calculation formats are used; these formats are used for illustrative purposes only and should not be construed as recommendations of this RP. The format of the error calculations, which sections are included, the sequence and title of the sections, and the content of the sections must be determined by the preparer of the calculation. In addition to these format issues, other specific administrative and technical content/ approach requirements such as the level of justification(s) required for any non-obvious assumptions, acceptable "round-off" techniques, etc., should be governed by the user's quality assurance, design control programs, or other applicable administrative procedures related to preparation of calculations.

The calculations are supplemented by a "COMMENTS" column to the right of the page. These comments are not intended to be part of actual calculations; the comments are intended to provide additional information or appropriate references to the RP and to highlight unique aspects of each calculation.

CALCULATION	SPECIFIC CONCEPTS ILLUSTRATED
Pressure Trip	Applicable to several channels Assumptions to cover missing data One M&TE value for entire channel Check calculation for allowable value
Flow Trip	Non-linear device Plant specific setpoint methodology used as reference Drift data extrapolation Independent and dependent M&TE
Level Trip	Multi-variable (temperature compensated) loop Accident condition errors IR losses Signal Conditioning module uncertainty propagation
Radiation Trip	Non-linear device Logarithmic error analysis Non-LSSS loop

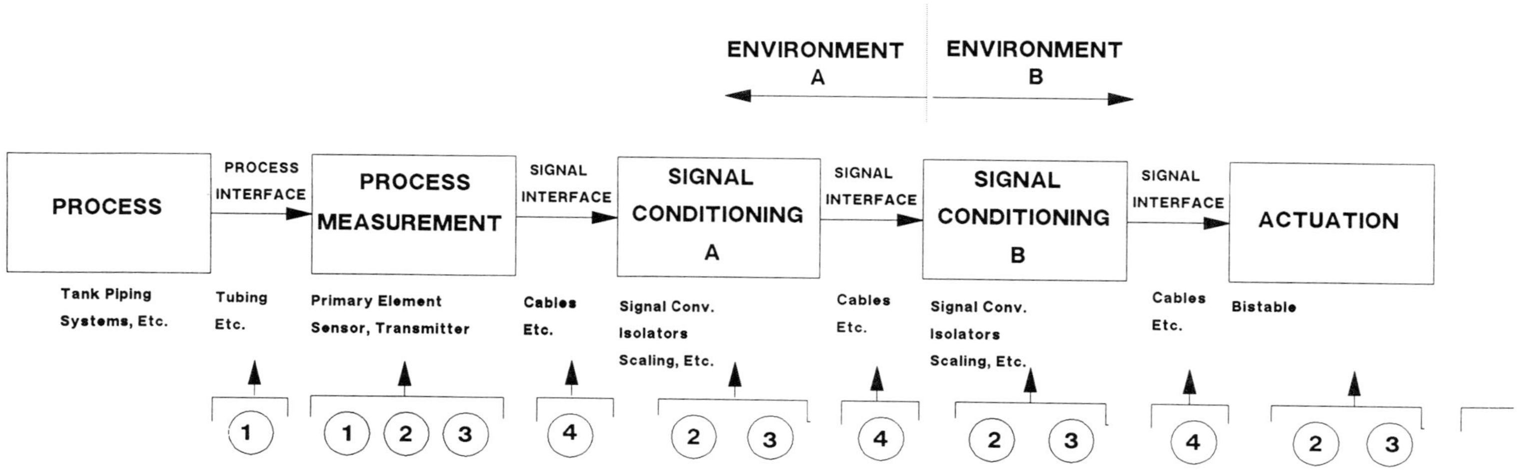

UNCERTAINTY ALLOWANCES

(1) Process Measurement Effects
(2) Instrument Uncertainty
(3) Calibration Uncertainty
(4) Other Uncertainties (IR, Leadwire Effects, Etc.)

FIGURE 5-1: Typical Instrument Channel Layout

FIGURE 1

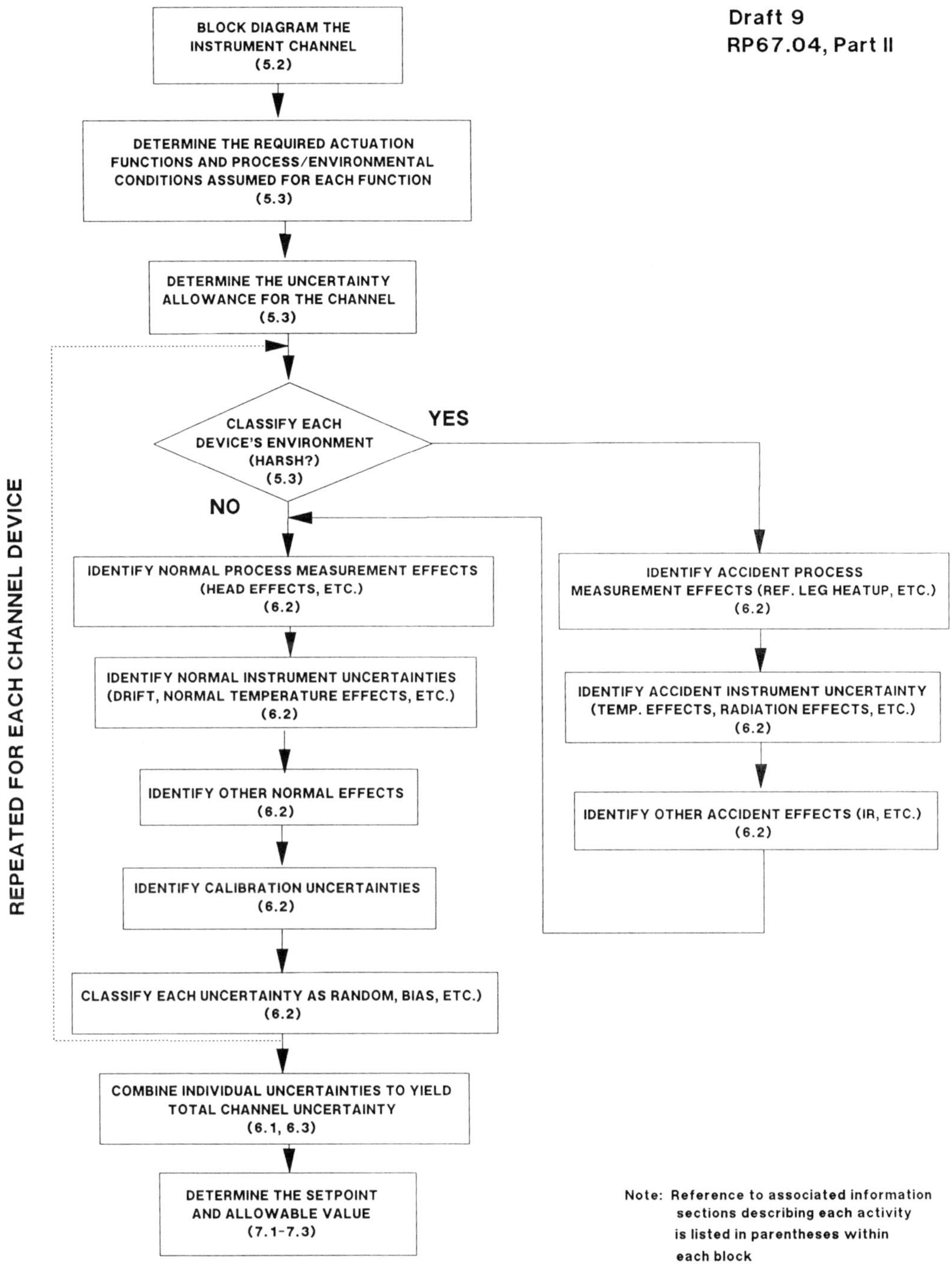

FIGURE 5-2: Setpoint Calculation Flowchart

FIGURE 2

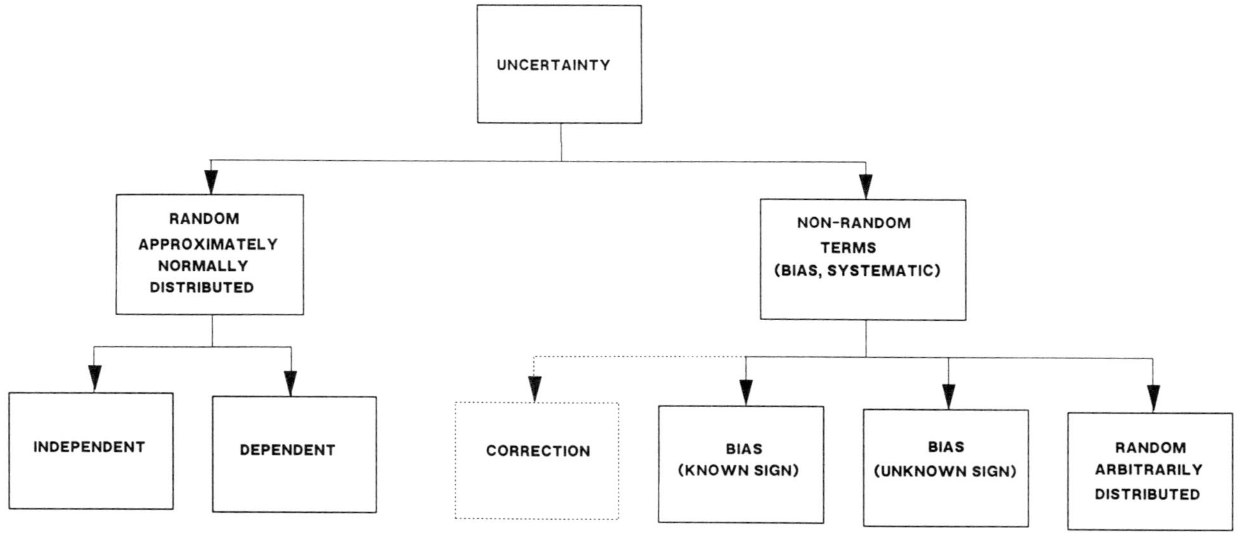

	INDEPENDENT	DEPENDENT	CORRECTION	BIAS (KNOWN SIGN)	BIAS (UNKNOWN SIGN)	RANDOM ARBITRARILY DISTRIBUTED
ATTRIBUTES	Variable Magnitude, Random Sign		Fixed and Known Sign and Magnitude	Variable or Fixed Magnitude and Known Sign	Variable or Fixed Magnitude and Unknown sign	Variable Magnitude, Random Sign
OTHER NAMES	Statistical, Accidental, Precision	Correlated	Offset	Systematic	None	None
COMBINATIONAL RESTRICTIONS	SRSS	SRSS After Linear Partial Summing	Not Used to Calculate Channel Uncertainty	Combine Like Signs Linearly	Absolute Value to Produce a Conservative Result	
QUANTIFICATION	Two Sigma(95%) Probability Limits		Constants	Estimated Limits of Error For Two Sigma(95%) Probability Limits		
EQUATION 6-2 TERMS	+/-A, +/-B, +/-C	+/-D, +/-E	None	+L, -M	+/-F	

FIGURE 6-1: Uncertainty Model

FIGURE 3

Safety Limit

Analytical Limit

Allowable Value

Trip Setpoint

Normal Operating Point

FIGURE 7-1: Setpoint Relationships

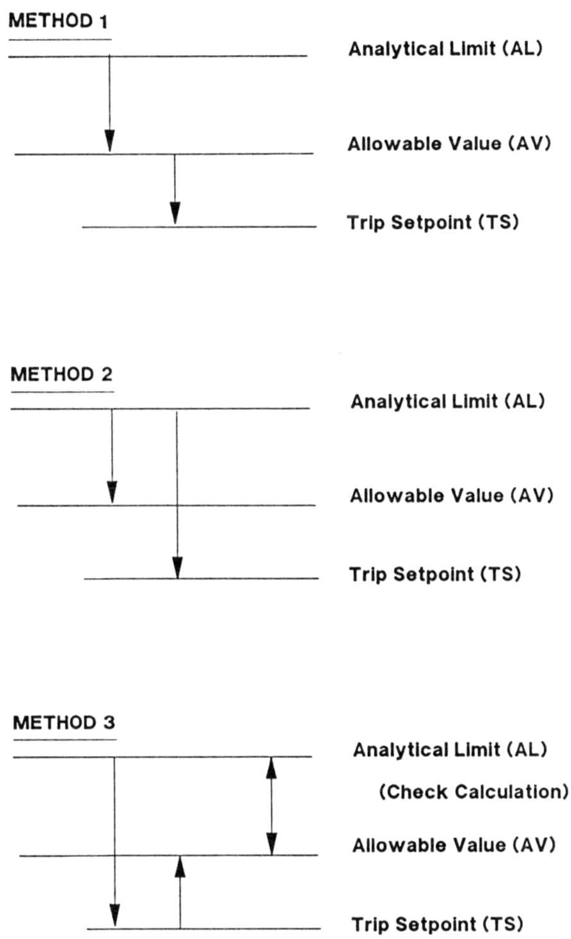

FIGURE 7-2: Methods For Determining The Allowable Value

FIGURE 4

0074-056X/91/189-215/$0 + .50pp

TENNESSEE VALLEY AUTHORITY'S DEVELOPEMENT OF AN ACCURACY CALCULATIONS PROGRAM AND SOFTWARE TO SUPPORT IT

Edward J Bradley
I&C Engineering Specialist
Tennessee Valley Authority
1101 Market St., COC
Chattanooga, Tn. 37402

Ronald A Jarrett
I&C Engineering Specialist
Tennessee Valley Authority
1101 Market St., COC
Chattanooga, Tn. 37402

Richard L Brehm
I&C Engineering Specialist
Tennessee Valley Authority
1101 Market St., COC
Chattanooga, Tn. 37402

Marvin R Belew
Sr I&C Engineering Specialist
Tennessee Valley Authority
1101 Market St., COC
Chattanooga, Tn. 37402

KEYWORDS

Instrument Accuracy Calculations, ISA Standard 67.04

The purpose of this paper is to discuss the calculations program implemented at TVA along with the historical background of how we got to where we are with accuracy calculations. First a brief description of TVA is needed to set the stage. TVA is a multi-site and multi-unit organization with a centralized Corporate Engineering office providing guidance and oversight to the onsite engineering organizations performing implementation. The onsite organizations consist of both TVA direct employees and contractor employees. Because the engineering has been decentralized it was vital to develop the guidance and tools needed to ensure consistent engineering quality. These tools have to aide the production aspect to ensure timely development of calculations and to support maintenance, operations and modifications.

TVA has been generating new calculations and maintaining existing accuracy calculations since the early 1980's. Over the past 10 years, TVA has learned through experience many of the pitfalls of preparing and maintaining calculations. Although the objective of calculations and setpoint verification has maintained constant through out the past decade the art of setpoint calculations has evolved from very simple and rudimentary justifications to detailed qualitative and quantitative analyses. The objective of setpoint calculations is twofold: 1) to prove the application of the instrumentation will not violate analytical safety limits, and 2) to ensure that the setpoints will not adversely impact the operability of the plant by reducing operating margins or increasing spurious trips.

The benefits of a calculations program are that you analytically prove the safe operability of a plant and establish the design basis documentation, but on a secondary level you may be able to remove overly conservative margin assumptions to gain added operating margin, reduced surveillance, and potentially reduced License Event Reports (LER) events due to more realistic Technical Specification limits. Either manufacturer or Technical Specification limits change or setpoint changes may result. The drawbacks of a backfit accuracy calculation program that requires rigorous quantitative analysis is that you may find that inaccuracy parameters that you previously considered insignificant may actually be significant factors thus requiring revising existing plant setpoints, technical specifications, limits, or hardware replacement. Even when a calculation shows that the analytical limits are being challenged there are many solutions to the potential problem. These consist of but are not limited to

- Revise accuracy calculation to remove any conservatism,
- Reevaluation of the analytic limits calculations to remove conservatism (TVA has been successful in moving the limits with the assistance of General Electric (GE) and Westinghouse (W)),
- Equipment changeout,
- Increased surveillance,
- Further device testing to better quantify effects,
- Moving setpoint and/or plant operating point,
- Modification of plant support systems to mitigate environmental extremes,
- Limit the application range of the instruments (ie., a flow loop should be used above 25% flow),
- Calibrate for different conditions,
- Prudent use of statistical analysis and probability amplifiers (combinational logic) to increase total channel confidence levels ...etc.

Accuracy calculations are required to be performed on all safety related parameters at TVA. These include pressure, temperature, flow, radiation monitoring, and time (eg., Diesel Generator load sequencing, line break detection delays, ...etc). Their purpose is to baseline the existing design's adequacy, analyze the effects of plant changes (hardware/environmental conditions), and analyze new equipment purchases prior to approval. Not all instrumentation in a nuclear power plant require a detailed quantitative accuracy calculation. Many of the instrument loop calculations can be performed from conservative generic accuracy parameters with only a few requiring detailed analyses and justifications. The calculations must be prioritized based first on their importance to safety and second on economics, consequences of inaccuracy, and existing margin between operating limits. Instrumentation with wide margins can use the more conservative accuracy analyses which yield wider inaccuracy bands resulting in much faster calculation preparation. Those loops that have narrow margins require a more detailed analysis of accuracy to minimize the predicted inaccuracies. This paper discusses the general accuracy methodology and the methods of refining inaccuracies where needed.

It has become apparent over the past 10 years that accuracy calculations must progress with the environmental qualification programs, pre-op test programs, surveillance programs, and simulator/operator training feedback. In other words, it is best to involve all parties that must utilize the plant instrumentation on the front end of calculations rather than produce the calculations in a vacuum. The plant maintenance personnel will have valuable qualitative and quantitative knowledge about the operability of the instrumentation and the operations/training personnel will have valuable information on the transient behavior of the plant systems. The use of the plant operating experience and maintenance can form the basis of reducing conservatism in accuracy and limits calculations. The usefulness of this information is important in two ways. First it allows the accuracy analysis to show that the plant can operate acceptably and secondly it can potentially be used to move the operating point for the systems based on increased margin.

The basic inaccuracy calculation consists of four categories of information. The loop function, operating and safety limits, the instrumentation application inaccuracy terms (repeatability, drift, span, environmental,... etc), and the process sensing and measurement physics and variations. The first step in preparing calculations is data gathering on process variations, installation and sensing physics, device inaccuracy parameters, calibration methods and equipment. The neededinformation is great, but the sources for the information are also many. Vendors provide many basic parameters such as accuracy and temperature effects and calibration requirements. Environmental qualification tests provide radiation and harsh temperature environmental effects. It should be noted that many of these Environmental Qualification tests are not statistically sound and leave a great deal to be determined by the calculation preparer. Seismic testing provides post and during effects. Field calibration records in conjunction with vendor specifications can provide drift variations as well as confirmation of basic repeatability and normal temperature variations. The Instrument Society of America (ISA 67.04) and United States Nuclear Regulatory Commission (Regulatory Guide 1.105) standards provide guidance on how the data is to be utilized to determine channel inaccuracy. The ISA standard discusses two methods of combining the individual inaccuracy parameters. These are the square root of the sum of the squares (SRSS) of each term and the Algebraic combination. The most commonly used method is a combination of the two methods. For the accuracy terms that are independent of each other and random in magnitude and sign, the SRSS method is used. For the terms that are systematic unidirectional or dependent, the algebraic method is used. It is the responsibility of the calculation preparer to decide which of the calculation methods the data supports. There are many parts to the numerical accuracy calculation which are used for different purposes. The numerical calculations calculate the expected inaccuracies during calibration, normal operation, accidents, and post seismic events. The allowable value, margin of safety, and calibrated spans (based upon normal process conditions, sensing offsets, waterlegs, and elevated static pressure effects) are also calculated.

To minimize the efforts of analyzing the existing data and to maximize the availability of multi-site data, the analysis of specific manufacturer and model number devices are handled by the Corporate Engineering offices. The Corporate Engineering office prepares generic calculations to document the inaccuracies associated with their qualification level of all devices and provides a listing of all associated accuracy parameters that apply. These generic calculations form the basis of the specific application calculations performed by the site organizations. Issuance of the generic calculations by a central organization eliminates the reanalysis by multiple organizations thus reducing the manpower costs. It also ensures consistent analysis of the same devices and a central point for reevaluation and dissemination of new information. In the generation of these generic calculations, the data is evaluated to determine its hardware dependencies and statistical significance. Plant historical data analysis and vendor testing data are analyzed to determine inaccuracy terms. As an example, we have evaluated the test results of drift testing performed by one of the Nuclear vendors on their nuclear qualified transmitters. This information indicates that standard specifications are often very conservative. Appendix A is an example of a typical data analysis performed on environmental qualification tests. It is an application of a multiple linear regression curve fit and statistical analysis of the raw data to determine device repeatability, time related drift, and normal temperature variation effects. The analysis was performed as an attempt to reduce the accuracy terms related to drift and repeatability such that an extended calibration interval could be justified. The results of the analysis indicated that the vendor standard specifications were conservative and that the test results supported a much narrower accuracy band thus allowing an extended calibration interval.

In addition to performing the generic calculations, Corporate Engineering is also responsible for establishing the methodologies to be used by the sites. TVA has standardized on the guidance given in RG 1.105 and ISA std 67.04 as documented in our own internal standards. With this guidance, we evaluate the equipment and determine the dependencies of the inaccuracy parameters. We then use statistical (SRSS) and algebraic (SUMMATION) combination techniques to determine the total inaccuracy of a device. This guidance also includes methods of analyzing the interaction of safety and operating limits and of multiple setpoint interactions such as the switchover of Emergency Gas Treatment System (EGTS) trains. In general, the typical accuracy calculation contains only one setpoint and it's two limits (HI and LO), but there are cases that contain multiple setpoints and control ranges with only two overall limits. One such example is analyzed in Appendix B. The basic item to remember in a multi setpoint analysis is that if the different devices are of the same type and exhibit known bias offsets then these offsets cancel when analyzing the overlap or separation of the setpoints from each other. It should be noted that if the accuracy terms are random and bi-directional then the Square Root of the Sum of the Squares statistical combination of accuracy of both devices can be used to statistically evaluated overlap of setpoints.

Corporate Engineering also provides the methodologies and procedures to analyze process sensing effects, waterleg environmental effects, and thermal time lag to reduce environmental extremes. Guidance is also provided for the use of statistics in the analysis of plant historical calibration data and test results.

The Site organizations utilize the generic calculations by applying them to specific conditions and the associated process effects are analyzed to form the application specific portion of the calculation. These application specific calculations provide a location for showing the conversion of plant operating condition limits to calibration values for static pressure compensated devices and capturing any special maintenance and or calibration requirements (ie., calibrate for hot conditions).

It was recognized early on that if we took the overly conservative worst case accuracy terms provided in Environmental test reports that our plants would not run. These conservative numbers would require setpoint changes to move the setpoints away from the analytic limits and into the operating regions. Because of this recognition, we have expended considerable time evaluating the raw test data from various environmental tests at various temperature and radiation levels and drawn conclusions concerning intermediate conditions. We attempt to determine if the temperature, seismic, or radiation effects are a function of the extreme so that we could interpolate from the tested extremes to conditions that are more representative of our plants. We do this by performing statistical analysis of the deviation data from pre to post test conditions for the various tested conditions. Care must be taken when drawing conclusions from this data because you need to consider the statistical confidence you need and the sample size of the data. If all you have is a single test on a single device, you can draw no statistical conclusions and should apply the maximum deviations as a bidirectional bias in calculations. If you have a few test results on a few test samples, then you may get unreasonably large inaccuracy terms when you apply tolerance and confidence limit multipliers. You should examine your data and it's distribution when compared to a normal distribution to determine if the distribution is normal before drawing conclusions to reduce these predictions. The optimal situation is where you have multiple test results from multiple devices (5 or 6 are usually sufficient) which allows for a reasonable prediction of the response of the general population of devices from the tested sample. The value of 5 to 6 samples is based on the tolerance limits tables which provide a reasonable multiplier for the standard deviation of the test sample. Where large variances occur in test results, it is often necessary to increase the sample size and retest. It is often necessary to test larger sample sizes to verify that the error distributions are indeed normal.

Another place where statistical analysis has been useful is in the analysis of plant calibration history. Here you can evaluate the deviation from as-left to as-found over known periods of time and normal seasonal variations. From the evaluation of this data by examination of deviation versus duration, or deviation versus month of year, you can draw conclusions about the time relation of drift and the temperature variation effect on devices. It should be noted that the evaluation of plant calibration history yields the combined effects of accuracy, normal temperature variation effects, calibration equipment, and drift; therefore, it can replace the normal hardware accuracy terms but not the accident terms in an accuracy calculation. With the statistical evaluation of drift you can eliminate the need to multiply the standard deviation by both a confidence interval and a tolerance interval if the devices you are evaluating are 100% of the population to which the data will be applied. Only the confidence interval is required. This is because you are drawing a conclusion about the total population from results of the calibration history of the entire population and not from the history of a sample of the population.

Process dynamic behavior must also be considered in either the operating limits determination or the accuracy response time analyses. Transients are normal operating system and system startup fluctuations. Timing delays on functions can be used to filter them out if the transients can be quantified (ie., delay isolation on hi steam flow to ride through startup transients on BWR High Pressure Coolant Injection systems).

Process effects are dependent on the sensing device (ie., orifice plate, thermal dispersion, vortex shedding, ...etc) and therefore will not be discussed in detail. Suffice it to say that the variations of process temperature and pressure should be considered in conjunction with the permutations of environmental extremes to determine sensing errors. Another type of process effect is the conversion of process limits, defined at a particular location in a system into limits at the sensor. An example would be the conversion of reactor vessel head pressure limits into equivalent Residual Heat Removal suction pressure limits. For these cases water legs, pipe flow loses, level variations, and even pressure drop across flow elements must be considered.

In addition to statistical analysis of equipment, it is also possible to use analysis or test of other devices if you can show similarity of fit, form, and function. However the similarity analysis must show similar materials, design (circuit analysis), and operating physics to be able to draw any quantitative conclusion.

The interdependence of accuracy parameters must also be addressed in the accuracy calculations to be able to apply the proper combinational methods. Some of the parameters of interest are:

- Device repeatability
- Linearity
- Hysterisis
- Deadband
- Power supply effect
- Calibration equipment
- Calibration technique
- Temperature effects (normal/accident)
- A/D software
- Seismic
- Time dependent drift
- Process sensing physics
- Senselines
- Dynamic behavior
- Process operating range
- Radiation effects

As stated in the ISA standard 67.04 the true nature of the interdependencies of these parameters must be known to properly combine them. Also stated in the standard, it is well recognized that empirical evidence suggests that the normal distribution provides a good representation of many physical variables including instrumentation errors. This view is also held by TVA with added clarifications for several effects that could be considered dependent. These are the accident temperature effects and radiation effects. The accident induced waterleg errors are always used as non-random bias values that are combined algebraically. When the accident temperature extremes and radiation extremes do not peak simultaneously they are considered independent and combined via the SRSS method as long as the associated accuracy parameters are not systematic biases. Timing of accidents can also plays a crucial role in determining if a parameter is applicable (ie., the Reactor Protection System scram functions that occur in less that 1 minute are not exposed to elevated environmental extremes of radiation or temperature). For instrumentation loops which are marginally unacceptable, a thermal lag analysis can be run to determine the actual temperature of the device rather than the average temperature of the environment.

The flowchart of the program implemented at TVA is shown in the following figure. In the figure, both the procedural and calculational methods developed to implement the guidance of ISA standard 67.04 and USANRC RG 1.105 are shown.

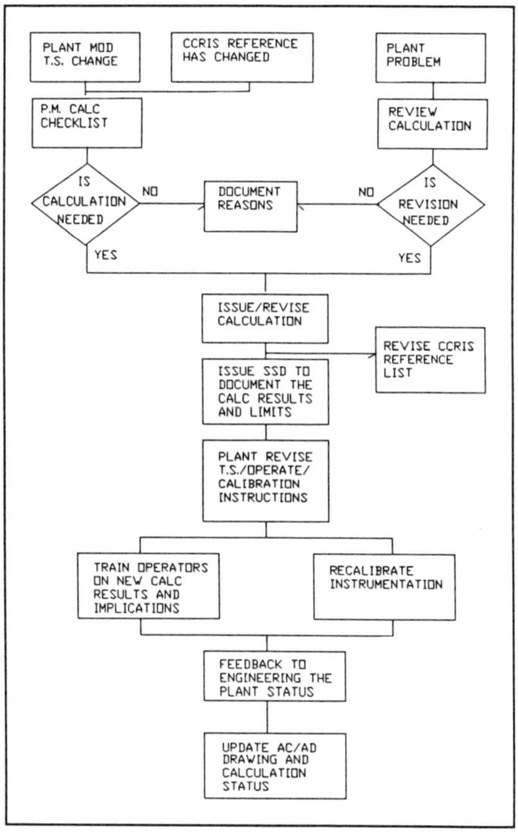

TVA has utilized many methods to implement the programs described. We have performed calculations manually, we have word processed them, used programs such as BASIC, Mathcad, and LOTUS 123, and we have used several AE software packages. Our conclusion has been that use of software languages like BASIC and LOTUS 123 are acceptable but require very computer literate individuals to take advantage of the software and reduce manpower and duration costs. The Architectural Engineering packages we examined did not live up to our expectations. They were not the user friendly packages that were expected nor were they extendable without extensive lead time and cost. We determined that since the mathematical requirements required to perform accuracy calculations is not overly complex that we would develop a software program on our own. The software program we chose is a marriage of FOXPRO and MATHCAD.

FOXPRO is a structured database language with popup menu and windowed interfacing capabilities and MATHCAD is a free format mathematical analysis language that lets the user program by typing in equations as if they were being written in a lab book. It takes advantage of the ease of developing userfriendly interfaces and cross referencing capabilities of FOXPRO and the ease of developing math templates for accuracy and process effects evaluations in MATHCAD. The development of the software inhouse meant that we own the rights to the source code and can modify it for minimal costs. TVA chose to develop a non-verified, non-validated software program so that it could be modified quickly and inexpensively. The printed calculation is a stand alone document with all the information to perform a complete manual verification of information and calculations. The software package is only a tool to aide the instrumentation engineer in the gathering of data and preparation of a consistent calculation. It does not take the place of good engineering thought. The calculation methodology can easily be changed by the user to account for unique differences in of the loop or process under evaluation.

Our calculation methodology has several separate and distinct parts that are mirrored in the software. These parts breakdown into three distinct categories: 1) generic manufacturer parameters, 2) environmental zones information (temperature, radiation), and 3) loop specific application information (calibrated span, calibration methods, setpoints, limits, ... etc). Generic calculations are performed on the manufacturer devices to support the software databases. Environmental response spectra are developed for qualification purposes and are input to the databases. Loop definitions are processed as the calculation building tool. This is accomplished by selecting generic devices and environmental zones and defining the span, setpoint, limits, calibration interconnection and process effects. After the entry of the loop information, a calculation can be directly produced. The logistics of the calculation processing are transparent to the user but consist of FOXPRO summarizing the parameters, environmental extremes, etc., and passing it to MATHCAD. Mathcad then analyzes the data via a standard math template that performs the accuracy calculation numerics. Mathcad is also utilized to perform scaling, evaluate waterleg effects, process effects, and perform statistical data reduction, again through the use of generic templates. New analysis methods can be developed and included in the software program as easily as saving the template file.

By using a combination of software packages we were able to take advantage of the strong points of the two programs and avoid their limitations. FOXPRO has excellent user interface and database security features but the math capabilities are not easily implemented, modified, nor validated. MATHCAD on the other hand has excellent math capabilities but limited wordprocessing capabilities and no database capabilities. With the math being performed in MATHCAD, it is easy to change the combinational methods utilized if needed. Also, with the math being performed in generic Mathcad templates, the extension of the Accuracy software program to include new process effects or statistical methods is as easy as creating a new mathcad file and saving it in the proper user directory. These new templates will be issued and distributed via Calculations Methods manuals and disks. Use of a database to store the bulk of the calculations allows for the quick and easy determination of calculations effected by changing environments or generic calculation revisions.

The software is a combination of a database manager and a mathematical processing engine that are connected via the disk operating system (DOS) batch files that accomplish the data transfer and running of the two programs (FOXPRO and MATHCAD). The accuracy parameters, environmental conditions, and loop definitions are stored in the database and accessed via the menu driven windowed user interface. The numerical processing of the parameters in the database are performed by standard MATHCAD templates. The actions requested by the user modify the DOS batch files that control program flow between the two software packages. The following sketch shows the menu structure and the program flow control between the FOXPRO and MATHCAD packages.

In conclusion, it must be remembered that every device in a loop must be evaluated for

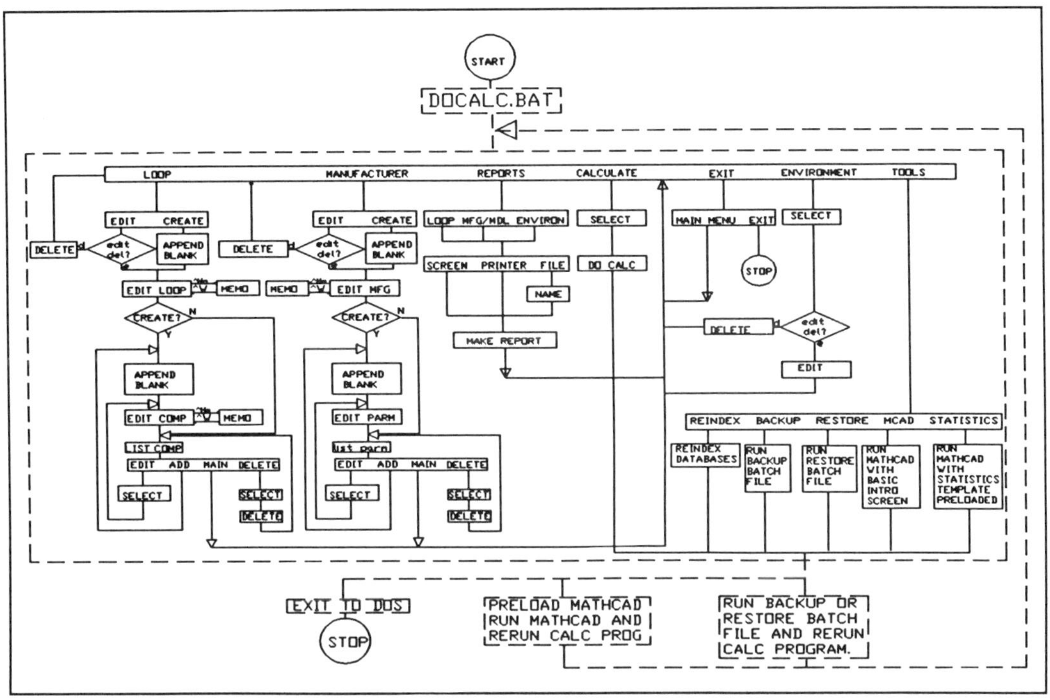

accuracy inclusive of senselines, cable, sensing and processing devices. These analysis should be performed in as efficient a manner as possible so as to minimize the initial manpower expenditure and rework effort. They should be performed in close connection to the Environmental Qualifications efforts, maintenance operations, and training/operations programs. They should be self explanatory stand alone documents that are easily checked, used, and revised because they will be revised many times over the life of the plant.

Appendix A

The following is an example of the data analysis performed at TVA. This evaluation is an analysis of the data provided by a transmitter manufacturer (which will not be mentioned due to the proprietary nature of the test results). This particular manufacturer has had a long running test setup to analyze drift. The test includes 12 transmitters that have been powered and pressurized. The data taken on these transmitters has been the as-found values for specific calibration points. The data including calibration temperature was recorded monthly for 63 months.

The analysis of the data is an application of a linear regression least squares curve fit of the dependent variable (deviation) from the two independent variables (time and temperature). Linear regression is described in NUMERICAL METHODS FOR ENGINEERS by Chapra and Canale, McGraw Hill 1985. Quattro Pro and Lotus 123 were both used to perform the linear regression. The analysis was performed to analyze the deviation data (dependent variable) as a function of independent variables (months since calibration) and temperature (deviation from calibration temperature). The raw data is shown in the following table:

Month	Delta Temp	Temp	Zero Shift From Ideal	Time/Temp Crvfit	Max Temp Limit	Min Temp Limit
0	0	64	0.030	0.03975271	0.18916237	-0.109657
1	2	66	0.080	0.03610713	0.1855168	-0.1133025
2	16	80	0.035	-0.0053773	0.14403232	-0.154787
3	21	85	0.000	-0.0184827	0.13092701	-0.1678923
4	20	84	0.000	-0.0126685	0.13674116	-0.1620782
5	-1	63	0.080	0.05621049	0.20562015	-0.0931992
6	12	76	0.010	0.01787925	0.16728891	-0.1315304
7	-38	26	0.150	0.17820226	0.32761193	0.0287926
8	-41	23	0.210	0.1903229	0.33973256	0.04091323
12	-14	50	0.090	0.11582899	0.26523865	-0.0335807
15	20	84	0.040	0.01660148	0.16601115	-0.1328082
18	-18	46	0.115	0.1444074	0.29381707	-0.0050023
21	-56	8	0.200	0.27221333	0.42162299	0.12280366
25	-13	51	0.090	0.14726755	0.29667721	-0.0021421
27	28	92	0.010	0.02330644	0.1727161	-0.1261032
30	-2	62	0.090	0.12588642	0.27529609	-0.0235232
34	-30	34	0.200	0.22482083	0.3742305	0.07541117
36	-16	48	0.160	0.18599726	0.33540692	0.03658759
38	18	82	0.010	0.08410884	0.23351851	-0.0653008
40	14	78	0.090	0.10204363	0.25145329	-0.047366
42	6	70	0.140	0.13259138	0.28200104	-0.0168183
44	-20	44	0.310	0.21989749	0.36930715	0.07048782
45	-26	38	0.330	0.24147785	0.39088751	0.09206818
48	2	66	0.260	0.16116979	0.31057946	0.01176013
50	22	86	0.060	0.10342677	0.25283643	-0.0459829
53	-18	46	0.290	0.23753917	0.38694884	0.08812951
55	-28	36	0.310	0.27439341	0.42380307	0.12498374
57	-22	42	0.250	0.26079577	0.41020544	0.11138611
62	4	68	0.150	0.19211602	0.34152568	0.04270635
63	8	72	0.140	0.18216395	0.33157362	0.03275429

The results of the regression analysis are shown in the tables that follow. Before examination of the results, the format of the tables must be discussed. There are several pertinent pieces of results that are to be examined. Since the curve fit is a fit of the dependent variable to two independent variables, there are four parts to the equation that represents it. These are:

Deviation = Constant + Time*Timeslope + Temp*Tempslope ± Std_y_error

Where the slope constants are shown in the table:

Regression Output:	
Constant	0.03975271
Std Err of Y Est	0.04725163 (repeatability)
R Squared	0.79933641 (goodness of fit)
No. of Observations	30
Degrees of Freedom	27
X Coefficient(s)	0.00266091 -0.0031532
Std Err of Coef.	0.00042617 0.00040676
	Time and Temp Coefficients

As can be seen from the goodness of fit indicator (R squared), the linear regression is good (79.93%) but not great (100% is perfect). To test to see if it is better than a fit of just time or just temperature, their regressions were also performed and follow. Their fits are not as good.

TIME

Regression Output:	
Constant	0.04798607
Std Err of Y Est	0.08333576
R Squared	0.35272024
No. of Observations	30
Degrees of Freedom	28
X Coefficient(s)	0.00292646
Std Err of Coef.	0.00074919

TEMP

Regression Output:	
Constant	0.11421308
Std Err of Y Est	0.07253651
R Squared	0.50960903
No. of Observations	30
Degrees of Freedom	28
X Coefficient(s)	-0.0033574
Std Err of Coef.	0.00062241

The time and temperature curve fit indicate that there is a high degree of correlation of the deviation to both time and temperature. The figure shows the raw data as X's, the time and temperature curve fit (horizontal line) and the repeatability band consisting of the standard error of y estimate (or repeatability) times the tolerance multiplier for 27 data points as vertical lines. The multiplier is a tolerance limit based on two sided statistics, 27 samples, and a desired 95% confidence level. It is assumed that the data is roughly normally distributed.

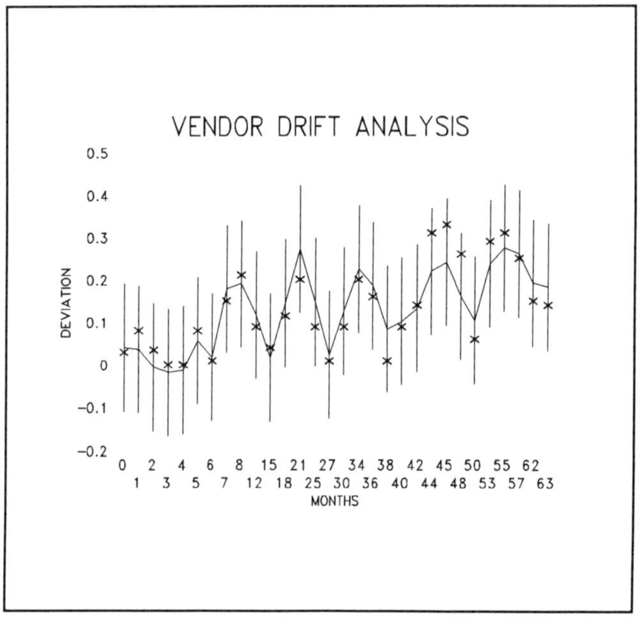

As can be seen in the figure the raw data is always enveloped by the predicted deviation limits represented by the vertical bars. Here the horizontal curve is the temperature and time (drift) effects and the vertical bars are the repeatability or uncertainty of the devices.

This analysis is only an excerpt from the complete analysis in which all 12 transmitters were analyzed to better draw conclusions concerning direction and magnitude of effects.

Appendix B

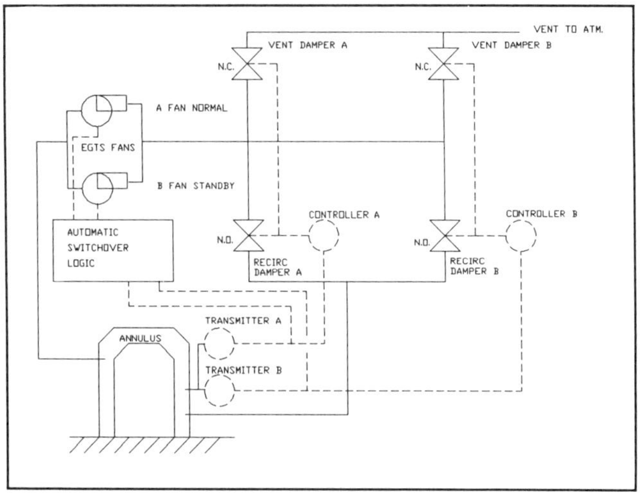

The following is a description of and the analysis of the Eemergency Gas Treatment System EGTS autotransfer system. It shows the overlap of multiple setpoints. EGT is utilized post accident to maintain a negative pressure in the Annulus of the Primary Containment structure to ensure that all outleakage of irradiated air is processed through filters. The system consists of two redundant paths. One starts automatically on the occurrence of an accident signal. The other is in standby and only starts if the first system fails to start and maintain the annulus within proper limits. The following page shows the system diagrammatically and shows the logic that initiates the transfer from the normal train to the standby train on detection of inadequate operation. It should be noted that the EGTS is required to maintain the annulus at a 0.5 inch water column (WC) vacuum while the normal vacuum system maintains the annulus at a 5 inch WC vacuum. As can be seen from the transfer logic, there are four pressure switches that arm and initiate the transfer of failure detection. These switches are designated A, B, C, & D. Their functions are described as follows: The A switch detects the total loss of normal and standby vacuum control by detecting a 0.2 inch WC decreasing vacuum indicating that the EGTS controller cannot maintain the 0.5 inch WC vacuum required. The D switch (4.0 inch WC increasing vacuum) is an arming function that indicates that the normal vacuum control system is functioning and the plant is running. The C switch (0.7 inch WC decreasing vacuum) works in conjunction with the D switch as an additional arming function. It indicates that the normal

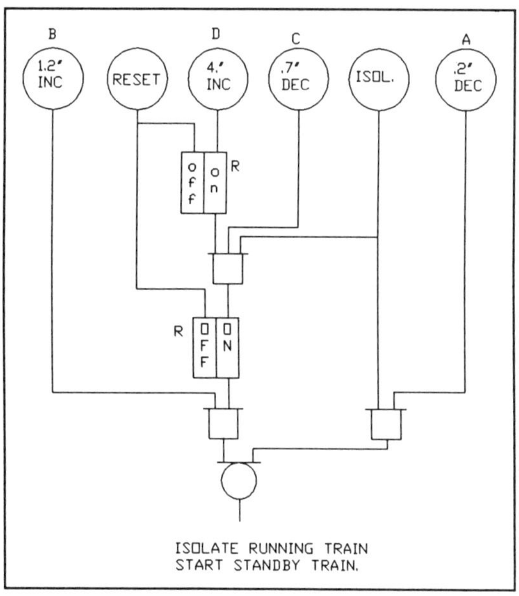

vacuum control system has been lost (normal control is at 5.0 inch WC vacuum) which indicates that the EGTS accident system is controlling the vacuum. The B switch (1.2 inch WC increasing vacuum) works in conjunction with the two arming switches (D and C) to indicate that the EGTS accident vacuum control system has failed and is pulling too much vacuum. The significance of too much vacuum is that there is too much exhaust going through the filtration system with too little holdup time for radiation filtration and absorption. The significance of too little vacuum is that the can be leakage directly to the environment without filtration. The interaction of the differing switch actuation points and the control ranges are shown on the figure following the logic.

The figure shows that the small range of vacuum causes a congested layout of switch actuation points and control ranges. This type of layout of switch actuation is more representative of actual plant situations than the simple examples of the standard with one setpoint and one limit. In this example it should be noted that biases effecting the different device setpoints cancel in the analysis of overlap. It also should be noted that for extremely congested setpoints statistical overlap can be considered to allow closer setpoint arrangement but still guaranty a 95% confidence that the setpoints do not overlap nor violate the safety and operating limits. This is possible by taking into consideration that the accuracies of each device is random and therefore can be statistically combined such that the overlap is in the tails of the normal distribution.

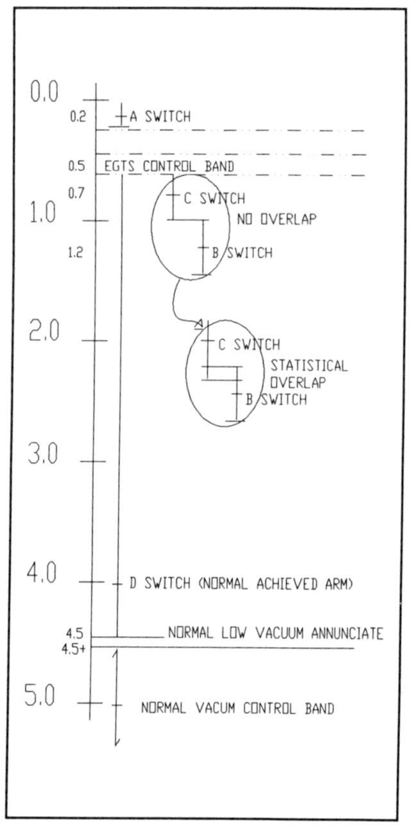

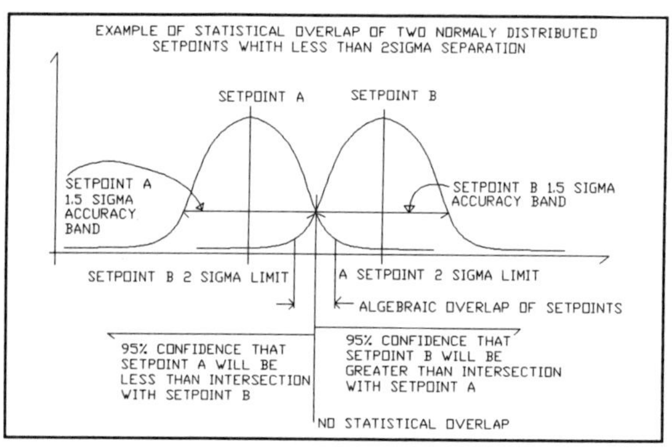

Appendix C

The following sections describe the use of the TVA software and show the menu system and data entry screens. However before the menu system and software are described, a brief background on the logistics of the calculation should be given. A basic calculation is performed as shown in figure below.

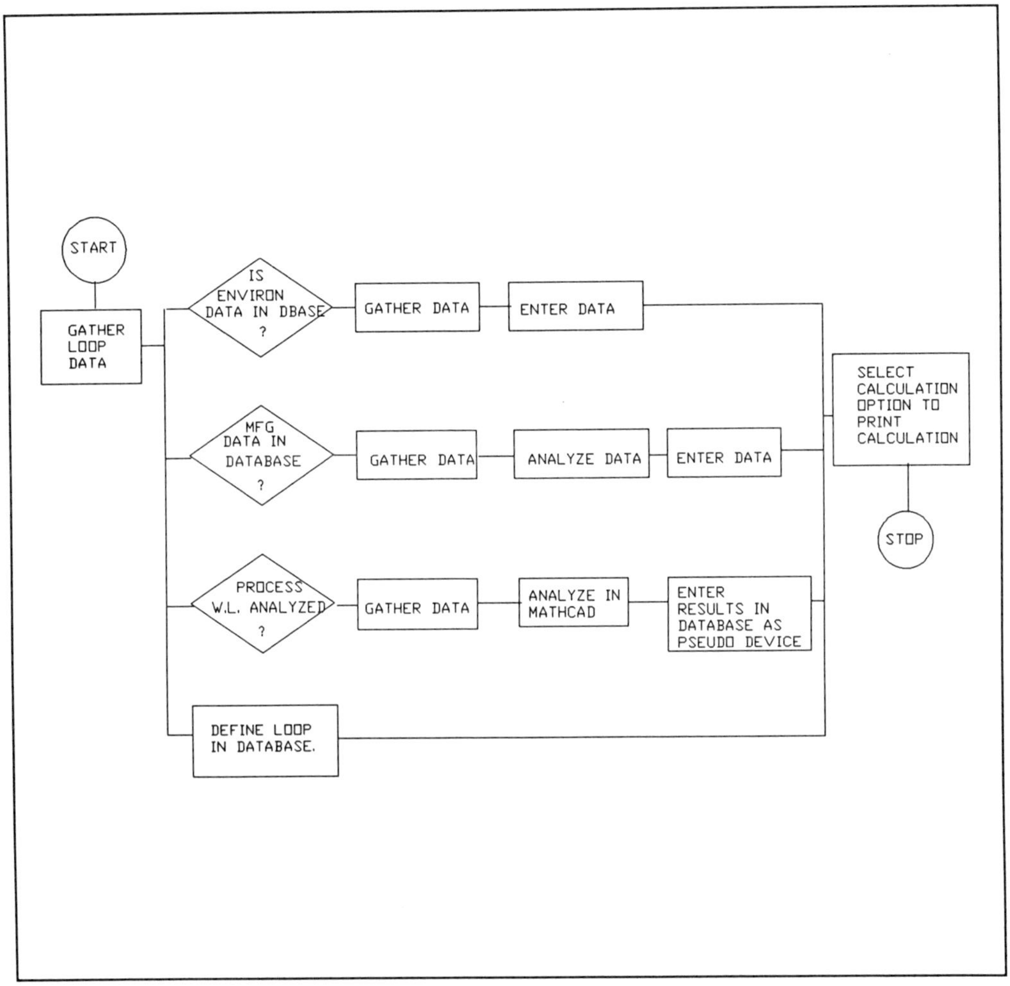

In the figure, an instrument loop is broken into it's subcomponents and their support information is developed. From this initial parsing of the loop, you can review the database in the computer to determine if the Environmental data for the component locations has already been entered. Also, you can determine if the particular manufacturer and model number devices have been generically analyzed previously.

If these portions of the loop data have already been gathered and entered in the computer, then only the loop specific data is required to be entered, thus reducing the data entry and gathering time and manpower.

Upon starting the software, you must login and enter a password to inform the computer of your security clearance and level. Upon successful password comparison, the following main menu is displayed.

Loop   Manufacturer   Reports   Calculation   Exit to DOS   Tools   Environ

On this menu, it can be seen that there are seven options which can be summarized as follows:

LOOP	allows the entry of loop specific data, listing of components, and calibration requirements
MANUFACTURER	allows the entry of data on specific manufacturer and model devices (qualification limits and accuracy)
ENVIRONMENT	allows the entry of the temperature and radiation levels of particular environmental rooms for various time frames
REPORTS	allows display/printout/file capture of summary level reports of the first three options data
CALCULATION	this option processes the accuracy calculation for a defined loop
TOOLS	allows the reindexing/backup/restore database maintenance features to be accessed and provides a bridge to MATHCAD for analysis of process and waterleg effects
EXIT TO DOS	a self explanatory option

The main menu and submenus are shown on the program navigation chart shown below. On this chart, all the menu choices are shown as well as the data entry and memo screens. The chart also shows the keys that are used to move between the menus, data entry, and memo screens.

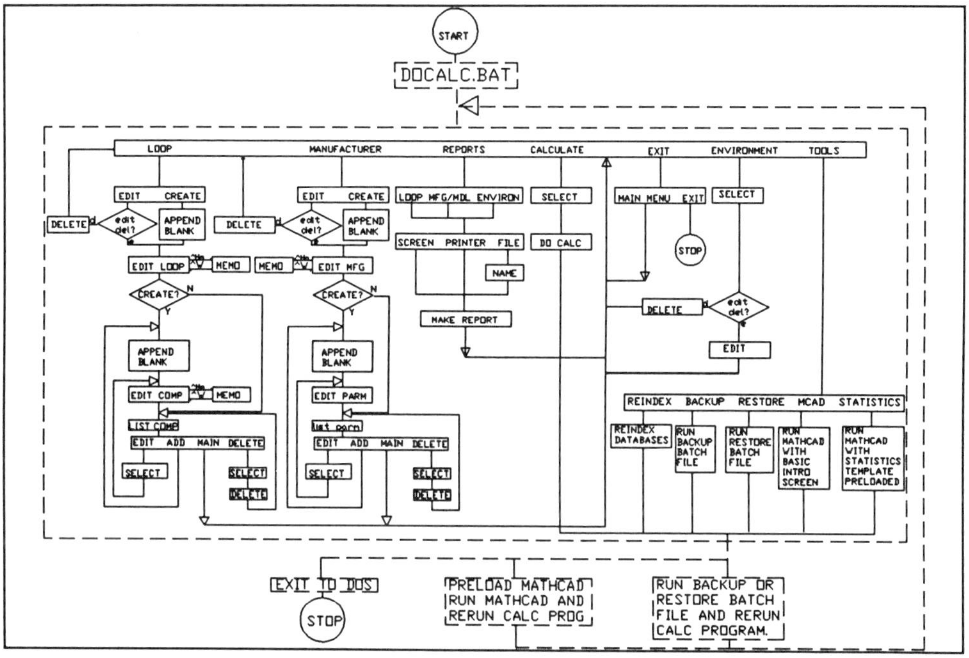

The following section shows the appearance of the screens for the above described software. It is typical of a session that would edit or create a loop calculation with no previous data. Before discussing the options on the main menu, it should be noted that the calculation program's help utility is available at all times by pressing the **Shift-F1** keys respectively. The first option on the main menu is the **LOOP** option which is selected by highlighting the selection using the cursor movement keys and pressing return.

Loop Manufacturer Reports Calculation Exit to DOS Tools Environ

<CREATE>
0-L-TEST
2-PS-73-1A
2-PT-1-81B

You then highlight the **loop** to edit or highlight the CREATE option to start a new loop definition. Pressing enter invokes the edit/addition and displays the loop info data entry screen.

```
Loop     Manufacturer   Reports   Calculation  Exit to DOS   Tools   Environ

<CREATE>
PT-3-1        LOOP NAME:  PT-3-1
PT-3-2        PURPOSE Memo
RON
              Setpoint and limits are in process units as defined as the
              input units of Device 01.


              SETPOINT              :    18.000
              ANALYTIC LIMITS (LO):    566.000
                              (HI):    568.000      VERIFIED
```

The loop data entry screen requires the entry of limits, setpoints, normal process effects, and provides an unlimited free format data entry **memo** field for entry of references and notes associated with the loop definition. It is accessed by placing the cursor on the **MEMO** field and pressing **ctrl-home** which yields the following display.

```
Loop     Manufacturer   Reports   Calculation  Exit to DOS   Tools   Environ

<CREATE>
PT-3-1        LOOP NAME:  PT-3-1
PT-3-2        PURPOSE Memo
RON
                              LOOP->PURPOSE
     PURPOSE: The purpose of this calculation is to determine the          ▲
              accuracy of the instrumentation and based on said            ◆
              instrumentation determine the ability of the loop to
              perform it's intended safety function.

    FUNCTION:
    LOOP ASSUMPTIONS:
    LOOP DESIGN INPUT:
    LOOP CALCULATION BASIS:
    COMPUTATIONS AND ANALYSIS:
                                                                           ▼
```

The memo field is exited by pressing the **ctrl-w** keys. After entry of all loop related information a list of all components in the loop is displayed along with a three item menu.

```
Loop   Manufacturer   Reports   Calculation   Exit to DOS   Tools   Environ

<CREATE>
PT-3-1          LOOP NAME:  PT-3-1

COMP_NAME          MFG                  MDL                        ROOM
WL-3-1             WATERLEG             WL-3-1                     0
PT-3-1             BARTON               763 (259F 1MR) LOT 2       A0
PI-3-1             WESTINGHOUSE         VX-252                     A6

                   RETURN TO MAIN MENU
                   EDIT A COMPONENT
                   ADD A COMPONENT
                   DELETE A COMPONENT
```

This submenu allows the addition of new components to a loop, the editing of an existing component or return to the main menu. The edit option when selected presents you with a list of components that you highlight to select.

```
Loop   Manufacturer   Reports   Calculation   Exit to DOS   Tools   Environ

<CREATE>
PT-3-1          LOOP NAME:  PT-3-1

                   MFG                  MDL                        ROOM
WL-3-1             WATERLEG             WL-3-1                     0
PT-3-1             BARTON               763 (259F 1MR) LOT 2       A0
PI-3-1             WESTINGHOUSE         VX-252                     A6
```

Using the cursor keys to select the component and pressing enter to access the edit, causes a data entry screen to appear. This screen has two submenus that popup to allow point and shoot access to change the manufacturer and model number of a component or to select another environmental zone and or time frame. The data entry screen looks like:

```
Loop    Manufacturer   Reports   Calculation  Exit to DOS   Tools   Environ

<CREATE>                 MANUFACTURER BARTON
PT-3-1             LOOP  MODEL 763 (259F 1MR) LOT 2

                         DEVICE NUMBER 02 DEVICE NAME PT-3-1
 COMP_NAME               CONTRACT 90744        LOCATION 25-5
 WL-3-1                  COMPONENT NOTES AND REFS Memo
 PT-3-1
 PI-3-1                           WORST APPLICABLE ACCIDENT
                         ROOM AO   Accident ABNORMHI TIME

                                      INPUT      OUTPUT
                         UNITS       IN WC         MA
                         LO            300.000        4.000
                         HI            336.000       20.000

                         Reading error    0.2500     0.0160
                         M&TE             0.2500     0.0160
                         SETTING TOLERANCE  (Ab):    0.0160
                         CALIBRATION INTERVAL (MONTHS) 18.0
```

Again, there is a **MEMO** field which can be accessed via the **ctrl-home** keys and exited from via the **ctrl-w** keys.

After entry of all component information, you are returned to the three option submenu from which you can add, edit, or return to the main menu.

The second option of the main menu is the **Manufacturer** selection where you choose to create new entries or edit existing manufacturer and model number devices. Again, selection is made via the point and shoot method where you highlight the desired item and press return to select.

```
Loop    Manufacturer   Reports   Calculation  Exit to DOS   Tools   Environ

        <CREATE>
        ASCO
        ASCO       288A
        BARTON     752
        BARTON     763 (259F 1MR) LOT 2
        BARTON             ▼
               ▼
```

After selecting a manufacturer and model device, you are presented with the data entry screen to enter the qualification limits, transfer function, and maximum limits of calibration.

```
   Loop     Manufacturer   Reports   Calculation  Exit to DOS   Tools   Environ

MANUFACTURER:  BARTON
MODEL: 288A
MINIMUM INPUT     0.000
MAXIMUM INPUT   600.000
IS THIS A SWITCHING DEVICE? (T/F) T
TRANSFER FUNCTION IIF(IN<SP,0,1)

MAXIMUM QUALIFIED TEMPERATURE 200.0
MAXIMUM QUALIFIED RADIATION         50000
NOTES AND REFERENCES Memo     verified
```

The manufacturer data has a **memo** field that allows the entry of references and clarifying notations as follows.

```
   Loop     Manufacturer   Reports   Calculation  Exit to DOS   Tools   Environ

MANUFACTURER:  BARTON
MODEL: 288A
MINIMUM INPUT     0.000
MAXIMUM INPUT   600.000
IS T                             MFG->PARM_NOTE
TRAN  PARAMETER NOTES:                                              ▲
                                                                    ◆
MAXI  ALL PARAMETERS ARE FROM GENERIC CALCULATION WBPEVAR9007013
MAXI  (B260726401), Refer to it for details of analysis as summarized
NOTE  below.

      PERTINENT NOTES ARE:

      RESPONSE TIME CONSIDERATIONS:  This device is a bellows actuated
      switch.  It will respond as quickly as the process and senseline
      can convey the pressure to the device.

      This device and the generic accuracy calculation only support use
      of this device in a MILD ENVIRONMENT.
                                                                    ▼
```

After entry of the manufacturer limits information, a list of all defined accuracy parameters is displayed and a three option menu. The menu allows the return to the main menu , the addition of new parameters, and the editing of existing parameters.

```
 Loop    Manufacturer  Reports   Calculation  Exit to DOS   Tools    Environ

MANUFACTURER:  BARTON
MODEL: 288A
M
I  NAME     RANDOM  POS_BIAS  NEG_BIAS  TYPE
T  DE       0.0000  5.4000    5.4000    P  M
M  SE       0.0000  0.5000    4.7000    S  SU

                    RETURN TO MAIN MENU
                    EDIT A PARAMETER
                    ADD A PARAMETER
                    DELETE A PARAMETER
```

Selecting the edit option displays a list of parameters from which you choose to edit.

```
 Loop    Manufacturer  Reports   Calculation  Exit to DOS   Tools    Environ

MANUFACTURER:  BARTON
MODEL: 288A
M
M
            RANDOM  POS_BIAS  NEG_BIAS  TYPE
DE          0.0000  5.4000    5.4000    P  M
RE          0.2000  0.0000    0.0000    S  M
SE          0.0000  0.5000    4.7000    S  SU

N
```

Highlight the parameter of interest and press enter to select it. The data entry window the opens for entry of the name and numerical value of the accuracy parameter. The code field allows coding to define the parameter as a value or percent of span, setpoint, or range as well as define it as a fixed or scaled variable based on temperature, calibration interval, or radiation level.

```
  Loop    Manufacturer   Reports   Calculation  Exit to DOS   Tools   Environ

MANUFACTURER:  BARTON
MODEL: 288A
M
M
I  NAME     RANDOM POS_BIAS NE PARAMETER NAME RE
T  DE       0.0000   5.4000    RANDOM    0.2000
   RE       0.2000   0.0000    POSITIVE BIAS   0.0000
M  SE       0.0000   0.5000    NEGATIVE BIAS   0.0000
M                              % Span
N
                               For all conditions.
                               Measurable during calibration
                               verified         Change Application code Y
                               Type  code: (S) % Span
                               Scale code: ( ) Value
                               Apply code: ( ) All conditions
                               Calib code: (M) Measurable
```

The Application code is presently a four digit code to represent the definition of the parameter. The first digit represents the type of number (ie., % of span, % setpoint, value). The second digit represents the dependency of the value and is used for scaling the number (ie., per 50°F, per month, per MRAD). The third digit represents usage clarifiers (ie., seismic only, accident only, etc). The fourth digit defines if the parameter is measurable during calibration.

After entry of the parameter data, you are returned to the three option menu where you can continue adding, or editing, or return to the main menu.

Back at the main menu you can select the reports option which allows you to display summary level reports of the loop data, manufacturer data, or environmental data.

```
  Loop    Manufacturer   Reports   Calculation   Exit to DOS  Tools   Environ

                         LIST LOOPS AND COMPONENTS
                         LIST MFG/MDL AND PARAMETERS
                         LIST ENVIRONMENTAL ZONES
```

After selecting the report of interest, you are prompted to select the destination for the report. It can go to the screen, the printer, or a disk based file.

```
Loop    Manufacturer   Reports   Calculation    Exit to DOS  Tools    Environ

                            ┌────────────────────────┐
                            │LIST LOOPS AND COMPO┌───────┐
                            │LIST MFG/MDL AND PAR│SCREEN │
                            │LIST ENVIRONMENTAL Z│PRINTER│
                            └────────────────────│FILE   │
                                                 └───────┘
```

After selecting the destination, the file is sent to it as well as the screen. As an example, one of the reports is shown.

```
Loop    Manufacturer   Reports   Calculation    Exit to DOS  Tools    Environ

                            ┌───────────────────────────┐
                            │LIST LOOPS AND COMPONENTS  │
                            │LIST MFG/MDL AND PARAMETERS│
```

Record#	MFG	MDL	NAME	RANDOM	TYPE	POS_BIAS	NEG_BIAS
1	ROSEMOUNT	1152	RE	0.250	S M	0.000	0.000
4	ROSEMOUNT	1153	RE	0.250	S M	0.000	0.000
5	ROSEMOUNT	1153	DE	0.000	SK M	0.020	0.010
6	ROSEMOUNT	1153	TNE	0.350	STN	0.000	0.000
7	ROSEMOUNT	1153	TAE	0.350	STA	0.050	0.000
8	ROSEMOUNT	1153	RADE	1.000	S A	0.000	0.000
9	ROSEMOUNT	1153	SE	1.000	S S	0.000	0.000

Press any key to continue...

The next main menu option is the **CALCULATION** option which allows you to select the loop of interest to prepare a calculation. Upon selection, the computer reviews the databases and prepares a calculation which is immediately printed for use. The calculation performs a Square Root of the Sum of the Squares combination of the random portion of the variables in conjunction with the algebraic summation of the positive and negative biases. These variables are scaled appropriately based upon the environmental extremes defined, the calibration intervals, and scaling. Several accuracy determinations are performed for calibration conditions, normal, accident, and seismic conditions. The calculation option displays a list of the defined loops that a calculation can be performed upon.

```
Loop    Manufacturer    Reports    Calculation    Exit to DOS   Tools    Environ
                                  ┌──────────────┐
                                  │0-L-TEST      │
                                  │1-L-1-1       │
                                  │2-PS-73-1A    │
                                  │2-PT-1-81B    │
                                  │PT-1          │
                                  └──────────────┘
```

The next item on the main menu is the **ENVIRON** option which displays the list of currently defined environmental zones and time frames from which you highlight and select for editing. You can select CREATE to define a new zone.

```
Loop    Manufacturer    Reports    Calculation   Exit to DOS    Tools    Environ

     ┌───────────────────────────────────────────────────────────────┐
     │<CREATE>                                                       │
     │EL 565'                    ABNORMHI      100.0        500      │
     │EL 565'                    ABNORMLO       40.0        500      │
     │EL 565'                    LOCA     1H   210.0    1000000      │
     │EL 565'                    NORMHI         90.0        500      │
     │EL 565'                    NORMLO         60.0        500      │
     │EL 540'                    ABNORMHI      104.0        500      │
     │EL 540'                    ABNORMLO       40.0        500      │
     │EL 540'                    LOCA     1H   180.0    1000000      │
     │EL 540'                    NORMHI         90.0        500      │
     │EL 540'                    NORMLO         60.0        500      │
     └───────────────────────────────────────────────────────────────┘
```

You highlight and select a zone of interest and are presented with a data entry window in which you define the time frame, and environmental extremes. There is a **memo** field that is used to capture clarifying notations and references.

```
Loop    Manufacturer   Reports   Calculation    Exit to DOS  Tools    Environ

            ROOM NAME    0
            ROOM DESCRIPTION EL 565 T5,6 / R11,12

            OPERATING CONDITION HELBIPC
            TIME INTO INCIDENT 1Y
            MAX TEMP SO FAR (F) 104.0
            RADIATION TID (RADS)      4000000
            REFERENCES AND NOTES memo
                        PRESS Ctrl-Home KEY TO ACCESS MEMO FIELD
```

You access the **memo** field by pressing **CTRL-HOME.**

```
Loop    Manufacturer   Reports   Calculation   Exit to DOS   Tools   Environ

              ROOM NAME 0
              ROOM DESCRIPTION EL 565'

              OPERATING CONDITION ABNORMHI
                           ENVIRON->EQREF
  3.0  ENVIRONMENTAL NOTES AND REFERENCES

     ___  The environmental information is from
          drawing ______________________________ Revision ___

     Special notes regarding the environmental excursions for
     this zone are as follows:
     _________________________________________________________
```

The **memo** field is exited by using the **CTRL-W** keys.

There are two remaining options on the main menu. The **EXIT-TO-DOS** option is self explanatory, and the **TOOLS** option. The **TOOLs** option allows database management (reindexing, backup, and restore) and provides a bridge to the MATHCAD math processing portion of the software package. MATHCAD is a free format spreadsheet that allows the use and display of complex mathematical formulas. It processes matrix and arrays of data with simple (+ - ÷ · ... etc) and complex ($\surd$ π Σ...etc) operators. The MATHCAD software is used to analyze the process and waterleg effects through the use of predefined templates which can be augmented with new templates as new methods or applications are developed and analyzed.

```
Loop   Manufacturer   Reports   Calculation   Exit to DOS   Tools   Environ
                                             +----------------------------+
                                             |COPY OFF                    |
                                             |COPY ON                     |
                                             |Database BACKUP             |
                                             |Database RESTORE            |
                                             |Mathcad (Statistics)        |
                                             |Mathcad (introduction)      |
                                             |Mathcad FLOW WATERLEG       |
                                             |Mathcad LEVEL WATERLEG      |
                                             |Mathcad PRESSURE WATERLEG   |
                                             |REINDEX                     |
                                             +----------------------------+
```

The selection of one of the MATHCAD options causes the MATHCAD software package to be loaded with an introduction screen describing what commands and files are available. An example of one of the screens follows.

Welcome to MATHCAD via the TVA Accuracy Calculations Program

You should erase this entry screen then either create your new process error

calculation or load your existing process error calculation via the F5 key.

The MathCad main menu can be accessed by pressing the F10 key and the MathCad

HELP screens can be accessed by pressing the F1 key. Please remember to print

and save your results before exiting back to the

Accuracy Calculation Main Menu.

CTRL-O will print your calculation and CTRL-Q will exit from MathCad.

Upon exiting from MATHCAD you are returned to the accuracy software via the login screen. The remaining option is the **Exit to DOS** option which is self explanatory.

© 1991 - Paper #91-715
0074-056X/91/217-224/$0 + .50pp

TOUCH-SCREEN INTEGRATED DCS TWO BOILER REFUSE DERIVED FUEL FACILITY

R. J. Gross

ABSTRACT

During the last ten years a virtual explosion in the computer industry has taken place. Industrial applications have lagged the marketplace in general because of the general conservatism of industrial users. Their predominant concern is to meet market demand and state of the art equipment often takes a back seat to known, reliable methods of production.

In the fall of 1986 Babcock and Wilcox embarked upon a project to supply a turnkey project for a refuse-to-energy plant. Bechtel Civil Inc. was our partner for this endeavor and they did the majority of the civil construction at the site.

B&W supplied a DCS system with touchscreen capability. This system was deemed to be a state-of-the-art control system which would illustrate B&W technology in the waste-to-energy market. B&W committed a large investment to the control system with the hope of using this plant to showcase B&W expertise and hardware in the turnkey marketplace.

INTRODUCTION

In December 1986, a contract was awarded by the Solid Waste Authority of Palm Beach County, Florida. This contract was obtained by B&W and Bechtel as partners. B&W's logic design responsibility boundaries were from boiler feed systems to the stack. Bechtel was responsible for the turbine logic and balance of plant logic design.

Babcock and Wilcox was responsible for supplying the control hardware. Because this plant was supposed to be a showcase for B&W refuse capability management decided to fund an advanced control system to highlight our advantages in the marketplace.

The plant specifics were as follows. The plant was designed to burn 2000 tons/day of refuse. By incinerating the refuse in two boilers and using the steam to operate a turbine-generator the plant was able to generate 60 MW of power.

This contract was originally awarded as a thirty-six (36) month project from date of signing until customer acceptance. Given the size of the plant (500,000 sq. ft. under roof) this was an ambitious schedule considering that the details of the control system hardware were not firm at the project start date.

PHYSICAL PLANT

The plant sits on approximately 200 acres of ground in northern Palm Beach County. In addition to the actual RDF facility there are also facilities for truck weighing and ash disposal.

The RDF processing equipment was provided by National Ecology Inc. They supplied the refuse shredders, classifiers, conveyors and tire shredding machines. In the latter stages of this project B&W purchased National Ecology and it is now a B&W subsidiary.

Heil Co. supplied the refuse feeders in the boiler building. In addition to the belt conveyors they also supplied overfeed augers which sent excess RDF to the RDF storage building for weekend operation when RDF operations were not running.

Babcock and Wilcox supplied the two boilers and their associated accessories. The boilers were designed to run at 650 psig and deliver 1.2 million lbs/hr steam to the turbine.

The turbine was a Westinghouse manufacture designed to produce 60 megawatts of power for resale to Florida Power and Light. In tandem with the turbine a full load condenser was installed capable of accepting full load steam indefinitely.

The stoker grates and drives were supplied by Detroit Stoker. The stokers on these units were of the traveling grate typewith an adjustable speed to vary burn rate on the grate. Detroit Stoker also supplied the feeder rams and the transport air systemwhich were used to introduce RDF into the furnace.

Joy Manufacturing supplied the scrubber and precipitator equipment. This system was well overdesigned to insure that adequate particulate removal would be available at all times.

PROJECT CONSTRUCTION

After contract signature, work was immediately begun to delineate work responsibilities between the partners. Tagging conventions for hardware were established and drawing interface points were decided.

Once work began in earnest, virtual daily contact with the partners and the control vendors took place. As the project began to gel, the parties were able to focus on the detail work and actual logic began to take shape.

Due to the partnership arrangement, it was necessary for all of the documents to flow between all of the parties. On this project, there were approximately six hundred (600) drawings issued by the control system vendor alone at the revision "0" level. Document control became a monumental task and in retrospect, it would be wise to set up a workable system to control documents up front. On this project, the volume of drawings forced B&W to establish a project control with a tight schedule minimizing drawing turnaround, which was imperative to maintain schedule.

SYSTEM ADVANTAGES

In the course of this project, many systems had logic designed without having full knowledge of the hardware supplied. The microprocessor based system made this possible. When logic systems which are based on hard station technology are designed without prior knowledge of the equipment to be controlled the result can be costly and time consuming.

provide a control room with all of the graphics this system has in a hard format. Even if hard stations and physical graphic representation were provided, they could not be interactive in the sense that a CRT based system is interactive.

When real time parameters are provided via an interactive CRT, the mechanics of operations are simplified. By optimizing the graphics screens, it is possible to have all of the pertinent indicators and control stations for critical loops on a single screen. It is not required to dance back and forth from station to station for an operator to observe the system dynamics, it can all be done on a single screen.

CRT interface also lends itself to space constraints. A full capability control system can be provided with a single CRT as the system interface. For purposes of redundancy and operator convenience, four (4) CRT's were provided on this project. This allows one operator to monitor both boilers, the fuel feed system and the turbine island. If the same system had been equipped with hard stations, the room would have needed to be at least twice the physical size and it is doubtful that all the displays now available could have been represented in that space.

Additionally it is doubtful that a single operator could adequately address all of the necessary control functions in such a large control room. The physical size would dictate additional personnel to man the critical functions of the boilers, turbine generator and multiple fuel systems.

In a further effort to increase reliability, the system logic was also carefully partitioned. This was accomplished by dividing certain critical logic systems and segregating them by hardware groupings.

For instance, Boiler #1 combustion control was on a separate set of modules and power supplies from Boiler #2. Burner management for each boiler was likewise partitioned. The turbine generator equipment had its own partitioning as well.

SYSTEM FLEXIBILITY

A third party system was also involved in supplying both hardware and software. This equipment and software was developed independently of the rest of the project by a subvendor to B&W's precipitator.

The third party software was developed to supply controls for the ash handling from the precipitators and the scrubbers. Even though the equipment was installed at a later time and did not have the benefit of involvement from inception, it was linked to the rest of the system with a minimum of fanfare.

Graphics were developed and implemented by a group separate from those involved in the logic development. With a minimum of personnel interface, the subsystem was able to be integrated as a viable portion of the project.

Communications between geographically distributed cabinets and the CRT's were greatly simplified. The advantages of geographic distribution are many fold but it all boils down to a few basics. Hard wire runs to the cabinets can be reduced and labor cost of the installation can be drastically slashed.

Using a control system with "hard" (physical) stations, recorders, gauges, etc. would dictate the field equipment and logic requirements to be conceived prior to purchase of the control system. When this type of situation occurs the lead time from design to implementation must be increased by about 50%.

A microprocessor based system with CRT interface alleviates many of these constraints. Work on many logic systems for this project was, by necessity, started without a clear understanding of the equipment to be controlled because of time constraints. In a number of cases, assumptions made for the control of this equipment were incorrect.

One typical example of this would be the operator interface with the feed belt equipment. Heil Company supplied the feed equipment and B&W assumed that Heil would control their own sequencing of the belts which flow the refuse from the storage building to the boilers. Unfortunately this was not Heil's understanding.

Each boiler had a completely redundant feed system with the capability of feeding the boiler from one of two belt feed systems. In the event that maintenance was required on one of the systems the redundant system could be brought on line while repairs were performed on the faulty equipment.

In addition the feed system was designed in such a manner that if feeder trouble were detected online an autostart of the redundant feed system was to occur this transfer would shift refuse feed to the other conveyor line without shutting down the boiler for lack of fuel.

When it came to light that Heil had not provided equipment to control their feed system the decision was made to incorporate the control and logic within the plant control system. For each boiler there were ten feed belts and six diverter gates that had to be controlled.

Including all of the feeder subsystems each boiler required twenty-seven digital and three analog stations for feed system control. If we would have been required to supply hard stations for all of this equipment at least two additional sections of control console would have had to be added just for the feeder controls. By utilizing the advantage of the CRT interface B&W was able to bypass costly delays on equipment ordering and avoid using additional space in the control room.

In addition the dual path the partners were following sometimes led to erroneous interface points. Input and output interfacing was sometimes incorrect and the system had to be revied during the course of ongoing work. On a microprocessor based system, this is much more manageable. When errors were made, it was possible for a small team of people to correct these errors quickly. Thanks to the flexibility of the system, a great number of field errors or misconceptions were dealt with in the software.

No changeout of equipment other than I/O cards was necessary to correct the majority of these errors. When control stations, start/stop pushbuttons, indicators, alarm stations and graphics are all software based, a number of potential problems are eliminated. Long lead times for hardware, misapplication, faulty equipment, shipping damage and pilferage, for example, are all eliminated.

During the course of this project, several iterations of graphic changes took place. At one point, this became a bottleneck on the project because of the touch screen capability, but overall this was a huge time and cost savings. It would be prohibitive space wise to

The main control room can be downsized because less hardware is accommodated in the control room. Wire pulls to the control room are also reduced and are actually limited to a small number of communication cables and power wire feeds to the DCS system.

Geographically distributed cabinets were the biggest savers of wire on the project. By linking all of the DCS system with a redundant data highway wire runs were reduced by approximately forty percent.

A forty percent reduction in wire, conduit, wire tray and labor required to install this equipment translated into major savings on this contract. Such a reduction favorably impacts the space allotment of the physical plant.

This control system was originally conceived to function with a single control room operator at the console. Initially this was not possible, too many malfunctions of equipment and day to day problems precluded a single person from manning the console.

At this point the majority of the equipment is in automatic and a single operator is able to oversee the control console. The ability of an operator to do this has come as something of a surprise. It was felt that a single operator would not be able to successfully monitor and maintain the functions of two boilers, twelve feeders, a turbine generator and two precipitators.

A great deal of credit must be given to the logic design in this regard. The fuel feed trips are now handled automatically whereas this function had previously been done manually because the system was not correctly tuned. It is not possible to overstate the importance of patience in tuning these types of control systems. It is all too easy to dismiss the control system as faulty and run the equipment in manual.

Every one in the process control industry has seen a number of installations where equipment is running in manual that should be automated. Taking the time to understand process is essential to utilizing the power of these systems.

Control system revision is an issue which must be addressed from the inception of the project. On this project drawings were issued by the vendor "as shipped." These were our baseline documents during the start-up phase of the project. During start-up, thousands of changes were instituted and the revisions were only recorded on the as shipped drawings but not on the CAD system.

After the start-up was completed, these revision markings were transmitted for updating in the CAD system. This was a clerical nightmare. Approximately two thousand manhours have been expended to date revising the drawings.

Until recently the jobsite did not have a set of drawings which accurately reflected the logic in the system. To compound this problem, logic changes went on for six months without the benefit of any hardcopy documentation.

This type of DCS systen offers a degree of flexibility that was a two-edged sword. It is possible to engineer revisions directly onto the DCS without revising the CAD drawings. This is a definite benefit during startup because the logic can be revised online without taking the system out of service.

The danger of this is that the entire system is then at the mercy of the revisor who may or may not be diligent about recording the alterations to the system he has made. Ideally, the revisions should be implemented in the CAD system. By doing this, the drawing record is automatically updated and the revisions can be printed out as hard copy for a record as well as being stored as electronic media.

If changes are made at the console it should always be backed up by the CAD system immediately afterward. If the changes are not quickly recorded and the team that initiated the revisions is dissolved, it is very difficult for the next group of people to decipher the intent of the revisions.

CONCLUSION

A CRT based control system for this application was an unqualified success. Despite concerns by management about cost for the system it has proven to be well worth the added cost. The money saved to initiate field changes alone could justify any added cost.

This type of system has proven to be well suited to large scale industrial projects where the jobs tend to be schedule driven. As long as the personnel are properly trained and supported and the system is maintained the control system will be very reliable. It also offers the added benefit of automatically updating drawing records as long as the operators are diligent in their data entry.

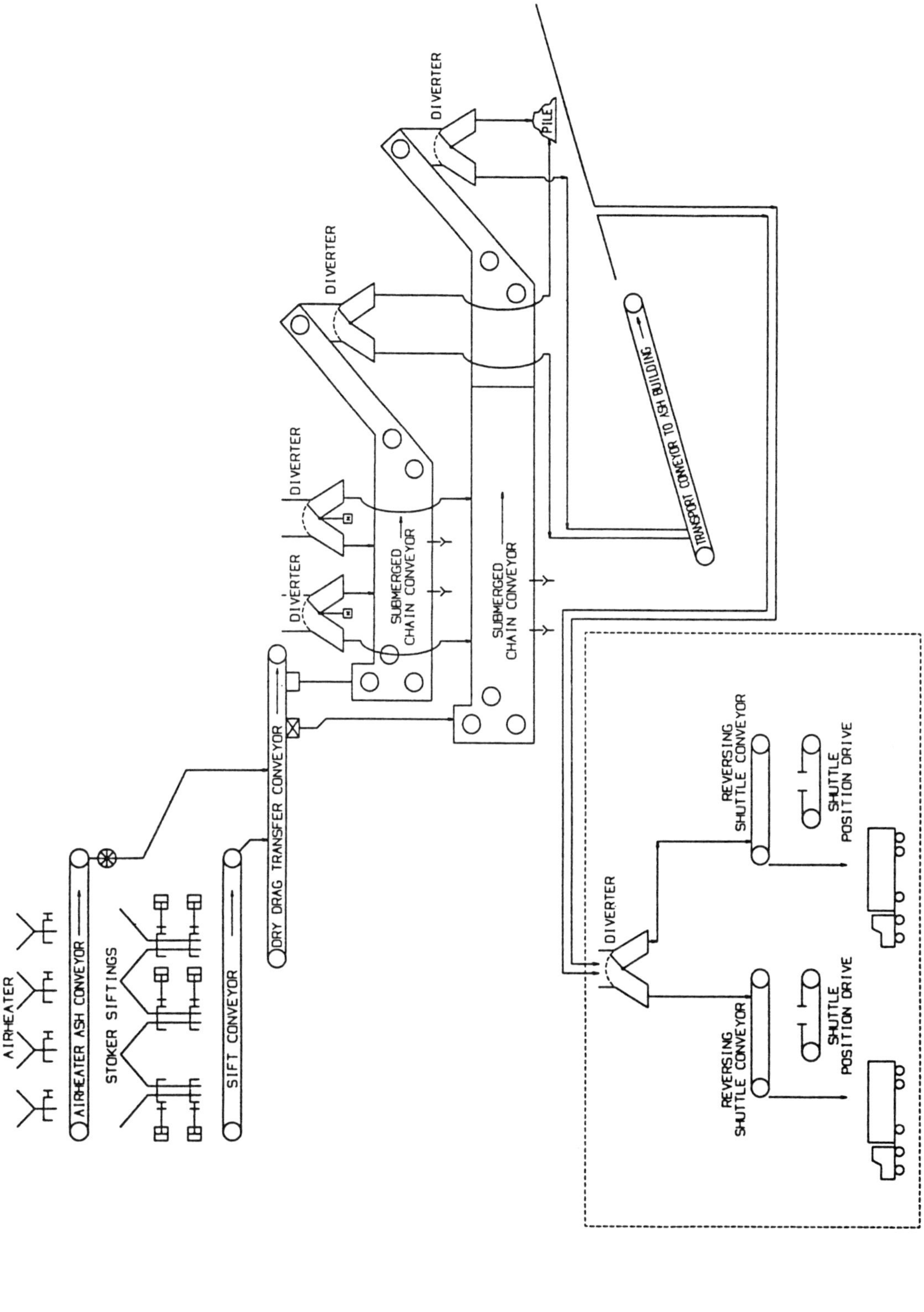
AIRHEATER
AIRHEATER ASH CONVEYOR
STOKER SIFTINGS
SIFT CONVEYOR
DRY DRAG TRANSFER CONVEYOR
DIVERTER
DIVERTER
DIVERTER
DIVERTER
DIVERTER
SUBMERGED CHAIN CONVEYOR
SUBMERGED CHAIN CONVEYOR
PILE
TRANSPORT CONVEYOR TO ASH BUILDING
DIVERTER
REVERSING SHUTTLE CONVEYOR
SHUTTLE POSITION DRIVE
REVERSING SHUTTLE CONVEYOR
SHUTTLE POSITION DRIVE

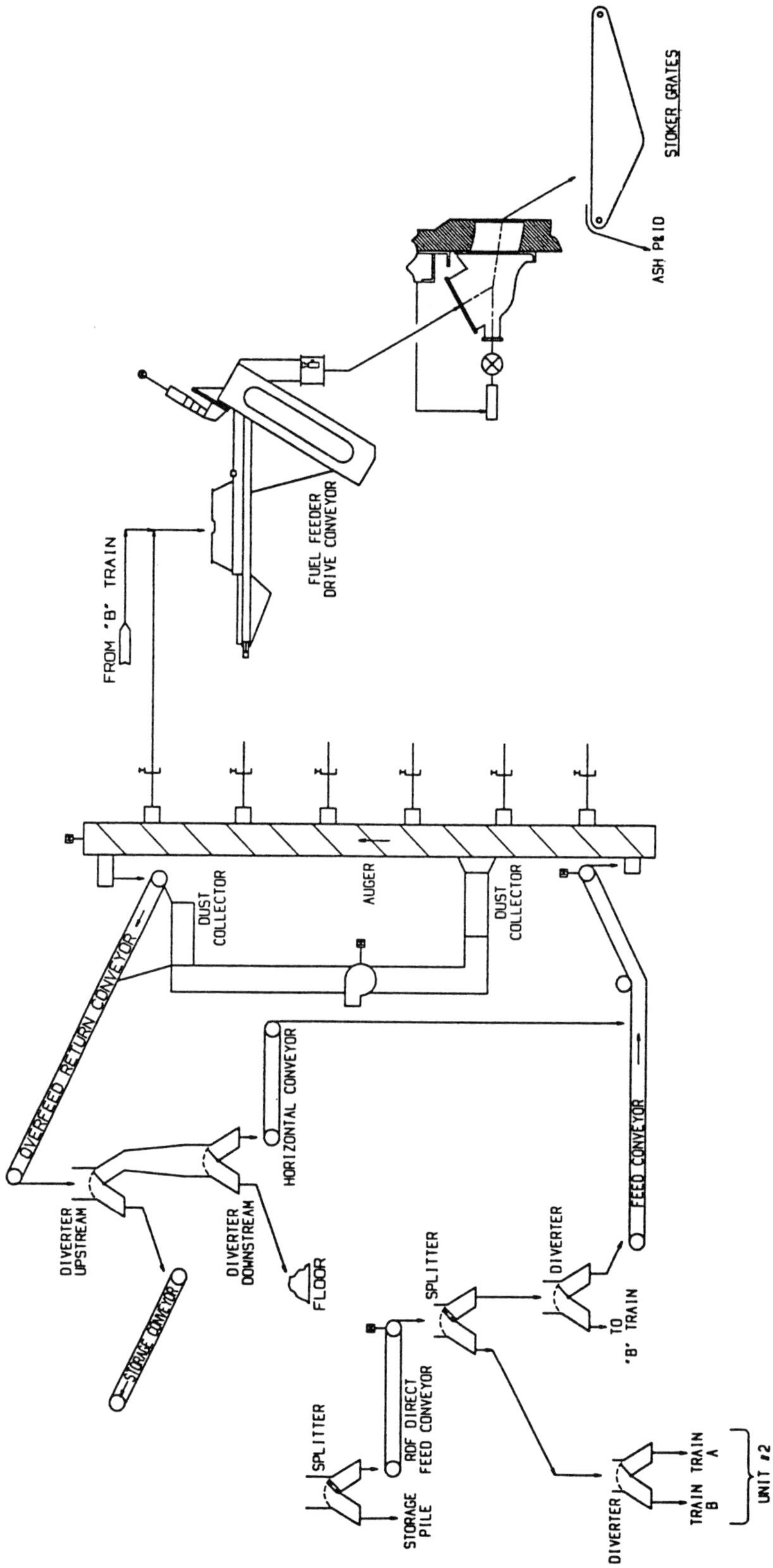
STOKER GRATES
ASH PIT
FUEL FEEDER DRIVE CONVEYOR
FROM 'B' TRAIN
AUGER
DUST COLLECTOR
DUST COLLECTOR
OVERFEED RETURN CONVEYOR
HORIZONTAL CONVEYOR
FEED CONVEYOR
DIVERTER UPSTREAM
DIVERTER DOWNSTREAM
FLOOR
STORAGE CONVEYOR
SPLITTER
DIVERTER
TO 'B' TRAIN
RDF DIRECT FEED CONVEYOR
SPLITTER
STORAGE PILE
DIVERTER
TRAIN B
TRAIN A
UNIT #2

©1991 - Paper #91-716
0074-056X/91/225-231/$0 + .50pp

SOME INTERESTING DCS START-UP EXPERIENCES WITH A WASTE TO ENERGY PLANT

P. Y. Keskar Ph.D. P.E.
Sandwell Inc.
Jacksonville, FL 32256

ABSTRACT

This paper describes a couple of interesting DCS start-up experiences gained during the startup of a 32 MW waste to energy plant. Two measurement related problems are identified, analyzed and the solutions are discussed. The paper can be treated as a tutorial in problem solving. An attempt is made to show that most of the start-up problems that seem difficult have simple practical solutions. Major emphasis should be placed on defining the problem correctly and then approaching the solution using engineering fundamentals and simple common sense.

INTRODUCTION

The start-up experiences described in this paper relate to a 32 MW waste-to-energy plant. The author supervised a team of engineers and programmers involved in the turnkey implementation and startup of a distributed digital control and monitoring system (DCS) for the power plant. The system included five distributed microprocessors, communicating with a redundant MICROVAX II host computer system. The system capacity exceeded 900 input/output points.

Out of the several problems encountered during startup, two measurement related problems have been selected for discussion here because of the interesting nature of the problems and relative simplicity of the solutions. The problems discussed have interdisciplinary flavor requiring understanding of fundamental engineering principles as well as simple common sense.

The first problem is related to measurement of steam flow due to waste burning and is encountered in the waste to energy plants, industrial cogeneration and small utility plants. The second problem discussed is related to measurement of boiler drum level, which is common to all boilers.

MEASUREMENT OF STEAM FLOW CONTRIBUTION FROM REFUSE BURNING

The waste to energy plant discussed here had three boilers, each designed to deliver 75,000 lbs/hr. of steam at 625 psig at a temperature of 700°F.

In a waste to energy plant, waste is the main source of fuel for the boilers. However, each boiler is provided with an auxiliary fuel oil burner which is used for startup as well as during normal operation to maintain desired steam flow by compensating for serious irregularities and periods of low refuse flow. Since the main purpose of a waste-to-energy plant is to burn refuse and not oil as fuel, the oil consumption has to be minimized. For this reason the vendor supplied boiler and burner management system required a signal proportional to steam flow from refuse, which could be used to adjust the refuse input as well as fuel oil flow. Another reason for the need to compute steam flow from refuse was that the plant management, regulatory agencies as well as the City wanted to document this data, to be used as one of the important performance indices for the plant. However, the steam flow attributed to refuse burning cannot be quantized, because of the highly irregular heat value (BTU/LB) as well as input rate.

There is a very simple and practical solution to this problem. Instead of working with unknowns (waste input), one should work with the known quantities. The following observations can be made with regard to the solution of this problem:

1.0 The total steam flow rate from the boiler is measured and known.

2.0 The oil flow into the auxiliary burner is a well behaved process variable, and is measured accurately (known quantity).

3.0 An experiment can be performed on the boiler with only auxiliary burner running, with no waste input. The steam flow from the boiler can be measured and recorded for several values of fuel oil flow. It is important to note that for each value of fuel oil flow rate, the steam flow reading should be noted only after steady state condition is reached. Based on the collected data, a graph can be plotted between steam flow and auxiliary burner fuel oil flow rate. This graph can be stored in the DCS either in the form of a look-up table or as several piece-wise linear functions. A look-up table will require a large number of data to be stored, with linear interpolation for points not covered in the table. However, a piece-wise linear approximation of the steam flow versus oil flow is a more practical approach. Usually only two or three linear regions can describe the curve, and one need to store only the slope, intercept and range for each linear region.

4.0 During normal boiler operation, when the refuse is being burnt as well as auxiliary burner is also running, the steam flow attributed to refuse at any given time can be computed in the DCS as follows:

4.1 Look at the fuel oil flow rate and steam flow rate.

4.2 Using the data generated in Item #3 above, compute the steam flow corresponding to the fuel oil flow rate. This value is the steam flow attributed to the fuel oil.

4.3 Steam generated from refuse is the difference between the steam flow rate of the boiler and the steam flow rate corresponding to the burner oil flow rate.

Figure 1 shows an actual plot of % fuel oil flow versus steam flow rate in thousands of lbs/hour. It can be readily seen that the data can be approximated with two linear regions, a low gradient and a high gradient region. Thus, the steam produced by the auxiliary burner can be approximated by the following equations:

$$Q_A = G_L * (F_O - 30) + 11.35 \tag{1}$$

$$\text{for } F_O < 74\%$$

$$Q_A = G_H * (F_O - 74) + 29.5 \tag{2}$$

$$\text{for } F_O >= 74\%$$

where Q_A: Steam flow from auxiliary burner

F_O: Fuel oil flow rate in %

G_H: High gradient = 0.663

G_L: Low gradient = 0.425

Note that the curve is divided into two linear regions with different gradients.

The steam flow from refuse is computed as follows:

$$Q_R = Q_T - Q_A \tag{3}$$

where Q_R = Steam flow from refuse.

Q_T = Total steam flow from the boiler, as measured.

Q_A = Steam flow from auxiliary burner, computed from equation (1) or (2).

Based upon the computed value of steam flow from refuse (Q_R), a proportional 4-20 ma signal was generated by the DCS, and provided to Boiler Vendor's Control system for the control of boiler and the burner management system. Q_R was totalized on shift and monthly basis and reported on shift and monthly reports.

It should be pointed out that the method described here for computation of steam flow from refuse is oversimplified, because accuracy of the result was not important. One of the major limitations of this approach is that the efficiency of the boiler with one fuel (oil only) is higher than the efficiency when oil and refuse are burnt simultaneously. Therefore, the graph shown in Figure 1 yields values of steam flow that are higher than the actual values generated for equivalent oil flows when both fuels are burning. A correction can be applied to Figure 1, and the graph can be "lowered" to account for the lower efficiency of actual operation, to improve the accuracy of the result. This correction was not applied in the case described here because the inaccuracy resulted in lower, more conservative values of steam flow from refuse.

MEASUREMENT OF BOILER DRUM LEVEL

The problem described herein is a common occurrence when a well meaning software engineer forgets to apply density corrections while developing engineering unit conversion routine for the boiler drum level signal.

The specific problem that occurred is described as follows:

> As discussed before, the plant had three boilers. For each boiler the control room had two separate means of drum level indications, as required by the ASME code. One was a hard wired visual level indication system and the other a display of drum level on DCS graphics, generated by converting a drum level transmitter signal into engineering units. During startup, one of the operators noticed that when the drum was at room temperature the reading from DCS and the hard wired level indicator were very close. However, during operation of the boiler the two level readings differed significantly. The operator correctly concluded that there was something wrong with the DCS level reading.

The DCS application programs were reviewed, and it was found that density correction was not applied while converting the signal to engineering units. A brief analysis of this problem follows:

> Figure 2 shows the typical level transmitter connections to the boiler drum. The net transmitter differential pressure (ΔP) is given by

$$(\Delta P) = h_3 - (h_1 * G_S + h_2 * G_W) \tag{4}$$

All quantities are expressed in inches of water at room temperature, and the variables are defined as follows:

$h_1 * G_S$ = equivalent head of a column of saturated steam having a length from the upper drum pressure connection to the water level. G_S is the specific gravity of steam.

$h_2 * G_W$ = equivalent head of a column of water at saturation temperature having a length from the water level to the lower drum pressure connection. G_W is the specific gravity of water in the drum.

h_3 = equivalent head of the water column in the wet leg at ambient temperature, equal to the distance between the two pressure taps. Note that the specific gravity of water in the wet leg is assumed to be 1.

It should be noted that the same boiler drum pressure is applied to both drum taps and thus it cancels out in equation (4). The distance between the taps, h_3 is fixed and the value is 30" in this specific case. Taking into account the specific gravities of steam and water, expressing h_1 as $(30 - h_2)$ and simplifying, equation (4) can be rewritten as:

$$\Delta P = 30\,(1 - G_S) - h_2 * (G_W - G_S) \qquad (5)$$

where G_S, and G_W are the specific gravities of steam and water respectively. G_S and G_W vary with drum pressure and temperature. The range on the transmitter can be decided by looking at ambient and maximum operating conditions. For ambient operating conditions set $G_S = 0$ and $G_W = 1$, and replace h_2 with h_{2A}, the drum level at room temperature. Equation (5) can be rewritten as:

$$\Delta P = 30 - h_{2A} \qquad (6)$$

ΔP varies from 0" to 30" of water, as the level at ambient temperature, h_{2A} varies from 30" to 0". The maximum operating conditions for each boiler were 760 PSIA at 512°F. For these conditions replace h_2 with h_{2M} and set G_W = 0.7734 and G_S = 0.0267, as read from steam tables. Equation (5) can be rewritten for maximum operating conditions as follows:

$$\Delta P = 30\,(1 - .0267) - h_{2M}\,(.7734 - .0267)$$

where h_{2M} is the drum level at maximum operating conditions.

$$\Delta P = 29.2 - 0.7467\,h_{2M} \qquad (7)$$

when $h_{2M} = 0$; ΔP = 29.2" of water

when $h_{2M} = 30$; ΔP = 6.8" of water

From above analysis, the transmitter calibration of 0-30" of water checks out to be adequate for maximum operating conditions. The transmitter produced an analog output of 4 to 20 mA proportional to 0 to 30" of differential pressure. Thus ΔP can be expressed in terms of milli-amp signal output as follows:

$$\Delta P = (30/16) * (X - 4) \qquad (8)$$

where X is the value of the signal within the range of 4 to 20 mA.

Equations (5) and (8) are equivalent. Equating the right hand sides of both these equations, an expression for h_2 can be developed as a function of the transmitter milli-amp signal.

$$h_2 = 30 * [(1 - G_S)/\,(G_W - G_S)] - 1.875 * [(X - 4)/(G_W - G_S)] \qquad (9)$$

In equation (9), h_2 is the actual compensated level of water in the boiler drum, and X is the transmitter signal. G_W and G_S are the specific gravities of water and steam. As a check, for ambient conditions setting $G_W = 1$, $G_S = 0$ and $h_2 = h_{2A}$, neglecting the effect of pressure and temperature we get:

$$h_{2A} = 30 - 1.875\,(X - 4) \qquad (10)$$

Equation (10) converts milli-amp signal to drum level at ambient conditions. As expected, according to equation (10), at 4 mA the drum level is 30", and at 20 mA the level is 0".

It is customary to set the drum level zero at half full, thus the 0 to 30" range is converted to -15" to +15" range by applying zero elevation. Equations (9) and (10) can be rewritten as follows:

$$h_2 = 30 * [(1 - G_S)/(G_W - G_S)] - [1.875/(G_W - G_S)] * (X - 4) - 15 \quad (11)$$

$$h_{2A} = 15 - 1.875 (X - 4) \quad (12)$$

Equation (11) was used to convert the mA signal from the transmitter to the actual drum level. The values of specific gravities G_W and G_S were stored in a look-up table for different drum pressure and temperature points. Values of G_W and G_S corresponding to a specific operating point were used to compute the correct compensated level. The signal conversion algorithm in the DCS was changed to include equation (11). After this change was made, the drum level indicated on the DCS graphic display closely followed the value shown by the hard wired level indicating system, and the controversy was resolved. To further emphasize the importance of density correction, let us compute the boiler drum level using equations (11) and (12) when the transmitter signal is 12 mA. Assume that the boiler is operating at maximum operating conditions, i.e., $G_W = .7734$ and $G_S = 0.267$.

The density compensated drum level computed from equation (11) is 4.024" whereas the uncompensated drum level computed from equation (12) is 0", which is in error because the effects of boiler pressure and temperature are ignored in computing this value.

CONCLUSIONS

No matter how many times you do it, a power plant startup is always a challenging event for an I&C engineer. Most of the troubles can be solved using simple practical solutions if the situation is correctly understood. A problem may be a computer hardware or software related, a field sensor malfunction, incorrect wiring or a combination of all, but one should concentrate on solving it rather than "passing the buck". As it turns out, both problems discussed here were software related and had very simple solutions.

ACKNOWLEDGEMENT

The author wishes to thank Mr. T. P. Davis of Sandwell Inc. for many helpful review comments that improved the quality of this paper.

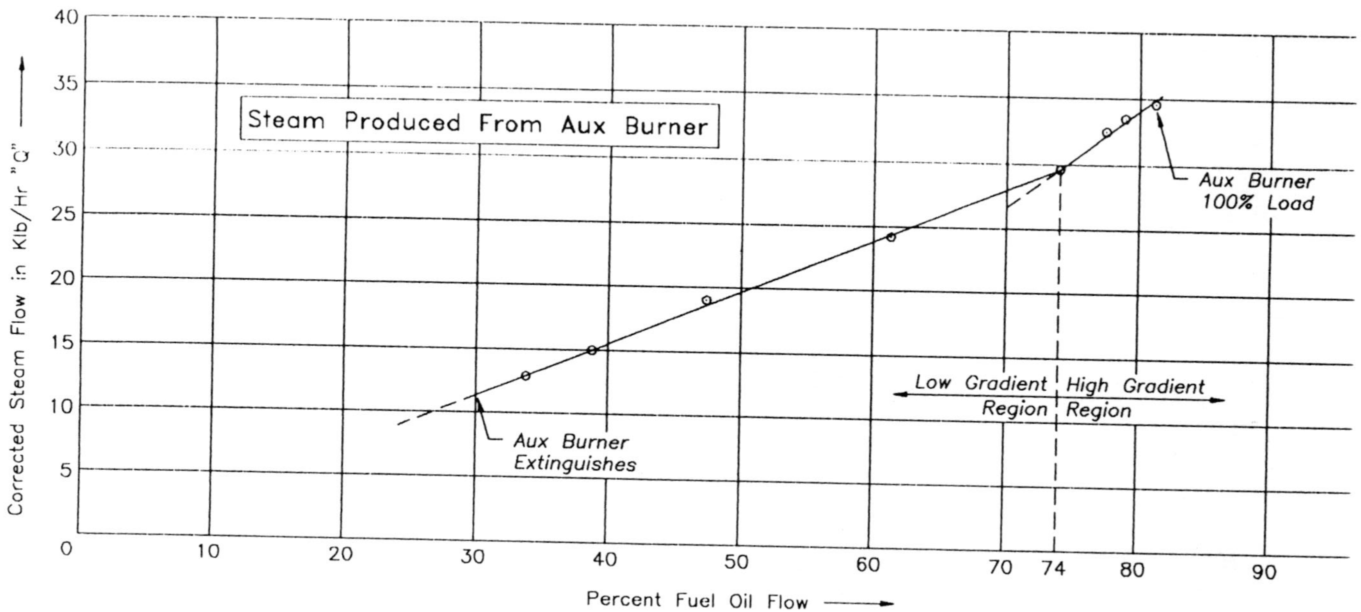

FIGURE 1 – STEAM PRODUCED FROM AUXILIARY BURNER

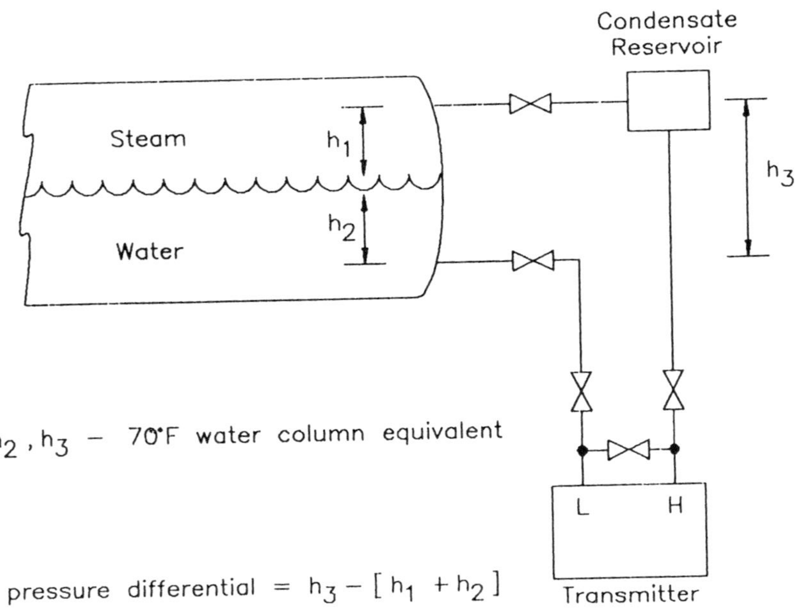

h_1, h_2, h_3 – 70°F water column equivalent

Net transmitter pressure differential = $h_3 - [h_1 + h_2]$

FIGURE 2 – DRUM LEVEL TRANSMITTER – CONNECTION AND CALIBRATION

© 1991 - Paper #91-717
0074-056X/91/233-265/$0 + .50pp

RESPONSE TIME TESTING AND CALIBRATION OF RTDs IN CONJUNCTION WITH BY-PASS MANIFOLD ELIMINATION PROJECTS

H. M. Hashemian
K. M. Petersen

Analysis and Measurement Services Corporation
AMS 9111 Cross Park Drive
Knoxville, TN 37923
(615) 691-1756

ABSTRACT

This paper presents a review of response time and calibration tests performed on thermowell-mounted Resistance Temperature Detectors (RTDs) that are installed in Westinghouse plants in conjunction with RTD by-pass manifold elimination projects. These tests involve using the Loop Current Step Response (LCSR) and the self heating methods for response time testing, and the cross calibration method to verify the consistency and assess the accuracy of the RTDs after installation. These tests are performed remotely on each RTD as installed in the plant in the normal configuration for service.

1. INTRODUCTION

In Westinghouse plants where RTD by-pass manifold elimination projects are implemented, direct immersion RTDs are replaced with thermowell-mounted RTDs which usually have a larger response time due to the thermowell. This often reduces the margin between the desired response time and the response time that can be achieved and maintained with thermowell-mounted RTDs. In addition to response time, the implementation of the by-pass manifold elimination projects affect the accuracy requirements of the primary coolant RTDs. More specifically, tighter tolerances are required of the new thermowell-mounted RTDs than were required of the old direct immersion RTDs. To insure that the desired tolerances are achieved and maintained, in-situ response time testing and on-line cross-calibration tests are performed on the RTDs upon installation and subsequently once every fuel cycle.

1.1 Response Time Testing Requirements

Thermowell-mounted RTDs offer many advantages over direct immersion RTDs but have an important disadvantage. That is, the increased response time and the sensitivity of response time to installation of the RTD into its thermowell. Therefore, it is crucial to ensure that these RTDs are properly installed in their thermowells to be able to achieve and maintain a suitable response time. This consideration has stimulated the following series of laboratory and in-plant response time tests for new thermowell-mounted RTDs that are installed in nuclear power plants:

- Laboratory testing of dynamic response characteristics of representative RTDs and thermowells to check for RTD/thermowell interchangeability problems and to verify that all the RTDs and thermowells to be installed in the plant have comparable response times and that the response times are suitable to meet the plant's technical requirements. In addition, some laboratory tests are usually performed, especially on newly designed RTDs or thermowells, to verify the testability of the RTDs with the Loop Current Step Response (LCSR) method and document the accuracy of the LCSR results for the RTDs.

- In-plant testing at cold shutdown upon the installation of the new RTDs to identify and resolve any RTD performance problems without an impact on the plant operation. Since the response time requirements in nuclear power plants are usually close to the best response time achievable with a thermowell-mounted sensor, there is usually little margin for any factor that can increase the response time.

- In-plant testing at hot standby conditions at the end of the outage to measure the actual in-service response times of the RTDs necessary to satisfy the plant technical specification requirements. The hot standby tests are needed because the response time of any temperature sensor depends on the fluid flow rate and fluid temperature in which the sensor is used. These effects can only be accounted for if the tests are performed under normal plant operating conditions.

The laboratory tests usually use a combination of plunge, LCSR, and self heating methods to characterize the dynamic response of the RTDs and provide baseline response time results. The tests are performed on the RTDs with and without their thermowells.

The in-plant tests at cold shutdown and hot standby conditions are performed remotely from the control room or cable spreading room where the RTD field leads reach their temperature transmitters. The LCSR and self heating methods are both used in performing the cold shutdown and the hot standby tests. The tests require approximately 30 minutes per RTD and two RTDs are usually tested at a time. If the work is to be completed quickly, more than two RTDs may be tested at a time. The details of the LCSR and self heating methods are described in Section 5.

1.2 Calibration Requirements

In light of temperature streaming problems inherent in the hot leg pipes of PWRs, the implementation of a by-pass manifold elimination project can affect the accuracy requirements of the primary coolant RTDs. More specifically, the new RTDs that are installed in conjunction with the RTD by-pass manifold elimination projects are often required to yield better accuracies than the old RTDs. This is because with the new system, the bulk temperature of the reactor coolant is obtained from an arithmetic average of three RTDs as opposed to the old system in which the reactor coolant was first well-mixed in the by-pass manifolds, and its average temperature was then measured. Since the uncertainties of arithmetic averaging of temperatures may be larger than that of natural temperature averaging by well-mixing of the water, it is important for the RTDs in the new system to be as accurate as possible to limit the uncertainties within an acceptable band.

In order to verify that a group of RTDs have adequate accuracies, the cross-calibration technique as described in Section 6 may be used. The cross calibration tests are performed at hot standby conditions before or after the response time tests. If the cross calibration schedule coincides with the "critical path" period, in order to avoid critical path delays, the tests can be performed at lower temperature plateaus or during heat up and the results extrapolated to normal operating conditions. The design of the cross-calibration test, the amount of data collected, and the details of the test procedure depends on the desired accuracy of the cross-calibration results.

2. ADVANTAGES OF RTD BY-PASS MANIFOLD ELIMINATION

By-pass loops are used in Westinghouse plants to correct for temperature stratification problems in the hot leg pipes and to provide for easy replacement of RTDs. This is accomplished by directing several samples of the reactor water through by-pass pipes to a manifold where the temperature of the water, which at this point is well-mixed, is measured with direct immersion RTDs installed in the by-pass manifold. The disadvantage of this system is that it adds 300 to 400 feet of piping with associated valves and other hardware to the plant's primary coolant system, and creates costs and maintenance problems. Table 1 shows some examples of the advantages of removing the by-pass manifolds in Westinghouse pressurized water reactors (PWRs).

In addition to creating maintenance problems, experience has shown that the by-pass loops are a source of nuclear radiation due to radioactive material that becomes trapped in the system. These problems have recently motivated the nuclear industry to remove the by-pass loops and install well-type RTDs directly into the reactor coolant pipes. As of the end of March 1991, fifteen nuclear power plants have successfully implemented RTD by-pass manifold elimination projects. A listing of these plants is given in Table 2 along with dates of response time tests performed at cold shutdown and hot standby conditions to verify proper dynamic performance.

Figure 1 shows a simplified schematic of a primary coolant loop in a PWR with and without the by-pass manifolds. The system with the by-pass manifolds uses direct immersion RTDs with fast response times to compensate for the time lag required for the primary coolant to reach the by-pass manifold in which the RTDs are installed. This time lag is about 2 to 2.5 seconds in most plants and is usually referred to as the transport time delay.

TABLE 1

Advantages of RTD By-Pass Manifold
Elimination Projects

Eliminates

- 300 - 400 feet of RCS pressure boundary piping
- 60 - 70 valves
- More than 100 hangers
- More than 60 snubbers

Reduces

- Man-Rem exposure
- Forced outages
- In-service inspection

TABLE 2

Plants Where RTD By-Pass Manifold Elimination Projects Have Been Implemented As of March 1991

Plant	RTD Manufacturer	RTD Model	Number	Response Time Tests Cold	Response Time Tests Hot
Salem I	Weed	N9004E-2A	16 Dual	11/87	01/88
Millstone III	Weed	N9004S-2B N9004E-2B	12 Single 4 Dual	01/88	02/88
Catawba II	RdF	21232-1	20 Single	02/88	02/88
McGuire II	RdF	21232-1	20 Single	07/88	07/88
McGuire I	RdF	21232-1	20 Single	11/88	12/88
Salem II	Weed	N9004E-2A	16 Dual	N/A	11/88
H. B. Robinson	Weed	N9004E-2B	12 Dual	01/89	02/89
Indian Point III	Weed	N9004S-2B N9004E-2B	12 Single 4 Dual	06/89	06/89
Beaver Valley I	Weed	N9004S-2B N9004E-2B	9 Single 3 Dual	N/A	12/89
Catawba	Weed	21232-1	20 Single	03/90	04/90
Sequoyah I	RdF	21465-1	20 Single	05/90	05/90
V. C. Summer I	RdF	N9004E-2B	12 Dual	05/90	05/90
Beaver Valley II	Weed	N9004S-2B N9004E-2B	9 Single 3 Dual	10/90	11/90
Callaway	Weed	N9004E-2A-SP	16 Dual	10/90	11/90
Sequoyah II	RdF	21465-1	20 Single	10/90	11/90

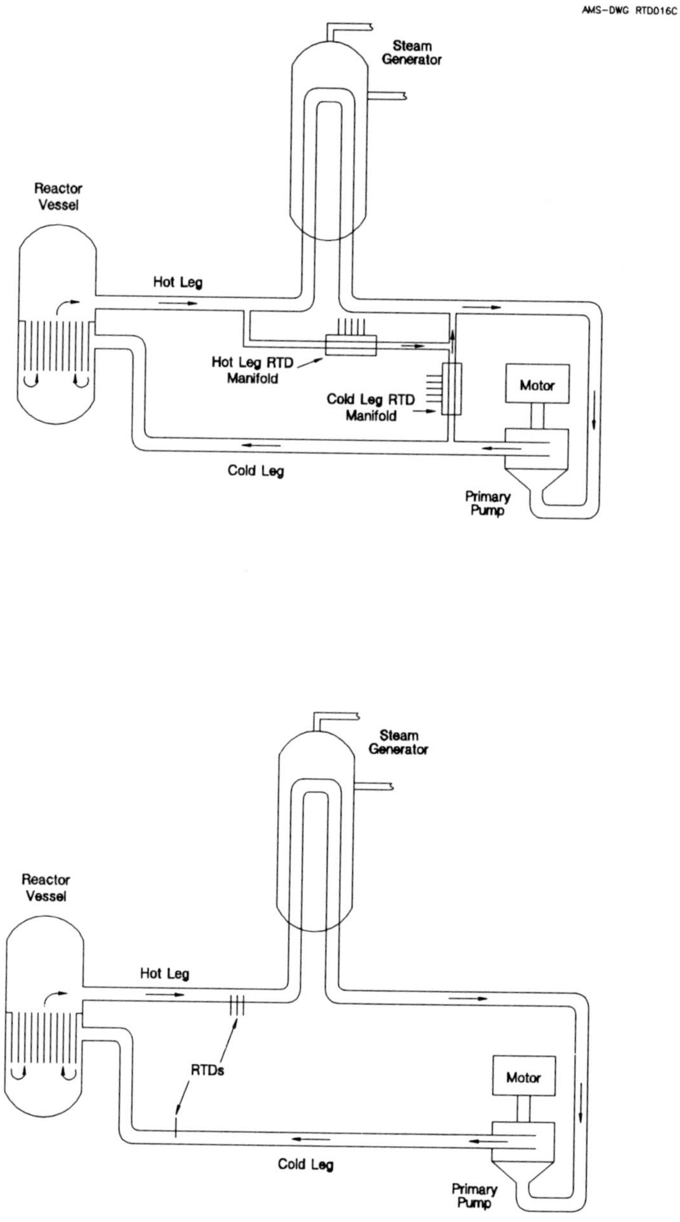

Figure 1. A Typical PWR Primary Coolant Loop With Without RTD By-Pass Manifolds.

Table 3 shows an example of typical system response times before and after RTD by-pass removal[1]. The RTD response time of 4.75 seconds shown in Table 3 includes a 10 percent allowance for accuracy of the Loop Current Step Response method. Note that the slow response of thermowell-mounted RTDs is partly compensated for by the transport time delay that is eliminated when the RTDs are installed directly in the reactor coolant hot leg and cold leg pipes. The 0.25 seconds shown in Table 3 is for the transport delay inside the hot leg scoops in which the RTDs are located. There are three scoops 120° apart around each hot leg pipe in Westinghouse plants (Figure 2). At least two of the three scoops are in the same plane in the pipe and the third one is either in the same plane around a ring in the pipe or somewhat displaced a few feet away from the reactor closer to the steam generator.

In the original by-pass manifold design, scoops were used to sample the reactor water from three locations in the pipe and direct them to the by-pass manifolds to provide a well-mixed water sample for an average temperature measurement. In the new system, this is accomplished by simply averaging the reading of the RTDs in the hot leg pipes. It is because of this simple averaging scheme that very high accuracy is often required of the new RTDs that replace the old direct-immersion RTDs. The high accuracy is necessary to ensure that the temperature measurement uncertainty with the new RTDs is not increased with implementation of the RTD by-pass manifold elimination project. More specifically, because of temperature streaming problems inherent in most PWRs, it is crucial to ensure that the bulk temperature identified from the average of three hot leg RTDs have the same or a better accuracy than the temperature measured in the by-pass manifolds where the water is well-mixed.

The temperature streaming occurs in the reactor outlet temperature when the reactor is at power. Due to radial power and flow distributions in the core, the reactor coolant temperature is often non-uniform as it leaves the core. As the reactor coolant flows through the hot leg pipe, mixing causes the non-uniformity to diminish. However, since the primary coolant RTDs are usually located only a short distance from the core, temperature stratification still exists when the reactor coolant reaches the RTDs. Typical temperature differences around the pipe at the plane of the RTDs vary from several degrees near the pipe wall to several tenths of a degree around the sensing tip of the RTDs. The temperature gradients across the hot leg pipes are usually not stationary or well defined.

3. PRE-INSTALLATION LABORATORY TESTING OF NEW RTDs

3.1 Response Time Testing

Prior to installation of new RTDs and new thermowells into a plant, it is often important to perform a series of independent laboratory tests to verify that the RTD/thermowell assemblies have a good chance of providing the desired response time. These tests are especially important and useful for new RTDs when they are first introduced to the nuclear industry or after a design or a product improvement change in the RTD or the thermowell.

TABLE 3

System Response Time Before and After Removal of RTD By-Pass Manifolds

Response Component	Time Constant (sec)	
	W/By-Pass	W/O By-Pass
RTD	3.0	4.75
Electronic	1.0	1.0
Transport or Mixing	2.0	0.25
Total	6.0	6.0

W/By-Pass: *Before removal of the manifolds*
W/O By-Pass: *After removal of the manifolds*

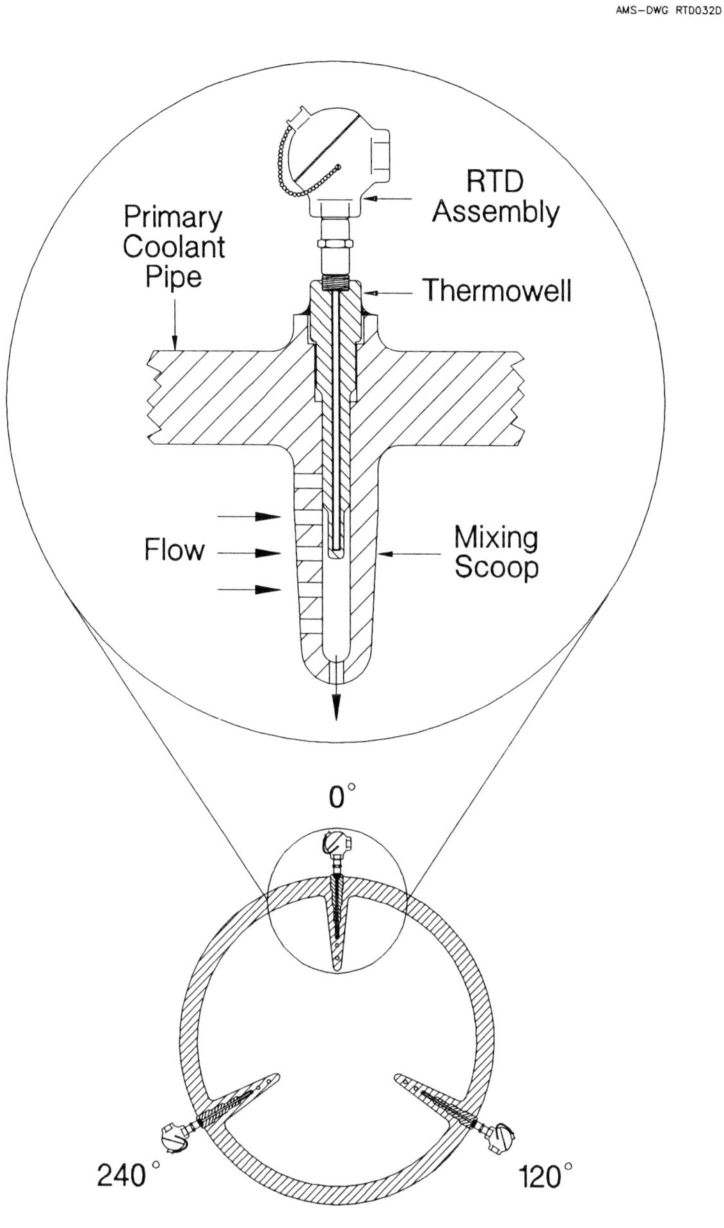

Figure 2. A Typical RTD Installation in a Hot Leg Pipe
After the By-Pass Manifold is Eliminated.

The laboratory tests should encompass both the calibration and the response time testing of the RTDs. For response time, the RTDs are often tested with and without the thermowells. A thermowell-mounted RTD is referred to as "bare" when it is tested without the thermowell. Table 4 shows response time results for three identical thermowell-mounted RTDs tested bare and in three identical thermowells. Note that the response time varies from one thermowell to another and that there is a not a strong correlation between the bare response time of the RTDs and corresponding response times in thermowells.

In addition to RTD/thermowell interchangeability tests that are performed in the laboratory using the plunge test method, the Loop Current Step Response testability of new RTDs to be installed in a plant must be experimentally verified and the accuracy of the method established for the type of RTD in hand. This is accomplished by performing laboratory plunge and LCSR tests on a sample of new RTDs to demonstrate that the response time results from the two tests, performed under the same temperature and flow conditions, agree to within better than approximately 10 percent.

3.2 RTD Calibration

In addition to response time measurements, the calibration of new RTDs should be checked prior to installation into the plant. This is especially important for those RTDs that have been in storage more than a year. It has been shown that RTDs can loose their calibration while in storage and should therefore be calibrated after any long period of storage[(2)]. A four point calibration in an ice bath and an oil bath is usually adequate for checking the RTD's calibration. The results should be compared with that of the manufacturer's to verify that the differences are small (e.g., less than 0.1° to 0.2°C). If the differences are large compared to the desired accuracy for the RTDs, then the RTDs should be recalibrated and new calibration tables generated. Figure 3 shows typical examples of the differences between the original manufacturer's calibration and the calibration performed several months later for two nuclear-grade RTDs. Each plot in Figure 3 represents the differences between the calibration tables generated with the Callendar equation for each RTD tested at two different times. The differences are plotted as a function of temperature over the range of 0°C to 300°C. The shape of the difference curves are often parabolic for calibration tables that are generated with the Callendar equation.

Generally, for any nuclear grade RTD that shows more than about 0.2°C (~0.4°F) difference anywhere in the temperature range of interest, a new calibration table should be generated to replace the old calibration table. If it is determined that a difference of more than 0.2°C is due to calibration shift rather than an error in the original calibration, it is best to anneal the RTD and then recalibrate it to generate a new table.

Annealing is usually a good remedy for reducing the chances of future drift. Figure 4 shows the stability of a group of nuclear grade RTDs before and after annealing. The annealing for these RTDs was performed in a furnace at a temperature of 400°C for about 24 hours. The data shown in Figure 4 represent the average calibration drift of the RTDs over a one week period before and after annealing. Each RTD was calibrated daily for a week and the drift results for the week were averaged to obtain the data shown in Figure 4.

TABLE 4

Laboratory Plunge Test Results From
A Study of RTD/Thermowell Interchangeability

RTD Number	Time Constant (sec)			
	Bare	Well #1	Well #2	Well #3
1	1.7	4.8	5.2	6.0
2	1.8	3.6	4.1	4.6
3	1.9	4.4	4.9	5.9

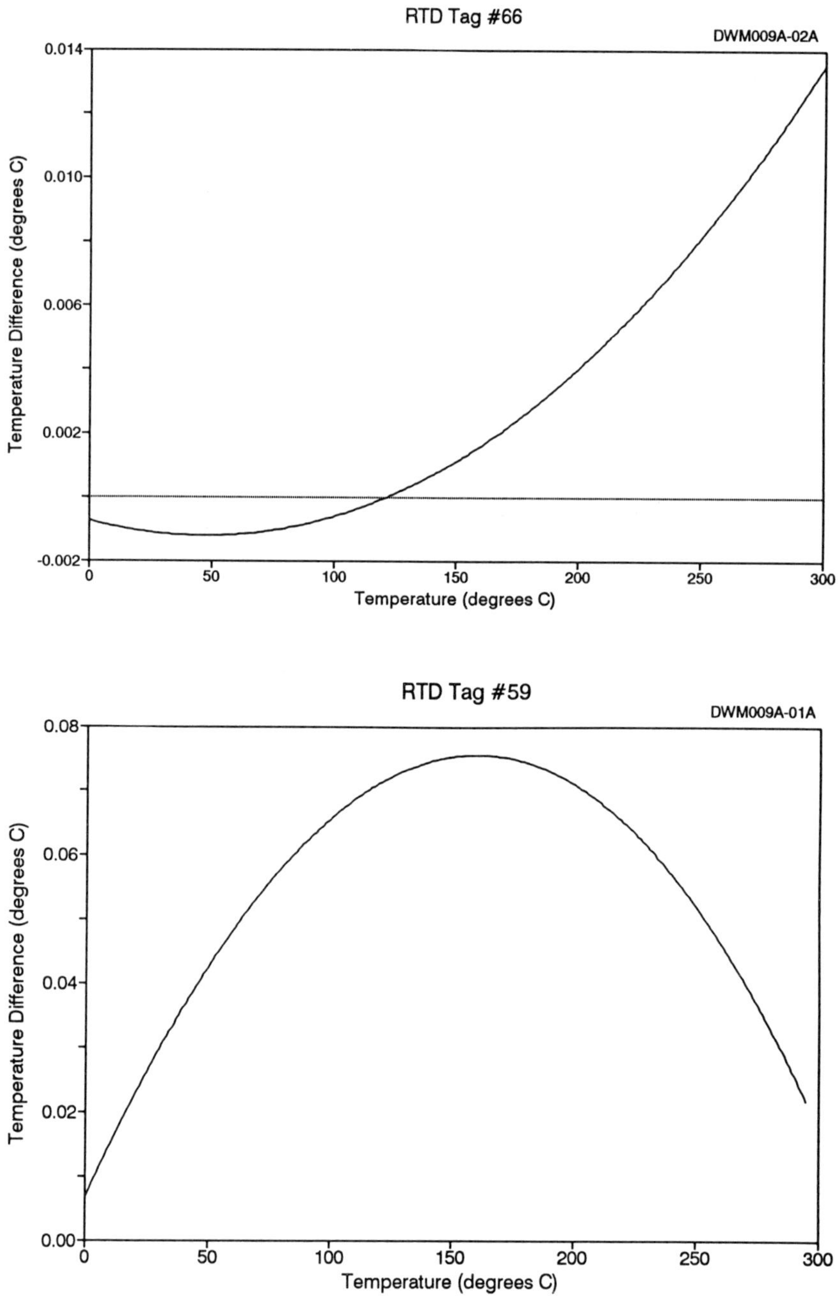

Figure 3. Examples of Calibration Difference Plots for Two RTDs.

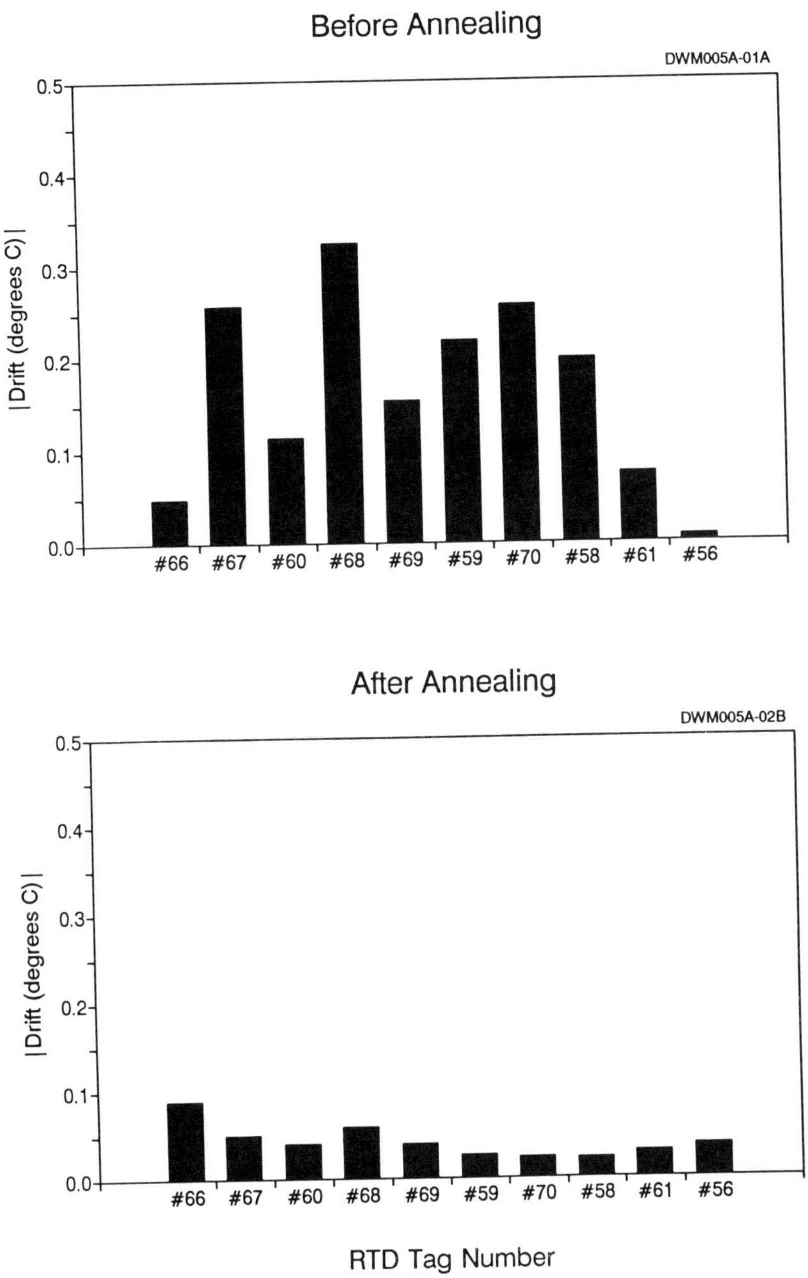

Figure 4. RTD Stability Before and After Annealing.

4. POST-INSTALLATION RESPONSE TIME TESTING OF NEW RTDs

Two series of tests using the Loop Current Step Response and self heating test methods are usually performed upon installation of new RTDs in nuclear power plants. These tests are referred to as cold shutdown and hot standby tests as described below.

4.1 Cold Shutdown Tests

The tests at cold shutdown are performed at the ambient temperature and flow conditions after the new RTDs are installed in their thermowells in the plant, their wiring is completed, and the primary coolant loops are filled. As the test results at cold shutdown depend strongly on the water flow rate, it is desirable that all RTDs are tested at approximately the same flow rate. Stagnant water or low flow conditions which are typical in the reactor coolant system during a refueling outage are usually acceptable for the cold shutdown tests. A small flow induced by the residual heat removal (RHR) system is often present when the plant is at cold shutdown. The flow induced in the primary coolant pipes by the RHR system sometimes helps and sometimes interferes with the cold shutdown tests. The test personnel must therefore be aware of the RHR flow and account for it in interpretation of the cold shutdown test results. The RHR system is normally operated to remove heat from the core during cold shutdown periods. It takes suction from one of the hot leg pipes, and after passing through a heat exchanger, discharges to one or more of the cold leg pipes. In some plants, the RHR flow configuration can be switched to take suction from different hot legs and to discharge to different cold legs. The changes that can be made in the flow configuration of RHR system are often useful in inducing flow around suspect RTDs to verify response time problems that are identified during the cold shutdown tests.

The cold shutdown test data are analyzed on-site and the time constants and self heating indices for all RTDs are intercompared to identify the outliers (if any). These tests have proven useful in identifying gross installation problems with thermowell-mounted RTDs in nuclear power plants. Examples of problems that are encountered with thermowell-mounted RTDs which have been detected in the past by response time testing at cold shutdown include: bent RTDs, bent thermowells, RTD/thermowell mismatch, insufficient RTD-in-thermowell insertion due to dimensional or mechanical problems, dirt in thermowells, and wiring problems that are not readily detectable by routine measurement of electrical loop and insulation resistances. Table 5 shows some examples of response time results from cold shutdown tests in a few PWRs.

Note that the cold shutdown results are useful only for verification of proper installation and not for measuring the actual in-service response times of the RTDs. The "in-service" response times are measured at hot standby conditions as described below.

In addition to RTD by-pass manifold elimination projects, the cold shutdown tests are used in plants where new RTDs are installed or existing RTDs are removed and replaced during an outage.

TABLE 5

Examples of Response Time Problems Identified and Resolved at Cold Shutdown

Time Constant (sec)		Problem Resolution
As Found	As Left	
11.6	4.7	Cleaned Thermowell
22.5	7.5	Cleaned Thermowell
14.7	5.9	Cleaned Thermowell
15.5	6.2	Cleaned Thermowell
37.4	13.0	Cleaned Thermowell
9.0	5.0	Reseated RTD
18.0	14.0	Reseated RTD
19.2	9.5	Reseated RTD
14.5	5.4	Installed New RTD
24.0	7.8	Installed New RTD
24.0	17.0	Debris Removed From Thermowell
27.9	5.8	RTD Moved to a Different Thermowell

Note: The response time results shown in this table are from response time tests performed at cold shutdown conditions in several PWRs. The large variation in the response time values given in this table are due to the fact that some of the RTDs were tested in stagnant water conditions and some in flowing water conditions. These results do not represent the typical response times of the RTDs.

4.2 Hot Standby Tests

The response time of an RTD depends on the temperature, pressure, and flow conditions of the process in which the RTD is used. Since these conditions can not be simulated in laboratory tests, the actual in-service response times of RTDs are measured by in-situ testing at or near normal operating conditions. The Loop Current Step Response (LCSR) and self heating test methods are used in performing the in-situ tests. For newly installed RTDs, these tests are performed at the hot standby condition at the end of the outage during which the RTDs were installed. In most plants, the hot standby condition during which the tests are performed is referred to as "Mode 3". It is required in most plants that the response times of newly installed RTDs be measured at hot standby conditions (Mode 3) to verify that the response times are acceptable before the plant can proceed toward power operation.

The response time results at the hot standby conditions are expressed in terms of a time constant for each RTD element that is tested. The time constant is defined as the time that would be required for the RTD to reach 63.2 percent of its final output following a step change in the reactor coolant temperature. Table 6 shows typical time constant results from in-plant testing of sixteen RTDs as installed in a PWR at hot standby conditions.

In addition to the time constant, the self heating index of each RTD element in terms of ohms per watt (Ω/watt) is recorded as a part of the response time test results. The self heating test results are not usually used in any acceptance criteria and are considered as useful additional data for explaining aging degradation and gross RTD performance problems.

5. DESCRIPTION OF LCSR AND SELF HEATING TEST METHODS

5.1 LCSR Test

The response time of nuclear plant RTDs are measured using the Loop Current Step Response (LCSR) method. This method is based on heating the RTD sensing element with a small electric current (40 to 80 milliamperes) applied to the RTD extension leads. The current causes a temperature transient in the RTD that can be analyzed to give the RTD time constant. A typical LCSR transient is shown in Figure 5. This transient is the average of twenty LCSR transients obtained in a PWR at hot standby conditions. Depending on the noise level of the signal, the LCSR test is repeated ten to twenty times and the raw data is then averaged and analyzed. The analysis of the LCSR data involves a computer fitting of the data to an equation of the following form:

$$T(t) = A_0 + A_1 e^{-t/\tau_1} + A_2 e^{-t/\tau_2} + \ldots \quad (1)$$

where

$T(t)$	=	RTD output from the LCSR test (called LCSR transient or LCSR data)
$A_0, A_1, A_2, \ldots$	=	constants referred to as modal coefficients
$\tau_1, \tau_2, \ldots$	=	modal time constant.

TABLE 6

Examples of a Set of In Service Response Time Results for Primary Coolant RTDs in a PWR

Item	Tag Number	Time Constant (sec)
1	412A	2.9
2	412B	3.5
3	412C	2.7
4	412D	2.7
5	422A	3.3
6	422B	2.6
7	422C	2.9
8	422D	3.0
9	432A	3.0
10	432B	2.8
11	432C	2.8
12	432D	2.6
13	442A	3.8
14	442B	2.8
15	442C	3.2
16	442D	2.7

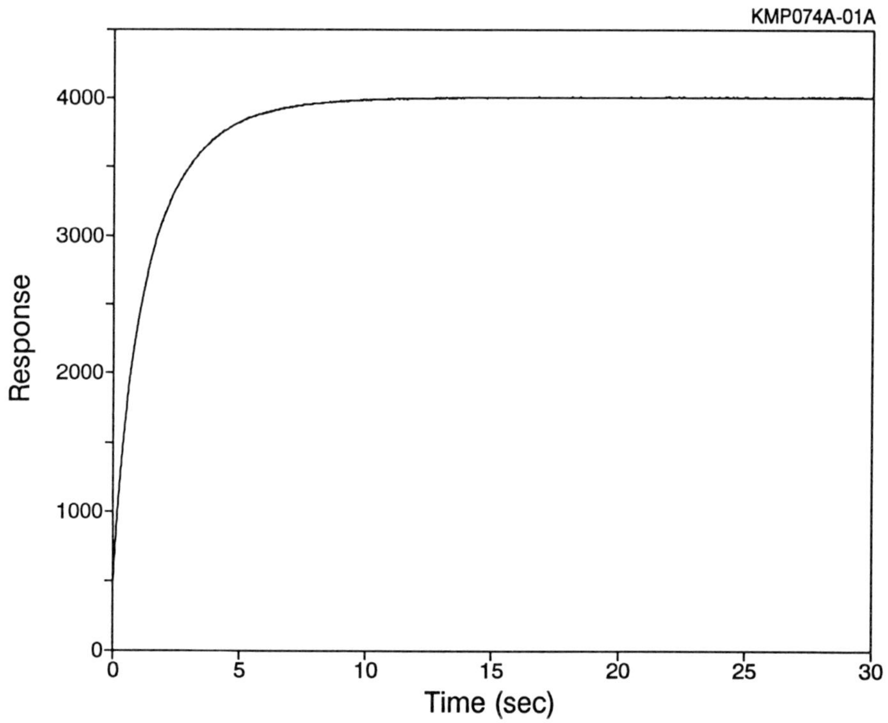

Figure 5. A Typical LCSR Transient From In-Plant Testing of an RTD in a PWR.

The fit of the averaged LCSR data to Equation 1 results in the modal time constants that can be used in the following equation to provide the overall time constant of the RTD:

$$\tau = \tau_1 \left[1 - \ln\left(1 - \frac{\tau_2}{\tau_1}\right) - \ln\left(1 - \frac{\tau_3}{\tau_1}\right) \ldots \right] \quad (2)$$

Although the LCSR data results from internal heating of the RTD, the above time constant (τ) is the same as the time constant that would be obtained for the RTD if the fluid temperature around the RTD undergoes a step change. The validity of the LCSR test has been established by numerous laboratory measurements with typical nuclear plant RTDs[(3)].

The advantage of the LCSR test is that it permits on-line testing of installed RTDs from outside the reactor containment and provides the actual "in-service" time constants of the RTDs. The test accounts for all installation and process condition effects on RTD response time.

The accuracy of the LCSR test results for the RTDs for which the LCSR method has been validated is ± 10 percent provided that the test data is properly analyzed. Any RTD to be tested by the LCSR method must first undergo a comprehensive series of laboratory tests to verify the validity of the LCSR method for measurement of the RTD time constant, and to establish the accuracy of the time constant results for the particular RTD design. If the RTD is thermowell-mounted, the LCSR validation must be performed with the RTD as installed in its thermowell. The comprehensive laboratory validation tests are necessary because the validity of the LCSR test depends on certain assumptions about the RTD heat transfer that must be satisfied to ensure that accurate results are obtained from the LCSR method.

5.2 Self Heating Test

The self heating test is often performed on nuclear plant RTDs in addition to the LCSR test. The self heating test is also based on heating the RTD with a small electric current. In this test, the changes in steady state resistance of the RTD are measured for different values of applied current, and plotted as a function of the applied electric power. For platinum RTDs, the plot is a straight line whose slope (called self heating index) is proportional to the RTD response time. Figure 6 shows a typical self heating curve for an RTD element in a PWR tested at normal operating conditions.

The self heating test is performed in conjunction with the LCSR test to provide a complete set of data on the dynamic characteristics of each RTD. This test is often useful for monitoring for gross response time degradation of RTDs. Baseline self heating data on new RTDs are useful for diagnosis of aging effects on RTD performance.

It must be pointed out that the correlation between the self heating index and response time depends on the design of the sensing element of the RTD and the condition in which the RTD is tested. For some RTDs, the correlation is strong and for others the correlation is weak. A weak correlation means that only large changes in self heating indices are indicative of possible changes in RTD response time and vice versa.

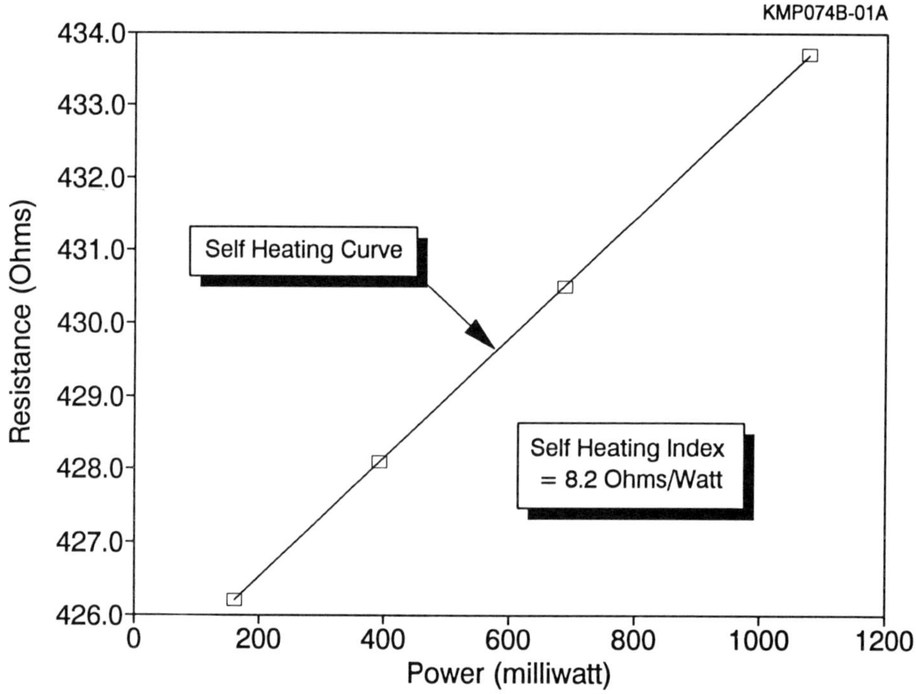

Figure 6. Typical Self Heating Data and Fit From an In-Plant Test.

5.3 In-Situ Response Time Testing Procedure

The basic steps in performing a LCSR or a self heating test in a nuclear power plant are:

1. Set up equipment in the control room or cable spreading room where the RTD field leads reach their signal conditioning equipment in the protection cabinets.

2. Disconnect the RTD field leads and connect them to the response time testing equipment to perform the LCSR and self heating tests.

3. Adjust the electrical current levels, the amplifier gains, and the analog to digital converter settings on the test equipment and execute the data acquisition software to perform the LCSR test and store the data. This is followed by performing the self heating test.

4. Analyze the LCSR data to identify the response time of each RTD that has been tested. The response time is expressed in terms of a time constant for each RTD tested. The self heating index is obtained by calculating the slope of the self heating curve in terms of ohms/watt. This is referred to as the self heating index (SHI) of the RTD.

A complete set of LCSR and self heating tests of an RTD as installed in an operating nuclear power plant usually requires about 30 minutes of testing. The best time to perform the response time tests is when the plant is operating normally at power prior to a refueling outage. ,This will provide the in-service time constants of the RTDs, and if any RTD is found to have a problem, it can be corrected during the outage and retested at the hot standby condition after the outage to obtain its time constant. Of course, for new RTDs such as those that are installed in conjunction with RTD by-pass manifold removal, the initial set of tests to measure the in-service time constants of the RTDs must be performed at the hot standby condition while the plant is coming out of the outage. All subsequent tests can be done at power except when an RTD is moved in the thermowell or replace, in which case, the test must be performed as soon as the plant is at or near normal operating conditions.

6. ON-LINE CALIBRATION OF RTDs

The steady state characteristics of RTDs are defined in terms of their calibration accuracy and stability. The initial accuracy of RTDs is determined from the laboratory calibration performed to generate the RTD calibration tables.

Once the RTDs are installed in the plant, a test called "cross calibration" may be performed on-line once every fuel cycle to verify that the RTDs are maintaining their consistency and to assess their installed accuracy. The best time to perform the cross calibration test is at the end of a fuel cycle when the plant is in the process of shutdown for a scheduled refueling outage, or at the end of an outage when the plant is at the hot standby condition.

The cross calibration test can be performed at one or more temperature plateaus at isothermal conditions when the reactor coolant hot leg and cold temperatures are at approximately the same temperature. To avoid performing the tests during "critical path" periods, the cross calibration test is sometimes performed at lower temperatures and the data are extrapolated to higher temperatures. Cross calibration can also be performed when the temperature is increasing or decreasing monotonically at a reasonably constant rate. The choice for when and how to perform the cross-calibration tests depends on the desired accuracy in the cross-calibration results.

Table 7 shows the results of a cross calibration test of sixteen single element RTDs. The test was performed in a PWR using a system which scanned the RTDs four times to measure their resistances. The resulting resistances were recorded as Pass 1 through 4 and the average of the four measurements were calculated for each RTD. The average resistances were converted to temperature and the resulting temperatures were averaged. This average temperature was then compared with the temperature indicated by each RTD and their deviations were calculated. Any RTD that deviated from the average by more than a predetermined criteria was flagged or excluded from the average depending on the magnitude of the deviation. In this case, the RTDs that deviated from the average by more than 0.3°C were flagged.

Table 8 provides the criteria used in seven nuclear power plants for excluding (from the average) or flagging of RTDs. The RTDs that are excluded (from the average) are usually replaced and those that have been flagged are often corrected by adding a bias in their temperature transmitter to null the deviation.

Cross calibration is sometimes performed at several plateaus to verify the accuracy of the RTD over a wide temperature range. In some plants, new resistance versus temperature charts are generated using the cross calibration data. A calibration chart generated with cross calibration data often has larger uncertainties than a calibration chart generated with laboratory measurements. The approach for generating a calibration chart using cross-calibration data is similar to performing a laboratory calibration without an ice point. The impact of calibration without an ice point was studied in the laboratory using six nuclear grade RTDs. A four point calibration was performed at 0, 100, 200, and 300°C, and the data were analyzed with and without the ice point. The results are shown in Table 9 in terms of the differences at ice point and 200, 280, and 300°C. These are the differences between the fitting results with and without an ice point. As expected, the differences are small at higher temperatures and large at 0°C.

Cross calibration is an effective method for verifying the consistency of a group of RTDs. The method does not account for common mode drift or any other systematic calibration problem unless one or more newly calibrated RTDs are included in the test. The decision on whether or not to include one or more newly calibrated RTDs in the cross-calibration, the number of plateaus at which cross-calibration data are collected, and other criteria for cross-calibration tests must be based on the accuracy requirements for the RTDs.

Until recently, cross-calibration was thought to be limited to a test of consistency between a group of RTDs. However, recent research[(2)] has proven that the drift of RTDs is random rather than systematic, and therefore

TABLE 7

In-Plant Cross Calibration Results

Temperature Plateau: 280°C

Tag	Resistance Measurements (Ohms) Pass 1	Pass 2	Pass 3	Pass 4	Avg. Res. (Ohms)	Temp. (°C)	Dev. (°C)
1	408.3075	408.3614	408.4015	408.3950	408.3663	278.070	-0.035
2	408.0508	408.1079	408.1378	408.1279	408.1061	278.243	0.137
3	408.0823	408.1500	408.1696	408.1619	408.1410	278.034	-0.072
4	408.4108	408.4825	408.4903	408.4813	408.4662	278.206	0.100
5	408.0488	408.1355	408.1286	408.1235	408.1091	278.161	0.055
6	408.2716	408.3567	408.3478	408.3457	408.3307	278.212	0.106
7	408.1202	408.1654	408.1850	408.1714	408.1605	277.759	#-0.347
8	408.3659	408.3917	408.4105	408.4059	408.3935	278.529	# 0.423
9	408.3339	408.1414	408.4245	408.4219	408.3986	278.047	-0.059
10	407.9606	408.0369	408.0359	408.0314	408.0162	278.053	-0.053
11	408.3245	408.3936	408.3777	408.3836	408.3698	278.061	-0.045
12	408.4128	408.4656	408.4527	408.4555	408.4466	278.124	0.018
13	408.2628	408.3532	408.3531	408.3370	408.3265	277.961	-0.145
14	407.9214	408.0120	408.0225	408.0074	407.9908	278.067	-0.039
15	408.5879	408.5910	408.5987	408.5848	408.5906	278.318	0.212
16	408.1690	408.1644	408.1806	408.1533	408.1668	277.847	-0.259
	Average Temperature (°C):					278.106	

 # Denotes that acceptance criteria of 0.3°C is exceeded.

TABLE 8

Examples of Cross Calibration Criteria
in Various Nuclear Power Plants

Nuclear Plant	No. of Repeats	No. of Plateaus	Acceptance Criteria (°C) Exclude	Flag	Max. Acceptable	Remarks
1	4	1	0.6	0.17	1.1	1
2	4	1	0.6	0.17	1.1	1
3	4	4	0.6	0.11	0.6	2
4	4	2	0.3	0.30	1.1	3
5	4	4	0.6	0.27	N/A	4
6	N/A	1	0.4	0.17	N/A	5
7	4	4	0.7	0.11	0.7	6

Plant stability criteria for RTD cross calibration is typically about ± 0.3 to ± 0.6°C.

Remarks:

1. *Cross calibration data is taken for any number of plateaus. However, only the data for 292°C is used to meet acceptance criteria and adjust the temperature transmitters as needed.*

2. *Data can be taken at a constant heat up rate. On sixteen RTDs, data is taken as follows: RTD number 1 to 16, reverse current 16 to 1, reverse current 1 to 16, etc. This presumably corrects for both the ramping temperature and for EMF effects (reversing the current). Data is taken around 95, 170, 230, and 292°C. This data is used to apply 2nd order fit to the data.*

3. *Two plateaus: 170°C and 292°C. For deviations greater than 0.17, the deviation at 170°C and 292°C is used to determine the error offset and the slope and apply the corrections to the transmitter.*

4. *Data is taken at four plateaus on sixteen RTDs sequentially, 1-16, 16-1, etc. The heat up rate is also measured.*

5. *The sixteen RTDs in this plant are tested one channel (four RTDs) at a time. Data is taken for 25 minutes at 5 minute intervals. This is repeated for all four channels. The plant stability requirement for the tests is 0.17°C, i.e., the temperature cannot change by more than 0.17°C from beginning to the end of any test run.*

6. *Tests are done at 120, 180, 230, and 275°C. Data is taken on sixteen RTDs, 1 to 16, 16 to 1, 1 to 16, 16 to 1. The four calibration points are used to determine a zero and a slope for the correction to temperature transmitters.*

   N/A: Data Not Available

TABLE 9

Calibration Errors Without Ice Point

Tag	Difference (°C) 0°C	200°C	280°C	300°C
15A	0.09	0.013	0.002	0.006
15C	0.09	0.013	0.002	0.005
16A	0.09	0.014	0.003	0.004
16C	0.07	0.011	0.002	0.004
17A	0.08	0.013	0.002	0.005
17C	0.09	0.016	0.004	0.005

Above are differences between the fitting results of a four point calibration with and without the ice point.

the accuracy of cross-calibration results can be made to approach that of a laboratory calibration especially if one or more newly calibrated RTDs are included in the cross-calibration process.

6.1 Cross-Calibration of RTDs in Eagle 21 System

The Eagle 21 system is a Westinghouse product that is used in the instrument cabinets in nuclear power plants for signal conditioning. In performing cross calibration with the Eagle 21 system, instead of measuring the resistance of the RTDs directly, a measurement of the voltage across the RTD element and the current in the RTD loop is made and the resistance is deduced from the measured voltage and current data. The advantage of this approach is that it does not require the RTDs to be disconnected during the cross-calibration process. The measurement of the current in the RTD loop in an Eagle 21 system is accomplished by measuring a voltage drop across a fixed resistor in the RTD current loop.

6.2 Cross Calibration of RTDs in Westinghouse 7300 Cabinets

In Westinghouse plants where the RTD signal conditioning is performed in 7300 cabinets, the cross calibration tests can be performed without having to lift any wires in the cabinets. The cabinets are equipped with test cards and electronic relays that can be used to switch a group of RTDs to a test condition for a few seconds so that their resistances can be measured. A suitable fixed resistor can be switched in to mimic the RTD for the few moments while the RTD resistance is being measured. This procedure is implemented on a computer controlled switching and simulation system used for cross calibration of RTDs that are used in Westinghouse 7300 cabinets.

6.3 Accuracy of Cross Calibration Method

The accuracy of cross-calibration results depends on the accuracy of the test equipment used, the procedure by which the work is implemented, and more importantly, by accounting for plant temperature fluctuations and any temperature differences between the hot leg and cold leg temperatures in different loops. Because of incomplete mixing and the differences in heat removal in steam generators, even at the best isothermal conditions, the hot leg and cold leg temperatures in the same loop or across the reactor are slightly different. Software has been developed to account for these differences and correct the data to improve the accuracy of the cross-calibration results. This software can also identify and correct for the inherent temperature fluctuations in the primary coolant loops.

The uncertainties of the cross calibration method are calculated from a knowledge of the following uncertainties:

- Uncertainty of Measured Resistances. This depends on the accuracy and drift of the resistance measurement equipment. The best resistance measurement equipment available has accuracy and short term drift characteristics in the range of 0.005 to 0.01°C.

- Stability of the Process Temperature. Plant temperature fluctuations shall be accounted for in evaluating the overall accuracy of cross calibration. The cross calibration can be performed when the plant temperature is ramping up or ramping down at a constant rate. The advantage of this approach is better stability and the disadvantage is the error which would result if the ramp rate is not constant.

Figure 7 shows in-plant cross calibration data for 16 RTDs. The data is shown before and after correcting for fluctuations. The correction involved implementing a curve smoothing procedure. Note that there is about 1°C of temperature fluctuation that has been reduced to about 0.1°C with the smoothing procedure. The data shown in Figure 7 is the result of four consecutive cross calibration repeats involving sixteen RTDs in each repeat.

Table 10 shows stability data at 3 plateaus with and without correction for fluctuations. The data given in Table 10 are presented in terms of the standard deviation of 64 data points representing four sets of cross calibration data on 16 RTDs. Note that the corrections improved the stability by about 0.06°C. This data along with results of other similar measurements have shown that the uncertainties that result from plant temperature stability range from 0.03 to 0.1°C depending on whether or not the data is corrected for temperature stability.

- Uniformity. Because of incomplete mixing and differences in loop heat removal, the temperature of primary loops are not necessarily equal at isothermal conditions. We measured as much as 0.5°C difference between the hot leg and cold leg temperatures in the same loop of a two loop PWR at isothermal conditions and as much as 0.2°C across its core (Figure 8). These differences are referred to as uniformity error. The error can be identified and subtracted from the data to achieve better accuracy. The range of uniformity errors determined from cross calibration data in a two loop PWR was 0.065 to 0.21°C

Among other factors that may need to be included in the evaluation of the cross-calibration uncertainities is the accuracy of the RTD calibration charts used to convert the RTD resistances to equivalent temperature. The RTD bias errors due to possible unidirectional drift is another factor to consider. Of course the bias error will not be of concern if newly calibrated RTDs are used in the cross-calibration for reference. To establish the validity of our cross calibration approach, we performed laboratory cross calibration tests in an oil bath on twenty newly calibrated RTDs at 300°C. The results are given in Table 11. Note that except for an unstable RTD (Tag 21) that was included intentionally here, most deviations are less than 0.05°C. Also note that the average bath temperature (i.e., 300.786°C) obtained by averaging the indication of all the RTDs is within 0.002°C of the true temperature of the bath as indicated by the Standard Platinum Resistance Thermometer (SPRT) that was included in this validation test to give the true temperature of the bath. Although the results in Table 11 represent a one point calibration check of the RTDs, it is apparent from the close agreement of the SPRT with the average value of the other RTDs that the cross-calibration has a solid foundation as a method of checking the consistency, and if done properly, the installed accuracy of a group of RTDs.

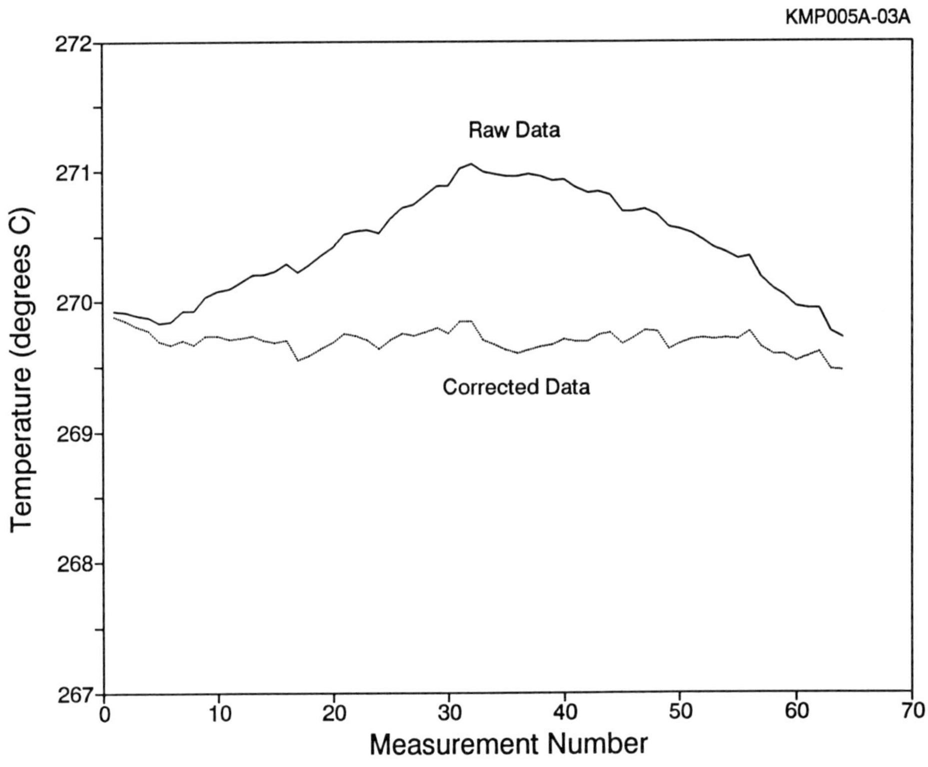

Figure 7. Cross Calibration Data Before and After Correction For Fluctuations

TABLE 10

Temperature Stability Correction

Temperature Plateau (°C)	Run #	Stability (°C) Raw Data	Stability (°C) Fluctuation Removed
280°C	1	0.10	0.03
	2	0.08	0.03
220°C	1	0.12	0.02
	2	0.06	0.02
170°C	1	0.09	0.03
	2	0.11	0.03
Average		0.09	0.03

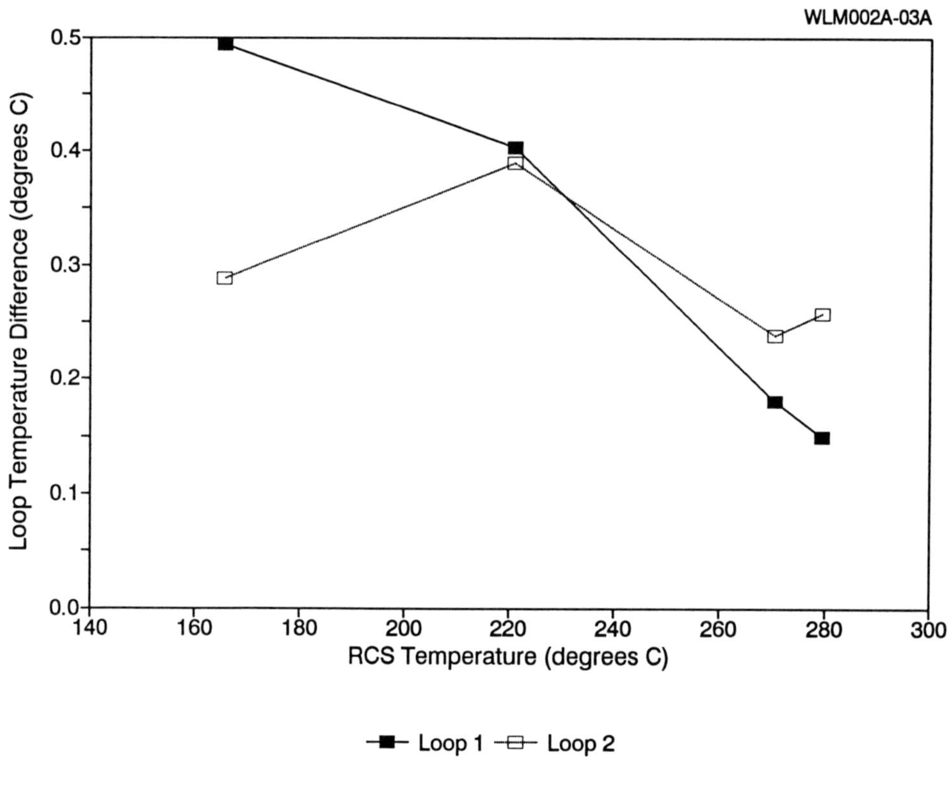

Figure 8. Differences Between Hot Leg and Cold Leg Temperatures in Each Loop of a Two Loop PWR.

TABLE 11

Laboratory Validation of Cross Calibration

Plateau: 300°C

Tag	Resistance Measurements (Ohms) Pass 1	Pass 2	Pass 3	Pass 4	Avg. Res. (Ohms)	Temp. (°C)	Dev. (°C)
21	212.2644	212.2643	212.2741	212.2634	212.2666	*303.537	# 2.751
19	429.4928	429.4910	429.5060	429.4940	429.4960	300.759	-0.028
12A	429.1156	429.1152	429.1298	429.1134	429.1185	300.811	0.025
12C	429.0084	429.0088	429.0224	429.0102	429.0125	300.807	0.021
03	424.6416	424.6324	424.6384	424.6494	424.6405	300.785	-0.001
18	429.8286	429.8200	429.8228	429.8394	429.8277	300.819	0.033
15A	424.6800	424.6818	424.6862	424.6988	424.6867	300.774	-0.012
15C	424.2652	424.2698	424.2740	424.2818	424.2727	300.774	-0.012
13A	428.9974	428.9962	429.0018	429.0144	429.0025	300.762	-0.024
13C	428.9734	428.9702	428.9726	428.9840	428.9751	300.771	-0.015
9C	430.2376	430.2382	430.2478	430.2404	430.2410	300.787	0.001
9A	430.1488	430.1532	430.1658	430.1480	430.1540	300.780	-0.006
17A	424.3216	424.3148	424.3264	424.3152	424.3195	300.840	0.054
17C	424.2266	424.2230	424.2322	424.2186	424.2251	300.833	0.047
16A	424.7866	424.7882	424.7912	424.7800	424.7865	300.806	0.020
16C	424.5208	424.5294	424.5258	424.5228	424.5247	300.806	0.020
07	430.0636	430.0732	430.0620	430.0628	430.0654	300.724	-0.062
20	430.3424	430.3492	430.3344	430.3448	430.3427	300.749	-0.037
SPRT	54.7764	54.7778	54.7761	54.7767	54.7768	300.788	0.002
	Average Temperature:					300.786	

** Not used in average*
 # Deviation limit exceeded

7. PERIODIC IN-PLANT TESTING OF RTDs

The response time and calibration of most nuclear plant RTDs are periodically tested to comply with regulatory guides 1.118 and 1.105, plant technical specifications, and various industry standards and requirements. Presently, these tests are performed once every fuel cycle and all the technical data that are available to date indicate that this is an adequate test frequency, considering an 18 or 24 month fuel cycle. The tests are performed on-line using the methods that have been described in this paper.

The best time to perform the response time and the calibration tests is toward the end of a fuel cycle. The advantage of this schedule is that if any RTD problem is identified, it can be corrected during the refueling outage. The response time tests should be done at normal power operation, and the cross calibration tests should be done at isothermal or ramp conditions at the end of the outage while the plant is being shut down. If any RTD is replaced during the outage or moved in its thermowell, its response time must be tested at hot standby coming out of the outage. The cross calibration test does not have to be repeated unless there is desire to ensure that the calibration of the newly installed RTD was not affected by any installation stress.

8. CONCLUSIONS

The response time and calibration of new RTDs that are installed in PWRs in conjunction with by-pass manifold elimination projects should initially be tested very thoroughly to ensure that the plant safety and economy is not adversely affected after the removal of the RTD by-pass manifolds. The initial response time tests to be performed should involve two series of tests. The first series of tests should be performed at the cold shutdown condition upon the installation of the new RTDs. This series of tests are needed to identify and resolve any installation or wiring problems that can interfere with optimum response time performance, or LCSR testability of the RTDs when the plant resumes normal operation. The cold shutdown tests should be followed by a series of response time measurements performed at the hot standby condition to measure the actual in-service time constants of the RTDs necessary to satisfy the plant's technical specification requirements. Since the response time of RTDs depends on the process temperature and flow conditions, the in-service response times can only be identified by testing at or near normal operating temperature and flow conditions.

The cross calibration tests may be performed at either temperature plateaus or temperature ramp conditions during plant heat up. The number of temperature plateaus and whether or not the data are collected at temperature plateaus or during a temperature ramp condition should depend on the desired accuracy for the cross calibration results, and the temperature range over which the accuracy of the RTDs is to be verified. For a one point calibration check at near normal operating condition, and a moderate accuracy requirement such as 1°C (~2°F), a set of cross calibration data collected at hot standby conditions is usually sufficient. For a wide temperature band and/or better accuracies, it may be necessary to take data at several temperature plateaus. To account for any bias in the calibration of the RTDs, one or more newly calibrated RTDs may need to be included in the cross calibration test of any group of RTDs.

The results of the response time tests should be expressed in terms of a time constant for each of the primary coolant RTD elements tested, and the results of the cross calibration tests should be expressed in terms of the temperature deviation of each RTD element from the average of all RTDs. To achieve maximum accuracy in the cross calibration results, the test data must be corrected for the plant temperature fluctuations and any differences between the hot leg and cold leg temperatures in each of the primary coolant loops and across the reactor.

Experience has shown that once a group of new RTDs are properly installed in the plant and thoroughly tested and all problems are properly addressed or corrected, the subsequent periodic tests performed once every fuel cycle are often routine.

REFERENCES

1. McGarry, J.T., Versluis, R.M., "RTD By-Pass Loop Elimination on Pressurized Water Reactors", International Conference on Operability of Nuclear Systems in Normal and Adverse Environments, Opera 89, Lyon, France, September 1989.

2. Hashemian, H.M., et al., "Aging of Nuclear Plant Resistance Temperature Detectors", U.S. Nuclear Regulatory Commission, Report Number NUREG/CR-5560, (June 1990).

3. Kerlin, T.W., et al., "In-Situ Response Time Testing of Platinum Resistance Thermometers", Electric Power Research Institute, Report Number NP-834, Volume 1, July 1978.

©1991 - Paper #91-718
0074-056X/91/267-269/$0 + .50pp

SENSING LOOP PERFORMANCE MONITORING IN THE SAFETY SYSTEMS OF NUCLEAR POWER STATIONS

Robert C. Colley
Marjorie Widmeyer

Joseph M. Weiss
H. Reed Wiegle

ABSTRACT

Plant technical specifications and NRC regulatory guides require testing of sensing loops to detect degradation and failure. Industry efforts have focused on specific manual testing to detect individual failure modes such as increased response time and calibration drift. Recent work performed by EPRI and by others using instrument loop data, failure modes, and effects analyses (FMEAs), and experience with utility on-line sensor health monitoring programs has established qualitative physical models of the sensing loop. This methodology has demonstrated that sensing loop cross comparison techniques can provide equivalent indication of sensing loop performance. It also provides more frequent sensing loop health indication than manual testing and reduces the requirement for manual testing.

INTRODUCTION

Surveillance and testing requirements were developed to assure that the process sensing loops in nuclear plants perform as described in the plant safety analyses. The sensing loops covered by these requirements include those in the reactor protection system, the engineered safeguard functions, and containment isolation system. When the surveillance and testing requirements were defined, the actual failure modes for these loops were not known: Nor was it known how accurately the measured parameters reflect the health of the sensing loop. In order to account for potential failures, a snapshot approach was taken requiring response time testing, loop functional tests, and calibrations to determine instantaneous sensing loop performance. The required tests are generally intrusive and have led to a number of plant scrams and Licensee Event Reports (LERs). Furthermore, it has been determined that these tests are not capable of detecting certain sensing loop failures.

Sensing loop health is determined by monitoring and trending sensing loop performance to detect existing or incipient failures. This paper reviews the current testing requirements in light of new information about sensing loop health monitoring. This new information is predicated on developing a qualitative physical model of the sensing loop obtained by examining past instrument problems, performing failure modes and effects analysis (FMEAs), and by gaining experience from on-line sensor health monitoring programs. The qualitative model is used to determine what failures are possible, what measurable parameters will indicate the failure, and the sensitivity of the parameters to the various failure modes. The programs monitor the entire process sensing loop daily in a non-intrusive manner and decrease the need for frequent manual testing. A significant benefit of the sensing loop health monitoring approach is that it is a non-intrusive means of interrogating sensor signals to determine the sensing loop's current and future performance with respect to the various degradation modes.

Sensing loop health monitoring will have additional paybacks in reducing the number of components in the sensing loop that require calibration, increasing the time between needed calibrations, and anticipating failures of the loop. This monitoring program will permit longer fuel cycles without requiring shutdowns during the fuel cycle to recalibrate instrumentation. More accurate sensor information can reduce instrument uncertainties which can provide increased plant operating margins. This methodology can also be used to eliminate certain manual testing such as response time testing of pressure sensors since work performed by EPRI and others indicates that these tests contain redundant information about the sensor's health.

CURRENT TEST REQUIREMENTS AND PROCEDURES

Surveillance and testing are expected to detect, but not predict, all failure modes in a timely manner. This includes:

- o Daily channel checks that are a cross comparison of like sensing channels to determine drift or other indications of sensing loop degradation;
- o Monthly or quarterly loop functional checks that monitor the electronics of the the sensing loop up to the final actuation device;
- o Calibrations performed at each refueling outage to compare the accuracy of the sensor to a benchmark standard; and
- o Response time measurements of the sensing loops.

Nuclear power plant sensing loop surveillance and testing requirements are specified by industry standards such as IEEE 338, ISA 67.06, and NRC standards such as Regulatory Guide 1.118. Each of these standards provides for the elimination of a specified testing requirement if other means can be shown to provide the same information. NRC Generic Letter 91-04(1) provides the required technical bases for instrument test interval extensions.

The design of safety systems in nuclear plants includes redundancy, per IEEE 279, to allow a plant to have inoperable safety loops and still meet the required system reliability. Loops may be inoperable due to either identified failures or testing requirements. This design allows the sensing loop to be tested while the plant is on-line. The testing is generally intrusive. It is performed by lifting leads, placing the instrument channel in the tripped condition, or inputting a step or ramp signal, and recording the response. This intrusive testing has led to numerous scrams and/or safety system actuations.

REVIEW OF PAST PROBLEMS

An informal review of NPRDS and LER data by the authors identified several scrams and safety system actuations caused by manual testing errors.

An EPRI survey of nuclear plant I&C managers(2) revealed the following were the highest concerns among I&C activities:

1. Documents and drawings:
   - inconsistences among various types of documents including procedures, manuals;
   - drawings, databases, etc., and between these and the plant as-built equipment; and
   - inadequate or erroneous information in documents, inadequate document control, lack of good retrievability, and time lags in updating documents.
2. Equipment testability:
   - absence of needed test jacks, switches, etc.; and
   - hard to test easily, properly, or risk-free because of built-in "traps."
3. Frequency of testing:
   - wear out by excessive testing; and
   - unnecessary operations.
4. System procedure content:
   - steps, precautions, prerequisites, necessary conservatisms, needed guidance, and other contexts inadequate or erroneous;
   - requirements from other documents not properly implemented;
   - inadequate verification, validation, and review before release
   - inadequate vendor information; and
   - not fully anticipating operating features or conditions.

All of these concerns are related to performing sensing loop surveillance and testing. Decreasing the number of intrusive testing procedures would ease these problems and should reduce the number of LERs.

CURRENT METHODOLOGY

The methodology for sensing loop health monitoring is based on guidance given in IEEE 338. This methodology includes the determination of failure modes and detection of these failures by analyzing plant process signals. The methodology is a combination of theoretical work and an evaluation of plant data. The theoretical basis utilizes FMEAs, analytical and hardware redundancy, and signal validation methods. The plant data is used to confirm the failure modes and to validate the non-intrusive test methods.

FMEAs are performed to determine potential failure modes and, consequently, assure that existing testing would detect these failures. ORNL(3) performed FMEAs for the three most common pressure sensor measurement types: strain gauge, capacitance, and force balance. The ORNL study did not examine electronics, relays, etc.

The EPRI response time testing program(4) evaluated extensive plant data and performed independent FMEAs on pressure transmitters and switches from 6 different sensor vendors to determine if all failure modes effecting response time would be detectable by other means such as sensor drift. The plant data included over 4,200 individual response time data points from 39 plants. Reference 5 identified the response time testing techniques in current use; ie; hydraulic step and ramp testing, and electronic white noise testing. The plant response time testing data did not identify any response time failures or response time degradation trends.

EPRI performed FMEAs on sensors from Rosemount, Barton, Foxboro, Statham, SOR, and Tobar (Veratrak). The FMEAs verified much of the previous ORNL work. That is, all previously identified failure modes, including changes in response time, would be detectable by sensor output drift. However, the FMEAs also identified one new failure mode that was not easily detectable by the existing required testing; slow loss of sensor fill fluid (oil loss). Comprehensive industry studies(6-10) determined that trending sensor drift to a more precise level could also detect oil loss. This would involve trending drift to approximately 0.5 percent drift accuracy rather than approximately 2 percent. Moreover, it was demonstrated that response time degradation could be detected from the same sensor drift trending that would detect all important sensor degradation modes including oil loss. If this drift trending accuracy is approximately at the sensitivity of the sensor, it effectively becomes an on-line calibration check and provides a basis for an instrument calibration reduction program.
In order to eliminate response time testing on the rest of the loop, it would be necessary to perform a study similar to that of the sensors. This would include performing FMEAs on each of the individual components in the loop such as electronics, relays, etc.

The EPRI Instrument Calibration Reduction Program (ICRP)(11) uses calibration of one redundant channel and channel cross-comparisons to the other redundant channels to infer calibration. Northeast Utilities Millstone II plant served as a host site for the development and initial testing of the ICRP. The testing to date has demonstrated that use of hardware redundancy is a viable means of signal validation, determining instrument loop drift, and verification of calibration. A program of similar intent has been employed at Southern California Edison(12).

At Millstone II, two types of data are collected: daily steady-state data and transient data that cover the operating range of the instrument. Daily steady-state monitoring of the instrument loop provides on-line fault detection and alarming. This steady-state data is logged and the average of the validated signals is used as the "truth" to determine the instrument loop drift of each instrument. Figure 1 shows steady state data for one parameter set.

Transient data is collected during plant transients, such as a cooldown, over the operating range of each of the instruments. The transient data from each instrument channel or sensor is compared to one instrument chosen as the reference and calibrated. This calibrated instrument loop is used as the "truth" to which each of the other channels is compared. If that instrument or channel is demonstrated as being in calibration, and the other instruments or channels (including the associated 95 percent confidence interval) do not deviate from this reference by more than the manufacturers' specified accuracy over the range of the transient, then the other redundant instruments can be inferred as calibrated. The initial tests of the method show the transient data signal estimates as indicated by the diamonds in Figure 2 have an accuracy of <1 psia over the measured range of 900 psia or about 0.1 percent with a 95 percent confidence interval as indicated by the triangles and boxes. The comparison to standard calibration methods as shown in Figure 3 indicates the methods are of comparable accuracy.

The utility has estimated that if the methodology were applied to 55 instruments at Millstone II the number of calibrations could be reduced to 16. This would result in a savings of 250 man-hours per outage of I&C technician time.

The current approach to extend calibration intervals for pressure sensors is to use on-line cross calibration techniques. This is the same methodology recommended for calibration of Resistance Temperature Detector (RTD) channels in PWRs(13). Therefore, the possibility of extending calibration intervals using cross calibration techniques for pressure sensors would also be applicable to RTDs. Consequently, the cross calibration methodology for pressure sensors and RTDs is consistent with the requirements given in Ref. 1 for extending calibration intervals from 18 to 24 months.

Another sensor trending program was installed at Millstone III to monitor Rosemount transmitters (which utilize capacitance as the measurement method) for sensing cell oil loss. This program has detected transmitter degradation in Veratrak transmitters (which utilize a strain gauge as the measurement method). This program confirms the Ref. 3 conclusions that sensor drift is an indication of transmitter degradation for any of the pressure transmitter types investigated.

Several industry initiatives are in progress dealing with instrument performance monitoring. ISA S67.06 (Ref. 4) is in the process of being revised with a scope change from response time testing to instrument channel performance monitoring. This standard will provide utilities an alternative to existing Technical Specification required testing that could reduce the amount of manual testing. The Boiling Water Reactor Owners Group (BWROG) is in the process of preparing a technical specification amendment to eliminate specific response time testing requirements. An EPRI-utility instrument channel performance monitoring group is in the formation process.

CONCLUSIONS/RECOMMENDATIONS

The basic requirements of a sensing loop performance monitoring system are becoming clear. The system would monitor and predict failures by performing cross channel checks, determining drift over the the instrument's calibrated range, and verifying functionality of the electronics.

In order to stabilize plant operations and maintenance (O&M) costs, it is essential that industry take advantage of testing and analyses efforts to reduce the amount of manual testing of instrument loops. This should be accomplished by working with the standards organizations (eg, ISA, IEEE) to modify existing standards or develop new standards. Technical specification amendment changes to incorporate these techniques should be pursued on an owners group or overall industry basis.

REFERENCES

1. NRC Generic Letter 91-04, "Changes in Technical Specification Surveillance Intervals to Accommodate a 24-Month Fuel Cycle," April 2, 1991.
2. Thie, J.A., "Surveillance of Instrumentation Channels at Nuclear Power Plants, Volume 2, An Approach to Classifying Problems and Solutions," EPRI Report NP-6067, June 1989.
3. NUREG/CR-4257,Volume 2, "Inspection, Surveillance, and Monitoring of Electrical Equipment in Nuclear Power Plants, Volume 2, Pressure Transmitters," Oak Ridge National Laboratory, August 1986.
4. Weiss, Joseph, et al, "Justification of Response Time Testing Requirements," EPRI Report NP-7324, in publication.
5. ISA-S67.06, "Response Time Testing of Nuclear Safety-Related Instrument Channels in Nuclear Power Plants," 1984.
6. "Rosemount Technical Bulletin 1," May 10, 1989.
7. "Rosemount Technical Bulletin 2," July 12, 1989.
8. "Rosemount Technical Bulletin 3," October 23, 1989.
9. "Rosemount Technical Bulletin 4," December 22, 1989.
10. Weiss, Joseph, "Guidance for Failure Detection of Rosemount Transmitters," EPRI Report NP-7121, December 1990.
11. Lord, R., "Instrument Calibration Reduction Program," EPRI Report NP-7207-CCML, February 1991.
12. Bockhorst, R., B.K. Basu, E.L. Quinn, "San Onofre Nuclear Generating Station Units 2 And 3 Instrument Drift Study Summary," ISA POWID 90, May 21-23, 1990, Toronto, Canada.
13. NUREG/CR-5560, "Aging of Nuclear Plant Resistance Temperature Detectors," Analysis and Measurement Services Corporation, March 1990.

©1991 - Paper #91-719
0074-056X/91/271-273/$0 + .50pp

JUSTIFICATION OF RESPONSE TIME TESTING REQUIREMENTS FOR PRESSURE AND DIFFERENTIAL PRESSURE SENSORS

J.M. Weiss C. Mayo V. Swisher

ABSTRACT

Response time testing (RTT) requirements were imposed on pressure, differential pressure sensors as a conservative approach to insure that assumptions in the plant safety analyses were met. The purpose of this project has been to identify the need for response time testing using the bases identified in IEEE Standard 338. A combination of plant data analyses, failure modes, and effects analyses (FMEAs) was performed. Eighteen currently qualified sensor models were utilized. The results of these analyses indicate that there are only two failure modes that affect response time, not sensor output concurrently. For these failure modes, appropriate plant actions, and testing techniques were identified.

Key Words: Response time testing, pressure transmitters, pressure switches, differential pressure transmitters, level switches.

INTRODUCTION

Previous industry efforts have not systematically evaluated pressure s sensor performance to determine what type of testing can best detect failures or sensor degradation. Rather, assumptions have been made as to what type of testing would be appropriate. Previous efforts dealing with pressure sensor RTT have been toward developing test methods assuming the testing was necessary. This EPRI program was to evaluate the basic need for performing RTT.

Safety system RTT requirements were established by IEEE Standard 338-1975, "Criteria for the Periodic Testing of Class IE Power, Protection Systems,"(1) presuming the need existed for this testing. This standard established guidelines for periodic testing to verify that loop response times of installed nuclear safety-related equipment were within the limits presumed by the design basis plant transient, accident analyses. The requirements covered all passive, active components in an instrument loop, including sensors. Individual components could be tested either in groups or separately to determine the overall loop response time. IEEE Standard 388-1975 requirements were adopted by NRC Regulatory Guide 1.118 "Periodic Testing of Electric Power and Protection Systems," Revision 1, issued in 1977(2). IEEE Standard 338-1977, "Criteria for the Periodic Testing of Nuclear Power Generating Station Safety Systems"(3) was adopted in the NRC Regulatory Guide 1.118, Revision 2 issued in June, 1978(4). Additional guidance is provided in Instrument Society of America Standard ISA S67.06, "Response Time Testing of Nuclear Safety-Related Instrument Channels in Nuclear Power Plants," August 29, 1986(5).

Recently, utilities began assessing the need for RTT requirements for pressure, differential pressure sensors. Experience and data from a significant number of plants indicated that, while RTT was both resource and exposure (ALARA) intensive, an insignificant number of pressure sensor failures had been detected through this type of testing. Various reviews of industry failure data showed most, if not all, pressure sensor failures were detected through calibration (typically performed prior to RTT), channel checks, and surveillance testing. The industry concluded from these reviews and related industry reports that:

- o RTT provided potentially little benefit in the identification of failed or degrading sensors;
- o RTT appeared redundant to other periodic tests;
- o By themselves, response time failures seemed unlikely; and
- o Verification of response time compliance to safety analyses could be met without RTT.

EPRI PROGRAM

EPRI established a program to evaluate the need for RTT of pressure and differential pressure (level and flow) transmitters and switches. The primary purpose of the investigation was to determine if deleting RTT could be justified per IEEE Standard 338 for specific pressure and differential pressure transmitters and switch models.

IEEE Standard 338 defines a basis for eliminating RTT. Section 6.3.4 states: "...response time testing of all safety related equipment, per se, is not required if, in lieu of response time testing, the response time of the safety system equipment is verified by functional testing, calibration check, or other tests, or both." The primary objectives of this project were to:

- o Evaluate existing RTT methodology using plant response time data;
- o Utilize FMEA methodology to determine if any response time degradation modes exist that would not be detected by periodic (non-RTT) methods; and
- o Provide RTT methods for sensors that exhibited response time degradation modes not detected by periodic methods.

This approach would not impact existing plant safety analyses and would be equally applicable to pressurized water reactors (PWRs) and boiling water reactors (BWRs).

DATA COLLECTION AND EVALUATION

The response time investigation was divided into separate data gathering and assessment activities. The first activity involved collecting and evaluating RTT data from utilities currently subject to RTT requirements. The second activity involved performing FMEAs on the major components of the qualified pressure sensor models currently used in safety-related applications.

Thirty-nine nuclear power plants supplied over 4,200 RTT measurements representing over 2,100 different pressure and differential pressure transmitters and switches. Three types of testing methods were found to be represented: ramp and step testing using a hydraulic test rig, and white noise analysis using an electronic test rig. Along with the test data, information regarding sensor type, sensor use, instrument range, trip set point, allowable response time, test method, and type of test rig were also provided. The plant data covered transmitters manufactured by 10 different vendors including Barton, Foxboro, Fischer & Porter, Tobar (Veritrak), GE, Barksdale, Static-O-Ring, ASCO, Magnetrol, and Rosemount. With the exception of Statham, all vendors supplying safety-related pressure and differential pressure sensors to the nuclear industry were represented by the plant data.

FMEAs were performed on the 18 sensor-types listed in Table 1. Based on the data collected from the industry for this investigation, these 18 sensor-types represent the majority of pressure sensor instrumentation currently installed or expected to be installed in safety-related systems in U.S. plants. These sensor-types also are representative of the various sensor designs (eg, bourdon tube, force balance, capacitance, strain gauge).

No response time failures were found in the over 4,200 measurements. Six measurements, each from a different sensor, exceeded the allowable response time provided on the plant data sheets, but were not actual response time failures. Discussions with personnel at the respective plants indicated that the allowable times in test procedures were set below technical specification requirements as a conservative measure. The response time measurements were compared to like sensors by plant personnel and found to be only slightly higher. As a result of this comparison and the fact that response times were below technical specification limits, the sensors were declared acceptable.

At the time the plant data was collected, there were more than 100 confirmed failures of Rosemount transmitters involving loss of fill fluid. Even though this failure mode results in significant degradation of response time, none of these failures were identified by RTT.

RTT RESULTS

The FMEAs indicate that RTT is redundant to other periodic testing for all cases but slow loss of fill fluid where the fill fluid is the working fluid and variable damping potentiometer misadjustment. For loss of fill fluid, drift trending will provide the necessary verification for the assumptions in the plant safety analyses. Using fixed rather than variable damping or procedurally assuring that variable damping is properly set will alleviate the other response time degradation concern. The conclusions drawn from the RTT investigation follow.

- o Based on the 4,200 measurements, RTT has not identified any sensors that have failed response time requirements using hydraulic or electronic white noise analysis techniques. However, calibrations and other tests have detected transmitters with excessive response times.
- o The limited number of test points per transmitter, the variances in RTT measurements due to type of test method, and differences in test performance minimized the usefulness of the data for predicting failure trends.
- o Sensor-specific response time of new sensors may not be known; pressure sensor vendors provide a nominal response time but do not provide sensor-specific RTT values to the utilities.
- o Hydraulic step and ramp RTT methods, as currently performed around the sensor set point, may not detect a fill fluid failure mode prior to exceeding response time limits.
- o Electronic white noise analysis RTT will not detect response time degradation due to loss of fill fluid failure prior to exceeding response time limits.
- o The FMEAs identified two failure modes having the potential to affect response time without concurrently affecting sensor output: (1) slow sensor fill fluid leak during pressurized operation; and (2) variable damping potentiometer misadjustment. The FMEAs also identified two manufacturing/handling defects having the potential to affect response time: (1) low sensor fill fluid from the manufacturing process; and (2) crimped capillaries from the manufacturing process or improper handling by the manufacturer, or in the field during maintenance or plant modifications.
- o Sensor failure modes associated with all Barton transmitters, models 763, and 764, and switch model 288/289, Tobar (Veritrak) models 32PA1, 32PD1, and 32PG1, SOR switches, and Foxboro (Weed) models N-E11GH, N-E11DM, and N-E11GM do not affect sensor response time without significantly affecting calibration and/or sensor response. Correlations to specify the amount of sensor output versus response time degradation currently do not exist for Tobar transmitters.
- o For Rosemount, Statham, and Foxboro E13 models, loss of fill fluid may significantly degrade sensor response time before impacting sensor output in a detectable manner using existing testing methods (eg, hydraulic RTT, electronic white noise analysis RTT, channel checks, surveillance testing, calibration).
- o Rosemount has developed correlations between drift and response time degradation that can be used by the industry to identify loss of fill-fluid prior to degrading response time. Periodic drift monitoring can be used as a method of detecting loss of fill fluid before response time is affected for Rosemount transmitters using the criteria in the Rosemount Technical Bulletins.
  - At the present time, hydraulic RTT is the most effective method for response time testing Statham transmitters even though it may not identify a slow loss of fill fluid in the early stages.
  - The power interrupt test is a viable method of RTT for force balance transmitters.
  - Variable damping potentiometer misadjustment can affect response time without affecting sensor output.
  - Hydraulic ramp and step, and electronic white noise analysis RTT methods can detect variable damping misadjustment (increased response time).

RECOMMENDATIONS

A summary of recommendations for modifying the current RTT program for pressure, level, and flow sensors is outlined. These recommendations are based on improving sensor failure detection, minimizing the potential for human error resulting in LERs, reducing personnel exposure, and reducing potential impact on plant down time. Utilities, NSSS owners groups, or NUMARC can utilize this report as a technical basis for eliminating RTT requirements from technical specifications.

- o Perform a hydraulic RTT prior to installation of a new transmitter/switch or following refurbishment of the transmitter/switch (eg, sensor cell or variable damping components) to determine an initial sensor-specific response time value; the power interrupt test is an alternate method to use on force balance transmitters.

- o Delete periodic sensor RTT requirements except for transmitters utilizing variable damping. If the utility can assure that variable damping is at its required setting, and not subsequently changed, there is no need to RTT. If RTT is necessary for transmitters with variable damping, test response time using hydraulic or electronic methods.
- o Perform periodic drift monitoring on all Rosemount pressure and differential pressure transmitters, models 1151, 1152, 1153, and 1154; guidance on drift monitoring can be found in EPRI NP-7121(6) and Rosemount Technical Bulletins(7-10); and drift monitoring intervals can be based on utility response to NRC Bulletin No. 90-01(11).
- o Utilities using Statham pressure, differential pressure transmitters models PD3200, and PG3000 should continue to perform hydraulic RTT until a drift criteria can be developed relating sensor drift to response time degradation.
- o Utilities using Tobar models 32PA1, 32PD1, and 32PG1 should continue to perform hydraulic RTT until verification is provided that sensor output significantly changes prior to response time degradation.
- o Perform power interrupt tests on all Foxboro (Weed) force balance transmitters, models N-E13DA and N-E13DM, at an interval to be determined by the utility.

CURRENT STATUS

The BWR Owners Group Subcommittee on RTT met with the NRC in January 1991, and has committed to submit a technical specification amendment to eliminate RTT during the fourth quarter of 1991 based, in part, on the EPRI report (12). ISA is currently in the ballot stage in revising the scope, purpose of ISA S67.06. The intent is to have S67.06 cover sensing loop performance issues including on-line drift monitoring for RTT, instrument calibration reduction techniques rather than only RTT.

REFERENCES

1. "Criteria for Periodic Testing of Nuclear Power Generating Station Safety Systems," IEEE Standard 338-1975.
2. US Nuclear Regulatory Commission Regulatory Guide 1.118, Revision 1, 1977.
3. "Criteria for Periodic Testing of Nuclear Power Generating Station Safety Systems," IEEE 338-1977.
4. US Nuclear Regulatory Commission Regulatory Guide 1.118, Revision 2, 1978.
5. "Response Time Testing of Nuclear Safety-Related Instrument Channels in Nuclear Power Plants," ISA S67.06, 1986.
6. "Technical Guidance for Incipient Failure Detection of Rosemount Pressure Transmitters", EPRI Report NP 7121, Electric Power Research Institute (EPRI), December 1990.
7. "Rosemount Technical Bulletin 1," Rosemount, Incorporated, May 10, 1989.
8. "Rosemount Technical Bulletin 2," Rosemount, Incorporated, July 12, 1989.
9. "Rosemount Technical Bulletin 3," Rosemount, Incorporated, October 23, 1989.
10. "Rosemount Technical Bulletin 4," Rosemount, Incorporated, December 22, 1989.
11. "Loss of Fill-Oil in Transmitters Manufactured by Rosemount," US Nuclear Regulatory Commission IE Bulletin 90-01, March 9, 1990.
12. "Investigation of Response Time Testing Requirements", EPRI Report NP-7423, Electric Power Research Institute (EPRI), in publication.

TABLE 1

SENSOR TYPES INCLUDED IN FMEAs

Vendor and Model	Source
Barton 288	Differential pressure indicating switch
Barton 763	Gauge electronic pressure transmitter
Barton 764	Differential pressure electronic transmitter
Foxboro N-E11DM	Differential pressure transmitter
Foxboro N-E13DM	Differential pressure transmitter
Foxboro N-E13DH	Differential pressure transmitter
Foxboro N-E11GH	Gauge pressure transmitter
Foxboro N-E11GM	Gauge pressure transmitter
Tobar 32PA1	Absolute pressure transmitter
Tobar 32PG1	Gauge pressure transmitter
Tobar 32DP1	Differential pressure transmitter
Rosemount 1151, 1152, 1153, 1154	Differential pressure transmitter
Rosemount 151, 1152, 1153, 1154	Pressure transmitter
Statham PD3200	Differential pressure transmitter
Statham PG3000	Pressure transmitter
SOR	Differential pressure switch
SOR	Pressure switch

©1991 - Paper #91-720
0074-056X/91/275-289/$0 + .50pp

RESPONSE TIME TESTING OF PRESSURE TRANSMITTERS IN NUCLEAR POWER PLANTS

H. M. Hashemian
K. M. Petersen

Analysis and Measurement Services Corporation
AMS 9111 Cross Park Drive
Knoxville, TN 37923

ABSTRACT

This paper presents a review of the response time testing practice in the nuclear power industry for the safety-related pressure, level, and flow transmitters. Two methods which have been developed and recently validated for response time testing of these transmitters are described. The advantage of the new methods is that they permit remote testing of installed transmitters from the control room while the plant is operating.

Also presented in this paper are experimental data on the contribution of sensing lines to the overall response time of pressure sensing systems, and the delays that can result from blockages or air entrapped in the sensing lines.

INTRODUCTION

Response time measurements are performed on safety related pressure, level, and flow transmitters (hereafter referred to as pressure transmitters) in a majority of nuclear power plants in the United States. The tests are performed in both pressurized water reactors (PWRs) and boiling water reactors (BWRs). A variety of test methods are used for these measurements. Two of the most recently developed methods and their advantages and disadvantages are presented in this paper. These methods are referred to as the power interrupt (PI) test, and noise analysis. Both methods have been validated for on-line measurement of response time of pressure transmitters in nuclear power plants.

The significance of the new methods is that they permit on-line measurements from the control room as opposed to the conventional methods which require physical access to each transmitter, and can therefore be performed on most transmitters only when the plant is at cold shutdown.

The PI test was conceived in early 1980's[1], but noise analysis has been mentioned as a candidate method for sensor response time testing since the mid 1960's[2]. In fact, noise analysis has been used to some extent in the nuclear power industry for sensor response time degradation monitoring and simple equipment has been developed for qualitative response time testing based on noise analysis[3]. Most recently, research completed on validation of noise analysis has shown that the method can readily be used for quantitative response time measurements of nuclear plant pressure transmitters[4].

The noise analysis technique has also been a candidate for response time testing of Resistance Temperature Detectors (RTDs) and thermocouples, but it has not been validated for these sensors and its accuracy is not currently known. A more exact method, called the Loop Current Step Response (LCSR) test is available for measurement of response time of temperature sensors[5]. The LCSR test is a simple method which is used in almost all nuclear power plants requiring response time measurements of temperature sensors.

The PI test is a simple method, but it is applicable only to force-balance pressure transmitters and has been validated only for the force-balance pressure transmitters that are made by Foxboro. The PI test can be performed on installed transmitters whether the plant is on-line or at cold shut-down.

The noise analysis method is not as simple as the PI test, but it is applicable to all transmitters of the types subject to response time testing in nuclear power plants. Another advantage of noise analysis is that its results include the contributions of the sensing lines to the overall response time of the pressure sensing system. The capability of the noise analysis method to account for the sensing line delays is important in testing of such transmitters as Barton. It has been determined by laboratory measurements that the response times of pressure sensing systems involving Barton transmitters are significantly affected by the length of the sensing lines and any significant blockages in the line that may occur, for example in PWRs, due to formation of sludge or solidification of boron in the reactor coolant system. The response time of other nuclear safety-related pressure transmitters such as those made by Tobar, Foxboro, and Rosemount companies are also affected by sensing line effects, but to a lesser extent than Barton transmitters.

RESPONSE TIME TESTING REQUIREMENTS

There is a wide variation in the response time testing requirements for pressure transmitters in nuclear power plants. Some plants test all of their safety-related pressure transmitters once every fuel cycle, some plants test only a portion of their transmitters once every fuel cycle, and some plants do not test any of their pressure transmitters. The variations in testing requirements usually depend on the age of the plant and whether or not standard technical specifications are in effect in the plant.

There is also a wide variation in response time requirements for pressure transmitters. The range of acceptable response times covers a 0.2 to 2.0 second span with a majority of applications requiring response time results in the neighborhood of 0.50 seconds including any contributions by the sensing lines.

The dominant range of achievable response times of most nuclear plant pressure transmitters is 0.05 to 0.5 seconds depending on the type, manufacturer, and the pressure range of the transmitter. For those transmitters whose response times are affected by sensing lines, about 0.1 to 0.3 seconds may be needed to account for the delay due to the length of the sensing lines. With 0.50 seconds as the average value for an acceptable overall response time, 0.20 seconds as the average value for the nominal response time of most pressure transmitters, and about 0.2 seconds for some sensing lines, only a small margin is sometimes left for any response time degradation due to aging of the transmitter or blockages in the sensing lines. It is because of this small margin that periodic response time tests are performed in nuclear power plants to ensure that transient pressure information is transmitted fast enough to mitigate the consequences of an undesirable pressure change.

About 10 to 60 transmitters are tested in most nuclear power plants once every fuel cycle, depending on the plant, and whether or not one or more safety channels are included in the periodic tests. Examples of pressure transmitters that are tested for response time in nuclear plants are given in Table 1. Table 2 lists the four most commonly used pressure transmitters in the safety systems of U. S. nuclear power plants and their nominal response times as specified in literature provided by their manufacturers.

CONVENTIONAL METHOD FOR RESPONSE TIME TESTING

Testing of response time of pressure transmitters in nuclear power plants began in the late 1970's when testing technology was developed under a project sponsored by the Electric Power Research Institute (EPRI). The EPRI project resulted in the development of a test instrument referred to as a hydraulic ramp generator[6]. As its name implies, this system provides a test signal in the form of a pressure ramp that is injected to the transmitter under test to measure its delay with respect to the response of a high-speed reference transmitter. Although this instrument performs well in providing a suitable test signal and is simple to use for testing of transmitters on the bench, practical considerations make it tedious to perform accurate tests in the field for those transmitters that are not readily accessible in the plant, such as those that are located in the containment. These considerations have led to the development and validation of new methods that are now available for on-line measurement of response time of pressure transmitters as described later in this paper.

TRANSMITTER AGING STUDY

Pressure sensing systems in nuclear power plants are exposed to conditions that can cause performance degradation over a long period of exposure. The conditions include:

- Heat, humidity, vibration, radiation, fluctuations in transmitter power supply voltages, etc.

TABLE 1

Examples of Pressure Transmitters Tested
for Response Time in Nuclear Power Plants

Pressurized Water Reactors

1. Containment Pressure
2. Pressurizer Level
3. Pressurizer Pressure
4. Reactor Coolant Flow
5. Refueling Water Storage Tank (RWST) Level
6. Steam Flow
7. Steam Generator Level
8. Steam Pressure
9. Turbine Impulse Pressure

Boiling Water Reactors

1. Drywell Pressure
2. Main Steam Flow
3. Reactor Vessel Pressure
4. Reactor Vessel Water Level
5. Reactor Water Clean Up (RWCU) Flow

TABLE 2

Manufacturer's Specifications for Response Time of Some Nuclear Plant Pressure Transmitters

Manufacturer	Model	Range	Nominal Response Time (sec)
Barton	763	All	<0.18
	764	All	<0.18
Foxboro	E11	All	<0.30
	E13	All	<0.30
	NE11	All	<0.30
	NE13	All	<0.30
Rosemount	1152	All	0.2-1.67
	1153	3	2.0
		4	0.5
		5 - 9	0.2
	1154	4	0.5
		All Others	0.2
Tobar/Veritrak	32 DP1	0-600"	0.08-0.15
	76 DP2		

Notes:

1. *Above results are general estimates of nominal response times of the transmitters at laboratory conditions. The in-service response times of identical transmitters may be significantly different than the values given in this table.*

2. *The above response time data from different manufacturers may not be based on the same definition. Therefore, these values should not be used for comparison of sensors from different manufacturers.*

- Stresses due to changes in normal operating conditions and plant trips that result in temperature and pressure cycling.
- Degradation induced during calibration, testing or maintenance such as over-pressurization, injection of test signals to the wrong side of the transmitter, etc.

These conditions can affect both the calibration and response time of pressure transmitters. Some examples of the effects that can cause response time degradation in typical nuclear grade pressure transmitters are listed in Table 3. In a transmitter aging study currently underway, the author is involved in performing experimental measurements with both artificially and naturally aged pressure transmitters to establish the response time degradation rates, the failure modes, and the failure rates. As a part of this study, individual components of pressure transmitters that have been identified as those that play a role in the speed of response of the transmitters are being tested for aging effects and their consequences on the response time of the transmitter system. This study is concerned with both the mechanical and the electronic systems of nuclear-grade pressure transmitters. The project is due for completion near the end of 1991 and is expected to provide objective results on the effects of aging on steady state and transient performance of pressure transmitters.

ON-LINE METHODS FOR RESPONSE TIME MEASUREMENTS

The response time of pressure transmitters in nuclear power plants can be measured on-line using the power interrupt (PI) method for force-balance transmitters, and noise analysis method for other transmitters. These methods are described below.

Noise Analysis

Noise analysis is based on monitoring the natural fluctuations that exist at the output of pressure transmitters while the process is operating. These fluctuations (noise) are due to turbulence induced by the flow of water in the system, random heat transfer in the core, and other naturally occurring phenomena. The noise is extracted from the sensor output by removing the DC component of the signal and amplifying the AC component. This signal can be analyzed to provide the response time of the pressure sensing system including the sensing lines. The analysis can be done in the time domain and frequency domain. In time domain analysis, the autoregressive (AR) method is used in which the noise data are fit to a general autoregressive model of order N to identify the coefficients of the model. These coefficients are then used to obtain such dynamic descriptors as the impulse response, step response and ramp response of the transmitters from which the response time is deduced.

In frequency domain analysis, the noise record for each transmitter is processed by Fourier transform analysis, and its power spectral density (PSD) is determined. The PSD is then fit to a mathematical model for the transmitter from which the response time of the transmitter is identified. Prior to any time domain or frequency domain analysis, the suitability of the noise data must be verified by a computer scanning and

TABLE 3

Examples of Effects That Can Cause Response Time Degradation in Nuclear Plant Pressure Transmitters

Degradation	Cause	Effect
Deformation of diaphragm	Pressure fluctuations pressure surges and mechanical vibration	Changes in stiffness of sensing element
Wear and friction of mechanical linkages	Pressure fluctuations and surges, corrosion and oxidation	Changes in system restoration ability
Partial or total loss of fill fluid	Manufacturing flaws Improper handling	Significant capacitance changes Nonlinear output
Degradation of fill fluid	Chemical changes of oil due to radiation and/or heat	Viscosity Changes
Failure of seals	Embrittlement and cracking	Moisture on electronics
Leakage of process fluid into cell fluid	Cracking of diaphragm	Capacitance changes
Changes in values of electronic component	Heat, radiation, humidity, voltage stresses, and maintenance	Changes in dynamic response and linearity of electronics
Hysteresis	Pressure fluctuations and surges and mechanical vibration	Distorted output
Setpoint drift	Calibration shifts	Increased time to reach setpoint

screening of the data. This is accomplished using data qualification algorithms that check for the stationarity and linearity of the data.

The noise analysis technique provides the response time of a pressure transmitter to small process perturbations while the response time of interest is often that of large process transients. Fortunately, the dynamic characteristics of almost all normal pressure transmitters that are subject to response time testing in nuclear power plants are predominantly linear. Therefore, the response times of these sensors are mostly independent of the magnitude of the excitation signals. That is, the response time test results for small input signals are equivalent to that of large input signals. This is evident in the results of numerous experimental measurements conducted with nuclear grade pressure transmitters tested with both noise analysis and conventional techniques involving a step or ramp pressure signal with much larger amplitudes than noise signals. A sample of these results are shown in Table 4 for three transmitters from each of the most common types currently used in the nuclear power industry.

It has been established by laboratory tests that the noise analysis results for pressure transmitters are generally conservative with respect to the response time results from the conventional ramp or step tests. That is, based on laboratory tests, the noise analysis results are often larger than the conventional test results. To date, the largest differences that have been observed in laboratory validation tests have been contained within less than ± 0.10 seconds. Therefore, the accuracy of the noise analysis technique is currently quoted as ± 0.10 seconds. It is expected that better accuracies will be achieved as more testing experience is accumulated.

In some instances such as when the process noise is narrowband, the noise analysis results may be excessively conservative. An exaggerated case in which the process noise is narrowband is the containment pressure for which the noise analysis technique can not be used . Another example of a pressure transmitter that can not be readily tested with noise analysis is the reactor water storage tank (RWST) level. These transmitters should be tested with the conventional ramp or step test methods unless the transmitters are force-balance type, in which case they can be tested with the power interrupt method described below.

Power Interrupt Test

The power interrupt (PI) test is applicable to force-balance pressure transmitters. The test is based on a momentary interruption of the normal electric supply power to the transmitter. The test is performed by turning the power to the transmitter OFF for a few seconds, and then ON. When the power is turned ON, the transmitter provides an output that can be analyzed to give the transmitter's response time. The method has been validated only for those pressure transmitters that are designed by Foxboro Company. The validation effort has not been extended to other force-balance transmitters, because Foxboro transmitters are the only force-balance type transmitters that are used for safety-related applications in nuclear power plants and are subject to response time testing.

The PI test accounts for the dynamic response of both the mechanical and the electronic components of the transmitter and thereby gives the overall response time of the complete electromechanical system of the

TABLE 4

Sample Results of Validation of Noise Analysis Methods for On-Line Measurement of Response Time of Pressure Transmitters in Nuclear Power Plants

Transmitter Manufacturer	Noise Analysis Response Time (sec) Ramp	Noise	Difference (sec)
Rosemount	0.05	0.05	0.00
Rosemount	0.32	0.31	-0.01
Rosemount	0.10	0.10	0.00
Tobar	0.08	0.09	0.01
Tobar	0.15	0.25	0.10
Tobar	0.33	0.35	0.02
Barton	0.17	0.18	0.01
Barton	0.12	0.12	0.00
Barton	0.11	0.13	0.02

transmitter. The few samples of the laboratory validation results listed in Table 5 show how well the PI test reproduces the conventional response times of the representative Foxboro transmitters.

The effect of sensor linearity is as important in the PI test as it is in noise analysis. Among numerous nuclear-grade pressure transmitters that we have tested for linearity effects, only a few were found to have slight nonlinearity and those were all Foxboro force-balance transmitters. Fortunately, for all these nonlinear transmitters, the PI test has produced conservative results under all test conditions that have been examined to date in the laboratory.

SENSING LINE EFFECTS ON RESPONSE TIME

The overall response time of a pressure sensing system is the sum of the response times due to the sensing lines and the sensor. The contribution of the sensing lines to the overall response time depends on the transmitter being used. For some transmitters such as Barton, the contribution of sensing lines is significant. Table 6 shows laboratory test results in which the sensing line delays were measured for four types of transmitters with 100 and 400 feet of sensing lines made of 1/2 inch (outside diameter) copper tubes. These results are the averages of test results for one Barton, one Rosemount, one Tobar, and three Foxboro transmitters. The results show that the sensing line contributions for Barton transmitters are larger than that of Rosemount transmitters by a factor of about 5. The reason for the difference is the difference in volumetric displacement of the sensing element of the two transmitters. As shown in Table 6, in Barton transmitters, a much larger volume of fluid must be moved through the sensing lines to flex the sensing element than in Rosemount transmitters. This effect is also important in the study of sensing line blockages. Any blockage that can restrict the flow can increase the time delay in the sensing line. Table 7 shows laboratory test results with a snubber in the sensing line to simulate blockages. Also shown in this table are the test results with the noise analysis method to demonstrate that sensing line blockages are accounted for in noise analysis results.

For the results shown in Tables 6 and 7, the response times of the pressure sensing systems in the laboratory were measured by applying a ramp signal at the input to the sensing line and measuring the delay of the transmitter/sensing line system with respect to two high-speed reference transmitters installed at the beginning and the end of the sensing lines. These results were regarded as direct test results to be distinguished from noise test results performed at the same time to examine the capability and the accuracy of noise analysis results for providing the overall response time of a pressure sensing system.

Air in sensing lines can also cause delays in the overall response time of a pressure sensing system. Table 8 shows the results of response time measurements with varying amounts of air in the sensing lines leading to a Foxboro force-balance pressure transmitter. The results shown are with nylon tubes and are therefore not representative of delays in stainless steel sensing lines used in nuclear power plants. Nevertheless, they demonstrate that air in sensing lines can increase the response time. In most cases, however, air dissolves in the water at high pressures and is therefore not a major problem.

TABLE 5

Sample Results of Validation of PI Method for On-Line Measurement of Response Time of Foxboro Force-Balance Pressure Transmitters in Nuclear Power Plants

PI Test

Transmitter Manufacturer	Response Time (sec) Ramp	PI	Difference (sec)
Foxboro	0.13	0.14	0.01
Foxboro	0.21	0.21	0.00
Foxboro	0.21	0.20	-0.01
Foxboro	0.21	0.22	0.01
Foxboro	0.20	0.21	0.01
Foxboro	0.16	0.17	0.01

TABLE 6

Sensing Line Delays as a Function of Sensing Line Length

Transmitter	Max. Volumetric Displacement (in ^{3})	Sensing Line Delay (sec)		Transmitter Response (sec)
		100′	400′	
Barton	0.13	0.08	0.29	0.14
Tobar	0.06	0.05	0.14	0.29
Foxboro	0.06	0.03	0.11	0.22
Rosemount	0.01	0.02	0.06	0.06

TABLE 7

Sensing Line Test Results

Transmitter Tested	Response Time (sec) Direct	Noise
Reference Alone	0.00	0.00
Reference & 100'	0.01	0.00
Reference & Snubber	0.34	0.27
Barton Alone	0.12	0.17
Barton & 100'	0.27	0.28
Barton & Snubber	3.00	2.94
Rosemount Alone	0.08	0.13
Rosemount & 100'	0.09	0.10
Rosemount & Snubber	0.07	0.13

TABLE 8

Response Time as a Function of
Air in Sensing Line

Length of Bubble (inch)	Response Time (sec)
0	0.12
8	0.13
20	0.16
30	0.19
60	0.39

1. *Above tests were performed with a 25-foot, 1/4-inch O.D. Nylon Tubing between the test unit and the transmitter under test.*

2. *The bubble lengths given above are approximate values.*

CONCLUSIONS

Two new methods, referred to as PI test and noise analysis, are now available for measurement of response times of pressure, level and flow transmitters as installed in nuclear power plants at power operation. The tests are usually performed at the end of a fuel cycle, a few weeks before the plant is to be shutdown for refueling. This provides the opportunity to resolve any response time problems at a convenient time during the outage.

The PI test of a pressure transmitter usually requires 30 minutes and one transmitter is tested at a time. Noise analysis requires about one hour for testing of a group of transmitters and there is no limit to the number of transmitters that can be tested at a time.

Both the noise analysis and PI tests have been validated by laboratory tests involving representative nuclear plant pressure transmitters. The validation tests involved measuring the response times of the transmitters with the conventional step or ramp test methods, and then with PI or noise analysis methods to demonstrate that the results of the new methods are comparable to those of the conventional method. These validation tests have shown that the noise analysis and the PI methods provide response time results with accuracies of better than ± 0.10 seconds.

REFERENCES

1. Soares, A.J., "Study and Dynamic Modeling of a Pressure Transducer That is Based on the Principle of Force-Balance", Ph.D. Dissertation, University of Tennessee, December 1982.

2. Thie, J.A., "Power Reactor Noise", American Nuclear Society, LaGrange, Illinois, 1981.

3. Hashemian, H.M., et al., "Apparatus for Measuring the Degradation of a Sensor Time Constant", U.S. Patent No. 4,295,128, October 1981.

4. Hashemian, H.M., et al., "Effect of Aging on Response Time of Nuclear Plant Pressure Sensors", U.S. Nuclear Regulatory Commission, NUREG/CR-5383, (June 1989).

5. Hashemian, H.M., et al., "In-Situ Response Time Testing of Thermocouples", ISA Transactions, Volume 29, Number 4, pp. 97-104, (1990).

6. Cain, D.G. and Foster, C.G., "A Practical Means for Pressure Transducer Response Verification", Nuclear Technology Vol. 36, December 1977.

0074-056X/91/291-306/$0 + .50pp

Portland General Electric Company

Trojan Nuclear Plant

CALORIMETRIC AND REACTOR COOLANT SYSTEM FLOW UNCERTAINTY

By

Lionel Bates and Tom McLean

ABSTRACT

This paper describes a methodology for the quantification of errors associated with the determination of feedwater flow, secondary power, and Reactor Coolant System (RCS) flow used at the Trojan Nuclear Plant to ensure compliance with regulatory requirements. The sources of error in Plant indications and process measurement are identified and tracked, using examples, through the mathematical processes necessary to calculate the uncertainty in the RCS flow measurement. An error of approximately 1.4 percent is calculated for secondary power. This error results, along with the consideration of other errors, in an uncertainty of approximately 3 percent in the RCS flow determination.

INTRODUCTION

The Trojan Nuclear Plant is required by Technical Specifications to maintain minimum Reactor Coolant System (RCS) flow. Compliance with this RCS flow requirement is determined by Plant tests (secondary calorimetric and RCS flow tests) which are similar to tests run at pressurized water reactor plants throughout the industry. Because the RCS flow calculated as a result of these tests has varied, the need arose to more rigorously quantify the uncertainty in the test results. The existing uncertainty calculations did not consider actual Plant instrument performance nor did they reflect the test methods or the effects of the mathematical processes used to calculate RCS flow.

As part of our effort to better quantify the RCS flow uncertainty, we have applied statistical analysis to Plant calibration data to determine actual installed instrument performance. This approach has been described in the draft ISA RP dS67.04, Part II, "Methodologies for the Determination of Setpoints for Nuclear Safety-Related Instrumentation" (RP). We have also used the methods described in ANSI/ASME PTC 19.1-1985 (ASME) and the ISA Measurement Uncertainty Handbook (Handbook) to propagate the instrument errors through the mathematical processes which define the tests.

The ISA RP and the ASME PTC give equations which can be used to propagate error through math functions. These equations are based on the general Taylor Series formula for the function Z = f(x,y):

$$\sigma_Z^2 = \left(\frac{\partial Z}{\partial x}\right)^2 \sigma x^2 + \left(\frac{\partial Z}{\partial y}\right)^2 \sigma y^2 + 2\,\frac{\partial Z}{\partial y}\,\frac{\partial Z}{\partial x}\,\alpha\,\sigma x \sigma y$$

The precision squared (σ_Z^2) of the function Z which results from the precision errors σx, σy of its variables can be calculated on the basis of the partial derivative $\left(\frac{\partial Z}{\partial x}, \frac{\partial Z}{\partial y}\right)$ of the function with respect to each variable, the error in each variable, and the correlation (α) between the variables x and y. There are several correlatable variables used in the calculation of secondary power and RCS flow. The ASME PTC states, "With complex relationships, the same measurement may be involved more than once in determining the result. Only when Taylor Series is used to relate the most elementary parameters to the result will there be a proper accounting for these linked relationships."

The ASME PTC classifies error as either being random (precision), which is essentially equivalent to the standard deviation of a normal distribution of error, and bias limit, B, which can be either symmetrical or nonsymmetrical and can be calibrated out in some cases. The bias limits are considered to be fixed because once they are set, they do not vary (in the absence of recalibration) from one measurement to the next. An example of random or precision error (designated S) is repeatability of an indication based on a known input. An example of fixed bias error (also called fixed random error) is the installed accuracy of an orifice plate. Assume that there exists a group of 100 orifice plates. The error of each plate is random, and the distribution of the error for the group of 100 plates is normal. Assume an orifice plate is chosen from the group and installed in a flow system. Once installed, the accuracy is more precisely described by a fixed bias limit.

The symmetrical uncertainty in a measurement is the combination of precision error and bias limits in the form:

$$U_{ADD} = \pm(B + t\bar{Sx})$$

where B is the total bias limit, $\bar{Sx}$ is the precision index of the mean, and t is the 95th percentile point for the two-tailed student t distribution. For samples of greater than 30 data points, t approaches 1.96 and is approximated as 2. The U_{ADD} expression results in an uncertainty interval which is characterized in terms of coverage because the B and S terms may contain non-normally distributed errors. The U_{ADD} interval defined by the expression will contain the true value 99 percent of the time. The square root of the sum of squares (SRSS) combination of the bias limit and precision errors results in 95 percent coverage. The expression is:

$$U_{95} = \pm[B^2 + (t\bar{Sx})^2]^{1/2}$$

In this paper, there will be some discussion of the results of the statistical studies related to RCS flow determination. We shall focus on the development of bias limits and precision errors and on the methods employed to propagate errors through the complex mathematical functions which are involved in feedwater flow, secondary calorimetric, and RCS flow determination. The results of the Trojan studies are also presented. Unfortunately, we are not able to show the development and propagation of all errors pertinent to the subject.

I. EQUATIONS FOR POWER AND FLOW

Determination of RCS flow in a nuclear power station requires that the steam power produced in the secondary system be determined. This determination is commonly called a secondary calorimetric. One of the parameters critical to secondary power (SP) determination is feedwater flow. Once calculated, the SP becomes the basis for RCS flow measurement.

In the Trojan Plant secondary calorimetric procedure, the SP is calculated by:

$$SP = 4 \times Mfw(0.9996 \times H_s + .0004 \times H_f - H_{fw}) - 0.9996 \times BD_{tf} \times (Hs - Hf) \quad \text{(Equation 1)}$$

Where the variables and typical values are:

SP = Secondary power in Btu/hour		$SP = 1.146 \times 10^{10}$
Mfw = Feedwater flow in lbm/hour		$Mfw = 3.683 \times 10^6$
H_f = Steam generator (SG) liquid enthalpy in Btu/lbm		$H_f = 518.562$
H_s = SG steam enthalpy in Btu/lbm		$H_s = 1197.97$
H_{fw} = Feedwater enthalpy in Btu/lbm		$H_{fw} = 414.76$
BD_{tf} = Total blowdown flow in lbm/hour		$BD_{tf} = 1.154 \times 10^5$

The variables in the secondary power formula are determined as follows:

A. The H_s, H_f, and H_{fw} values are look-up variables determined from steam tables on the basis of SG pressure and feedwater pressure and temperature.

B. The blowdown flow is read in gpm and converted to lbm/hour.

C. The M_{fw} variable is calculated as will be shown in Equation 3. Also note that the feedwater term in the equation is multiplied by four to account for the four SGs and their individual feedwater lines.

The RCS flow is calculated using the input from the secondary power calculation plus others in the following manner:

$$RCS_f = (SP - RCPh) \times (.1247) \times (\mu)/(T_{he} - T_{ce}) \quad \text{(Equation 2)}$$

Where:

RCS_f = RCS flow in gal./min. $\quad RCS_f = 3.731 \times 10^5$

SP = Total secondary power in Btu/hour	
RCP_h = Reactor coolant pump heat in Btu/hour	$RCP_h = 48 \times 10^6$
μ = Reactor coolant specific volume in ft^3/lbm	$\mu = .021389$
T_{he} = RCS average hot leg enthalpy in Btu/lbm	$T_{he} = 630.92$
T_{ce} = RCS average cold leg enthalpy in Btu/lbm	$T_{ce} = 549.34$

For the calculation of RCS flow, SP is calculated from Equation 1, and the RCPh is a calculated value with an assumed ±5 percent error. The Thot and Tcold enthalpies and RCS specific volume μ are look-up values from the steam tables made on the basis of average Thot, Tcold, and pressurizer pressure readings.

Mfw is calculated for the secondary calorimetric by the formula and typical values as follows:

δp = 399.32 in inches of water and is the feedwater venturi δp

ρ = 52.381 in lbm/ft^3 and is feedwater density

F_a = 1.00634 is dimensionless and is the venturi expansion factor

K_{ave} = 25305 is the average feedwater venturi discharge coefficient

$$M_{fw} = K_{ave} \times F_a \times \sqrt{\delta p \times \rho} \qquad \text{(Equation 3)}$$

$M_{fw} = 3.683 \times 10^6$ lbm/hour/loop or feed line

The feedwater density ρ is a steam table look-up value made on the basis of feedwater pressure and temperature indications. The K_{ave} value is the average of the four feed line venturi discharge coefficients and F_a is determined from vendor data (it is a function of feedwater temperature).

II. DISCUSSION OF INSTRUMENT ERRORS

Two Trojan Plant tests are run in order to determine total RCS flow. The Plant secondary calorimetric is run daily, and the RCS flow test is run annually. Feedwater venturi δp (three readings are taken of each of the four venturis), SG pressure (three indications of pressure are read for each of four SGs), feedwater pressure and temperature, and SG blowdown flow (one reading for each SG) are the Plant indications from which secondary power is calculated. RCS Thot and Tcold (three readings of Thot and one reading of Tcold for each loop) and pressurizer pressure (three indications) are read in order to calculate RCS flow.

The approaches described by the ASME PTC and the ISA RP were used to quantify the error in each of the individual indications. The Plant tests are run during normal Plant conditions; therefore, harsh environment effects did not have to be considered. The instrument temperature effects, calibration setting tolerances, drift, accuracy, readability, maintenance and test equipment effects, and Plant process computer inaccuracy (where applicable) were determined from vendor, Plant calibration, and industry data. The NUREG/CR-5560, "Aging of Nuclear Plant Temperature Detectors" was used as a source of some data for the resistance temperature detector (RTD)-related indication uncertainties. The bias and precision error applicable to the feedwater flow venturi δp indicator are shown below as an example of the instrument errors.

δp Indicator Errors

	Bias (B)	Precision (S)	Source
Accuracy		±0.5%	Assumed random portion of drift.
Stability		±0.577%	Observed value.
Readability	±0.2%		One-half of smallest division of scale.
Static pressure	±0.17%		Manufacturer's specification.
Temperature effect	±0.5%		Assumed bias because indicator is purely mechanical device.
Setting tolerance	±1.0%		From calibration records.
Drift	±0.5%		Assumed bias portion of drift.

The stability error for feedwater flow δp is a process dynamic effect because two of the Trojan feedwater flow control valves are unable to maintain a steady-state position and thereby introduce a variation in indicated feedwater flow δp. A steady-state variation in indication is also present in the RCS Thot which is measured with a high precision bridge.

The Plant procedures require multiple readings (a common technique) of some indications in order to reduce the precision error introduced into the calculations. Three readings of δp are recorded for each feedwater line (four) flow. The dynamic error has a magnitude of approximately 1 percent of δp indicator span. The precision error effect in δp indication due to this dynamic error after the three readings are averaged is calculated by:

$$S\delta p_{re} = \frac{1.0}{\sqrt{3}} \qquad S\delta p_{re} = 0.577 \text{ percent span}$$

The feedwater flow δp indicator bias (excluding readability) and precision errors are:

$$B\delta p = \sqrt{0.17^2 + 0.5^2 + 1.0^2 + 0.5^2} \qquad B\delta p = 1.2365 \text{ percent span}$$

$$S\delta p = \pm\sqrt{0.5^2 + 0.577^2} \qquad S\delta p = 0.7635 \text{ percent span}$$

Both the bias limits and precision errors are combined using SRSS in the ASME methodology. The ASME basis for this approach is that it works (gives reasonable results) and that, for example, the probability is very low (10^{-8}) that four biases would be within 1 percent of their limit at the same time. (This probability was calculated assuming rectangular distributions for the four bias limits.)

III. STATISTICAL ANALYSIS OF DRIFT

The bias and precision errors for the feedwater δp indicator did not contain separate errors for measuring and test equipment (M&TE), Plant vibration, normal radiation, and calibration instrument accuracy. The drift and accuracy errors which were included were determined by the statistical analysis of Plant calibration as-left and as-found data which includes the listed effects.

The feedwater flow δp indicators are calibrated at six points over their span. We have analyzed the calibration as-left and as-found data at each transmitter calibration data point. Trojan Plant secondary calorimetrics are performed at 100 percent power which corresponds to approximately 80 percent of δp span. Therefore, the 80 percent of span data is discussed here. The calibration data was loaded into a spreadsheet where the mean and standard deviation of the data were determined. A histogram of the data set of actual drift (in percent of span) at 80 percent of scale is shown in Figure 1 with a normal curve drawn over the data. The characterization of the data as a normal distribution appears to be reasonable and conservative due to the concentration of data around zero.

The data was imported into a SYSTAT software program and was checked for correlation with time. The results indicated that this drift data does not have a significant correlation with time. This result is consistent with all other data sets we have checked. The drift derived from calibration data has not exhibited time dependency at Trojan. The sample data was tested for normality by use of the Kolmogorov-Smirnov test at the 95 percent confidence level. The analytical result of the test was that the data does not show conformance with the normal distribution with a high level of confidence. This result is due to two influences. The histogram in Figure 1 displays a characteristic we have already mentioned which is a large group of data at zero drift. The normal curve is conservative due to this concentration of drift values around zero, but this behavior also is one cause for a non-normal analytical result.

A PPLOT (Figure 2, which is the data plotted on probability paper) of the data reveals the other reason for the non-normal analytical result. If the distribution were perfectly normal, the data would lie along the straight line drawn in the figure. Instead, most of the data is grouped at drift values of 0.0, ±0.2, ±0.4, ±0.6, etc., percent of span. These groups of data correspond exactly to the readability of the δp indicator. The 500 inches of water δp output scale is divided into 2-inch divisions. One-half of one division is exactly 0.2 percent of span. Hence, the grouping of the drift data is due to the readability bias.

The data set for the indicator drift at 80 percent of scale is again plotted in Figure 3 against the days (INTERVAL) between calibrations. The drift is modeled as a combination of random and bias errors. The standard deviation (random error) of the drift is approximately one-half percent. The bias limit is also assumed to be one-half percent. The 99 percent interval of drift is therefore ±1.5 percent. A readability bias of ±0.2 percent of span is also assumed in the uncertainty calculation. As the figure illustrates, more than 99 percent of the sample data lies within the ±1.5 percent interval.

IV. UNCERTAINTY IN STEAM TABLE LOOK-UP VALUES

Several of the parameters in Equations 1 through 3 are ASME steam table look-up values whose error depends on the error in the indications used as the basis for look-up. The H_s, H_f, H_{fw}, ρ, T_{he}, T_{ce}, and μ are the steam table look-up values determined from the Plant indications of SG pressure, feedwater pressure and temperature, RCS hot and cold leg temperature, and pressurizer pressure. An example of the error introduced into these look-up values by the errors in the indicated parameter is the bias limits introduced into H_s and H_f by the error in SG pressure. The individual SG pressure indication has a bias limit of ±2.477 percent over a 1,200-psi span. The error in four independent averaged indications (one per loop) is given by:

$$SG_{pe} = \frac{2.477}{\sqrt{4}} = 1.24 \qquad SG_{pe} = 1.24 \times \frac{1200}{100} = 14.88 \text{ psi}$$

Use 15 psi

There are actually 12 SG pressure indications (three per loop). Four was used as the average for conservatism because we have averaged bias limits which may have been calibrated with the same equipment (which would have the same bias; therefore, the bias would not be random).

The applicable values for interpolation and error determination from the steam tables are (average Plant SG pressure is 851 psia):

Saturated Pressure	SG Liquid Enthalpy	SG Steam Enthalpy
870	521.8	1197.3
860	520.1	1197.7
850	518.4	1198.0
840	516.7	1198.2

The bias limits for Hs and Hf were determined as follows:

$$BH_s = \frac{1198 - 1197.7}{10} \times SG_{pe} \qquad BH_f = \frac{520.1 - 518.4}{10} \times SG_{pe}$$

$$BH_s = 0.45 \text{ Btu/lbm} \qquad BH_f = 2.55 \text{ Btu/lbm}$$

V. ERROR PROPAGATION THROUGH MATH FUNCTIONS

The bias limits (B) and precision errors (S) in the Plant indications and ASME steam table look-up values are propagated through the equations for feedwater flow (Equation 3), SP (Equation 1), and RCS flow (Equation 3). The equations are listed in the order of propagation. The ANSI/ASME PTC 19.1-1985 contains the derivation of the formulas applicable to the propagation of errors through such math functions (see Section 6), "Propagation of Errors by Taylor Series". Similar formulas are derived in the ISA Measurement Uncertainty Handbook Appendix B. The ISA RP dS67.04 draft recommended practice also contains similar formulas (in a slightly different format) in Table 6-1. The SRSS formulas are used to propagate both bias and precision error. These formulas assume that the propagated errors have continuous distribution functions which are smooth and which have continuous derivatives around the point of interest. We previously looked at the feedwater flow δp indicator whose drift behavior we are able to model using a normal distribution by adding additional bias limits for drift and readability.

As an example of the propagation of error, the bias limits associated with the feedwater flow will be propagated through Equation 3. The propagation formula for these errors is taken from Page 37 of the ASME PTC 19.1-1985 Equation 5.3. The formula contains a non-squared term that accounts for the dependency between the feedwater density and venturi expansion factor errors which are both functions of feedwater temperature. Formula 5.3 is:

$$\frac{Bmfwt}{mfw} = \left[\left(1 \times \frac{BK}{Kave}\right)^2 + \left(\frac{B\rho}{2\rho}\right)^2 + \left(\frac{B\delta p}{2\delta p}\right)^2 + \left(\frac{BF}{Fa}\right)^2 + 2\left(\frac{1}{2} \times 1 \times \frac{B\rho}{\rho} \times \frac{BF}{Fa}\right) \right]^{1/2}$$

Where $\frac{BK}{Kave} = \frac{\pm.0075}{\sqrt{N}} = \frac{\pm.0075}{\sqrt{2}} = \pm.0053$ = bias limit in average venturi discharge coefficient,

and N is the number of loops (which is actually four, but two is used for conservatism).

$\frac{B\rho}{\rho} = .00163$ = bias limit of feedwater density due to temperature error

$\frac{B\delta p}{\delta p} = \frac{3.263}{399.32} = .00817$ = bias limit of venturi δp

$\frac{BF}{Fa} = \frac{3.5 \times 10^{-5}}{1.00634} = 3.48 \times 10^{-5}$ = bias limit of venturi expansion factor

$\frac{Bmfwt}{mfw} = .00676$ $\qquad \frac{BmfwF\rho}{mfw} = .00091$

The equation would also normally contain an error term for the bias limits associated with the feed line inside diameter (BδD) and venturi throat diameter (Bδd). The BδD and Bδd terms are zero in the Trojan application because the venturis have been calibrated. The use of calibrated venturis simplified the quantification of the BK error which is equal to the laboratory accuracy plus an installation error taken from ASME-MFC-3M-1989, "Measurement of Fluid Flow in Pipes Using Orifice, Nozzle, and Venturi".

The quantification of the K_{ave} bias limit at Trojan was also made easier because the feedwater flow venturis are periodically inspected and cleaned of any fouling. The venturi expansion factor error (BF) is a function of the error in feedwater temperature and is due to the thermal expansion of the venturi. This error can be graphically determined from the ASME standard. The total bias limit Bmfw and bias limit due to feedwater density and expansion factor errors ($BmfwF\rho$) were both calculated to aid in the quantification of correlation coefficients which are needed for further error propagation.

Going back to Equation 1, it can be seen that precision and bias limits for each term of the equation contribute to the final error interval - U_{ADD}. Assume that these errors have been determined and that we wish to combine the flow errors (feedwater or blowdown) with the enthalpy errors of Equation 1. The bias limit for the 0.9996Hs + .0004Hf combination of enthalpies is computed as the sum of:

$$BH_{sf} = 0.9996(BH_s) + .0004(BH_f)$$

$$BH_{sf} = 0.9996(.45) + .0004(2.55) = 0.4508 \text{ Btu/lbm}$$

For conservatism this bias limit was added to the bias limit of the feedwater enthalpy (BHfw) to determine the bias limit of the enthalpy difference.

$$BHD = BH_{sf} + BH_{fw} = 0.4508 + 1.985 = 2.436 \text{ Btu/lbm}$$

The relative bias limit of the enthalpy difference is given by:

$$\frac{BHD}{HD} = \frac{2.436}{.9996_{Hs} + .0004H_f - H_{fw}} = \frac{2.436}{782.94} = .00311$$

The relative bias limit for the feedwater term of the secondary power equation is calculated as the SRSS of the feedwater flow and enthalpy relative biases plus the contribution due to correlation between the two terms.

$$\frac{BSP}{SP} = \left[\frac{Bmfwt}{mfw}^2 + \frac{BHD}{HD}^2 + 2\,\alpha\,\frac{Bmfwt}{mfw} \times \frac{BHD}{HD} \right]^{1/2}$$

where α is the correlation between the feedwater and enthalpy relative bias limit terms. The correlation is due to the feedwater temperature error which affects both terms. For the enthalpy term, the feedwater temperature bias limit affects the bias limit in the feedwater enthalpy or BHfw. The portion of BHD due to BHfw is BHfw/BHD. This ratio is:

$$\frac{BHfw}{BHD} = \frac{1.985}{2.436} = .815$$

The feedwater temperature error affects the feedwater flow error by affecting the errors in venturi expansion factor (Fa) and feedwater density (ρ). This error (relative bias limit) was previously calculated as BmfwFρ/mfw. The total feedwater flow relative bias limit was calculated as Bmfwt/mfw. The ratio of feedwater flow error due to feedwater temperature error to total feedwater flow error is:

$$\frac{BmfwF\rho/mfw}{Bmfwt/mfw} = \frac{.00091}{.00676} = 0.134615$$

The correlation coefficient α is the product of the two ratios:

$$\alpha = \frac{BHfw}{BHD} \times \frac{BmfwF\rho/mfw}{Bmfwt/mfw} = 0.815 \times 0.134615 = 0.1097$$

Then the relative bias limit $\frac{BSP}{SP}$ is:

$$\frac{BSP}{SP} = [.00676^2 + .00311^2 + 2(.1097 \times .00676 \times .00311)]^{1/2}$$

$$= .007745, \text{ or in percent, } .7745 \text{ percent of calculated power}$$

The relative bias limit for the blowdown term of the secondary power equation is calculated in the same fashion as for the feedwater term. The two terms are combined using a weighted formula to account for the two orders of magnitude difference between the power of the feedwater term (greater) versus the power in the blowdown term (lesser). When this is done, it is found that the blowdown term may add an additional .02 percent to the relative bias limit. The resulting sum (.7745 + .02) is rounded to 0.8 percent.

The precision errors of the Plant indications were similarly used to calculate the precision errors for each of the terms of the secondary power equation. These errors were combined and propagated in a manner similar to the bias limit errors we have included for illustration. The magnitude of the total relative precision error of the secondary power equation was calculated to be ±0.2543 percent. For 99 percent coverage, the uncertainty in the SP calculation is (t = 2):

$$\text{UNCERTAINTY} = \pm(.80 + t \times .2543)$$

$$= \pm 1.309 \qquad \text{Use } \pm 1.4 \text{ percent}$$

The value of t was taken as 2 on the basis that the errors used in the calculation are from vendor data, drift analyses, and industry standards. These sources either have or can be assumed to have greater than 30 degrees of freedom. Therefore 2 is the correct multiplier.

VI. RCS FLOW UNCERTAINTY

The RCS flow was calculated from Equation 2 to be 373,100 gal./min. The uncertainty of this calculation can be quantified by continuing the propagation of errors using the ISA RP or the ASME PTC formulas. The sources of error to the RCS flow calculation are the secondary power, reactor coolant pump heat, RCS specific volume, Thot enthalpy, and Tcold enthalpy errors.

The Plant SP error has already been quantified. The RCS flow error is calculated using a Plant average basis as was done for the SP errors. The error estimates for the individual parameters (Thot, Tcold, and pressurizer pressure) that are read and converted to process values (specific volume and enthalpies) include both bias limits and precision errors. Because the bias limits and precision errors can be combined by SRSS in the ASME methodology, the RCS flow uncertainty error is calculated on a combined (bias limit and precision error) basis.

The RCS specific volume and core enthalpy rise are determined from ASME steam tables on the basis of pressurizer pressure, narrow range Thot (Th), and narrow range Tcold (Tc) measurements. The errors in Th and Tc were calculated by using the ISA RP dS67.04 methodology.

The Th error (T_{he}) and Tc error (T_{ce}) were calculated to be $\pm 2.47°F$ and $\pm 1.314°F$, respectively. Four independent T_h and T_c measurements are recorded. The error in the average T_h and T_c measurements are calculated by:

$$T_{ha} = \frac{T_{he}}{\sqrt{4}} \qquad T_{ha} = 1.235°F \qquad T_{ca} = \frac{T_{ce}}{\sqrt{4}} = 0.657°F$$

The error in the pressurizer pressure reading was calculated to be ± 25 psig. This error and the average T_h and T_c errors were applied to the ASME steam table values for enthalpy and specific volume as was illustrated in the secondary power calculation. The results of these error calculations are:

$V\mu = 2.037 \times 10^{-5}\ ft^3/lbm$ = error in RCS specific volume

$T_{hen} = 1.896$ Btu/lbm = error in hot leg enthalpy

$T_{cen} = 0.827$ Btu/lbm = error in cold leg enthalpy

The reactor coolant pump heat input is assumed to be 48×10^6 Btu/hour with an error of 5 percent of the value or 2.4×10^6 Btu/hour. The error contribution of reactor coolant pump heat (RCPh) is minor in comparison to the secondary power error. The error in the RCS flow calculation (refer again to Equation 2) can now be determined by the combination of the errors in the calculation input parameters. The error in the SP-RCP_h difference is treated as the sum of the two errors because the SP error is much larger than the RCP_h error and the sum approach is conservative:

$$SP_{err} = SP_e + RCP_{he} = 1.651 \times 10^8$$

Where SPe = 0.014 x SP and

$SP = 1.162 \times 10^{10}$ Btu/hour (100 percent power)

The error in the RCS flow calculation is calculated by first determining the error in the numerator of the RCS flow Equation 2. This is accomplished by application of Equation 6.14 of the ISA RP. This equation combines the parameter errors by SRSS to calculate the propagated error through the multiplication function. The value of the numerator of the RCS flow equation is also needed to propagate the errors. It is calculated as NUM below:

$$NUM_{err} = .1247[([SP - RCP_h] \times V\mu)^2 + (\mu \times SP_{err})^2 + (SP_{err} \times V\mu)^2]^{0.5}$$

$NUM_{err} = 4.413 \times 10^5$; $\quad NUM = .1247 \times (SP - RCP_h) \times \mu = 3.086 \times 10^7$

RCS flow is calculated as the quotient of the numerator over the denominator of $T_{he} - T_{ce}$. The error in this quotient is calculated by application of Equation 6.16 of the ISA RP as follows:

$$RCS_{fe} = \frac{\left[([T_{he} - T_{ce}] \times NUM_{err})^2 + \left[NUM \times (T_{hen}{}^2 + T_{cen}{}^2)^{0.5} \right]^2 \right]^{0.5}}{(T_{he} - T_{ce})^2 - [(T_{he} - T_{ce}) \times (T_{hen}{}^2 + T_{cen}{}^2)^{0.5}]}$$

$RCS_{fe} = 1.13 \times 10^4$ gallons per minute error

VII. RESULTS AND CONCLUSIONS

We have described the process used at Trojan for the calculation of feedwater flow, Plant secondary power, and RCS flow. The Plant indication errors impact the uncertainty associated with the calculation of these parameters. Examples of this impact have been given to demonstrate the methodology employed to propagate the individual errors through the mathematical processes. The secondary power error of approximately ± 1.4 percent is indicative of the advantage in accuracy achieved by averaging Plant indications to minimize uncertainty. The ± 1.4 percent error also illustrates the facility the Taylor Series error propagation technique gives the analyst to track the effect of each error and dependencies between errors. Care must be exercised to identify where dependencies exist between parameters. The RCS flow uncertainty of 11,300 gpm is approximately 3 percent of actual flow. Both the secondary power and RCS flow uncertainties meet Trojan Technical Specification requirements.

VIII. REFERENCES

1. ISA RP dS76.04, Part II, "Methodologies for the Determination of Setpoints for Nuclear Safety-Related Instrumentation".

2. ANSI/ASME PTC 19.1-1985, "Measurement Uncertainty".

3. ISA Measurement Uncertainty Handbook, 1980.

4. NUREG/CR-5560, "Aging of Nuclear Plant Temperature Detectors".

5. ASME-MFC-3M-1989, "Measurement of Fluid Flow in Pipes Using Orifice, Nozzle, and Venturi".

FIGURE 1

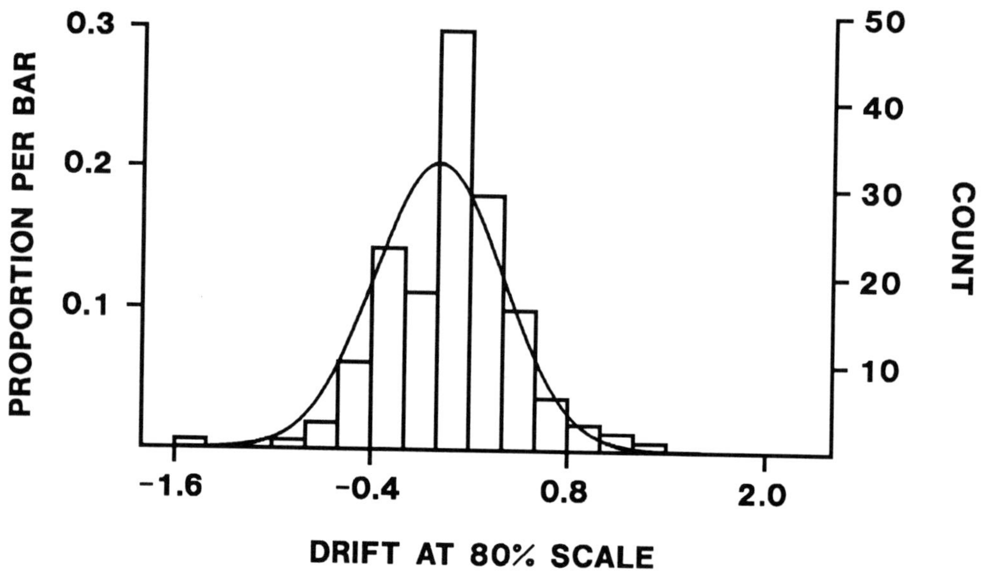
0.3
0.2
0.1
PROPORTION PER BAR
50
40
30
20
10
COUNT
-1.6
-0.4
0.8
2.0
DRIFT AT 80% SCALE

FIGURE 2

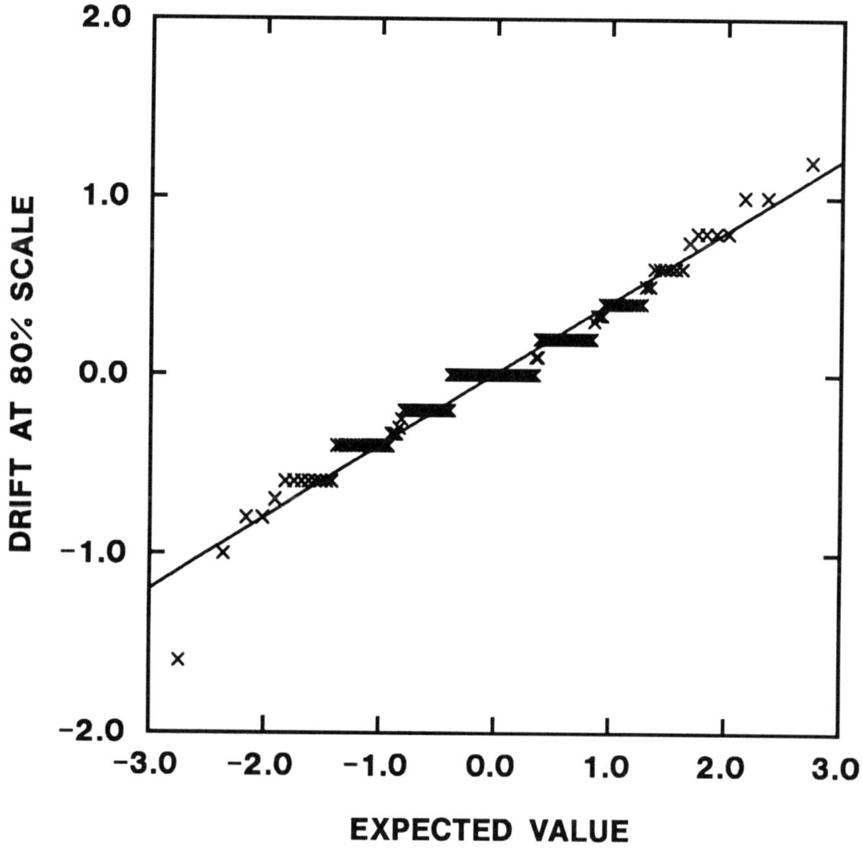

2.0
1.0
0.0
-1.0
-2.0
DRIFT AT 80% SCALE
-3.0
-2.0
-1.0
0.0
1.0
2.0
3.0
EXPECTED VALUE

FIGURE 3

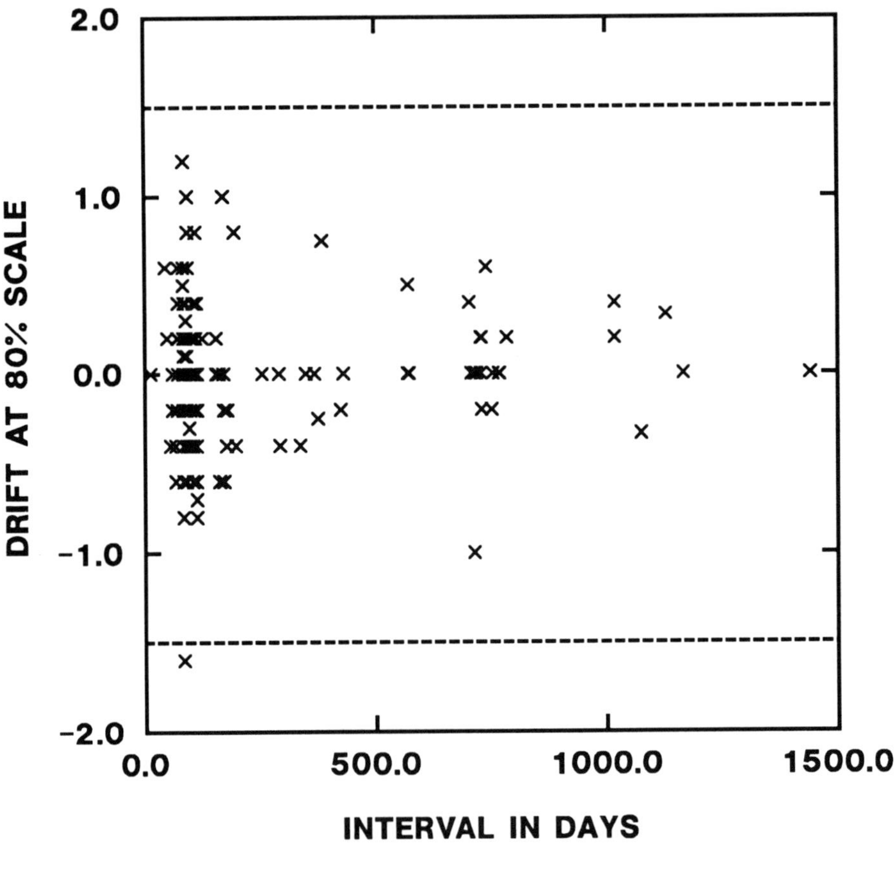
2.0
1.0
0.0
-1.0
-2.0
0.0
500.0
1000.0
1500.0
DRIFT AT 80% SCALE
INTERVAL IN DAYS

©1991 - Paper #91-722
0074-056x/91/307-313/$0 + .50pp

DELTA-P LEVEL MEASUREMENT SYSTEMS

Glenn E. Lang
James P. Cunningham

ABSTRACT

Delta-P level measurement systems are used extensively in Pressurized Water Reactor Systems to measure water level in various vessels. The measured level signals are used as inputs to plant protection and control systems and also for providing indicated level to the operators for use during normal plant operations and following postulated design basis events. If these level measurement systems are designed properly, they provide a reliable indication of actual water level in a vessel. However, if a few basic design details are ignored, several potential discrepancies become evident and may result in an ambiguous level indication. This paper briefly summarizes several discrepancies associated with a delta-P level measurement system that are inherent to the design of the systems and that are due to common design oversights.

Key words: Delta-P, measurement systems, reference leg, condensate pot.

INTRODUCTION

Water level measurement in numerous tanks or vessels is required to operate a Pressurized Water Reactor efficiently. Examples of tanks or vessels in which level is measured include, but are not limited to, the Volume Control Tank, Pressurizer Relief Tank, Boric Acid Tank, Refueling Water Storage Tank, Auxiliary Feedwater Storage Tank, Reactor Vessel, steam generators and the pressurizer. The level is usually measured in each of these tanks or vessels utilizing a delta-P level measurement system; however, for Westinghouse plants only the measured levels for the steam generators and the pressurizer are used as inputs to the Reactor Protection System for initiating reactor trip functions and engineered safeguard features. Because of the importance of these two measured levels, the discrepancies associated with these levels that are due to the inherent operating characteristics of the measurement system and to common design oversights are discussed in the following paragraphs.

UTILIZATION OF LEVEL MEASUREMENTS

The utilization of steam generator water levels and pressurizer water levels can be effectively subdivided into three areas: (a) inputs to the Reactor Protection System (RPS); (b) inputs used for normal plant control; and (c) indicated levels used for post accident monitoring. Each of these areas is discussed below.

1. Inputs to the Reactor Protection System

Steam generator water level is utilized as an input to the RPS for the following protection functions:

a. Initiates a reactor trip when a specified low steam generator water level setpoint is reached.
b. Initiates the Auxiliary Feedwater System (AFWS) when a specified low steam generator water level setpoint is reached.
c. Initiates steam generator blowdown line and sample line isolation when a specified low steam generator water level setpoint is reached.
d. Initiates a feedwater isolation and turbine trip when a specified high steam generator water level setpoint is reached.

Pressurizer water level is utilized as an input to the RPS that initiates a reactor trip function when a specified high pressurizer water level setpoint is reached.

2. Inputs Used for Normal Plant Control

Steam generator water level is used as an input to the steam generator water level control system in conjunction with other plant variables for controlling steam generator water level.

Pressurizer water level is used as an input to the pressurizer level control system in conjunction with Reactor Coolant System (RCS) average temperature for controlling pressurizer level. The level signal is also used to automatically isolate RCS letdown and to turn off pressurizer heaters when a specified low pressurizer water level setpoint is reached.

Finally, indicated pressurizer water level and steam generator water level are used during plant heatup and cooldown and other planned plant operations.

3. Indications Used for Post Accident Monitoring

The indicated pressurizer water level and indicated steam generator water level are used extensively throughout the plant Emergency Operating Procedures (EOPs) in several areas, which include the following:

a. Safety Injection System (SIS) termination and reinitiation
b. Indication of loss of secondary heat sink
c. Indication of loss of RCS inventory
d. Indication of restoration of normal steam generator water level
e. Indication of restoration of normal pressurizer water level

In general, indicated steam generator water level is used to monitor the status of secondary heat sink. The indicated pressurizer water level, in conjunction with reactor vessel level, is used to monitor the status of RCS inventory.

ERROR SOURCES ASSOCIATED WITH A DELTA-P LEVEL MEASUREMENT SYSTEM

Several areas have been identified which result in measurement errors in the delta-P level measurement systems. Each of these potential error sources is discussed in the following paragraphs.

1. Process Pressure Variations

   After the level measurement system is installed in the plant, each system is calibrated at a specific reference temperature and process pressure applicable to the system. If the process pressure changes as a consequence of plant operating conditions, errors are inherent in the level measurement system. An approximation of the measurement error due to changes in process pressure (assuming the temperature of the fluid in the vessel is at the saturation temperature corresponding to the specific pressure) is given by the following expression, as a fraction of the level span:

$$E_P = \{[(H_L/H)(\rho_g - \rho_{gc}) + (L/H)(\rho_f - \rho_g)]/(\rho_{fc} - \rho_{gc})\} - (L/H) \qquad (1)$$

   where

L	=	actual water level in vessel above lower level tap (feet)
H_L	=	maximum vertical distance from lower tap to water level in condensate pot at upper tap (feet)
H	=	vertical distance between upper and lower taps on vessel, i.e., the level span (feet)
ρ_g	=	dry saturated steam density at the process pressure (lbm/ft^3)
ρ_f	=	saturated water density at process pressure (lbm/ft^3)
ρ_{gc}	=	saturated steam density at the calibration process pressure (lbm/ft^3)
ρ_{fc}	=	saturated water density at the calibration process pressure if the system is hot calibrated or water density at the calibration process pressure and temperature if the system is cold calibrated (lbm/ft^3)

   The measured level signal can be electronically compensated for the error due to process pressure variations by inputting the measured process pressure into an algorithm, calculating the applicable level error, and correcting the measured level signal by subtracting the error term. This type of compensation can be easily implemented in a digital computer system.

2. Reference Leg Temperature Variations

   In addition to assuming a process pressure when calibrating the level measurement system, reference leg temperature is assumed. If the temperature of the reference leg changes (e.g., due to changes in normal containment temperature, containment heatup due to a high energy line rupture, etc.), errors are inherent in the level measurement system. An analytical expression for approximating the resulting error due to changes in reference leg temperature is given by the following expression, as a fraction of the level span:

$$E_T = (H_L/H)(\rho_{Lc} - \rho_L)/(\rho_{fc} - \rho_{gc}) \quad (2)$$

where

ρ_L	=	water density in reference leg (lbm/ft^3)
ρ_{Lc}	=	water density at the process pressure and reference leg temperature at which the level measurement system was calibrated (lbm/ft^3)

The measured level signal can be electronically compensated for the error due to reference leg temperature variations. This can be accomplished by installing strap-on RTDs on the reference leg to measure the leg temperatures, inputting the temperatures into an algorithm, calculating the applicable level error, and correcting the measured level by subtracting the error term. This type of compensation can easily accomplished using a digital computer system.

3. Fluid Velocity Effects

When calibrating the level measurement system, the fluid velocity near the tap locations is assumed to be negligible such that a differential pressure is not introduced due to the fluid flow. An approximation to the error introduced by fluid flow past the lower tap (the top tap is assumed to be in the steam space) is given by the following expression, as a fraction of the level span:

$$E_V = -(W/A)^2(1.0 + K_f)/2Hg_c\rho_{WT}(\rho_{fc} - \rho_{gc}) \quad (3)$$

$$K_f = 288g_c\rho_{WT}\Delta P\ (A/(W_R{*}CR_R))^2 \quad (4)$$

where

W	=	fluid flow rate normal to lower tap (lbm/sec)
A	=	flow area at lower tap (ft^2)
g_c	=	gravitational constant (ft/sec^2)
ρ_{WT}	=	water density in vicinity of lower tap (lbm/ft^3)
K_f	=	friction and form loss factor
ΔP	=	pressure drop in downcomer to lower tap (lbm)
W_R	=	steam flow at rated plant conditions
CR_R	=	circulation ratio at rated plant conditions

Conceptually, this error could be subtracted from the measured level if the downcomer velocity is measured or calculated. However, for some steam generator designs, it is beneficial to correct the control and high level setpoints, based on the calculated water velocity at rated plant conditions.

4. Downcomer Subcooling Effects

For a steam generator vessel, another source of measurement error is due to subcooling of the fluid in the downcomer region in conjunction with a saturated mixture around the steam generator U-tubes. The magnitude of the subcooling in the downcomer is dependent upon the following process conditions: main feedwater temperature, circulation ratio, and location of the feedwater nozzle with respect to the lower level tap. The measurement error introduced by downcomer subcooling is given by the following expression, as a fraction of level span:

$$E_S = (\rho_{WT} - \rho_f) \min (H_S/H,\ L/H)/(\rho_{fc} - \rho_{gc}) \qquad (5)$$

where

H_S = maximum height of water column above lower tap that is assumed to be subcooled (ft)

5. Condensate Pot And Instrumentation Line Design Features

Many areas have been identified that can result in the delta-P level measurement system not performing within its expected system accuracy requirements. The comments in this section assume a system configuration with a condensate pot designed to assure a known constant water level in the reference leg (Reference 1). Several of these potential areas are discussed in the following paragraphs:

a. The instrumentation line from the vessel to the condensate pot must not permit a water trap to occur. Design features that could result in a water trap include the following:

   1. A non-monotonic slope or inadequate slope in the line. The instrumentation line should have a slope of at least 1 inch per foot to facilitate the return of condensed steam to the vessel. In addition, the line must exhibit a sufficient slope to overcome the surface tension present in the instrumentation line.

   2. Insufficient line diameter to facilitate steam flow to the condensate pot and condensed steam flow to the vessel. The instrumentation line should have at least a 1/4 inch inner diameter to facilitate the steam and water flow.

   3. A water dam existing in the design of the root valve utilized. The installed root valves should not feature a design that exhibits a water dam and the valve should have a stroke distance of at least 1/4 inch. A particular installed orientation (e.g., horizontal, vertical, etc.) may preclude a water dam.

b. The instrumentation line from the vessel to the condensate pot must assure that steam is condensed in the pot. For example, if the line is too long, the steam will condense prior to reaching the condensate pot. Therefore, the condensate pot must be closely coupled to the vessel. However, the line must be of sufficient uninsulated length to permit sufficient heat transfer to ambient air to ensure that condensation occurs. For example, insulating the line may keep the condensate pot too hot to condense steam.

c. The condensate pot must be of sufficient volume to minimize the time period required to refill the reference leg if fluid is lost. Fluid may be lost due to dissolution of hydrogen following a system depressurization or due to a plumbing leak.

A recommended minimum size of condensate pot is a one and one half inch Tee.

d. If a film of water is postulated to run down the inside of the vessel shell normal to a flush penetration of the upper tap instrumentation line, the condensed steam may be precluded from returning to the vessel. An associated potential problem area is, if a component in the vessel (e.g., a moisture separator) is causing a water jet to impact the penetration of the instrumentation line inside the vessel, the resulting impingement may preclude the condensed steam from returning to the vessel. Eventually the instrumentation line from the vessel to the condensate pot will become full of fluid, thereby introducing an error in the measured level.

   The design for preventing this phenomenon is to extend the instrumentation line inside the inner shell of the vessel a short distance and to install a deflector such that jet impingement cannot impact the end of the line.

e. The high pressure and low pressure instrumentation lines from the delta-P transmitter to the condensate pot and the vessel, respectively, should exhibit a positive slope. A negative slope could result in air or non-condensable gasses becoming trapped in the instrumentation lines, which would introduce an error into the measured level.

   A generic guideline to follow is that all instrumentation lines associated with a delta-P level measurement should exhibit a positive slope from the delta-P transmitter to the condensate pot and the lower tap on the vessel. Even the existence of horizontal segments of an instrumentation line could require a significant period of time for non-condensable gasses to migrate out of the line (e.g., hydrogen coming out of solution, air trapped during calibration, etc.). In particular, when flexible hosing is used to connect the bottom of the condensate pot to the rigid instrumentation line, the hoses must be installed or supported such that they cannot sag and form a trap.

CONCLUSION

Delta-P level measurement systems are used extensively for measuring levels in various vessels in a Pressurized Water Reactor. If designed properly, these level systems will adequately measure levels that are used for the following: (a) inputs to the Reactor Protection System; (b) inputs utilized for normal plant control; and (c) indicated levels utilized in post accident monitoring. However, if these level systems are not designed properly, several potential sources of error exist that can result in the level measurement system providing ambiguous information. Because of the importance of reliable measured level signals, special attention should be devoted to these systems during their design, installation and maintenance. This paper summarized some of the commonly experienced discrepancies that have occurred in operating plants associated with delta-P level measurement systems. Recommended design features were also presented to address the areas discussed above.

REFERENCES

1. ISA-S67.02 - 1980 Nuclear Safety Related Instrument Sensing Line Piping and Tubing Standards for Use in Nuclear Power Plants.

©1991 - Paper #91-723
0074-056X/91/315-325/$0 + .50pp

DESIGN CONSIDERATIONS FOR A COGENERATION ENERGY MANAGEMENT OPTIMIZATION SYSTEM

D.J. Ahner
C. Kona

ABSTRACT

A cogeneration Energy Management Systems (EMS) should be designed to continuously control plant energy flows, insure the simultaneous satisfaction of process heat and power requirements and energy supply contractual commitments and protect plant personnel and equipment. Differences in the thermal performance characteristics of the plant's power generation, heat recovery, and process boiler equipment, in addition to the use of various fuels, and contractual commitments for power and/or process steam, provide a wide range of considerations which must be factored into the overall energy management and its optimization. In addition, other EMS considerations include unit commitment and interlinking of closed loop energy management control with plant process controls.

The availability of low cost, reliable microprocessor based plant control, monitoring and computational hardware, in addition to powerful optimization software techniques permits the practical application of advanced energy management concepts to these cogeneration applications. The authors are currently developing an EMS of the complexity described herein. The considerations discussed in this paper were used to initiate this development.

This paper develops approaches to this problem, applying concepts to specific application situations. A discussion of some of the general system aspects to be considered and their implementation is presented. The intent is to provide information and insight useful in the development of a specification for a comprehensive cogeneration energy management system.

INTRODUCTION

Cogeneration plants are generally characterized by multiple sources of energy, various types of prime movers (e.g. boilers, waste heat recovery, steam and gas turbines, etc.), and varying requirements for process heat and electrical power, particularly if bulk power is being dispatched to a utility grid. The ability to coordinate the operation of these various components to meet all the requirements is a formidable task.

Energy management systems for these plants have several simultaneous objectives. These include 1) dynamic control of energy flows, 2) maintaining the operating security of the energy supply system and equipment, and 3) optimization of energy use to meet given loads and constraints at the lowest costs.

It is important to understand the process characteristics and design requirements of these individual functions to develop an EMS which has the appropriate speed, data acquisition, computer software capabilities and control system interface to accomplish the intended objectives.

As shown in Figure 1, process controls may be considered a primary element in an EMS which interfaces with the plant equipment and is designed to accommodate the dynamics of the various process loops, providing responsive and stable control in maintaining specified process state conditions. Data gathering and reduction (i.e. averaging) and control logic and response requirements of process control are dictated by the process dynamics. The process control is typically overlayed with a hierarchy of control logic which provides system security protection in the event of contingency situations and optimization of energy use for power and process steam supply. These security and optimization functions impose certain speed of response and control logic depending on process needs. The optimization function is relatively slow acting in contrast to typical process control.

For example, where critical loads sensitive to electrical load frequency or steam supply conditions are involved, system security may require data acquisition and response times which do not permit significant deviations in process variables, such as frequency or steam pressure, which a conventional (underfrequency or underpressure) security system would utilize to initiate protection action. In such cases, load and generation imbalances must be immediately detected, and immediate control action with accurate priority load shedding must be initiated. Such requirements impose unique variable sensing, computer and control response requirements. Calculation of individual unit output capability, under all unit system trip scenarios may be necessary to establish the proper load shed for balanced operation. This procedure requires current on line unit information to keep load shed tables accurate and timely. Such calculations may be quite complex for industrial cogeneration type systems.

The Energy Management Optimization System (EMOS) can impose its own unique requirements in terms of data acquisition and response. A description of the various functions and implementation of process optimization with EMOS is the subject in this paper.

ENERGY MANAGEMENT OPTIMIZATION SYSTEM (EMOS)

The implementation of an EMOS involves the interaction of several main modules. It is helpful to examine each of these modules in terms of their functions and how each component interfaces and contributes to the overall solution of satisfying load demands at minimum cost. These modules are software packages which perform various functions required of a complete EMOS. Figure 2 is a schematic indicating the functional relationships of these modules.

Performance Monitor (PM)

This module interfaces directly with the plant process control. Its function is to collect, validate, average, store and output the information needed for the optimization calculation on a timely and consistent basis. The PM provides a consistent data base for current and historical data. It calculates and stores performance factors for the major equipment and continually calculates and updates the incremental heat rate curves for the equipment in the dispatch. Its data base provides information for graphic and supervisory reports, and alerts operators as to the status of the dispatch and the equipment.

The PM is key to insuring that the dispatch reflects the current state of all the equipment by continually updating equipment efficiencies and other plant performance factors. This allows the PM to also provide diagnostic capability concerning trends and unit operating changes. The calculated data reflects equipment degradation and

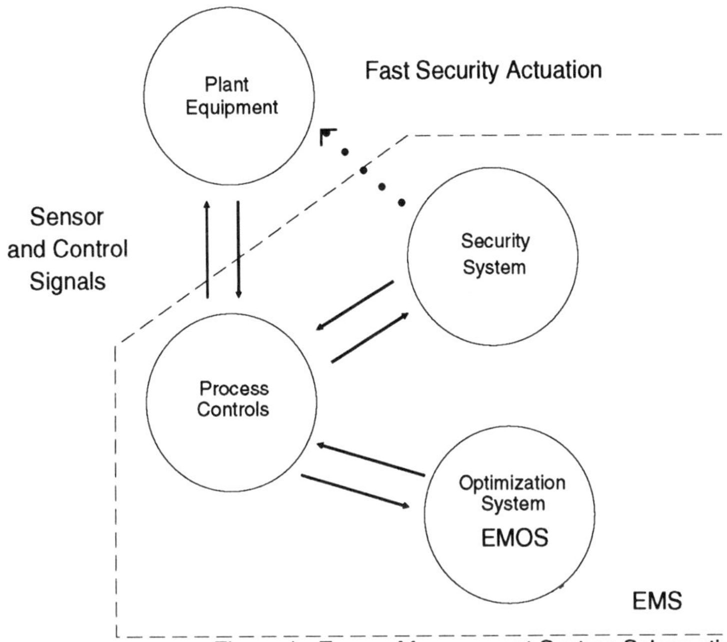

Figure 1. Energy Management System Schematic

offers the opportunity to reduce operating costs by also identifying required equipment maintenance concerns. In this respect, the PM not only provides information on the current state of the equipment for accurate dispatch, but also provides information on changes and how to improve the equipment status.

Dispatch

This function contains the logic for how equipment operation will be optimally changed to bring the system from its existing state to the optimum state for the existing power and thermal loads. The optimum state is determined by minimum fuel cost or under existing plant power and/or steam constraints. This function determines the set points, participation factors, etc. for the specific equipment which have been identified for dispatch. If a particular equipment is not responding to the desired state changes, the condition is displayed for operator review and that equipment may be removed from the automatic dispatch list and the optimum dispatch recalculated. The logic also includes such functions as system trend determination to insure that state changes are indicative of desired optimum system loading trends and to satisfy any limits on changes in equipment operating conditions as the system is moved to the optimum state. This function may also contain logic to establish changes in system configurations. This includes such items as changes in steam paths, loss of steam distribution components, availability of new units coming on line and/or equipment being removed from service. Additional functions may include dynamic or steady state distribution system simulation to insure that projected optimum states do not violate operating constraints.

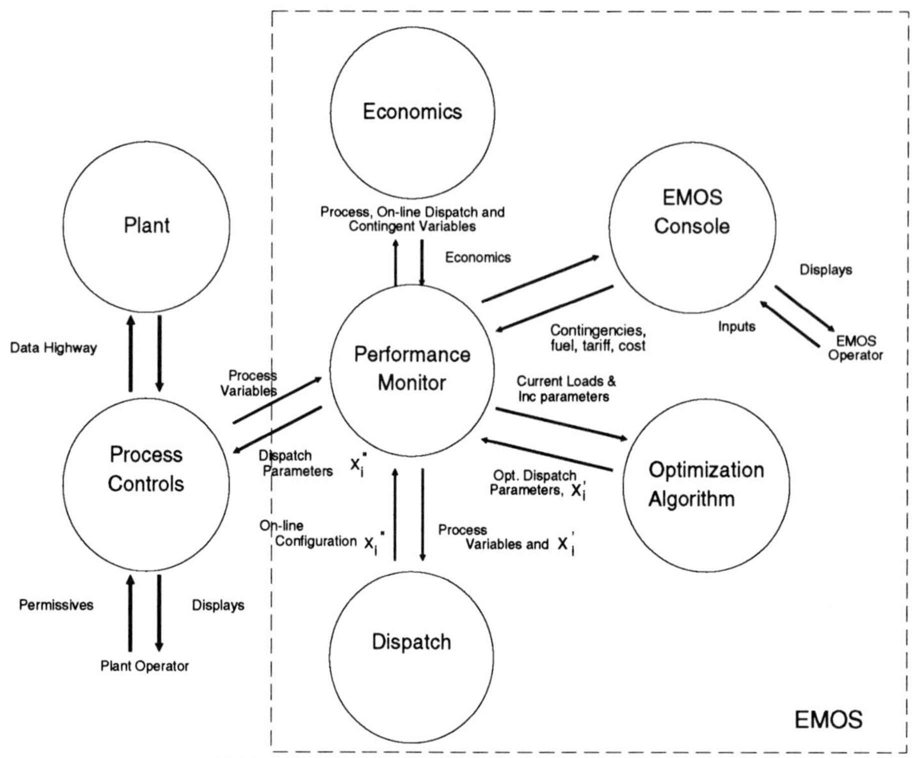

FIGURE 2. EMOS Functional Arrangement

Of concern, is the need to optimize the dispatch on an incremental basis. Due to equipment loading limitations, it is not possible to instantaneously move from one optimum state to another. By continuously optimizing on an incremental basis, the path from one optimum state to another is established. This means that if the target optimum state is not fully realized, the system, in moving towards that state along the most optimum path will be operated at minimum cost. This is especially important when considering optimum dispatch for continually changing power and process steam needs.

Economics

This module calculates the costs of operation for defined configurations. It determines the on-line cost of operation using the current costs and measurements of fuel and purchase power flows. This will include the energy costs of dispatched generators, steam boilers and power coming in through a tie line.

It may be desirable to establish multiple simultaneous cost calculations: 1) cost of online operating equipment configuration satisfying current loads, 2) cost of current configuration optimum unit dispatch, which satisfy current on line system loads, 3) cost of current configuration optimum dispatch to satisfy forecast loads, 4) cost resulting from user specified alternate equipment configurations which satisfies current on line loads, and/or 5) forecast of system loads.

On line calculations require the use of process measurements (power, flow, etc.). Cost calculations require operator inputs but utilize EMOS optimization and economics software.

Various calculated economic scenarios are used which: a) compare actual operations to optimum economic performance, b) determine economic loss due to process dynamics and/or control system response, c) determine economic impact of using alternate equipment and d) determine economic profile of forecasted loading and/or an alternate plant load condition.

In general, the economics function is the accountant for determining the plant operating costs resulting from current and forecast loads and to establish the economic impact of the alternate system configurations or equipment performance. Power and steam sale billing transactions are also part of its module's functions.

The economics function would include evaluation of unit availability and calculate associated startup and shutdown costs to be included in the economics of a particular operating scenario.

Interface

This is an operator interface to the EMOS, which provides the capability of display and review of any of the information from the other modules. It provides the user with the ability to change inputs such as power tariff schedule changes, fuel costs, or in general, to alter the EMOS calculation when necessary to account for changes to the operating environment. In addition, the user has the capability of pulling up data to perform off-line studies using the same information and software available to the EMOS algorithms and to review results and input permissives for EMOS implementation activity. This user interface provides the capability to perform "what if" analysis with respect to equipment, system configuration or load forecast situations.

Since the interface can communicate with all of the EMOS modules, it provides a window with which a performance engineer can evaluate the equipment operation. This can range from viewing on-line operation to pulling up historical data and determine equipment trends. On-line viewing screens can be graphs or tables or a combination with values changing continuously as plant data is updated. Historical data can be viewed as time plots, cross plots, tables and charts. This capability gives the performance engineer valuable information with which to diagnose the equipment and observe trends which may require attention.

Optimization Algorithm

This module determines the optimum loading dispatch of power and process steam energy sources based on the existing plant load and operating configuration information.. This module solves the problem of how to satisfy the energy demands at the lowest operating cost. It uses the incremental performance characteristics of the plant equipment recognizing limits and constraints in the cogeneration system.

OPTIMIZATION ALGORITHM RELATIONSHIPS

The thermal dispatch of power system boilers and turbines establishes the plant operation economic target. Economic power dispatch is widely applied in electric utilities and in industrial cogeneration applications. This establishes the economic dispatch of process steam and power output to the various boilers and prime movers in the specified configuration.

To establish this optimization, the configuration of the system components and their individual equipment load thermal performance characteristics must be known. This data is used in an optimization algorithm to calculate the optimum dispatch to meet a given system load at minimum cost. The optimum solution is determined from the solution of sets of energy coordination equations. These equations include the equipment incremental performance and any power or steam constraints. They are developed from the thermal performance characteristics, fuel costs and the system configuration. The calculations require an algorithm to find the optimum solution of these equation sets.

In contrast to the conventional utility power plant optimization, the algorithm for cogeneration plants is much more complex. The incremental relationships for a power plant may be expressed simply as plant fuel input versus power output. However, cogeneration applications require the consideration of multiple energy sources, individual plant equipment incremental performance, the simultaneous demands for power and process heat at various energy levels and complex system configurations generally associated with cogeneration plant design. The resulting solution and implementation is more complex than that associated with utility power plants.

The optimization equation for a cogeneration system are developed for each component supplying power and/or steam to the various plant headers. As an example, the coordination equations for the system shown in Figure 3 are:

COORDINATION EQUATIONS

Unit 1 Process Steam

$$\frac{\lambda_o}{\eta_{g1}} = \lambda_s \qquad (1)$$

Unit 1 GT Power

$$\frac{\lambda_o h_{g1} - \lambda}{\eta_1 (h_{g1} - 3413)} = \lambda_s \qquad (2)$$

Unit 1 ST Power

$$\lambda_s h_{s1} = \lambda \qquad (3)$$

Unit 2 GT Power

$$\frac{\lambda_o h_{g2} - \lambda}{\eta_{g2} (h_{g2} - 3413)} = \lambda_s \qquad (4)$$

Unit 2 ST Power

$$\lambda_s h_{s2} = \lambda \qquad (5)$$

Inc. Performance Characteristics

$$\begin{aligned} hg_{1,2} &= f_{1,2}\,(kW_{g1,2}) \\ \eta_{g1,2} &= g_1\,(kW_{g1}, q_{S/F}) \\ \eta_{g2} &= g_2\,(kW_{g2}) \\ \eta_{s1,2} &= m_{1,2}\,(kW_{s1,2}) \end{aligned} \tag{6}$$

Load Constraint Conditions

$$\begin{aligned} kW_p &= kW_{g1} + kW_{g2} + kW_{s1} + kW_{s2} \\ S_p &= S_1 + S_2 \end{aligned}$$

(7)

where $\lambda_0, \lambda_s, \lambda$ = *fuel, incremental steam and incremental power costs*

h_g, h_s, η_g, η_s = *unit power and process energy incremental heat rates and steam rates*

kW_n, S_n = *unit power and individual steam header energy outputs*

kW_p, S_p = *plant power and individual steam header loads*

Development of relationships governing the optimization are discussed in Reference 1.

In contrast, system optimization equations for power dispatch only, are of the form:

Coordination

$$\lambda_i H_i = \lambda \tag{8}$$

Plant Performance Characteristics

$$H_i = f\,(kW_i) \tag{9}$$

Load Constraint

$$kW_p = \sum^{n} kW_i \tag{10}$$

where H_i = *Individual plant incremental Heat Rate*

A comparison shows that, equations (1) through (7) contain multiple incremental cost parameters associated with power and individual process steam header supply. The form of the coordination equation depends on the type of unit and its configuration. In addition, the solution depends on multiple individual component thermal performance characteristics and the power and process steam constraint conditions. Equation (7) must be satisfied for all generators and for each steam header in the plant.

The form of the coordination equation also depend on the type of unit, its configuration in the cogeneration plant and the method of power and process steam control which is utilized. The thermal optimization may be modified by constraints in the power or process steam distribution system. Electrical equipment outages may require a redistribution of generator loadings to avoid bus or equipment overloads and/or violation of interconnection power or VAR flows. Similarly, flow limits may exist for specific plant steam headers and equipment outage may modify steam piping configurations. Minimizing electrical distribution system loss may also be desirable. Such considerations may also be introduced and solved in the optimization algorithm.

The degree of complexity in the solution of this optimization algorithm often discourages attempts at its solution and implementation. However, analytical techniques can be applied which can solve this complex system of equations and do not require sophisticated system models or extensive program development for implementation.

The optimization can result in widely varying results depending on equipment configuration and characteristics. That is, the optimization depends on each of the unit's current performance and availability. These are defined on an online basis through the performance monitor, the dispatch module and the economics evaluation. Figure 3 is a plant configuration schematic which can be used to illustrate an example of the differences in optimum equipment loading between a power and cogeneration plant application. The economic dispatch of this configuration is shown in Figure 4 as independent power plants on a utility grid. Figure 5 shows the economic dispatch operating in an (IPP) cogeneration plant supplying process steam to industrial users and power to a utility grid.

The individual plant utilizes similar gas turbines but one is equipped with a supplementary fired heat recovery steam generator (HRSG) and a corresponding larger steam turbine. The individual capacity of the gas and steam turbines relative to the combined plant capacity, is indicated in Figure 3.

The power plant configurations could be separate remote power plants dispatched to a given grid. A single process steam connection for an IPP cogeneration application is indicated. In a cogeneration plant application, cycle modifications such as a single autoextraction steam turbine might be included which would further modify the results. However, to illustrate the basic differences in optimum equipment loading sequence solely due to the power vs. cogeneration plant application, the plant equipment configuration is assumed identical in both applications.

Solution of relations 1) through 7) results in a cogeneration criteria for optimum dispatch:

$$\eta_{g1} = \eta_{g2} \tag{11}$$

This is distinctly different than a power only optimum solution which results in

$$H_1 = H_2 = H_i \tag{12}$$

Figure 4 indicates the optimum loading sequence for a noncogeneration power plant. It will be noted that the supplementary fired HRSG unit is started first. As system load demand increases beyond about 48% of total demand, the second unit is started and the first unit is backed off. Both units are then loaded in parallel and fully loaded by equal incremental cost, without supplementary firing, to 80% total system demand where the gas turbines are fully loaded. Supplementary firing is then applied to achieve full capacity at 100% demand.

The optimum dispatch situation in a cogeneration application is very different. Figure 5 indicates the dispatch of the same units in satisfying a given process steam and system load demand at minimum total operating cost. The loading sequence of Figure 5 will be different for various levels of process steam demand. For the sequence shown in Figure 5, the process steam demand is equivalent to about 40% of the total steam generation capacity. At this process demand, operation of both units are required in addition to supplementary firing of unit one. The

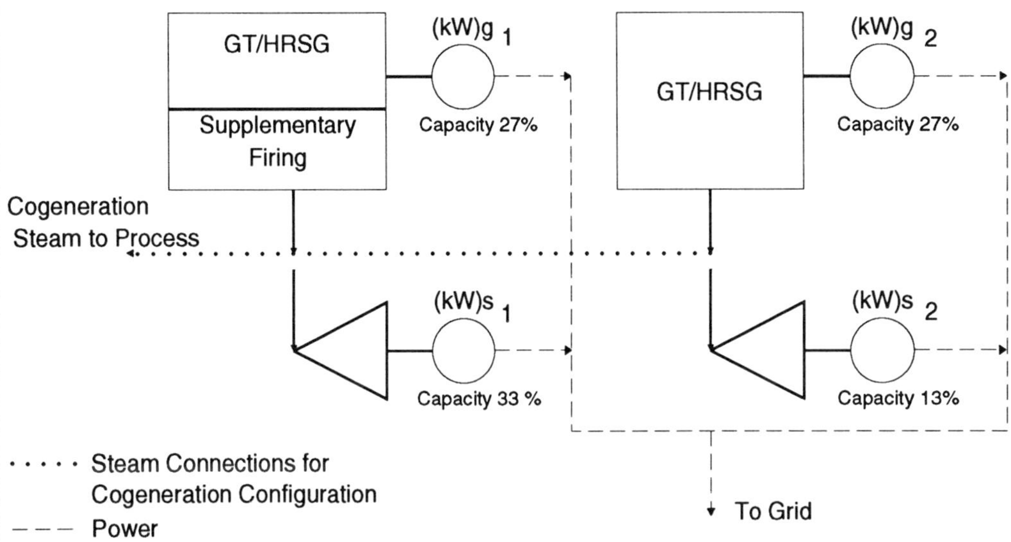

Figure 3. Load Dispatch Schematic

minimum system demand power to meet this process demand is 20%. In this application, the unfired HRSG unit is started and loaded first. The fired HRSG unit is then started and supplementary firing is used to equally load the HRSG's to satisfy the process demand at equal steam loading. With increasing system demand, the gas turbine with supplementary firing is loaded and the other gas turbine and both HRSG's are maintained at constant output, until both gas turbines are equally loaded at 40% system load without any supplemental firing. Parallel gas turbine and steam turbine loading proceeds to 53% system load where the gas turbines are at base load. Supplemental firing is reinitiated and brings the plant to full load, which is 80% of full capacity at this process steam demand. Although each loading sequence for the same equipment is optimum and meets the constraint conditions, the loading patterns and the incremental energy costs are very different for each of these applications.

Combined cycle power plants may typically load at an average incremental heat rate of about 6300 Btu/kWHR (industrial turbines, constant geometry and speed) and supplementary fired steam turbine power at an incremental of about 12,000 Btu/kWH. These incrementals depend on the equipment type and cycle configuration.

Thermal dispatch determines when minimum or maximum flows and powers are reached and the need to retain specific equipment in meeting the thermal and power load demands. The dispatch displays a flag to indicate which equipment may be removed from or placed into service.

The development of the proper coordination equations for the specific application and plant equipment configuration is essential to determine the optimum economic dispatch function and solution algorithm.

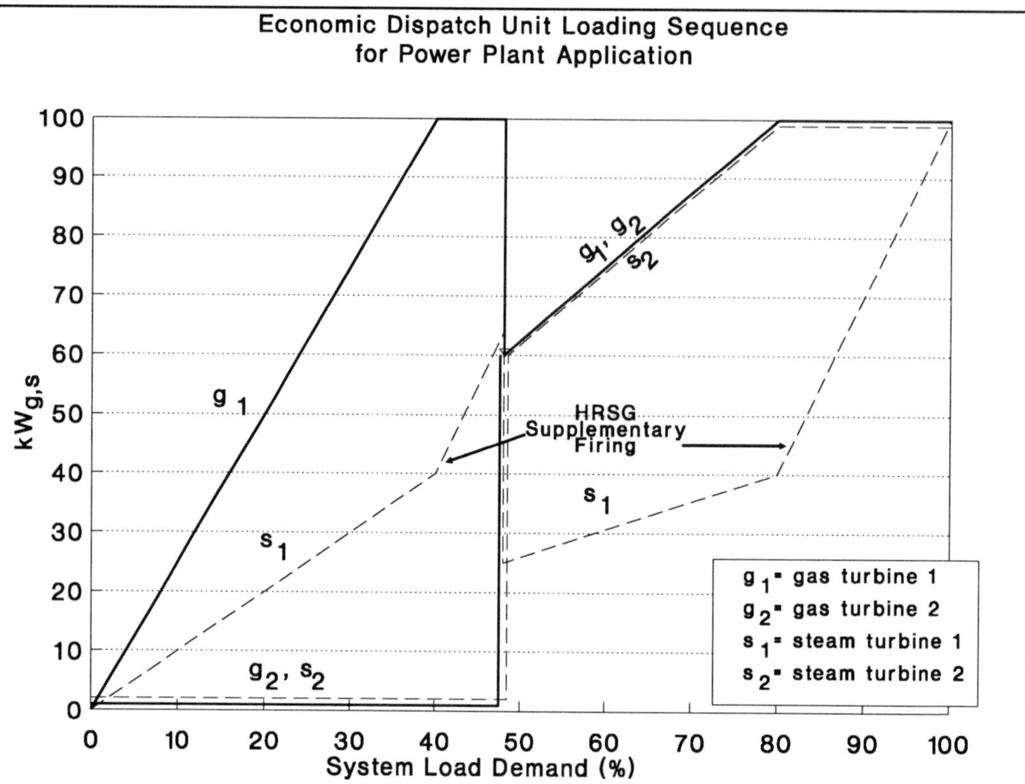

Figure 4

IMPLEMENTATION

EMOS optimization can be implemented as an off-line supervisory program, where process changes are infrequent and may occur over extended periods (e.g. base load). However, where process load changes are more frequent (e.g. several times a day), closed loop control with continual on line calculation is necessary. Additionally, interaction with process dynamics and control stability is required as part of any closed loop EMOS implementation.

The control action of EMOS may be implemented through the plant process control loops. Unit set points and participation factors, calculated in EMOS, are supplied to the plant process control. Consideration of such constraints as unit rate limits are factored into the EMOS dispatch and optimization calculations. Implementation through the plant process control preserves the control configuration and integrity of the individual process control loops and the plant security system.

SUMMARY

In cogeneration systems, the requirement to optimize steam load as well as power generation, must be taken into account. Consideration of the use and cost of multiple fuels or alternate process energy supply and tie line tariffs

Economic Dispatch Unit Loading Sequence
Process Steam Demand - 40% of Total Steam

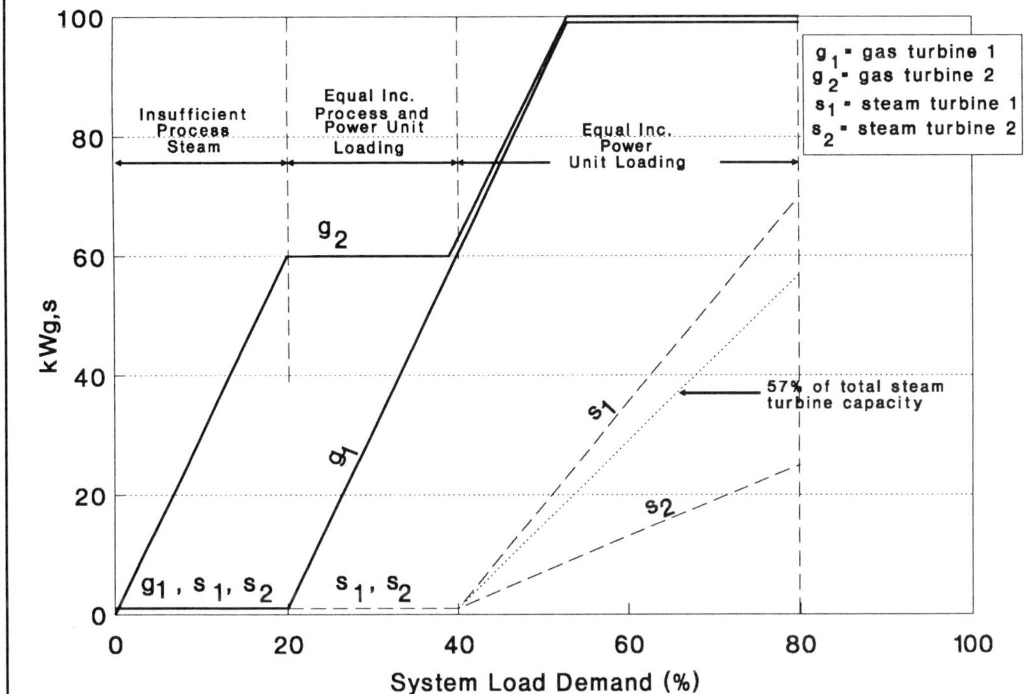

Figure 5

and economics are important in reducing overall energy operating costs. This was illustrated with an example showing the different dispatch requirements between a power plant providing only the electric power and one which must meet steam energy demands as well. This example illustrates the importance of understanding and using the differences in performance characteristics particularly for plants competing for incremental power on the same grid.

The tools and techniques required to develop an Energy Management Optimization System of this complexity are currently available. The various components of this EMOS were discussed along with considerations which should be taken into account to develop a valid specification for building this type of a system.

REFERENCES

1. Ahner, D.J., "Economic Optimization of Cogeneration Energy Management", ISA Paper #85-0988.

2. Crim, H.G., de Mello, R.W., Leyse, R.H., "EPRI ENGINEERING WORKSTATION - AN OBJECT ORIENTED APPROACH TO PLANT PERFORMANCE EVALUATION", presented at the EPRI Heat Rate Improvement Conference in Knoxville, TN, September, 1989.

©1991 - Paper #91-724
0074-056X/91/327-335/$0 + .50pp

COMBINED-CYCLE POWER PLANT PERFORMANCE MONITORING

G. Borrows* R. DesJardins**

ABSTRACT

Maximizing plant efficiency while meeting requirements of the Public Utility Regulatory Policies Act, (PURPA), is an important aspect of operating combined-cycle cogeneration power plants. A real-time computer program for monitoring and reporting the performance of combined-cycle cogeneration facilities has been developed and installed at the Indeck-Yerkes Energy Center in Tonawanda, NY.

Using real-time plant data, the program monitors and displays equipment performance information to assist plant personnel in maintaining optimum plant efficiency and profitability. Performance indices for the gas turbine, heat recovery steam generator, steam turbine, and condenser are calculated and permanently stored. This information is used to study changes in the operating condition of plant components.

The program calculates and trends PURPA qualification indices such as Thermal and Efficiency Standards on daily, weekly, monthly and yearly bases. Printed reports containing historical information on steam and electric energy production, gas turbine operation, PURPA qualification indices, auxiliary boiler operation, and supplementary fuel use are generated on request.

The program includes a controllable parameters screen that displays target values for parameters the operator can affect, along with their actual values. The hourly cost associated with each off-target parameter is calculated and displayed on the screen. This provides the operator with the information needed to optimize operating performance, thereby providing the most economic means of generating electrical and thermal energy.

The program has been successful in increasing operator awareness of plant efficiency and profitability, improving plant management's ability to track and maintain PURPA qualification, and providing a long-term data base of plant equipment performance.

*Operations Engineer, Indeck Energy Services, Inc.
**Senior Engineer, General Physics Corporation

INTRODUCTION

A computer program has been developed for on-line monitoring of combined-cycle power plant performance. The program is installed and operating at the Indeck-Yerkes Energy Center in Tonawanda, NY. Indeck-Yerkes is a 60 MW combined-cycle power plant providing electrical energy to Niagara Mohawk Power Corporation and process steam to the DuPont Company. The COMPAS program, (COmbined-cycle Management and Performance Advisory System), receives real-time data from the power plant control system and calculates performance indices for the plant's major components. The program also provides operational targets for operator controllable parameters, diagnostic flowcharts for efficiency-related problem resolution, and calculation of PURPA, (Public Utility Regulatory Policies Act), indices Efficiency and Thermal Standards.

Indeck Energy Services, owner and operator of the Indeck-Yerkes plant, desired real-time efficiency and qualification information to help plant personnel maintain peak operating efficiency, maximize plant profitability, and assure PURPA qualification status. The COMPAS program, jointly developed by Indeck and General Physics Corporation, accomplishes each of these functions on a personal computer interfaced to the plant control system. The program provides real-time targets for efficiency and PURPA-related parameters, real-time equipment performance indices, interactive diagnostic flowcharts for equipment problem resolution, and extensive data base management and trending capabilities.

SYSTEM OVERVIEW

The monitoring program operates on a personal computer under a multi-tasking, multi-user operating system. The system performs four tasks independently and simultaneously: 1) User Interface, 2) Data Acquisition and Calculations, 3) Data Base Management, and 4) Remote Communications. The flow of data and information is shown in the figure below:

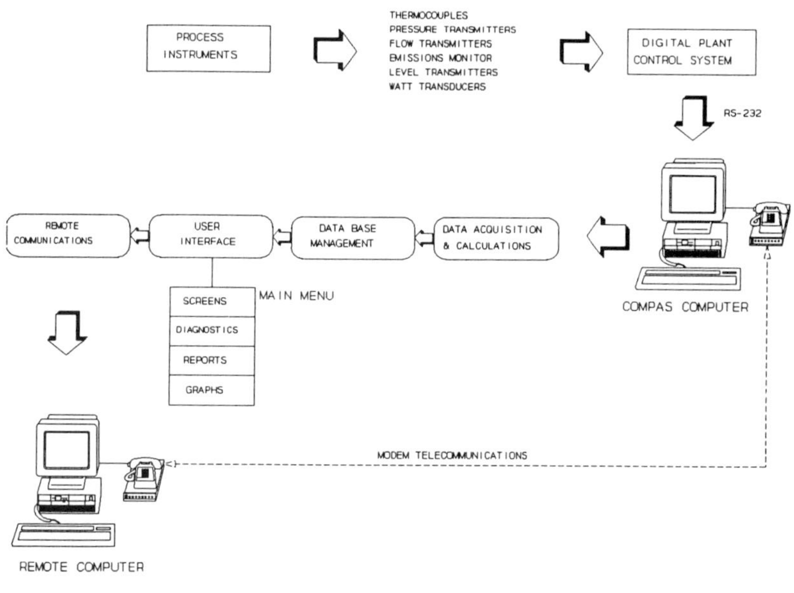

INFORMATION FLOWPATH

The Data Acquisition and Calculation task receives raw data from the plant control system and performs all program calculations including equipment performance indices, controllable parameter targets, off-target operating costs, data limit and integrity checking, and PURPA qualification indices. This information is then supplied to the User Interface task for display on the screen and the Data Base Management task for historical trending. The fourth task, Remote Communications, allows a remote or satellite computer to access the data base and screen information through the use of a modem. The User Interface is the primary task through which the system is operated. From this task, a main menu is displayed with the following selections:

- SCREENS
- REPORTS
- GRAPHS
- DIAGNOSTICS
- CUSTOMIZATION
- EXIT

The SCREENS selection invokes the three main operating displays: the Controllable Parameter Screen, the Management Monitor Screen, and the Equipment Performance Screen. These screens are updated automatically with new plant data every five minutes.

CONTROLLABLE PARAMETERS

The Controllable Parameter Screen displays plant parameters over which the control operator has a degree of control:

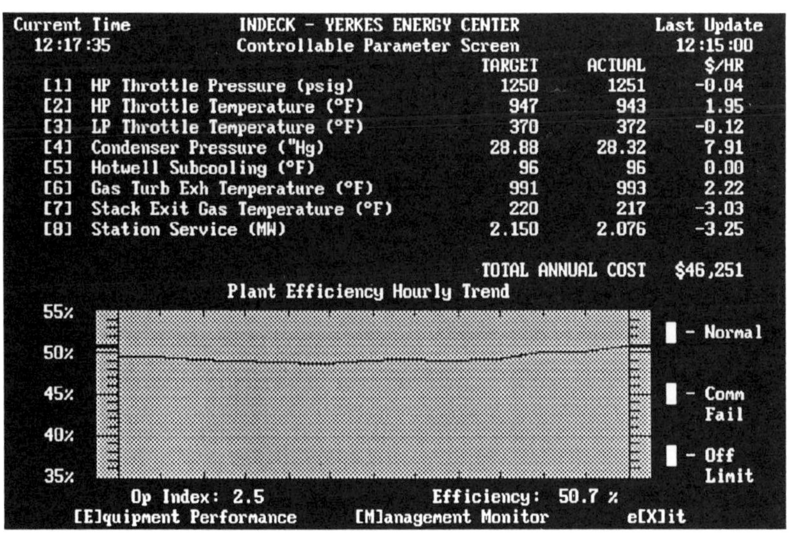

CONTROLLABLE PARAMETER SCREEN

Each parameter displayed on the Controllable Parameter Screen has an effect on overall plant efficiency and profitability. The screen displays the ACTUAL value of each parameter along with a TARGET value and hourly COST penalty for off-target operation. Using this information, the plant operator is able to rank off-target parameters based on cost and take action to resolve the highest cost items first. With real-time cost information, the operator may perform trade-off analyses by comparing the cost of maintaining one parameter over the savings gained from another. For example, comparing the cost of operating cooling tower fans in the high position to the savings of operating the steam turbine at a lower backpressure would constitute a trade-off analysis. The Controllable Parameter Screen also includes a rolling trend plot of Overall Plant Efficiency for the previous hour to assist in performing trade-off analyses.

MANAGEMENT MONITORING

The Management Monitor Screen displays information pertaining to plant outputs (steam and electric energy), and their relationship to PURPA guidelines for a qualified cogeneration facility:

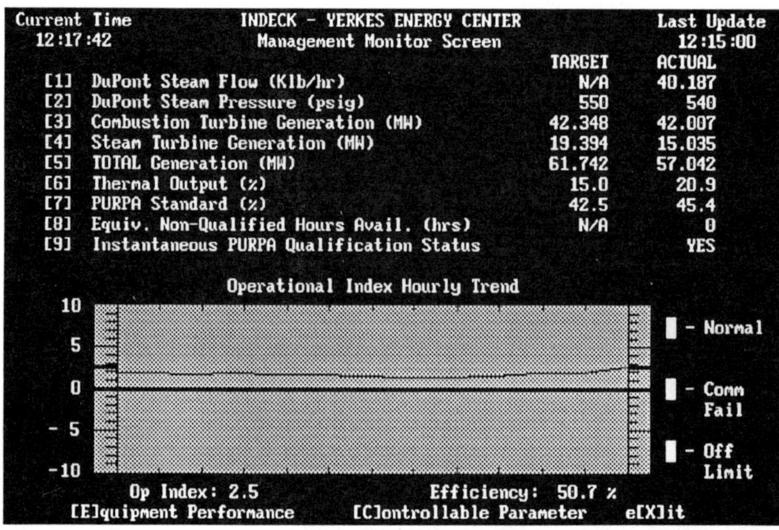

MANAGEMENT MONITOR SCREEN

Thermal Standard and Efficiency Standard are PURPA qualification indices defined by the Code of Federal Regulations. The Thermal Standard is the ratio of steam energy output to total plant energy output, (steam and electricity), expressed as a percentage. The Efficiency Standard is the ratio of electrical output plus one-half steam energy to fuel input, also expressed as a percentage. PURPA guidelines require that a qualified facility maintain the Thermal Standard and Efficiency Standard at or above certain limits defined in the Code of Federal Regulations. The Thermal Standard of a qualified facility must be greater than or equal to 5% over the period of a year. In addition, if the Thermal Standard is less than 15%, a qualified facility must have an Efficiency Standard greater than or equal to 45% measured over a year. If the actual Thermal Standard is greater than or equal to 15%, then the minimum required

Efficiency Standard drops to 42.5%. The program calculates the plant's Thermal Standard and Efficiency Standard, and based on the actual Thermal Standard, displays the appropriate targets for each. During periods of abnormal operation, the plant may not meet the Efficiency and Thermal Standards defined by PURPA. However, the regulations recognize that such periods of operation are possible, if not probable. As such the regulations require that the Efficiency and Thermal Standards bc maintained on a calendar year basis, not an instantaneous or continuous basis. The Equivalent Non-qualified Hours Available is the number of hours that are available for running in a non-qualified mode without jeopardizing the plant's qualification status. The last parameter displayed on the Management Monitor Screen is an indication of instantaneous qualification status. This parameter will display "YES" when the Thermal Standard and Efficiency Standard are at or above their corresponding target values. The indicator is instantaneous only and not cumulative over time.

The Management Monitor also includes an hourly trend plot of Operational Index. Operational Index is a dimensionless factor used to trend gross operating profit, (steam and electric sales less fuel costs). An objective of the monitoring program is to help the operator maximize the Operational Index on a continuing basis while maintaining PURPA qualification.

EQUIPMENT PERFORMANCE

The Equipment Performance Screen displays information about the performance of the plant's primary components, including the combustion turbine, steam turbine, heat recovery steam generator, and condenser:

```
Current Time            INDECK - YERKES ENERGY CENTER               Last Update
  12:17:48              Equipment Performance Screen                  12:15:00

 MISCELLANEOUS PARAMETERS :
                    Dupont Steam Flow (Klb/hr)                         40.187
                    Combustion Turbine Generation (MW)                 42.007
                    Steam Turbine Generation (MW)                      15.035
 COMBUSTION TURBINE PERFORMANCE :
                    Ambient Air Temperature (°F)                           30
                    Compressor Efficiency (%)                            84.9
                    Combustion Turbine Efficiency (%)                    33.6
 HRSG HEAT LOSS EFFICIENCY (%) :                                         81.4
 STEAM TURBINE PERFORMANCE :
                    Turbine Section Efficiency 1st Extract. (%)          59.9
                    Turbine Generator Efficiency (%)                     31.6
                    1st Stage Steam Pressure (psig)                       701
                    Feedwater Flow / Steam Flow                          2.05
 CONDENSER PERFORMANCE :
                    Circulating Water Inlet Temperature (°F)               62
                    # Circulating Water Pumps                               1
                    Cleanliness Factor                                  42.56
 OVERALL PLANT (THERMAL) EFFICIENCY (%) :                                50.7

          █ - Normal           █ - Comm Fail        █ - Off Limit
      [C]ontrollable Parameter    [M]anagement Monitor        e[X]it
```

EQUIPMENT PERFORMANCE SCREEN

Combustion Turbine Performance - This category displays the following parameters related to the gas turbine: 1) Ambient Air Temperature, 2) Compressor Efficiency, and 3) Combustion Turbine Efficiency.

Compressor efficiency is the ratio of the ideal temperature rise across the compressor to the actual temperature rise, (the ideal temperature rise is the rise that would occur if the compressor were 100% efficient). It is expressed as a percentage and is an indication of the physical condition of the compressor blades. As fouling occurs, the efficiency gradually drops off. A wash may be performed to restore "clean" performance. The impact of washing will be seen as improvements in compressor efficiency, gas turbine efficiency, gas turbine output, and overall plant efficiency.

Gas turbine efficiency is the ratio of gas turbine electrical output (MW) to gas turbine fuel input expressed as a percentage. It encompasses the efficiencies of the compressor, turbine, and generator.

HRSG Heat Loss Efficiency - The efficiency of the heat recovery steam generator (HRSG) is defined as the ratio of heat absorbed by the working fluid (water) to the heat added to the HRSG (gas turbine exhaust gases and supplementary fuel to duct burners). It is expressed as a percentage. The efficiency is not measured directly, rather the losses are measured and the efficiency calculated from those losses. The largest loss in the Heat Recovery Steam Generator is the Stack Loss. This loss accounts for the sensible heat in the flue gas leaving the HRSG which is not transferred to the working fluid. The Stack Loss is calculated as follows:

$$\text{Stack Loss (\%)} = (h_{SGT} - h_{AMB})/(h_{EGT} - h_{AMB}) \times 100\% \qquad (1)$$

where:

h_{SGT} = Enthalpy of Gas at Stack Temperature
h_{AMB} = Enthalpy of Gas at Ambient Air Temperature
h_{EGT} = Enthalpy of Gas at Gas Turbine Exhaust Temperature

Combustion gas properties are calculated in accordance with procedures set forth in ASME Power Test Code 4.4, Gas Turbine Heat Recovery Steam Generators. Based on the Stack Loss relationship shown in Equation (1) above, higher-than-normal stack exit gas temperatures will result in increased operating costs.

Steam Turbine Performance - Performance of the steam turbine is measured with two indices: 1) steam turbine section efficiency, (measured from the throttle inlet to the first extraction point), and 2) steam turbine cycle efficiency. Turbine section efficiency is an indication of the physical condition of the turbine steam path while cycle efficiency is an indication of overall turbine cycle performance. Turbine section efficiency is defined as the ratio of the measured enthalpy drop across the turbine section to the ideal enthalpy drop, (the ideal enthalpy drop is the enthalpy drop which would occur if the turbine section were 100% efficient). The actual enthalpy drop, in combination with the quantity of steam flowing through the turbine, determines the work performed by the turbine.

The overall steam turbine cycle includes these components in addition to the turbine itself: the condenser, the vacuum deaerator, the steam jet air ejectors, the steam seal regulator, the LP and HP feedwater pumps, and the steam turbine electric generator. The individual efficiencies of these components contribute to the overall efficiency of the turbine cycle. The efficiency of the turbine cycle is defined as the ratio of useful work performed by the cycle, (MW), to the heat added to the cycle for electrical generation, (heat absorbed by the working fluid in the heat recovery steam generator). As the efficiencies of the individual components decrease, the overall cycle efficiency will trend downward. For example, if the condenser pressure were to steadily increase, the turbine cycle efficiency will decrease.

Condenser Performance - Of all controllable parameters, the condenser pressure has the greatest impact on steam turbine cycle efficiency. A high condenser pressure will decrease the cycle efficiency and will lead to higher operating costs. Condenser performance is measured in terms of Tube Cleanliness Factor. During the design stage, the condenser manufacturer assumes a cleanliness factor of 85% for "as-new" condenser performance. For a given condenser arrangement and condition, the pressure maintained is a function of three parameters: 1) cooling water inlet temperature, 2) cooling water flow, and 3) condenser heat loading, (steam turbine exhaust plus returns). The program calculates the design performance expected for the actual cooling water inlet temperature and flow and the condenser heat loading. This expected performance is based on 85% cleanliness. The same calculation is run with varying cleanliness factors to achieve the actual condenser pressure. The result is the actual cleanliness factor displayed on the Equipment Performance Screen.

Overall Plant Thermal Efficiency - The Overall Plant Thermal Efficiency is an indication of the combined efficiencies of all plant components, including the steam turbine cycle, the gas turbine cycle, the heat recovery steam generator, and the auxiliary boiler. This efficiency is simply the ratio of steam turbine cycle output (MW and steam) and gas turbine cycle output (MW) to total fuel input. It differs from the PURPA Efficiency Standard in that the entire quantity of steam energy output is included in the numerator. Similar to the Efficiency Standard, it is expressed as a percentage.

REPORTING AND GRAPHING

The REPORTS function of the program allows the user to print hard copy reports of the Controllable Parameter Screen, the Management Monitor Screen, and the Equipment Performance Screen. In addition, a fourth report detailing cumulative performance on daily, weekly, monthly, and yearly bases is available. With the GRAPH feature, the user may produce graphical plots of both measured and calculated values. Parameters may be plotted versus time or versus one another. This feature is useful in studying equipment degradation and performing controllable parameter trade-off analyses.

DIAGNOSING EFFICIENCY-RELATED PROBLEMS

The Diagnostic Flowcharts are used to diagnose common problems that contribute to inefficient plant operation. By answering a series of YES or NO questions, the diagnostic flowchart provides the plant operator direction for resolving problems such as high condenser pressure, unsatisfactory cooling tower effectiveness, and low steam temperatures. Each of these items contributes to higher operating costs if not maintained as closely as possible to their "as-new" operating condition. The following Diagnostic Flowcharts are presently available:

- High Condenser Pressure
- Low Hotwell Temperature
- Cooling Tower Problem
- Low Compressor Efficiency
- Low Steam Temperatures

As questions are asked by the diagnostic and answered by the operator, the diagnostic will work toward a solution to the problem. Along the way, the diagnostic will recommend that certain actions be taken to either identify or eliminate a cause(s) of the problem. For example, when diagnosing high condenser pressure, the operator may be asked to check the operation of the steam jet air ejectors to make sure that non-condensibles are being removed from the condenser and not causing air blanketing. The diagnostic flowcharts provide the operator with a systematic and consistent approach to resolving efficiency-related equipment problems.

CUSTOMIZING THE SYSTEM

The monitoring system has been designed to allow on-line updating by the user of the constants used by the calculations. Equipment design information, PURPA-related constants, fuel composition and cost, steam and electric revenue rates, and constants associated with the raw plant data may be modified or updated as required through a series of menus. The system continues to collect and log data while the updates are made.

REMOTE COMMUNICATIONS

The system is equipped with remote communications features that allow direct access by another computer to the system's data base and screen data through use of a modem. The fourth task on the system computer acts as the server, providing unattended, password-protected access by a remote satellite computer. The satellite computer is configured similarly to the server computer and uses the same User Interface and Data Base Management system files. As such, the satellite has capability equal to the server to display and automatically update the main screens displayed on the satellite every five minutes. Changes may be made from the satellite to the server system constants using the CUSTOMIZATION option. The updated files may then be transferred to the server. This feature enables the System Administrator to administer several systems at different plants from a single remote location.

SYSTEM HARDWARE

The monitoring system is implemented on an 80386 personal computer having a 100 MByte hard disk drive, 8 MByte RAM, and an 80387 math coprocessor. Data transfer from the plant control system to the COMPAS System takes place through a stand-alone 80286 personal computer imbedded in the 80386 case. Raw data is serially transferred from the 80286 PC to the 80386 PC.

SYSTEM SOFTWARE

The system functions under The Software Link's PC-MOS 4.0, a multi-tasking, multi-user operating system capable of supporting many DOS applications. Calculation and data acquisition routines are written in Microsoft's QuickBASIC 4.5. The Data Base Management and User Interface tasks were developed using Borland's Paradox 3. Remote communications are implemented using The Software Link's LANLink 5X.

FUTURE ENHANCEMENTS

Future enhancements to COMPAS include development of an automatic compressor efficiency diagnostic alarm, operator guidance for improved cooling tower operation at low loads, and multi-unit generation economy evaluation. Presently, gas turbine compressor efficiency is calculated and archived by COMPAS. Evaluation of this trend with respect to time and ambient conditions is performed manually through manipulation of the data base. The automatic diagnostic will compare present compressor efficiency to past data to detect changes in performance levels and will alert the operator to the need for a compressor wash.

Economic operation of multiple cogenerating units will be investigated and a methodology for determining individual machine loading to achieve optimum economy, (economic dispatch), will be developed.

CONCLUSIONS

The COMPAS program has been successful in increasing operator awareness of plant efficiency and profitability, improving plant management's ability to track and maintain PURPA qualification, and providing a long-term data base of plant equipment performance. The remote communications feature has resulted in more efficient use of technical resources since a single performance engineer is shared among several plants. On-line data and calculations for the entire plant are achieved without need for data takers and time-consuming hand calculations. It should be stressed that the monitoring system is not a substitute for special component performance tests. Instead, the trending of actual performance over time with a full data base of coincident data provides the performance engineer with the information needed to spot declining component performance. The engineer may then schedule more detailed performance tests as required to confirm actual performance levels.

The Controllable Parameter Screen allows the plant operator to compare the economics of one operating scenario against another. These trade-off analyses lead the operator to the most economic decision, while maintaining PURPA qualification and equipment safety limits. Design-based parameter targets help to prevent off-design performance from becoming accepted as normal. The diagnostic flowcharts provide plant personnel with a systematic approach to resolving efficiency-related equipment problems.

ACKNOWLEDGEMENTS

The authors would like to thank Mr. John Drennan, Mr. Joseph Jaskulski, Mr. Joseph Nasal, and Mr. Michele Pontalti for their support of this project.

0074-056X/91/337-342/$0 + .50pp

FUTURE INDEPENDENT POWER GENERATION AND IMPLICATIONS FOR INSTRUMENTS AND CONTROLS

Dr. James H. Williams
RCG/Hagler, Bailly, Inc.
Suite 400
50 California Street
San Francisco, CA 94111

ABSTRACT

The independent power producers market is comprised of cogeneration, small power generation, and independent power production (IPP) segments. Shortfalls in future electric supply are expected to lead to significant growth in this market. The opportunities for instruments and controls will shift from traditional electric utility applications to the independent power market with a more diverse set of needs. Importance will be placed on system reliability, quality of power and increased demand for "clean" kWh.

BACKGROUND

The independent power market has grown dramatically during the last decade reaching almost 38 GW of operating capacity. If the projects representing 60 GW currently under development (i.e., planned and under construction) in this industry are any indication, the prospects for the future growth in the demand for instrumentation and controls will be dramatic. The type of instruments and controls required in the future will be dictated by many factors including, but not limited to, the future mix of prime movers installed and fuels used in independent power industry, and changing federal and state environmental requirements and power sales contractual obligations facing the industry.

THE STATUS OF INDEPENDENT POWER INDUSTRY IN 1990

1990: The Year in Review

The year 1990 was a year of unequalled growth. In fact, the year established a new record for the largest amount of capacity coming on line while new project announcements exceeded all expectations. A total of 224 independent power projects came on line in 1990, adding 6,356 MW of independent power capacity additions. This is 50% more capacity than in 1989 or 1988, and almost double that in 1987.

A principal reason for this increase in 1990 can be attributed to the increased success rate of the independent power industry in bringing projects on line compared to planned. The National Electric Reliability Council (NERC) indicated its last report that the success rate of actual commissioned capacity was 90% of the planned commissioned capacity of independent power projects, up from a 74% success rate in 1989.

By the end of 1990, there were over 38 GW of independent power capacity in operation--involving nearly 2,875 projects with an average project size of roughly 13 MW. Of this total, 45% (or over 17 GW) are natural gas-fueled, 22% (or over 8 GW) are coal-fueled, and 33% (or over 12 GW) use renewable energy and waste fuels.

During 1990, over 35 GW of capacity was announced as under development. This surpasses all previous years with over twice the capacity announced in either 1988 or 1989 and about 20% percent more projects than in either year. Although many of the newly announced projects were competitively-bid projects, the amount of new capacity announced in sole-source projects increased by more than 50 percent.

The dramatic increase in project announcements between 1989 and 1990 was fairly evenly distributed among the major fossil energy sources; the of amount of natural gas-fueled and coal-fueled project capacity announced were nearly equal.

Between 1989 and 1990, the largest regional increases in project announcements occurred in the South Atlantic region (three times that between 1988 and 1989); and in the Mid-Atlantic and East North Central regions (a two-fold increase in each case).

Notably, the largest five developers have increased their market share from 24% in 1989 to 29% in 1990. More generally, the 30 largest developers control about 71% of the market comprised of operating and under development projects.

The Status of Independent Power Industry

At the end of 1990, there were a 60 GW of independent power capacity under development. During the 1991-1993 period, nearly 24 GW (or 40%) of the developing independent power capacity is planned to become operational with the remainder becoming operational in 1994 and beyond.

The current backlog still shows a very strong market for cogeneration with 300 projects comprising

34 GW with an average capacity of slightly more than 110 MW. These projects represent 55% and 57% of the total independent projects and capacity respectively under development.

Correspondingly, there are 200 small power projects representing 11 GW of capacity under development, or about 37% of the total number of projects and nearly 18% of the total capacity. Small power is no longer "small" with an average project size reaching nearly 55 MW.

More profoundly, there are some 40 IPPs for a total of 15 GW with an average project size of 375 MW. Although they represent only 8% of the total number of independent power projects, they represent 25% of the total independent power capacity currently under development.

Of the total independent power capacity under development by the end of 1990, 69% (or 32 GW) is fossil fueled with natural gas-fueled projects amounting 47% (or 29 GW) and coal-fueled projects amounting to 22% (or 13 GW).

Both run-of-the-river and pumped storage hydroelectric independent power projects represent 14% (or less than 9 GW) of the total under development. Other renewable and waste fuel projects amount to 17% (or less than 10 GW).

Of the 50 states and the District of Columbia, six states possess 60% (36 GW) of the total independent power capacity under development at the end of 1990. Two of these states, New York and New Jersey, ranked number 1 and 2 respectively, represent over 28% (or more than 17 GW). A resurgence of activity in Florida makes it number 3 with currently 10% (or 6 GW) followed by Massachusetts and Virginia with 7% (or 4.4 GW) each. The changing geographical focus of the independent power market is exemplified by the fact that California which was ranked number 1 for many years is now ranked number 6 with less than 7% (or 4 GW) of the total market under development.

THE OUTLOOK FOR THE INDEPENDENT POWER MARKET

The Primary Market Drivers

The margins achievable in the independent power market are getting smaller for the following reasons:

- Electric utilities which purchase much of the electricity generated by independent power producers are negotiating "tougher" agreements. Some of the reduced returns (and comparatively lower revenue streams) are a result of competitive bidding between independent power producers.

- Environmental penalties are becoming more onerous resulting in project development delays and failures. Without the ability of independent power projects to engage in environmental cost "pass throughs," the environmental costs are detrimental to the bottom line.

- Consequently, independent power developers are paying the price of competitive bid delays and longer negotiations; in essence, they are financing for a planning activity that a utility normally expenses and passes through to its rate payers.

- The acquisition of the necessary financing is becoming more difficult for independent power producers, since several major lending firms are less active in project finance than in previous years.

- Finally, market saturation is occurring in 15 to 20 electric utility service territories affecting about one third of the national potential for independent power. As a result, independent power developers will incur the costs to become competitively positioned in changing geographical markets.

Based on the experiences of the 1980's, many of these projects will fail with less than 50% of the independent power capacity currently under development successfully reaching operation. However, we expect that another 35 GW or more of independent power capacity will be announced during the remainder of the decade replacing these casualties.

In the long term, the independent power will continue to undergo transitions due to changing attitudes and the bargaining power of electric utilities:

- Competitive bidding will become part of the emerging integrated resource planning imposed on utilities by state regulators which will pit independent power projects against utility-built projects, and against demand-side management alternatives.

- Balanced risk sharing between utilities and independent power producers is yet to be realized. Independent power developers are willing to assume the risks associated with project completion and operations, but want utilities to assume the risks of changes in utility system demand and public utility regulation, and share in fuel availability and environmental risks.

- More and more utilities are imposing very selective criteria as part of the independent power project competitive bidding process, and using differentiated pricing related to differentiated project features as part of the power purchase agreements.

In addition, the independent power industry must deal with the uncertainties associated with the following:

- Availability of primary fuels such as natural gas, coal and some waste fuels.
- The timing and substance of the expected reform in the Public Utility Holding Company Act (PUCHA) and regulations governing the transmission of independent power generated electricity.
- The impact of the Clean Air Act on host electric utilities and on the independent power projects.

The Forecast for Independent Power Market

As a result of these factors, we expect that between 47 GW and 57 GW of new independent power capacity--between 950 and 1,150 projects--will become operational during the remainder of the decade. As a result, independent power will represent between 35% and 40% of all new power generation projected to be operational by the end of the century. The investments to finance this development will be in the $80 to $100 billion range. To operate and maintain these projects, the independent power industry will need to acquire the services of between 20,000 and 30,000 new employees compared to the estimated 11,000 current labor force.

By the end of the year 2000, there could be over 4,000 operating independent power projects representing 90 GW of electric capacity selling a substantive amount of the generated electricity to nearly 220 electric utilities.

The expected character of the forecasted development:

- Project sizes will continue to increase over the next 5 years with nearly 70% of the capacity representing projects exceeding 150 MW.
- Fuel choice for projects will change dramatically for the remainder of the decade with the portion of capacity of new natural gas-fueled projects declining by nearly half of it current portion as new coal-fueled projects becoming more dominant owing to improved technologies and techniques such as circulating fluidized bed (CFB) to generate "clean" kWhs.
- The geography of the market will be altered significantly with roughly 45 new electric utilities located in a dozen states entering the independent power market.

IMPLICATIONS FOR INSTRUMENTATION AND CONTROLS MARKETS

Obviously the forecasted development and changing character of the independent power market has dramatic implications for the suppliers of instruments and controls:

- The number of independent power projects expected to become operational during the remainder of the decade will create opportunities for sales of products and services

- The changing mix of fuel types--proportionally less natural gas and more coal--will increase the demand for a broader range of products and services.

- The implementation of Clean Air Act regulations will necessitate a greater demand for controls and instrumentation.

- The increasing number of projects selling higher-quality electricity to local electric utilities will require more and more sophisticated instruments and controls.

- The increasing demand for more reliable independent power projects will create an increased demand for monitoring devices and controls to respond to operating variances and anticipated maintenance problems.

- The greater number of projects under the control, i.e., utility dispatchability, will require more remote automation.

- The increasing average size of projects, many of which being comprised of multiple prime movers and thermal generators, will necessitate more automated and sophisticated operations

- The potential shortfall in the number and quality of manpower required for operations and maintenance will require more automation.

Clearly, the instrumentation and controls required by the independent power industry in new projects will approach and possibly surpass the size, sophistication, diversity and quality required by electric utility industry. In addition, this industry is expected to continue to repackage and renovate older independent and utility power projects which will require improved instrumentation and controls. As a result, the research and development of innovative instrumentation and control systems currently being sponsored by technology developers and manufacturers, electric utilities, and EPRI becomes more compelling for the comparatively new and growing independent power industry. For the remainder of the decade the sales and installation of instruments and controls could easily exceed $5 billion in the independent power market.

©1991 - Paper #91-726
0074-056X/91/343-348/$0 + .50pp

COMBINED CYCLE PLANT CONTROLS RETROFIT CASE HISTORY - 2 YEARS LATER

Dennis Tenney
Operations Supervisor
Comanche Power Station
Public Service of Oklahoma
Lawton, Oklahoma

Tim Pieszchala
Regional Service Manager
Europe
Tri-Sen Systems Ltd
Kenilworth, England

ABSTRACT

The Comanche Power Station, Public Service of Oklahoma's combined cycle generating facility, underwent a controls and operator panel retrofit at the end of 1988. The plant consists of two gas turbines, two heat recovery boilers and a steam turbine along with three generators. This paper examines the extent to which the original goals and specification were met. Costs, operating principles and modifications since the original installation are discussed. Operating procedures are compared with the original system. The future of the plant is discussed and the impact on the power system grid is analysed.

INTRODUCTION

Public Service Company of Oklahoma's (PSO) Comanche Power Station in Lawton, Oklahoma is the site of a Westinghouse PACE combined cycle plant that became fully operational in 1974. This site consists of two Westinghouse W501-B gas turbines with afterburners, two heat recovery boilers and a Westinghouse non-reheat, condensing, axial exhaust steam turbine. Each gas turbine drives a Westinghouse 94 KVA, hydrogen-cooled generator and the steam turbine drives a Westinghouse 133 KVA hydrogen-cooled generator.

When PSO decided to build a new plant, they looked at the present economic and technological solutions. Natural gas prices and availability during the early 1970s was a leading factor in deciding on a plant type. In general, a natural gas fired plant is clean-burning, lower in pollution and requires less maintenance than an oil-fired plant. This, combined with low heat rate design and quick modular construction, less than 2 years to initial power production, made the PACE combined cycle plant attractive.

Comanche Station was commissioned and over the first few years of operation numerous control system modifications were required to improve the availability. The removal of the trip circuits that were not required for safe operation was one of the largest areas of modification. The availability grew from 52% in 1976 to 89% in 1979, see figure 1. The station output factor also increased, reaching 96% in 1979, see figure 2. Then, mechanical and control system problems started to increase, lowering the average availability from 1980 to 1985 to 77%. The output factor dropped to 71% in 1981, with a 5 year average between 1980 and 1985 of 81%.

Originally, the control system operated with different control modes. In coordinated control mode, the plant was to be treated as one load producer. The individual operational setpoint changes for the gas turbines, boilers and steam turbine were to be managed by the control system, with the operator selecting plant output level and load rate. Another mode was the automatic control mode. The individual systems, gas turbine, boilers and steam turbine, were managed by the control system as independent units, but the plant output was controlled by the operator. The third mode was manual, where each system was under full control by the operator.

The control system was not initially installed with automatic dispatch. Automatic dispatch would allow the central load dispatcher to set load levels for the plant to match the requirement for the grid. As control system problems increased, automatic dispatch was not practical because the plant spent less and less time in coordinated and automatic control. The plant eventually ended up running base loaded. Base load only allowed for a fairly constant output of load. The plant was therefore unable to meet increasing needs to cycle the load to meet swings in the system grid.

Over the years, the control system experienced more and more failures. The heat rate deteriorated due to the manual operation of the plant. Optimum firing temperature on the gas turbines was not maintained as well as automatic control. The heat rate increased 1.7% in 1986 and 2.3% in 1987 at an estimated cost of $1000 per day while operating in manual control.

The control system automatic mode failures contributed to an average loss of 4,256 Megawatt hours per year in the late 1980s. In the five years between 1982 and 1987, the control system automatic operation was responsible for 30 failures which affected plant load and operation. 10 of these resulted in plant or unit trips. In 1988 there were no unit trips attributed to control system failure since the plant operated the majority of the time in manual control.

As a direct result of the successive control system failures, system service hours increased. The increase of hours tended to be amplified by the scarcity of parts and service personnel to maintain the system. The late 1980s showed labor costs for service work averaging $ 25,000 per year, which did not include replacement and spare parts. During repairs, the control computer had to be removed from service increasing the cost of service by lowering the plants efficiency.

The increased costs due to control system failures, operation of the unit on manual control, and service time to repair faults in the control system prompted the investigation of a replacement system.

ORIGINAL GOALS AND SPECIFICATIONS

When it was decided to investigate the control system replacement, the station and engineering staff explored the present technology available which could be used for Comanche Station. The philosophy incorporated features such as fault-tolerant design, increased diagnostics and fault finding capabilities, along with automatic dispatch capability.

Fault-tolerant design was selected to improve overall plant availability. Faults in the system should not affect the plant performance and they should be easy to find and correct. A system which followed these guidelines could be maintained by the station personnel, allowing for quick turn-around and lower service charges from the control system vendors.

In addition to the fault-tolerance, the control system required fault finding features. Sequence of event recording instrumentation and alarming capabilities were needed to help the maintenance personnel to diagnose and fix problems quickly and efficiently.

The system needed to run in automatic mode reliably to acquire the best heat rate, one of the original design features of a combined cycle plant.. In addition, the automatic dispatch was required to help the system grid to maintain load swings. The project feasibility study pointed to the use of automatic dispatch capability as the major economic justification for the project.

One of the other goals was to retain as much of the field cabling as possible. Minimal disruption of operating procedures was a goal to bring the plant back into operation without large retraining expense. In addition, more information was required for the operators to increase their efficiency in running the plant, but not overloading the operator during upsets.

The challenge was to incorporate all these features into a control system that could be engineered and brought on-line in the shortest amount of time. The station needed to be retrofitted quickly to take advantage of the window provided by the steam turbine major overhaul in the fall of 1988 which was initially scheduled for eight weeks duration.

INSTALLATION - THE NEW CONTROL SYSTEM

PSO selected Tri-Sen Systems of LaMarque, Texas to supply their TS3000 Triple Modular Redundant (TMR) turbine control system. The scope of supply included the control system, the operator interface components and auxiliary items for the station retrofit. Other components would be supplied by the PSO engineering department, such as the sequence of events system.

The TMR controller allows the plant to continue operating in a safe condition if a fault is detected in the control system. The diagnostic tools available to the maintenance personnel would permit them to find and determine the best course of action for repairs. On-line replacement of failed components increase the plants availability, since the system does not have to be taken out of service to repair the control system.

Field hardware was reviewed and modified. The transmitters for the combustion shell pressure were duplicated to alleviate a single point of failure of the main control element on the gas turbines. The new control system provided the start-up sequencing logic. This allowed the original relay sequencer to be removed, along with all the interposing relays. On the steam turbine, additional throttle pressure and valve position transmitters were also added to extend redundancy to the field.

Single fault contingency was the main objective followed in the new system, along with the field equipment, to provide availability. No single fault should cause a trip or outage. This continuous operation philosophy included the control system, primary control elements where feasible, and the operator display and information system. The secondary objective was to allow the maintenance and operations personnel to have enough information to properly diagnose and repair problems.

The inclusion of the Rochester 3800 sequence of events recorder, in combination with the control systems alarming and event recording capability, allows the maintenance personnel to quickly decipher problems. Usually, the alarm gives the station personnel enough time to sort out the problem before it become a serious one. The alarm panels in the control room were reduced from 432 alarm windows to 144 alarms windows and are grouped according to function. Some alarm windows point to a detailed CRT display where more information can be viewed when required.

The control room operating panel was enhanced. Color graphic CRT displays mimic the process and provide direct visual feedback. The standard PC-type keyboards were decided upon, instead of the custom process entry keyboards to facilitate training and acceptance of the controls since the PC is now almost a household feature and the operators would readily accept them.

The original control was accomplished by 'walking-the-board', with each major control process is managed from a station. The station management decided to keep this approach since the operators were familiar with it. So, advanced computer interfaces were integrated with the established control board environment. This principle integrated the manual stations, status lights and push buttons into the same panel as the CRT. This idea is similar to the aircraft industry's cockpit upgrades, which allows the pilots to make the gradual transition to new controls.

INSTALLATION - THE PROBLEMS

Of course, any installation is not without its problems. From the time when the decision to retrofit the plant was made in late 1987, to the time the vendor was chosen, took about six months of valuable time. The system specifications had to be defined, written down and agreed upon by all the parties involved. The schedule for the shut-down was ambitious with only eight weeks scheduled, but was extended to 12 weeks just before the actual outage began.

The work order was issued in January of 1988. Five months of preliminary engineering work was done with site visits and design review meetings. Training for plant staff and technicians took place in July and August. At delivery time, the vendors factory was threatened by a hurricane which forced early shipment of the system. This caused some internal wiring errors on the control system which were responsible for three plant trips. The diagnostic facilities finally pointed out these problems and were repaired early after start-up.

This problem did point out one oversight in the planning of the project. The operators were not required by the control panel design to make a large leap in technology. But, the maintenance personnel were not as fortunate. The people who have to maintain the system need to be thoroughly trained and familiar with the equipment so they can start maintenance as soon as possible. This can be alleviated by having initial training and refresher courses on-site every year.

The plant shut-down commenced on September 9th. The control room was powered down and the old control system removed in five days. About one half of the components and cabinets, and all the wiring was to be removed and replaced. At the same time, the mechanical components on all the turbines and generators were checked. The gas turbines had complete hot-path and compressor section maintenance performed while the steam turbine major overhaul was taking place.

The first control software load was accomplished on October 24th. The cable terminations were completed November 16th. The longest time spent and not adequately scheduled was the point-to-point wiring checks. After the wiring checks were complete the first unit started for test on the 9th of December, 1988.

On February 1, 1989, all three units were released to the dispatcher. In the end, the installation required 20 weeks because of mechanical hold-ups and point-to-point wiring checks.

OPERATING PHILOSOPHY

The plant originally was run on base load since the old control system lacked automatic dispatch capability and the plant ran very efficiently. This meant the power output was fairly constant and little load following could be accomplished. The plant could only be called on as a normal power producer. The output factor typically was in excess of 80%. To manage the maximum output of the plant, the afterburners had to operate more than 60% of the time.

Now the plant operates on automatic dispatch greater than 95% of the time. The output factor is reduced to 75% and the afterburners are scarcely on-line. This allows close load following for the system dispatcher to allow him to optimize overall system fuel costs.

The operators required very little retraining. The majority of operating personnel find the panel very logical. The CRTs give the operators more information, making their job easier and provide a better insight to what is happening in the process.

COST SAVINGS

The cost saving for the entire grid system and not just the station has yet to be fully realized. Because the plant now operates under the direction of the central dispatch, load is shifted on the plant quickly and reliably. Also, the potential plant power can be utilized in the systems spinning reserve, without wasting fuel and time requirement for coming on-line.

In the plant, most of the cost savings comes from the lower maintenance costs. Since the availability is higher, less hours are used trying to diagnose trips. The start-up cost will also be less since the restarting of the units is more reliable after the machines are down for routine maintenance. The afterburners will not have to be serviced as often since the hours utilized on the burners are fewer. The other cost savings in the plant comes from the heat rate which has been typically 1.5% below expected since the upgrade.

Savings to the system grid are realized mainly from the current use and reliable operation of load following under the central dispatch. The efficiency of the plant is higher than most coal-fired units, so lower production costs have been realized by designating Comanche Station as a base load station with load following and allowing the coal units with lower fuel costs to maintain a constant load output.

The spinning reserves can be maintained by the extra power available from the heat recovery units by increasing gas turbine firing or lighting the afterburners. This also allows for quicker turn around to the grid and lessens the need for replacement power costs and start-up of other reserves.

FUTURE DEVELOPMENTS

The station has been running well. Some problems, however, still exist and require attention. The display system used for operations was customized to meet the original requirements. This led to problems with system revisions. Another problem that did occur was an inconsistency with the printing of the automatic shift reports. To solve both problems, the display system will be updated to a standard version which incorporates the basic functionality of the customized system. This solution will also take advantage of the normal product evolution and enhancements.

For the system, lower minimum loads are being studied to allow the dispatcher better cost control during low load periods. Also, methods are being investigated to decrease automatic dispatch turn-around time for better response.

An enhancement to the present system is black-start capability. Since the gas turbines do not have black-start capability currently, two diesel-powered generator sets will provide power for the electric motor starters. This will allow quick restart of the gas turbines for rapid re-establishment of a black-out system grid. This project should be completed by June of 1991. Also, to utilize the extra capability, the selling of waste heat or steam is being investigated.

CONCLUSION

Since the installation, station output capability has been stable. In 1989, the Comanche Power Station was named the most efficient fossil-fired generating station in the United States by the Utility Data Institute. This was accomplished in the first year of operation with the new control system. The station on-line trips are currently 50% lower in 1990 than the previous four year average. Availability was 94% for 1990, the highest achieved for Comanche Station.

The decision to retrofit the Comanche Power Station is paying off. The control system enhancement allows for proper operational requirements of the station and the system grid. The availability has increased with the use of TMR technology and implementing a fault-tolerant control philosophy.

Figure 1 - Equivalent Availability Factor 1976 Through 1990

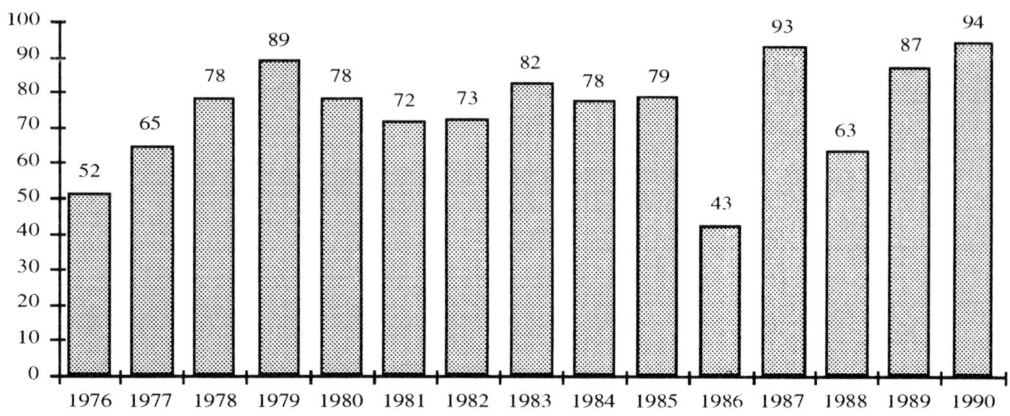

Figure 2 - Output, Availability and Heat Rate

Output Factor indicates how highly loaded the plant was when on-line. Declining figures indicate increased load following in automatic dispatch and is calculated:

Output Factor = (Actual Net Generation / Net Rated Output Capability) * 100

Equivalent Availability (EA) Factor indicates the amount of time the plant is available to run combined with load actually generated and is calculated:

EA Factor = ((Period Hours * Rated Plant Net) - Actual Net Generation) / (Period Hours * Rated Plant Net)) * 100

Heat Rate is the comparison to a known test standard for a given load range.

Year	Output Factor	EA Factor	Deviation from Expected Heat Rate	
1986	91.64	42.51	+ 145	BTU/KWH
1987	92.77	92.77	+ 191	BTU/KWH
1988	87.75	63.25	+ 37	BTU/KWH
1989	78.12	87.33	- 160	BTU/KWH
1990	75.56	93.74	- 117	BTU/KWH

©1991 - Paper #91-727
0074-056X/91/349-371/$0 + .50pp

UNIT STARTUP AND SHUTDOWN AUTOMATION DESIGN CONCERNS

N. Y. Michael
Sargent & Lundy

K. E. Hill
Bailey Canada, Inc.

ABSTRACT

The computational capability of microprocessor-based control systems can be extended into development of an automation program for unit startup and shutdown. This paper describes the design process of such a program for a 600-MW pulverized coal, once-through supercritical unit. It also describes the development of the design documents (flowcharts, time-based charts, control hierarchy logic diagrams, and graphic display); program structure; program language; and program mechanics/features.

INTRODUCTION

In October 1987, Sargent & Lundy (S&L)--as the leader of a consortium composed of Asea Brown Boveri (ABB), Combustion Engineering (CE), and Sulzer Brothers (Sulzer)--signed an agreement with Huaneng International Power Development Corporation (HIPDC) to design two 600-MW units.

Asea Brown Boveri's scope included the design, manufacturing, and delivery of two tandem-compound steam turbine generator sets and their auxiliary equipment, including a low-pressure (LP) bypass system for each turbine and the associated control and instrumentation systems. The design, manufacturing, and delivery of the two supercritical, coal- and oil-fired, spiral tube boilers, and their auxiliary equipment was divided between CE and Sulzer: CE's scope included the fuel, air, and flue gas parts; and Sulzer's scope included the pressure (steam and water) parts and a high-pressure (HP) bypass system.

Sargent & Lundy's scope encompassed the design and procurement of the balance-of-plant equipment for the two units and station-common systems. Bailey Canada, Inc., as one of the balance-of-plant contractors, was responsible for the control and instrumentation systems, including the implementation of the unit startup and shutdown automation program (UAM).

The commercial operation dates are October 1991 for Unit 1 and March 1992 for Unit 2.

DESIGN BASIS

The control and instrumentation systems (see Figure 1-1) use distributed microprocessor-based redundant controllers. As configured, no single fault will cause the unit to trip or limit its capability. Each unit has six operator consoles (CRTs and keyboards) located in the central control room. The design of the consoles allows assigning the monitoring and/or controlling functions of any equipment in the plant to any of the six consoles. In addition, a hardwired control board contains the essential control switches, indicators, and recorders that are required for maintaining the unit on line or, in the event of losing all six consoles, shutting it down safely.

The control hierarchy consists of the following levels.

Drive Level--Drive level control includes the capability of controlling individual drive units such as pumps and fans. The interlocking and protection functions are implemented in the drive unit controller. Drive level control is available from the CRT-based consoles and from the hardwired control panel for the essential services.

Group Level--Group level control includes the capability of controlling a group of functionally associated equipment. For example, the induced draft (ID) fan motor, ID fan bearings lube oil pumps, and ID fan speed change are controlled as a group. The group level controller supervises the implementation of the correct startup or shutdown sequence when commanded by the operator. The group level control can also select the sequence of starting the A and B groups of the equipment for the same function, e.g., selecting A-B or B-A sequence for the ID fans. Furthermore, provision is made for the operator to select automatic changeover of sequence (i.e., A-B to B-A) after every start or stop operation of a group. The group level control and the sequence selection is only available from the CRT-based consoles.

Unit Level--Unit level control is the highest level in the control hierarchy. The equipment start and stop commands are generated by the UAM.

UAM STRUCTURE

The UAM is divided into six master sequences. The unit master sequence (UMS) is an overall master sequence that includes the highest level logic for the entire plant and activates the other sequences. The other five sequences that relate to the five major equipment areas in the power plant are the following:

- BMS - boiler master sequence,
- TMS - turbine master sequence,
- BOPMS - balance-of-plant master sequence,
- BFPTAMS - boiler feed pump turbine A master sequence, and
- BFPTBMS - boiler feed pump turbine B master sequence.

Each of the five master sequences coordinates the startup and shutdown commands to and from the related equipment. Each master sequence is subdivided into sections and each section is subdivided into functional groups. The commands from the unit master are channeled to each of the master sequences and to the associated sections and functional groups.

UAM DESIGN DOCUMENTS

The basis for the design of the UAM is a set of flowcharts that describe the unit startup sequence and the unit shutdown sequence. The flowcharts present a complete definition of the process logic by which the control system will perform the automatic startup or shutdown of the unit. Figure 1-2 provides a sample of the process flowchart. The development of flowcharts by S&L required several iterations. A number of meetings were held with HIPDC, ABB, CE, and Sulzer to ensure that the sequence of start and stop operations is coordinated in a logical sequence and the program mechanics were adequately designed to provide the operator with proper guidance when the UAM is used.

Each master sequence consists of hundreds of steps and each step includes the following functions:

- step status that must be verified to allow the program to proceed automatically;
- output (command) to another system to perform a function;
- identification of the graphic or operator guide that must be displayed to monitor the execution of the command to another system;
- time delay to allow for the execution of command;
- feedback to verify execution of command to another system; and
- program abort in the event of failure to execute the command to another system within a predetermined time.

The flowcharts define the sequence of startup and shutdown sequence of the equipment associated with each master sequence. However, due to the different durations involved in preparing and putting the boiler, turbine, and balance-of-plant equipment in or out of service, the startup or shutdown operation is not an entirely linear process and consists of several parallel paths. To coordinate between these parallel paths, a set of time-based charts were prepared. The time-based charts (see Figure 1-3) define the time of startup or shutdown sequence initiation for each piece of equipment associated with any master sequence in relation to the equipment associated with the other master sequences.

Several factors impacted the development of the flowcharts and time-based charts:

- The duration of turbine shutdown determined the number of hours that the turbine has to be kept on turning gear before initiating the speed ramp.
- The boiler startup mode (cold, warm, or hot) is determined by the duration of boiler shutdown and water separator pressure and temperature. The selection of each of these modes determines the extent of boiler cleaning prior to light-off, the steam pressure and temperature ramp rates, and fuel firing rate and sequence.
- Pulverizer availability determines the light-off and shutdown sequence. During the engineering phase it was determined that the UAM will be permitted only when at least four pulverizers are available. This decision was necessary to minimize the number of probabilities; otherwise, there would have been 720 (i.e., 6!) possible combinations for each boiler startup mode.
- The duration of a turbine-driven boiler feedwater pump shutdown determined when it had to be put on turning gear before operating the turbine-driven boiler feed pump.

UAM CONFIGURATION

The UAM design criteria was based on the following:

- the ability to transfer from manual to automatic and automatic to manual operation;
- the use of operator guides and automatic displays; and
- the flexibility of the configuration.

The UAM configuration is a hybrid between graphical logic element BLOCK logic and a program written in a language similar to BASIC.

The BLOCK logic is used for communications and making step feedback available to the BASIC program. TRUE or satisfied condition for a particular step can be a simple boolean input from the data highway, a composite of several boolean inputs, or a comparison of an analog value. Certain steps of the program are required only during a certain phase of the startup or shutdown sequence. For example, the turning gear should be on until the initiation of turbine speed ramp. At that point, an appropriate logic derived signal or "LOOPBACK" is paralleled with the original feedback signal.

If the feedback to a step is lost before its "LOOPBACK" becomes active, a "STEPBACK" occurs. The basic program then identifies both the step which caused the "STEPBACK" and the last step that was active before "STEPBACK."

The BASIC language programming is used to augment BLOCK logic for the following three reasons:

- It allows the creation of a generic step-state processing function which operates essentially as a special purpose BLOCK (enabling both a larger area of the plant to be controlled by a single processor and a more sophisticated monitoring program).
- It simplifies configuration of branching and parallel segments of the sequences.
- It deals effectively with the complexity of driving the operator guide windows.

The program is layed out in two parts: the first part is the core logic module with several generic step processing functions and windowing functions; and the second part is the user implementation.

OPERATOR GUIDE

The operator guide is a single display with dynamic text elements. It is subdivided into six windows, one for each master sequence. As the startup or shutdown sequence progresses, the text in each window scrolls independently to display the states of currently active steps (see Figure 1-4).

For each dynamic text element there is a status indicating symbol and a remaining time indicator. The remaining time indicator is used only for configuration of HOLD steps, such as turbine heat soak.

Operation of the text elements requires the display itself, a list of text strings associated with index numbers in the operator console memory, and text index exception reporting blocks configured in the UAM processors.

The backbone of the operator guide is two arrays which are maintained by the step-state processing program. When the sequence is starting up, it first runs through the list of steps and stores the operator guide information for each step in the first array. On each subsequent cycle the list of active steps is stored in the second array. At the end of the cycle, another section of the program converts the information in the two arrays into a visible window on the operator console. This window, which consists of eight lines of text with step-state and remaining time indicators, provides the operator at a glance all the activity in the sequence. For a detailed description of the processing of operator guide window, refer to Appendix A (Visible Window Processor).

AUTOMATIC DISPLAYS

The automatic display feature consists of three parts: a mode of operation of the operator console; a list of display names and associated addresses of boolean exception reporting blocks; and a configuration in the UAM processors to drive the exception reporting blocks.

There are three lists of automatic displays in the operator console memory corresponding to the BMS, TMS, and UMS/BOP/BFPT. Each list may be assigned to any one of the six operator consoles by the operator. Once a list has been assigned, the automatic display of that list is made active on that operator console. The UAM will call displays to the console at the appropriate times by forcing the exception reporting block associated with the automatic display to the "ON" state and all the other blocks on the list to the "OFF" state. Therefore, at any given time only a maximum of three automatic displays in the UAM system can be displayed.

USER SEQUENCE IMPLEMENTATION

The order in which steps become active is determined by the CASCADE-IN parameter. "CASCADE-IN" is "CASCADE-OUT" from the previous step. Therefore, multiple parallel sections of a sequence can be made active simultaneously. See Figures 1-5 and 1-6 for an example of a section of the BASIC configuration and an explanation of the parameters in the function calls.

Step-state processing logic operates on the principle that, for any given state in which a step finds itself, it may make an immediate transition to only certain other states and that the criteria for those transitions can be explicitly defined. In this case, the full set of states in which a step can find itself is dependent upon the UAM AUTOMATIC, UAM MANUAL, and UAM OFF modes. It is also dependent upon the principle of STEPBACK. See Figures 1-7 and 1-8.

For a detailed description of the UAM sequence implementation, refer to Appendix B.

CONCLUSION

In the 1960s, when mainframe computers were introduced to the power industry, a number of utilities attempted to automate the unit startup and shutdown operations. These attempts were somewhat less than successful because the utility industry was limited in their capabilities to implement, use, and maintain the computer technology.

With the proliferation of digital technology has come the widespread use of microprocessor-based equipment, personal computers, and in general, computer literacy. Programming languages and application software for these computers has become more user friendly, making implementation much easier. Use of computers and application of computer technology to utility plant operation is widely accepted and operating personnel are knowledgeable and experienced in its use. Maintenance personnel are trained and experienced in minor programming and maintenance work required for reliable operation of microprocessor-based systems. The unit startup and shutdown automation program for utility plants is an excellent application for industry-proven digital technology.

During the last 10 to 15 years, the power industry has applied state-of-the-art control concepts in new plants and in controls retrofit projects. For example, the use of CRT-based control strategies has gained considerable acceptance and is replacing the conventional hardwired technology. Therefore, applications such as the UAM are the obvious next steps in the current trend of advanced control strategies application.

The use of a hybrid configuration method between BLOCK and BASIC programming languages is effective for the requirements of the UAM. BLOCK logic is employed for communications, AND-OR logic, and comparison logic. BASIC language is used for sequence branching, step-state processing, and driving a highly flexible operator guide and automatic display system. The method has produced a clear, maintainable configuration which makes an effective use of the processor time and memory.

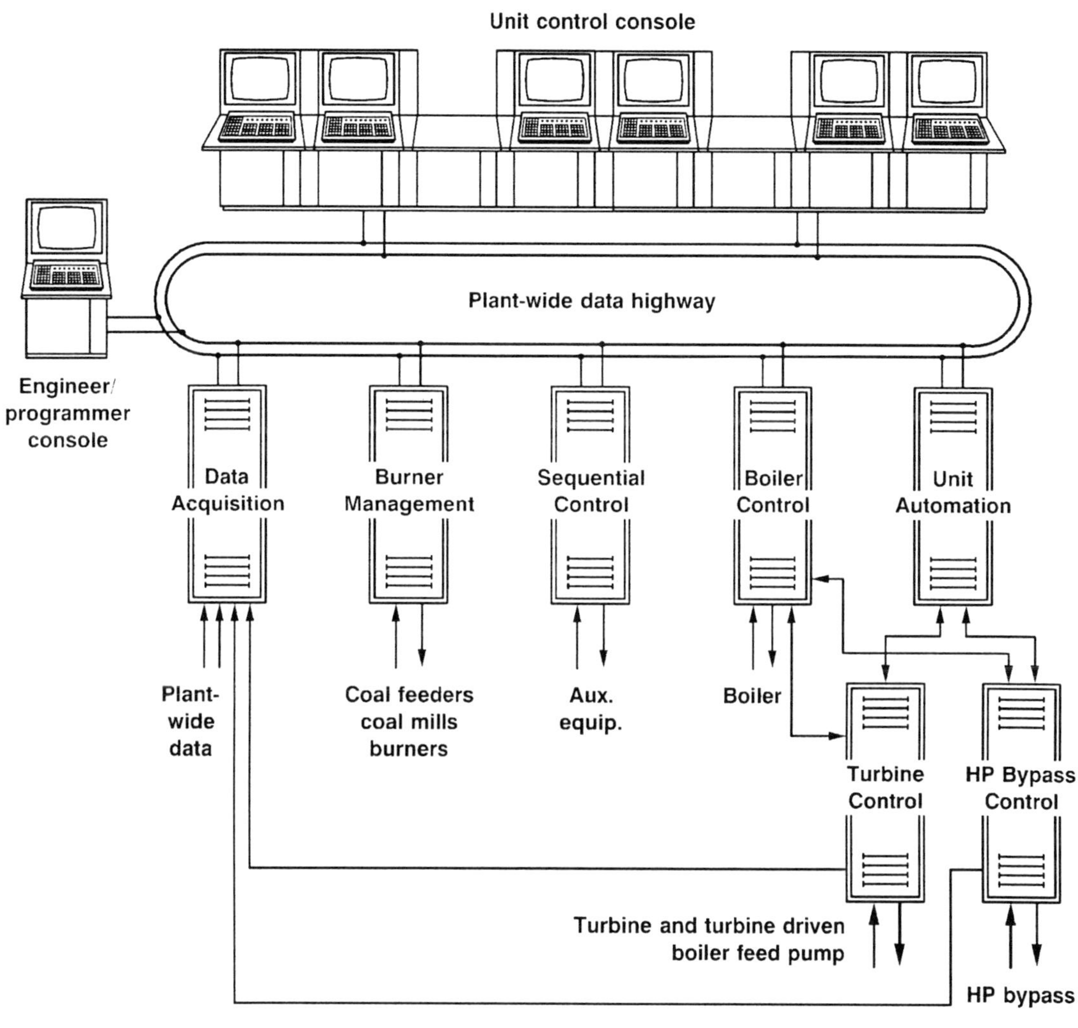

Figure 1-1—Control System Configuration

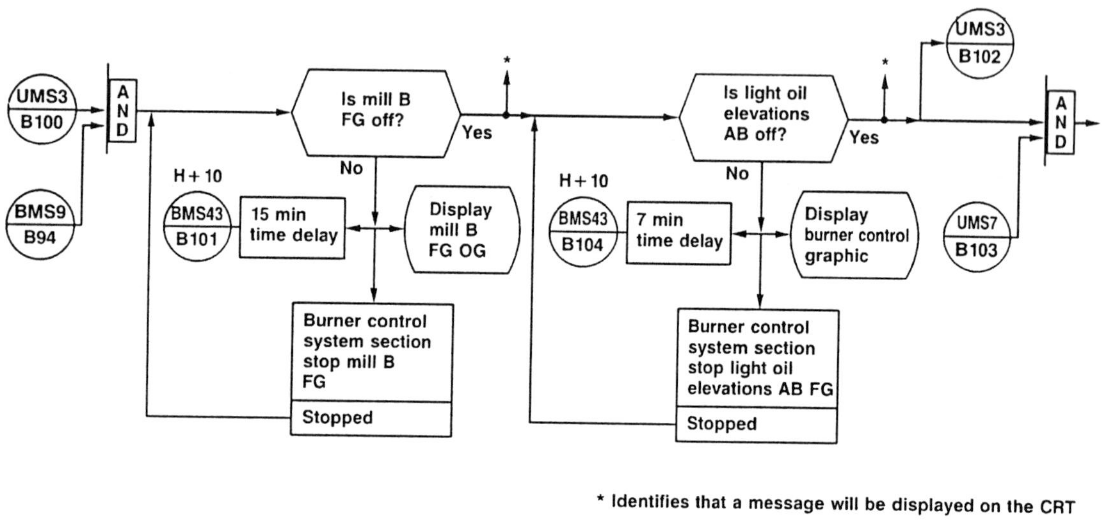

Figure 1-2—Sample Flowchart

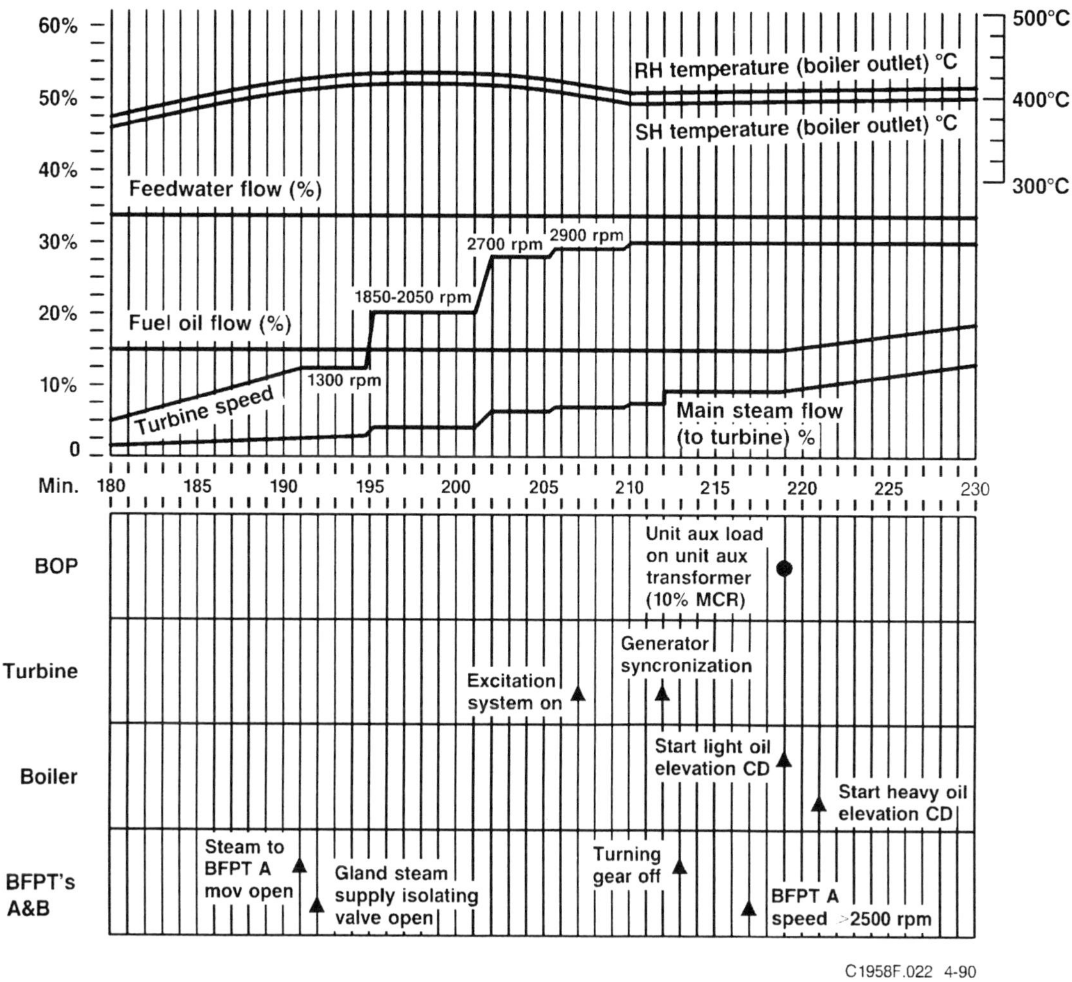

Figure 1-3—Sample Time-Based Chart

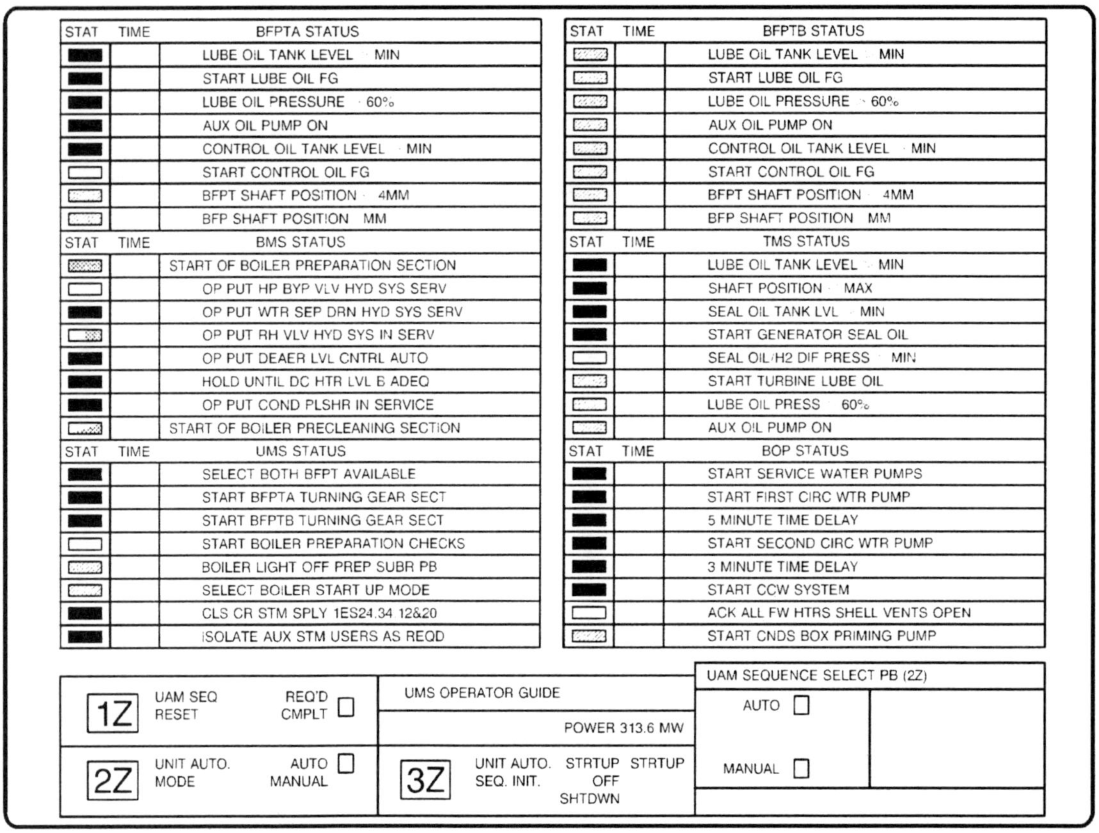

C1958F.040 4-90

Figure 1-4—UAM Operator Guide CRT Display

```
{START # PAGE 64}
{BFPTA TURNING GEAR SECTION}
     STEP_IDX = STEP IDX + 1
     SEQAR(CURS, STEP_IDX, 3) = 0
     SEQAR(CURS, STEP_IDX, 1) = 792
(IS LUBE OIL TANK LEVEL > MIN}
     CALL SEQ_START( P640R0159, P64S01, .5, 10, P64DI0161,\
             P64R2360, P64B2380, PA, 801)
{IS LUBE OIL SYSTEM ON}
     CALL QUICK RETURN (P64S01, P64S02, .5, 10, P640R0410,\
             P64R2361, P64B2381, PAG1, 802)
{LUBE OIL PRESSURE > 60%}
     CALL QUICK RETURN (P64S02, P64S04, 1, 30, P64DI0164,\
             P64R2363, P64B2383, PA, 806)
{CONTROL OIL TANK LEVEL > MIN}
     CALL SEQ_START( P640R0159, P64S05,.5, 10, P64DI0165,\
             P64R2365, P64B2385, PA, 804
{CONTROL OIL SYSTEM ON}
     CALL QUICK RETURN(P64S05, P64S06, .2, 10, P64DI0166,\
             P64R2366, P64B2386, PAG1, 805)
P64S07 = OFF
IF (P64S03 AND P64S04 AND P64S06)\
P64S07 = ON
{END PAGE 64}
```

NOTES:

```
"P64S01"       - S  =  CASCADE, THIS VARIABLE BEING ON INDICATES
                       STARTUP  PROGRAM  HAS  PROGRESSED  AT  LEAST
THIS FAR
"P64DI0161"    - DI =  DIGITAL INPUT
"P64R2360"     - R  =  REAL OUTPUT
"P64B2380"     - B  =  BOOLEAN OUTPUT
  {  }         - COMMENT LINE TO IDENTIFY STEP
```

Figure 1-5—Batch Language Program Sequencing Logic

```
                                ---- CASCADE_IN:  Enabling step cascade state.  Must be ON to
                                |      allow this step to proceed automatically.
                                |
                                |          -- CASCADE_OUT:  This step cascade state.  Is set to ON
                                |          |   when this step is complete and P64S01 is ON.
                                |          |
                                |          |          | ---Alarm timer value.
                                |          |          |
                                |          |          |    ---Abort timer value.
                                |          |          |    |
                                |          |          |    |        ----Step Feedback signal, may
                                |          |          |    |        |      be direct from Superloop
                                |          |          |    |        |      or may include a Loop-back
                                |          |          |    |        |
                                |          |          |    |        |          ----REAL output used
                                |          |          |    |        |          |      to report step
                                |          |          |    |        |          |      status to Event
                                |          |          |    |        |          |      Log.
                                |          |          |    |        |          |
CALL QUICK_RETURN  (P64S01,P64S02,.5,10,P640R0410,P64R2361,P64B2381,PAG1,802)
                                                                  |      |    |
                                                                  |      |    |
    BOOLEAN Output to drive---------------------------------------       |    |
    automatic command to                                                 |    |
    another system.                                                      |    |
                                                                         |    |
                         Auto Display identifier,------------------------     |
                         used to call the appropriate                         |
                         Auto Display (Control Graphic)                       |
                                                                              |
                         Step Status Message identifier,----------------------
                         used to drive the UAM Operator
                         Guide window on the UMS CRT.
```

Figure 1-6—Step Function Call Parameter Definition

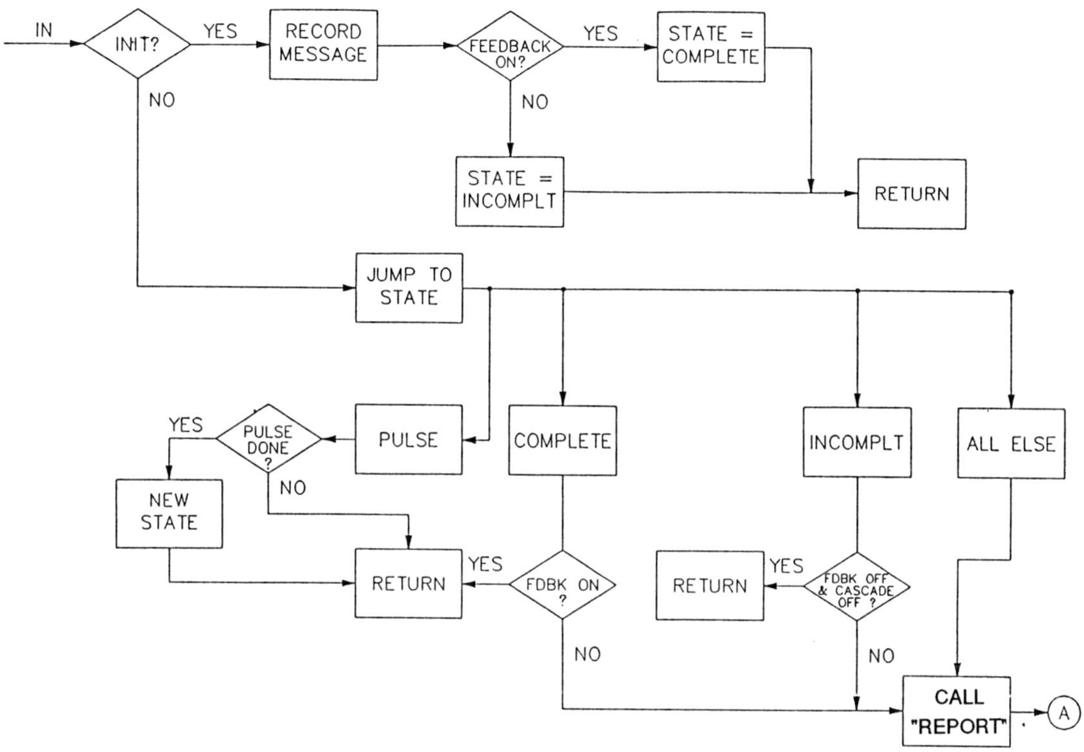

Figure 1-7—Core Logic Step-State Transition

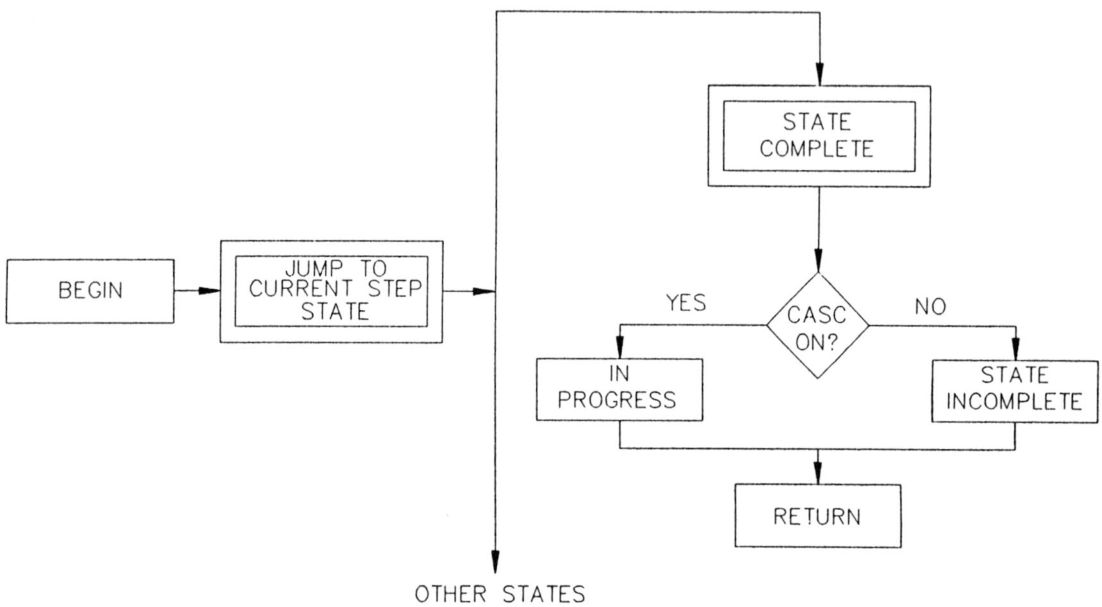

Figure 1-8—Core Logic Step-State Transition "Reports"

APPENDIX A
VISIBLE WINDOW PROCESSOR

The text index, step status indicator, and remaining time exception reporting blocks which make up the operator guide visible window for a sequence are updated every cycle. The sequence of accomplishing this function is as follow:

1. The number of active and inactive steps is found.

2. The number of regions and their starting indices and widths is found. A region is a section of a sequence in which one or more step is active. A second region is defined as an area with active steps which is too far removed from the first active region to display both sets of active steps and all intervening inactive steps within the eight lines of the operator guide window.

3. If there is only one region and the width of the active region is less than four, the bottom of the active region is placed on visible line four of the operator guide window.

4. If there is only one region and the width of the active region is greater than three, the top of the active region is placed on visible line one of the operator guide window.

5. If more than one region is active and the total width is less than nine, equal numbers of inactive steps are added to the bottom of all active regions, with the remainder added to the uppermost regions. For example, if there are two active regions with one encompassing two steps width and the other encompassing one step width, then, there are five inactive steps (five divided by two active regions is two with a remainder of one). The result is that the visible window will show the first two active steps of the first region, followed by the next three "future" inactive steps. Below that will be the one active step of the second window, followed by two "future" inactive steps. A region or subwindow with a width of more than one is where a number of actions which are required in parallel are in progress simultaneously.

6. If there is more than one region and the total width is greater than eight, a large composite window is built in column one of "TEMP-ARRAY." It is simply the result of concatenating all the active windows together. It is reduced to a width of eight by starting at the bottom and removing inactive steps as required while transferring the visible steps to column two.

   In the following example where there are three active regions and a total width of 14, subwindowing occurs when two or more sections of the sequence which are widely separated within the program are active simultaneously:

Line	Region	Column 1	Column 2 (Visible)
1	1	STEP INDEX 5 X	STEP INDEX 5 X
2	1	STEP INDEX 6 O	STEP INDEX 6 O
3	1	STEP INDEX 7 X	STEP INDEX 7 X
4	1	STEP INDEX 8 O	STEP INDEX 9 X
5	1	STEP INDEX 9 X	STEP INDEX 20 X
6	2	STEP INDEX 20 X	STEP INDEX 23 X
7	2	STEP INDEX 21 O	STEP INDEX 37 X
8	2	STEP INDEX 22 O	STEP INDEX 41 X
9	2	STEP INDEX 23 X	
10	3	STEP INDEX 37 X	
11	3	STEP INDEX 38 O	
12	3	STEP INDEX 39 O	
13	3	STEP INDEX 40 O	
14	3	STEP INDEX 41 X	

Note: X = Active
O = Inactive

Subwindowing mostly occurs in the unit master sequence and in the balance-of-plant master sequence.

7. The index number for the top of the window of the last cycle is memorized so that if there are no active steps on the next cycle, the program can "freeze" the visible window where it was at the last active cycle.

APPENDIX B
UAM SEQUENCE IMPLEMENTATION

The following is a detailed description of the UAM sequence implementation.

STEP-STATE PROCESSING

Each step in the flowcharts is represented as a call to a function such as "QUICK-RETURN" or its variations: "SEQ-START" and "ET-TIMER." Most of the time it can be determined that this step requires no more processing and it is on to the next step. Only a handful of steps require more complex processing during any one cycle. This greatly speeds execution of the monitoring program.

The results of monitoring a step are influenced by two external signals: "FEEDBACK" and "CASCADE-IN." FEEDBACK is the boolean signals that indicates a step is satisfied or can be ignored. CASCADE-IN is the CASCADE-OUT for the step which logically just proceeds this one. It signifies that all the logically preceding steps are satisfied.

DEFINITIONS OF MAJOR STATES

State	Definition
COMPLETE	FEEDBACK is on.
INCOMPLETE	FEEDBACK and CASCADE are off.
IN PROGRESS	FEEDBACK is off and CASCADE is on.
ALARM	An in-progress step which has had its alarm timer timed out.
ABORT	UAM forced to manual by a stepback abort or timeout abort.

AUTOMATIC COMMAND AND STEP-STATE LOGGING BLOCKS

Each step which can issue an automatic command has a boolean exception reporting block on the right-hand side of the page. In the automatic mode these blocks issue a three-second pulse when each step becomes in progress (see Figure B-1).

Each step has an analog exception reporting block. On major state transitions these blocks will cause a "STEP-STATE-CHANGE" message to be printed on an event printer.

TRANSFER FROM AUTOMATIC TO MANUAL AND MANUAL TO AUTOMATIC OPERATION

The ability to transfer from manual to automatic mode implies the ability to detect the state of the process. This is done with the principles of "AND CASCADE" and "LOOPBACK."

For a given set of conditions there is a point (or a number of parallel points) in the progression of the flowcharts for which all previous feedbacks are satisfied or can be ignored. Each step logically "ANDed" the preceding CASCADE input with its own feedback to produce its "AND CASCADE" output. The field feedback for the step may be logically "ORed" with a set of conditions occurring later in time which allow the feedback to be ignored (LOOPBACK).

The step which is the first in the sequence not to be satisfied (or able to be ignored) defines the current state of the sequence. Therefore, if the UAM system is active it is providing a sharp definition of the current state of the startup or shutdown sequence. In manual mode this is apparent to the operator as the current steps being shown on the operator guide and the current automatic displays.

The transfer to automatic mode activates the automatic command functions and alarm and abort timers.

CONTINUOUS MONITORING OF PERMISSIVES AND DETECTION OF FAULTS

Once a second the feedbacks which contribute to the "AND CASCADE" are re-examined. If a required feedback from a step previous to the currently active one is lost, there is a "STEPBACK." The active step makes the transition to "LAST-STEP-BEFORE-STEPBACK" and the step for which the feedback was lost makes the transition to "STEPBACK-ABORT." Visually, the operator is shown both the cause of the automatic sequence abort and the last active steps in the sequence. This information is also logged on the event printer.

DOCUMENTATION METHOD

Each processor sequencing logic is documented as a program which includes the core logic program. Core logic itself is a fourth document. See Figures B-2 and B-3.

Blocks which are defined as the feedbacks for a step and automatic output for the step are located on the same block page. Feedbacks and associated loopback and AND-OR logic are on the left. The outputs of these blocks typically do not feed into any other block (the exception being where the signal is a loopback or some other special circumstance). They are logically tied to the BASIC program through the step function call. Typically the digital and analog output for the same step is located to the right at the same vertical level on the page. See Figure B-4 for an example of the BLOCK configuration.

In general, the order of pages and the I/O blocks on the pages follows the flow of the UAM flowcharts. Of eight available standard step locations on a page, two have been left spare. Each of these pages is followed by a blank page number and a spare set of block numbers. The spare locations and pages are left in order to facilitate additions or changes in the future.

The BASIC program documentation works considerably different. Unlike on the CAD drawings there is no fixed page number associated with a given function call. The nature of the programming environment is that upon the addition of steps or documentation, the word processor simply bumps the following text down to the next page.

In order to provide continuity between the two, the steps are called in the same order in which they appear on the CAD drawings. Text is then commented at the beginning of the group of steps associated with a CAD page and again at the end to identify it. A single line of comment text proceeds each function call to fully describe its function. See Figures 1-5 and 1-6.

FLEXIBILITY OF CONFIGURATION

Two basics principles have been incorporated into the configuration method to facilitate maintainability of the control system and its documentation. The "LOOPBACK" method of requiring a permissive to continue to be present until explicitly overridden is a simpler alternative to explicitly defining the states of all relevant feedbacks at each step. The BASIC program automatically reorders the operator guide text display and the order in which automatic displays are called when the sequence is changed. A change to the control logic configuration automatically updates the operator interface.

Text calls and automatic display calls are done with index numbers. In the case of automatic displays, the selected number is fed out to a set of demultiplexed blocks. Each of the outputs of the demultiplexed blocks is driving one of the automatic display boolean exception reports. The index numbers used are declared in the program as constants of the same name as the automatic display itself. This improves the readability of the documentation.

The index number is reported to the operator console by a specialized exception reporting block. This number has the effect of selecting one of a list of hundreds of text strings residing in the console memory.

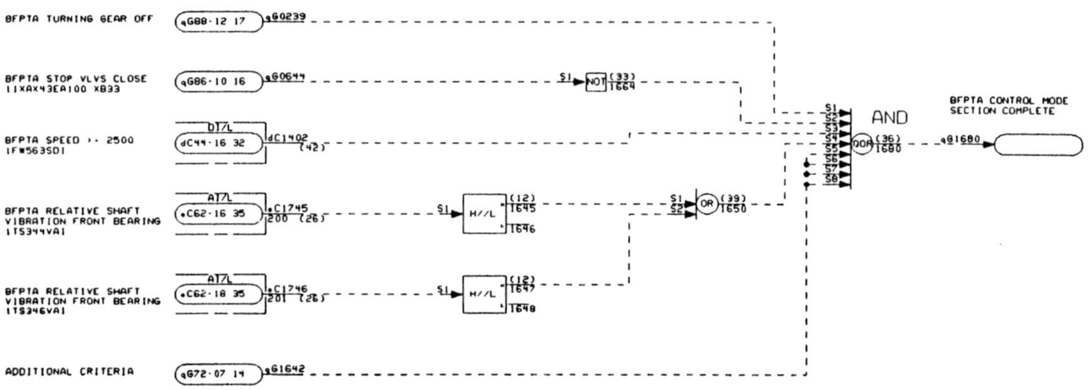

Figure B-1—Step Feedback and Loopback Logic

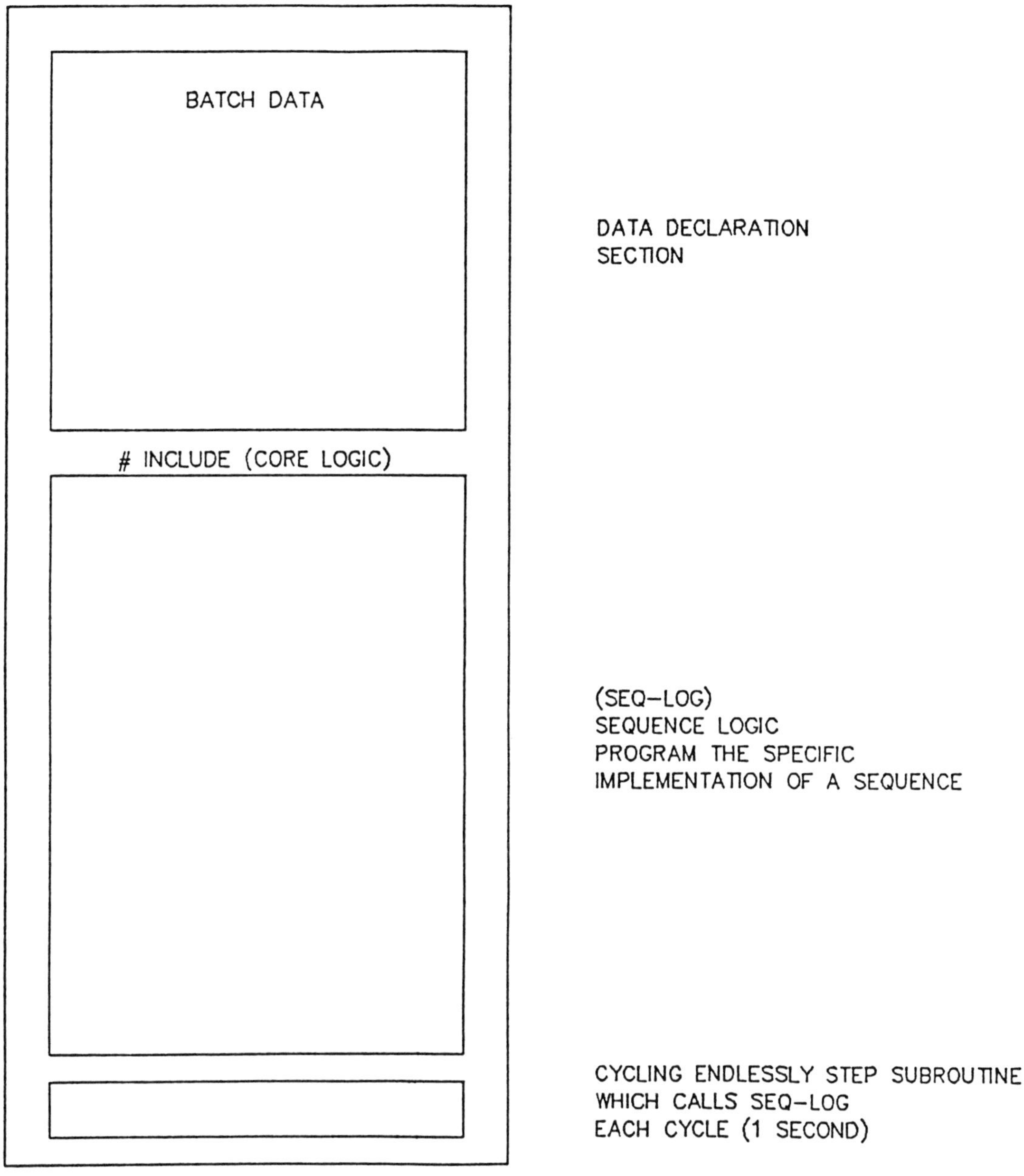

Figure B-2—Module Program Before "Include Is Activated"

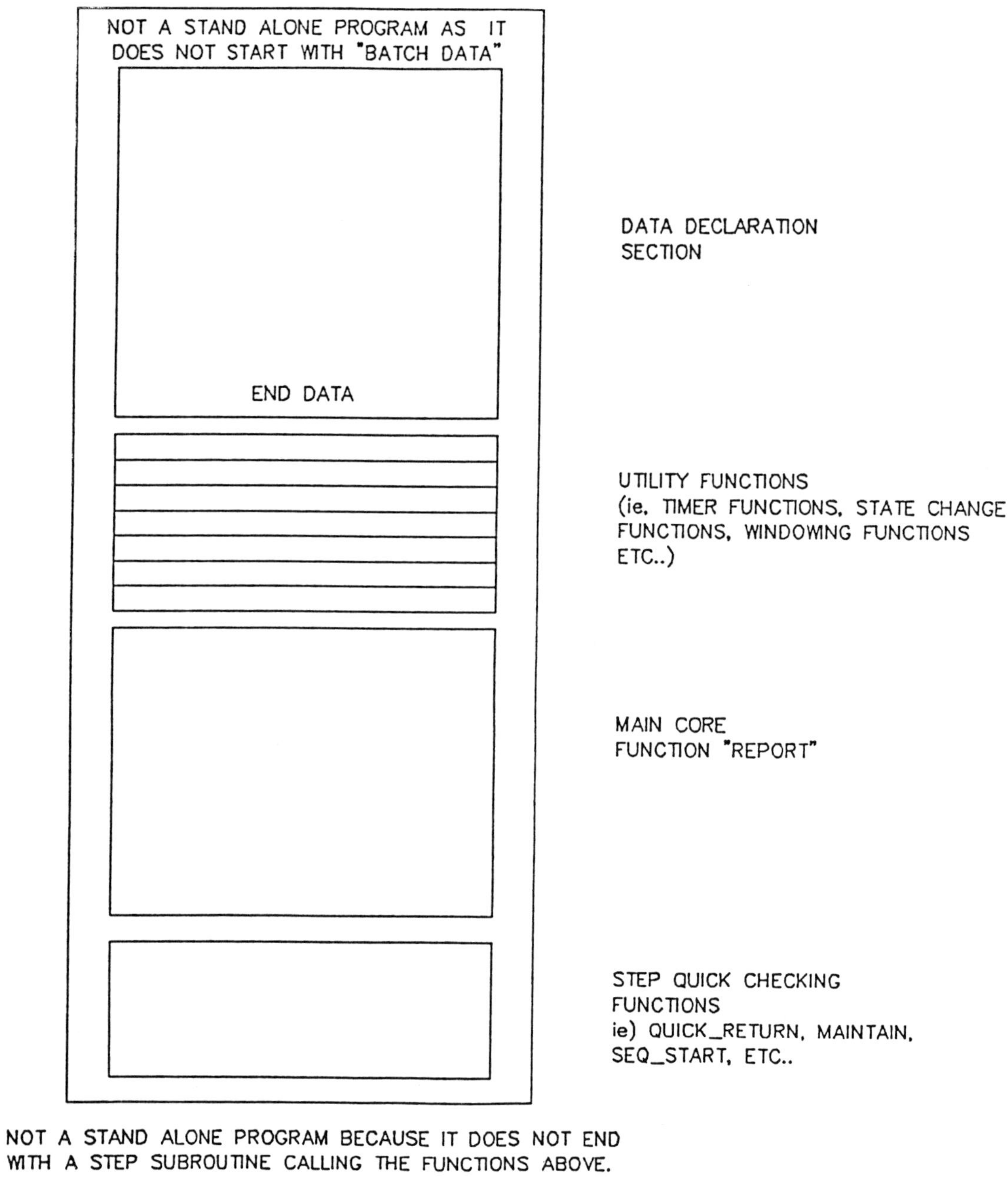

Figure B-3—Core Logic Sections

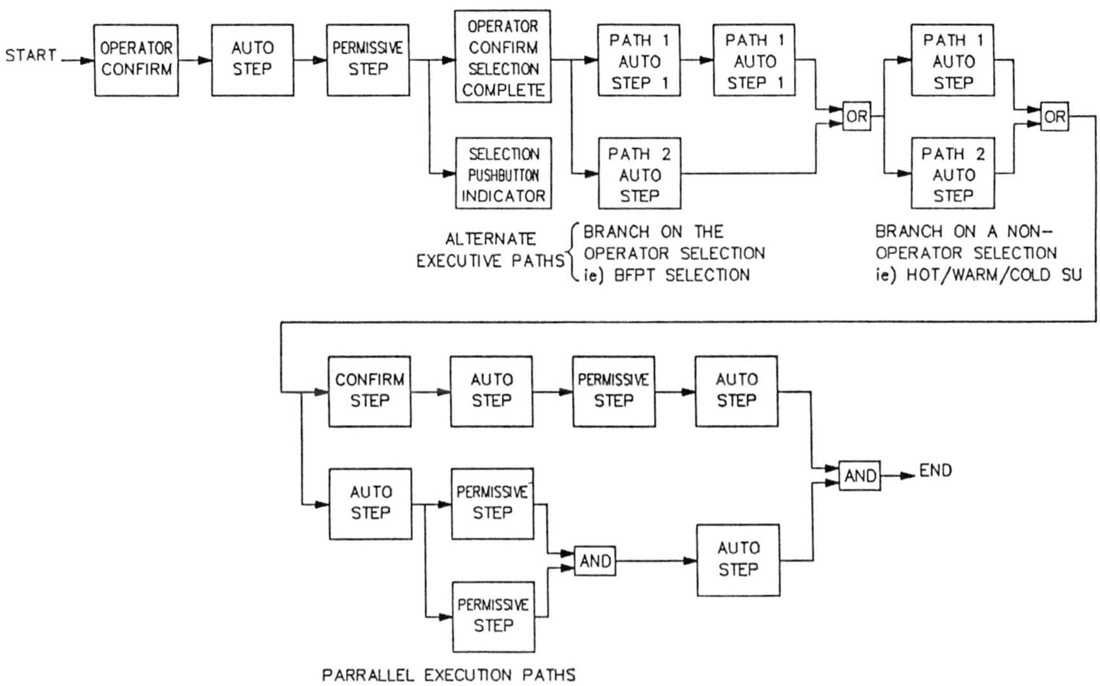

Figure B-4—Example Sequence Showing Use of Step Types and Branching

©1991 - Paper #91-728
0074-056X/91/373-385/$0 + .50pp

AUTOMATION SCOPE AT ROXBORO GENERATING STATION

Ali Ipakchi
Gary Chrisler
Scott Damesek
Asea Brown Boveri

Ervin W. Fields
Bernard C. Cook
Carolina Power and Light Co.

Murthy Divakaruni
Electric Power Reserach Institute

ABSTRACT

Many fossil fired power plant have replaced their aging unit control hardware with modern DCS technology. This paper describes a jointly sponsored project by EPRI, Carolina Power and Light Co. (CP&L) and ABB, to implement plant automation functions which would complement DCS technology, and result in improvements in plant operations in terms of heat rate, availability, and operations cost. These will be demonstrated at CP&L's Roxboro Station, a 4 unit coal fired plant with total capacity of nearly 2,477 MW. The scope of this four year project includes installation of a plant-wide information management system, automation of unit testing, implementation of advanced controls, intelligent alarms, operator advisory functions, and a plant-wide simulator, as well as support for plant unit economic dispatch.

1 INTRODUCTION

This paper presents the objectives of the EPRI-sponsored Roxboro Automation Project (EPRI RP2922-8). This project is intended to provide cost-effective and useful plant automation technologies for installation at CP&L's Roxboro plant. Successful implementation of these technologies will aid the plant in achieving their operational goals of reduced heat rate and improved availability.

1.1 Background

New competitive realities, economic pressures as well as changes in technology have caused many utilities to consider new methods for improving fossil power plant operations. One approach is the introduction of automation in an effort to increase efficiency and availability, and to improve operational flexibility. The rapid advancements of computer technology has opened new opportunities, in all areas of the plants to realize these improvements. Many utilities have initiated fossil plant modernization programs with a major emphasis on data acquisition and controls retrofit, and some are installing performance monitoring and diagnostic systems. However, much more needs to be done before utilities realize the full benefits of automation.

The application of automation in fossil plants, so far, has been through the use of a deluge of independent, specialized systems dedicated to specific needs. The utilities' quick adoption of these systems for solving specific problems has created "Islands of automation". These islands cannot easily pass information or communicate with one another, and require their own separate computer and user interface. The redundancy of interfaces and data for each independent system has added to the confusion in the plant, and has limited the potential benefits of automation. Very few power plants have addressed the integration issues.

To address these issues, and to demonstrate automation at a fossil-fired power plant, EPRI has initiated this Research Program. A project team composed of Carolina Power and Light, Co. (CP&L) and Asea Brown Boveri (ABB) has been assembled. CP&L's Roxboro Generating Station has been identified as the host site to this program.

1.2 CP&L Roxboro Electric Power Plant

The Roxboro Electric Power Plant is located on the Hyco Lake Reservoir, approximately nine miles northwest of Roxboro, N.C. It is the largest plant on the CP&L system and provides half of the Company's coal-fired generation. Roxboro is one of the largest power plants in the world, and has placed in the top ten in fossil production in the United States several times in the last 10 years.

Electricity was first produced commercially at Roxboro in 1966. By 1980, Roxboro had four generating units with a total capacity of nearly 2500 Megawatts. Unit 1 is a Riley Boiler serving one tandem-compound G.E. turbine Generator capable of generating 390 Megawatts. Unit 2 consist of a Combustion Engineering Steam Generator and G.E. tandem-compound 675 MW trubine generator. Units 3 and 4 have a similar design and incorporate Riley twin boilers and G.E. tandem-compound trubine generators. Unit 3 is rated at 707 MWs and unit 4 is rated at 700 MWs. Unit 4 also has one ICTG rated at 15 MW. A summary of plant unit characteristics is illustrated in Table 1.

The plant receives coal by "unit trains" or 100 car trains, at an average of two trains per day from West Virgina and Kentucky. Each coal car averages 100 tons of coal.

	#1	#2	#3	#4	I.C.	PLANT
GROSS GENERATION						
SUMMER (MW)	385	670	707	700	15	2477
WINTER (MW)	390	675	715	710	17	2507
COAL BURNED						
MAXIMUM (TONS/HR)	150	260	288	288		986
AVERAGE (TONS/HR)	128	208	208	208		749
COMMERCIAL OPERATION	5/66	5/68	7/73	9/80		
SMOKE STACKS						
HEIGHT (FT.)	400	400	800	800		
DIA. AT TOP-FT.	12.75	19.0	24.0	24.0		
UNIT (EQUIVALENT)						
AVAILABILITY)						
(5-YEAR AVERAGE)	87%	79%	89%	92%		

Table 1: A summary of Roxboro plant characteristics

The plant has made major investments on such items as electrostatic precipitators, dry flyash storage, and cooling towers to improve environmental conditions. A dry flayash storage system was completed in 1990.

All four Roxboro units are equipped with MOD 300 distributed data acquisition and control systems from ABB. The individual control systems are tied together via an Ethernet link. The controls modernization program at Roxboro started in 1986. A new data acquisition system was installed on units 3 and 4 in 1989, followed by similar installations on units 1 and 2 in 1990 and 1991, respectively. The changeover of unit 1 analog controls to MOD 300 was completed in 1990. All four units will be under digital control by 1993. The field instrumentation on all four units is being upgraded. The units are also being equipped with the OTIS performance monitoring system from ABB. Figure 1 is a the layout of Unit 3's DCN network. Other units are similarly equipped.

Electric generation at Roxboro is already very efficient, with the plant being ranked favorably among the leading national plants of similar design for many years. Despite this excellent record, CP&L has set aggressive performance improvement targets for Roxboro, which mandate significant cuts in unit heat rates, and gains in unit availability.

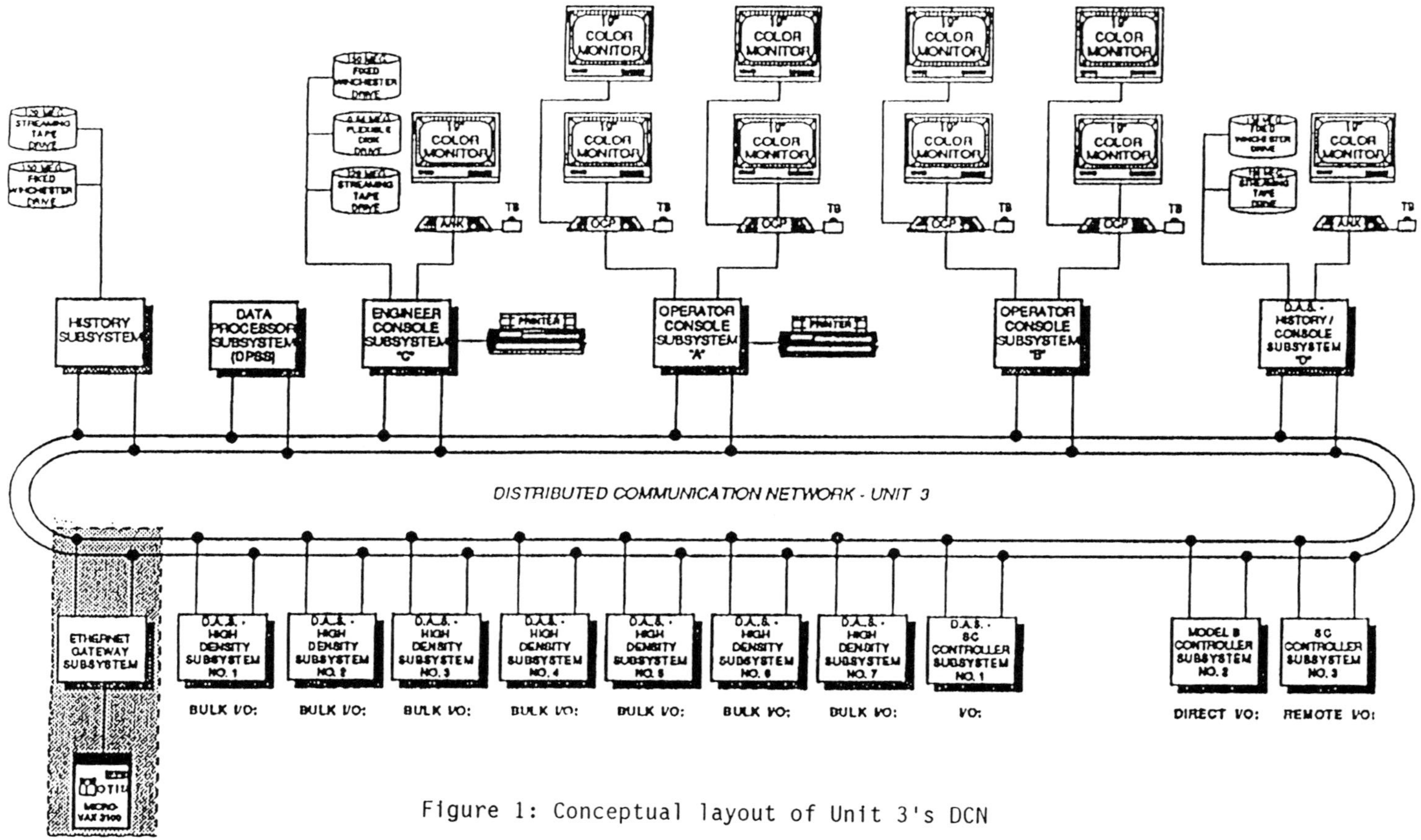

Figure 1: Conceptual layout of Unit 3's DCN

Power Automation Inc. A Division of ABB ASEA BROWN BOVERI	TITLE CAROLINA POWER & LIGHT ROXBORO UNIT 3 DCN NETWORK DIAGRAM	DRAWN M. KANE CHECKED D. McFARLAND APPROVED	DRAWING NO.: FIGURE 3 DATE: 1-9-91 REVISION: 0

2 PROJECT OBJECTIVES

The overall intent of this program is to demonstrate the application of state-of-the-art automation technologies at CP&L's Roxboro Station to achieve the following goals:

- Improved heat rate throughout load range (reduce current rate by 200 BTU/KwHr);
- Reduced forced outage rate;
- Improve availability percentage by 1.4 percent;
- Improve maneuverability throughout load range;
- Reduce O&M cost.

These goals will be achieved through implementation of specific projects which while fully utilizing the power of an installed distributed data acquisition and controls equipment, will integrate monitoring, diagnostics, on-line testing, and advisory and information management functions into a plant-wide automation system.

The scope of work will cover functions beyond those required for basic plant instrumentation and controls. The emphasis will be on vertical integration of plant-wide automation functions, as well as implementation of advanced control algorithms and specific diagnostics, operator advisory, process optimization, and information management systems to meet the stated goals.

Methods will be developed to transform the large quantities of data collected and generated by individual systems into useful information for access and effective use by plant personnel.

More specifically, the initial phases of this project will include the following activities:

- o Implementation of an advanced plant-wide information system.
- o Automation of plant equipment testing and data capture/analysis/reporting.
- o Implementation of advanced control strategies on selected loops.
- o Implementation of Operator Advisory systems, including on-line expert systems for alarm processing, optimum load-dependent setpoint selection, and diagnostic monitoring.
- o Implementation of improved techniques for plant unit dispatch.
- o Installation of an advanced simulator for support of plant operations, engineering, and control strategy development.
- o Development of Fossil Plant Automation Guidelines.

The tasks listed above are part of the work scope for the initial phase of the automation project. During the performance of the current Feasibility Study, which will define the details of implementing these automation technologies, additional automation goals may be formulated, which will be addressed during a follow-on phase of this project. This phased approach is being taken for several reasons, among which are the following:

a. New developments in automation technology can be utilized as they occur.

b. Budget expenditures can be moderated.

c. Adverse impacts upon an operating power plant can be minimized.

d. Implementation plans and goals can be adjusted, based upon the degree of success of the initial installations.

e. Unforeseen synergies between implemented technologies can be exploited.

The long-term goal of this project is to boost the efficiency of the Roxboro plant, by creating a state-of-the-art automation system. We intend to create a "Center of Excellence" of automation technologies at Roxboro, making Roxboro not only a more efficient unit, but also making Roxboro an automation showcase, demonstrating the benefits of advanced automation technology to other U. S. utilities.

3 PROJECT TASK STRUCTURE

This project is structured as eight tasks. They are defined as follows:

3.1 TASK 1: Feasibility Study

This task defines the work to be performed under the auspices of this automation project at Roxboro. The automation project entails automating operations and testing procedures which currently are time-consuming and expensive to perform, or are not performed at all. For example, equipment diagnostic testing is currently performed by spending up to 40 man-hours to construct special data acquisition and recording hardware, so that a few hours of data can be gathered. The data acquisition hardware is then disassembled. The laborious nature of this process severly limits the frequency of equipment diagnostic testing.

To automate this example process, required data for current and planned diagnostic tests must be examined, additional sensor hardware must be specified if required, data acquisition and storage software must be implemented, and generic software to perform configurable performance and diagnostic calculations must be developed and installed. The results of these calculations will be issued in operator-definable reports.

The Feasibility Study task will analyze and define the requirements of all the tasks currently defined in the project work-scope, including the example. During execution of this study, plant needs will be examined, available technologies will be evaluated, and execution schedules and resources will be allocated for accomplishing the goals of the other seven tasks. In addition, the study may recommend further tasks to be accomplished in subsequent projects, with the concurrence of the Roxboro members of the project team, which are beyond the scope and budget of the current effort. The findings and recommendations of the Feasibility Study will be reflected in this report.

3.2 TASK 2: Process Information Management System

The Process Information Management System to be designed and implemented during execution of Task 2 will be the backbone of the Roxboro Plant Automation effort. It will provide the platform upon which the other automation tasks will be implemented. The goal of Task 2 is to design, procur and implement a plant-wide data network and information processing capability, to support the needs of the Roxboro automation program. Where possible, the system will also demonstrate EPRI's Utility Communications Architechture (UCA), at a power plant level.

Each of the Roxboro units are equipped with ABB's MOD 300. Units are also equipped with other stand-alone computers for specific monitoring or data processing tasks, such as the PC-based OATIS emmissions opacity monitor, and the On-line Thermal Information System (OTIS) plant performance monitoring system, running on MicroVAXes. However, all these systems run on separate machines, require different commands to operate, and give access to their data only at consoles driven by the separate computers.

At this time a plant-wide information system to share resources, to access each unit's data bases, to generate plant reports, and to perform integrated analysis and system management is not available. Since all required data are available in digital form, however, such a system is feasible. The purpose of this task is to specify and develop such a system.

There are many advantages to such an integrated Information Management system. One such advantage is that all data, from all acquisition systems, application programs, or off-site processing centers connected by digital link, will be available from all terminals driven by the integrated system, regardless of physical location. No longer will operators have to walk from console to console, trying to locate the data they require. Another advantage is that such an integrated system will provide unified and consistent methods for accessing this diverse data, eliminating the specialized commands that were required by the individual stand-alone systems.

Another advantage inherent in integrating these heretofor separate applications lies in the area of efficiency and accuracy. Each application has access to current, up-to-date information as prepared by other applications, so the calculations prepared by each application are as accurate as possible, without any delay involved with transferring data between systems.

But the biggest advantage is in the entirely new applications that become possible through information integration, via the synergy of combining different capabilities. By integrating these diverse applications, we allow each application to access and use information generated by other applications, making possible new functions and services. For example, by combining the Dispatch Control Center near-future power demands with information about each unit's current equipment configuration and diagnostic results, an operator advisory system (task 5) can issue warnings to operators that equipment configuration changes will be required, and operators can decide whether these changes are feasible, or if Dispatch must be notified that requested generation changes cannot be met. The advisory system can also recommend that periodic maintenance services which lower unit efficiency while they are being performed, such as sootblowing, be initiated while central power demands for the unit are low, thus keeping unit efficiency high in periods of peak demand.

The approach to this systems integration task will be through applying recognized industry standards, such as OSI, SQL, and EPRI's UCA. The platform developed will be an "open system", fully expandable to meet Roxboro requirements for future growth. The initial hardware for this system will be manufactured by Digital Equipment Corporation, but by proper application of standards, this open system will accept connections to other hardware/software which utilizes these standards, regardless of vendor. This means that the development and implementation efforts for new applications will be minimized, as will specifying new applications. Furthermore, such a system provides the capability for a "phased implementation" approach, since it permits applications to be developed and installed one-by-one or all together, as budget and manpower constraints allow.

A conceptual diagram of the Process Information Management System at Roxboro is shown in Figure 2.

3.3 Task 3: On-line Unit Testing

The objective of Task 3 is to to provide high accuracy real time plant performance data through automated on-line equipment testing, analysis, and reporting. Current test and diagnostic methods at Roxboro, which are typical for the industry, are expensive, time-consuming, and labor-intensive. The difficulty of acquiring data for equipment diagnostics limits the frequency at which these diagnostics are performed, which could impact equipment health and unit down-times. By automating and simplifying this process, equipment diagnostics can be made easier, less expensive, and can be run more frequently, thus spotlighting degradations in equipment performance earlier, when more options for performing and

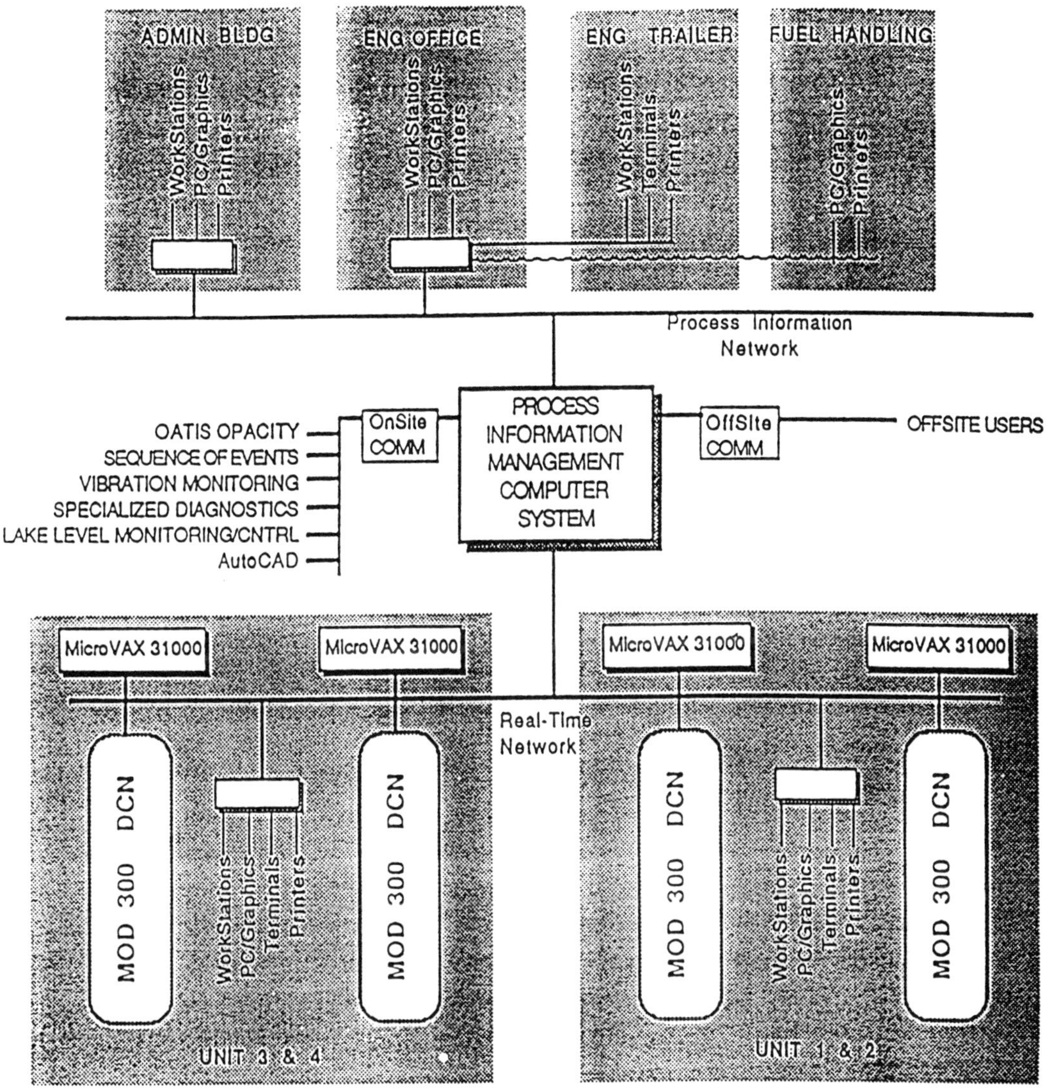

Figure 2: Conceptual diagram of the PIMS

scheduling repairs exist.

Roxboro units are equipped with a modern DCS capable of acquiring and processing all plant data required for these diagnostics. By taking advantage of the available digital processing hardware, this project will automate on-line real time testing of equipment through the use of high accuracy transmitters, and by providing data analysis functions, which can identify problem areas, and advise appropriate corrective action. This task includes both testing of systems, and individual equipment performance evaluations. The test calculation and reporting software will be configurable, and will allow easy implementation of both existing equipment tests, and new tests as they are defined. The Diagnostic Test function will permit scheduled, on-demand, and continuous monitoring of system and component health.

This method of testing will provide vital real time data to support future tasks such as strategic dispatching and performance optimization.

3.4 Task 4: Advanced Control Strategies

The objective of Task 4 is to evaluate the control loops of the Roxboro units, to recommend which loops can benefit from advanced control algorithms and strategies, and to design and implement the advanced control strategies on the selected loops.

Roxboro's distributed digital control systems, which have been installed as part of an ongoing controls upgrade program at the site, are capable of handling complex control algorithms. However, in most cases, the control strategies implemented parallel the analog control strategies they are replacing, and do not fully utilize the advanced capabilities available from their new DCSs.

Many vendors offer advanced control strategies which exploit the capabilities of DCSes. Adaptive, self-tuning and multi-variable controls, variable setpoint, and other algorithms have been used to improve the performance and responsiveness of the control systems under diverse operating conditions. EPRI has also initiated several projects in this area, including RP1403-26 to identify new control algorithms for fossil fired power plants. As part of the Feasibility Study for this task, existing research, performed by EPRI and others, and operational experiences with advanced control technologies at other domestic and international plants, will be examined, with an eye toward applicability to Roxboro controls.

Topics such as Sliding Pressure Setpoints versus Fixed Pressures during unit ramping, and Multi-variate Boiler Protection algortihms will be examined, and recommendations for modifications to Roxboro's control systems, both during the course of the current project and during possible subsequent porjects, will be made.

3.5 Task 5: Operator Advisory System

The objective of Task 5 is to develop and install computer-based operator assistance functions for:

a) Optimal Setpoint Adjustment

b) Operator Advisory Displays

c) Alarm Processing (Including Alarm Filtering and Alarm Diagnostics).

Under this task a set of specific operator assistance functions will be developed and integrated with the plant controls (MOD 300) and the process information management system installed as part of Task 2. This effort will be focused on:

- providing the operator with the information and assistance required to optimize the controllable parameters that are required to enhance the unit's performance,

- Computing and recommending optimum setpoint levels for controllable parameters, and

- implementation of alarm filtering and alarm diagnostics capabilities.

3.6 Task 6: Plant Unit Dispatching

Task 6 will develop and implement methods for optimal plant unit dispatching, and will research methods for improving plant performance under cyclic load control.

Advanced dispatching capability would allow each fossil generating unit to meet the load requirements of central dispatching while minimizing the effect on unit efficiency, reliability and unit availability. This will allow optimum unit performance while meeting system load demands. Utilizing the Process Information Management System, operators can also examine the effect that these load swings have on the unit, as compared to steady state operations.

Under the aegis of this project, no changes will be made to dispatching methods. Dispatching will be researched and examined, and results will be implemented, from a plant perspective. Recommendations will be made, however, for eventual improvements to dispatching methods, including direct data links between plant performance monitoring computers and dispatch centers.

3.7 Task 7: Simulator Based Training And Engineering Support

Task 7 will provide software and hardware facilities to:

a) Capture, archive, and replay plant data for operator and other staff training, and for engineering analysis.

b) Support a fossil plant simulator capable of handling multiple units, and support training and engineering tasks.

These simulator facilities will be in support of both training and engineering analysis applications, and will be used during the evaluation, installation, testing, and tuning portions of Advanced Control upgrades (Task 4).

During the performance of this portion of the Feasibility Study, Roxboro's training needs will be defined in detail, Engineering Analysis capabilities will be defined, and requirements for the simulator's support of ongoing controls upgrades will be spelled out.

These capabilities will include capturing a complete set of a unit's operating data, for replay of events and operator actions, evaluation of transients and other operating conditions of interest. This feature will utilize history archives to replay incidents for operation training and analysis of proper responses. This will also provide engineering staff with the ability to analyze disturbances to see if a modification to equipment, controls, or procedures can avoid future occurrences.

The simulator's control system models will be modifiable by Roxboro personnel, so that new control configurations can be tested and tuned in a simulated environment before they are installed in the actual plant.

The study will also evaluate possible benefits from combining a Dispatch Center simulation system with the plant simulator, to allow evaluation of plant dispatch responses vs. dispatch commands, and too quantify benefits of Strategic Dispatch enhancements.

4 PROJECT PLAN

Figure 3 presents the overall project schedule. The feasibility study activities were started in November of 1990, and are scheduled to be completed by June of 1991. The current planned phase of the project is scheduled to to complete by the end of 1994.

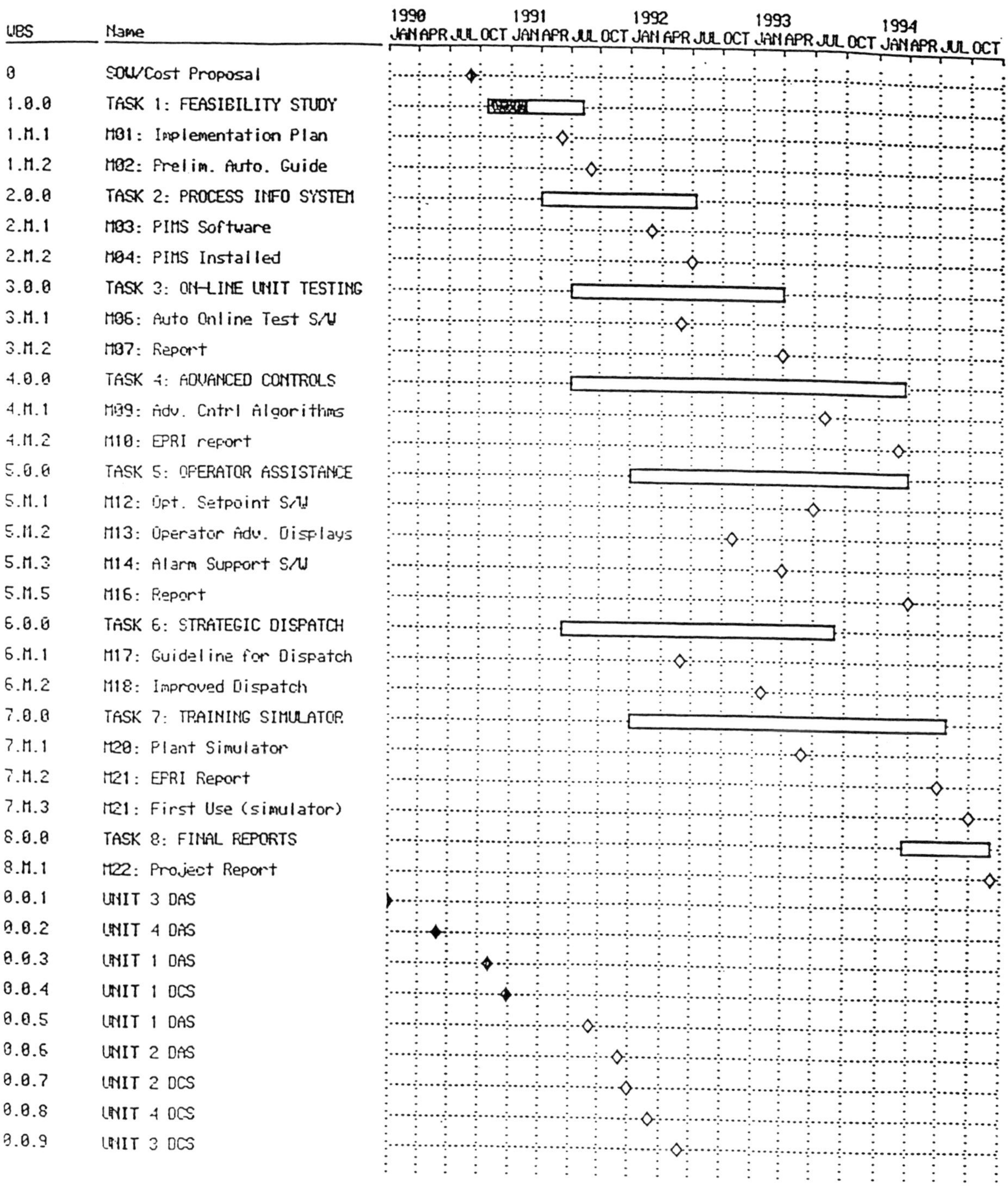

Figure 3: Roxboro Automation Project schedule

0074-056X/91/387-402/$0 + .50pp

FUTURE POWER PLANT NETWORKS

Michele A. Blanco, George L. Touchton,
S. Murthy Divakaruni, John R. Scheibel
Electric Power Research Institute

Presented at the ISA/EPRI Joint Controls and Automation Conference
St. Petersburg Beach, Florida
June 3-5, 1991

ABSTRACT

The advanced intelligence provided by power plant diagnostic and monitoring systems is key to maintenance strategies and improved operations aimed at heading off performance losses, availability losses, forced outages, and unnecessary capital investments in equipment replacements. The full potential of these systems, however, can only be realized by integrating this information into networks that support the complete plant operating environment and insure a "transparent" operator interface. Integration of monitoring and control into a comprehensive information network offers dramatic potential for improved operations and productivity. The power plant network is only one element of the overall integrated utility network that would serve an entire utility. This paper describes the essential, transforming, role information networks will play in future utility strategies, and gives an overview of EPRI-sponsored work to accelerate utility use of power plant and system level network technology.

Key words: Power plant information networks, monitoring and diagnostics, power plant control systems.

INTRODUCTION

Managing assets will be an overarching concern of the electric utility industry as we move into the 90's (1). Optimizing future operation including accurately assessing the value of existing resources and planning for future operation scenarios will be necessary in an environment of competition and increasing environmental pressure. This task will be complicated because the forces which shaped the acquisition and design of existing assets are giving way to a new set of imperatives. This means that these assets will be operated in a foreign environment and may be called upon to support operations for which they are ill- suited.

Information and intelligence are vital, but intangible, assets of all businesses. They are assets which are essential to managing plants, fuels, people, and other tangible assets, and more subtly, they are assets which require management. Surely, this concept has been readily accepted by the electric utility industry. Many utilities have installed sophisticated computer systems at the plant and office level. However, if not done carefully, these systems can simply replicate in silicon and software pre-existing solutions — including pre-existing constraints. For example, word and information processors rapidly generate memoranda and reports which are then put into the office mail system for distribution by pushcart. At the plant level, digital control systems mimic algorithms based on analog electric or eletro-hydraulic technology. As time passes, however, computer based technology is breaking down old barriers and fundamentally new processes are emerging, as illustrated by recent quantum advances in micro-chips, lasers, and fiber optics, which are the enabling technologies for fully distributed computing.

Fully distributed computing, where multiple computers are connected by a network and specialized "server" computers perform particular jobs such as managing data bases, printers or electronic mail systems, will force a complete rethinking of the mainframe computer for all businesses (2). These computing networks, coupled with sophisticated software, have the potential to make data, information, and knowledge available at any level of the organization — and in the format needed for decisions at that level. This paper describes network technology as it applies to the electric utility industry, beginning at the plant level.

CONTROL NETWORKS: POWER FOR PLANT OPERATIONS

Control systems were one of the earliest applications of computer technology, from crude magnetic core memories to the latest micro-processors. Modern control system requirements for sophisticated computational algorithms, high data flow rate, and reliability are now being fulfilled by distributed computing power.

Networks are integral to all of today's distributed digital control systems. Since the recognition that centralized computing schemes suffered from a lack of scalability and expandability, emphasis has shifted to distributed systems for real-time control. Explosive growth of micro-processor power has lead this decentralization, and today, many localized field controllers have the power of mainframes of a few years ago. Figure 1 illustrates a common configuration, where a plant network links several high-level user interface stations and a central computer, individual controllers, and controller subloops. A subnetwork allows communication between high-level elements without interfering with process traffic. Figure 2 illustrates another configuration with the following elements: (1) A local subnetwork that

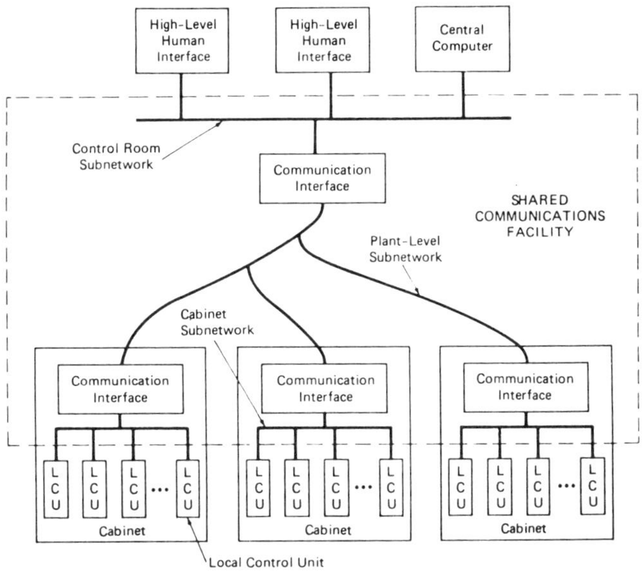

Figure 1. A common communication system for distributed plant control partitioned into subnetworks that follow the natural pattern of message traffic, while providing a mechanism for the subnetworks to intercommunicate as needed.

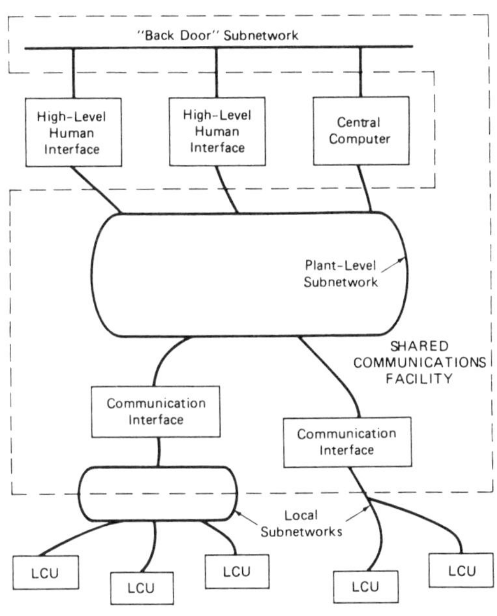

Figure 2. Another communication system for distributed plant control which allows for significant communication rates between the high-level elements without interfering with the process area traffic.

allows individual controllers in a process area to intercommunicate; (2) A plantwide communication system that connects high-level elements with the distributed process control elements; and (3) A "backdoor" subnetwork that allows for rapid high-level device communication, for example, rapid dumps of large databases, without interfering with process traffic (3). These types of systems allow the control operator transparent access to information at all levels of the plant.

The vastly increased information through-put and partitioning of control responsibility of a distributed digital control system also makes possible advanced control techniques such as full multi-variable control; enhanced archiving and data storage including on-line access to plant drawings; and fully automated operation.

DIAGNOSTICS AND MONITORING: INTELLIGENCE FOR PLANT ASSET MANAGEMENT

The advanced intelligence provided by diagnostic and performance monitoring is key to maintenance strategies and improved operations aimed at heading off performance losses, availability losses, forced outages, and unnecessary capital investments in equipment replacements. The masses of data generated by modern diagnostics and performance monitors are a processing and management challenge.

Incorporating diagnostic monitoring technologies into predictive maintenance strategies has helped utilities realize savings due to reduced plant downtime, reduced forced outages, reduced labor and equipment costs, and improved availability (4). Diagnostic information can be used to determine what maintenance is required before the equipment is taken out of service, to determine equipment inspection intervals, and to establish repair and replacement schedules. Maintenance staff and engineers no longer must solely rely on a preventive maintenance schedule that considers time-in-service over actual equipment condition.

In the same respect, performance monitoring technologies have revolutionized the way performance engineers conduct their business. Number crunching and adjustments to actual conditions can be performed instantaneously allowing for timely awareness and correction of performance problems by control operators. Other monitoring technologies also improve upon traditional ways of determining that an operational problem exists. For example, acoustic steam leak detection systems can detect leaks in plant equipment that would otherwise go undetected.

The advanced intelligence provided by monitoring technology can be utilized within the power plant at the process level for determining if a component or instrument is performing properly, for equipment condition assessment, and for determining sub-system or overall performance. Heat rate is fundamental to controlling plant costs, and to assessing the value of a given plant in operational scenarios. Extensive resources are available from five EPRI heat rate conferences (5,6,7,8,9) and four EPRI Incipient Failure Detection Conferences (10,11,12,13). Figure 3 depicts an assortment of monitoring technologies adopted by surveyed utilities.

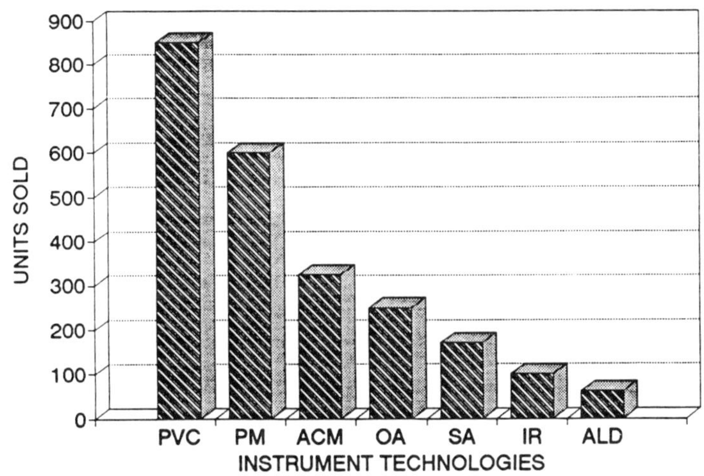

Figure 3. An EPRI survey of monitoring systems sold or installed since 1980 shows the relative market penetration of portable vibration collectors (PVC), performance monitors (PM), advanced continuous vibration monitors (ACM), ferrographic and spectrographic oil analysis (OA), stress analyzers for turbines, boilers, and piping (SA), infrared cameras (IR), and on-line acoustic leak detectors (ALD).

Power Plant Networks: The Future Today

Networks are the key to managing the flood of data from monitoring and diagnostics applications, processing it into information, and presenting it in appropriate formats to utility personnel of various functional groups. EPRI has taken the lead in integrating diagnostic and performance monitoring technologies into networks for improved operations in three major plant demonstration projects and is just launching a fourth. Each of these projects demonstrate how information networks can be used to improve operations and together they are pioneering a changing plant culture.

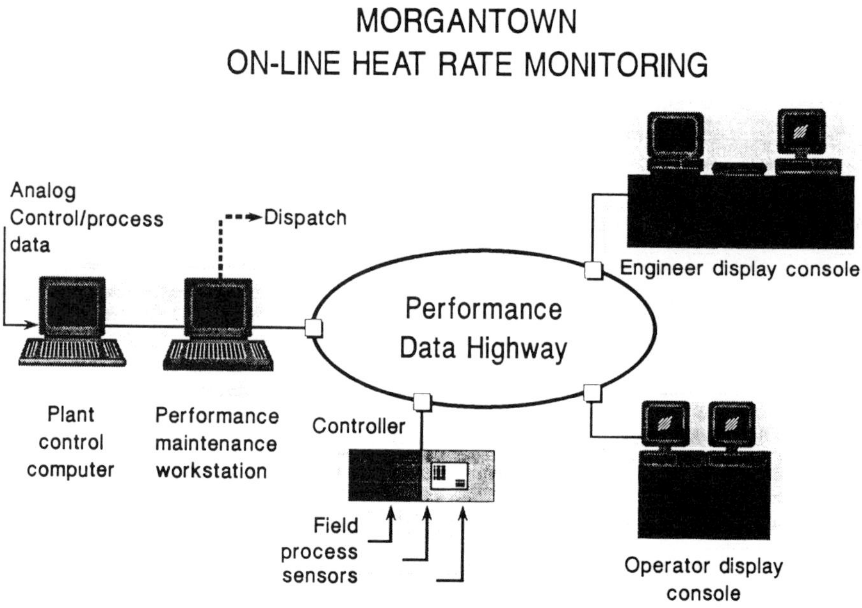

Figure 4. Morgantown performance monitoring system uses advanced monitoring devices, highly accurate instrumentation and sophisticated software to measure on-line plant heat rate.

Morgantown Performance Monitoring System

This Potomac Electric (PEPCo)/EPRI demonstration is the test bed for advanced thermal process instrumentation and on-line calculation of plant heat rate. 19 advanced monitoring devices and other process instrumentation on Morgantown Unit 2 are analyzed by the Performance Monitoring Workstation (PMW) (14). A data highway is used to interconnect sensor data, display consoles, and a single large performance computer as shown in Figure 4. The performance computer is linked to the plant data acquisition computer to obtain necessary operating parameters. This monitoring system combined with careful attention to instrument location, calibration, and redundancy has resulted in a 0.5% accuracy in measured heat rate. The workstation contains sophisticated software including thermal-hydraulic models used to compare measured and optimal performance, determine component degradation, and examine operational changes on plant performance. Cost of a full set of steam cycle instrumentation typically ranges between $300,000 to $500,000, including engineering, installation, data acquisition, and computer. Implementation has already improved maintenance planning, unit capacity and heat rate by over 100 Btu/kWh. Near real-time performance calculations at Morgantown will soon be used for on-line unit dispatch, allowing system operators to better determine Morgantown's actual incremental heat rate. Ideally, all plants in the system would be fitted with this capability to optimize system production costs.

PEPCo has installed the PMW on a second power plant unit, and additional installations are planned for single or multiple sites at four other EPRI member utilities: Public Service of Colorado, TU Electric, Pacific Gas & Electric, and Philadelphia Electric.

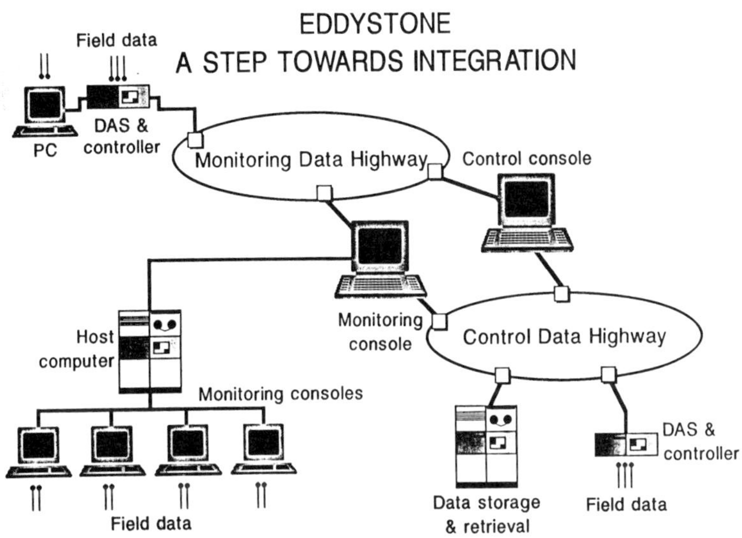

Figure 5. Eddystone diagnostic monitoring network integrates 24 on-line and periodic diagnostic technologies through a computer - data highway network.

Eddystone Diagnostic Monitoring Network

Demonstration and integration of diagnostic monitors and the development of predictive maintenance practices is the focus of this Philadelphia Electric/EPRI project. Twenty four on-line and periodic diagnostic technologies comprising over 2300 monitoring points are integrated through a computer - data highway network at the Eddystone Station (15). Data is drawn directly onto the highway, over gateways from the distributed control system, and from networked monitoring computers with their own dedicated data acquisition as shown in Figure 5. Detailed information from the monitoring computers located in an equipment room is displayed in an adjacent predictive maintenance analysis center. Key diagnostic information relevant to immediate plant operation is communicated between the predictive maintenance center, control room, and EPRI M&D Center located 1/2 mile from the plant. Through a combination of advanced maintenance scheduling, avoidance of extensive mechanical damage, and improved operating procedures to slow component wear, Eddystone is well on its way to achieving a 2-3 year payback on its $3 million monitoring system investment.

The monitoring network continues to grow with new instrumentation and diagnostic expert systems, including the Performance Monitoring Workstation described above.

INTEGRATED OPERATIONS AND MAINTENANCE NETWORKS: THE ADVANTAGE

Power plants have traditionally seen diagnostic and performance monitoring as separate functions from plant control. Reflecting this, stand-alone, single-purpose monitoring and diagnostic computer applications were developed as solutions to individual problems. Stand-alone systems create individual and unique islands of information. They may not be as effective as they could be because information is presented through an oftentimes unfamiliar man-machine interface, or because the system is physically remote from the potential user. The full potential of power plant monitoring and diagnostic technologies can only be realized by integrating this new information into networks that support the complete plant operating environment and insure a "transparent" operator interface.

Integration of monitoring technologies with the DCS has many advantages: (1) All plant data and calculated values reside in one common database. (2) Data developed by one diagnostic or performance monitoring application can now be shared with or accessed by another application. (3) Calculated values from monitoring programs can be integrated with the control operator's display, providing the operator with a "single window" on the process, and can be used in an advanced control strategy. (4) Plant sensors and instrumentation can be shared for both control and monitoring functions. (5) Maintenance logs, work orders and clearance tags can be automatically generated. (6) A single, plant-wide man machine interface can be developed for all users.

Yet, integration of power plant information systems, including control and monitoring, is hampered by the plain fact that the various DCS and other software and hardware suppliers employ proprietary communication protocols within their systems. The process of establishing communication between systems from various vendors is not straightforward. In general, many utilities have found that the proprietary solutions offered today prove to be far more difficult to integrate than initially envisioned. The early pioneers in this area have found that technically complex and costly custom bridges, gateways and routers are oftentimes needed. In fact, the cost to the electric utility industry for interconnection and translation devices is estimated to be $1 billion per year (16). Further, these custom interfaces generally require modification whenever an aspect of a system it is interfacing with changes.

EPRI sponsored a workshop in 1986 to assess utility interest in integrated utility communications and to identify and prioritize needed R&D (17). As a result, EPRI launched the Integrated Utility Communications (IUC) initiative to develop a utility information infrastructure that would allow for systems from various vendors to interoperate. Further, the infrastructure was to be designed in such a way that it could be easily expandable and should solve the above stated problems. Four major projects comprise the IUC program: (1) the development of a Utility Communications Architecture (UCA), meeting utility requirements, and based on the Open Systems Interconnection (OSI) Reference Model (18,19,20); (2) the development of a Database Access Integrated Services (DAIS) specification for easy exchange of data between diverse database types (21); (3) the demonstration of UCA and DAIS capabilities at two or more EPRI host utilities; and (4) the development and enhancement of software tools as needed for the planning, design, and monitoring of integrated utility communication systems.

EPRI and the utility industry are working with power plant test beds to make comprehensive power plant networks a reality, as evident by the following two demonstrations.

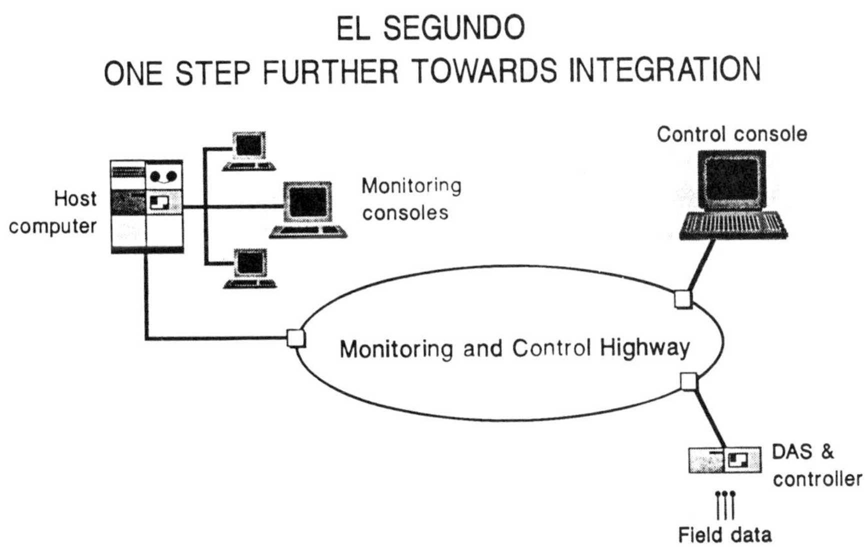

Figure 6. El Segundo control - diagnostic retrofit is a step towards fully integrating monitoring technologies with distributed digital control systems. The goal is to create a single display window on all plant operations and equipment condition.

A Step Toward Integration: The El Segundo Demonstration

EPRI is working with Southern California Edison to integrate diagnostic and performance monitoring, on-line advisory systems and advanced controls at El Segundo Power Plant, Units 3&4(Figures 6 and 7). This project is unique in that it has required an industry control system supplier to integrate distinct diagnostic monitoring systems from various vendors and expert systems with the control system. The project will address the design of the user interface and screen displays to deliver operationally pertinent diagnostic intelligence to the control operator for his action as well as allow for detailed analyses by others whether at the plant or remote. The criteria established by Southern California Edison for this project is to use a common sensor data acquisition platform, distributed processing, one plant monitoring computer for the diagnostic monitoring applications, one large integrated database, and a single set of screen displays accessible through the DCS consoles or any of the networked PC's providing a full and unrestricted window on all aspects of plant operations. To meet these criteria has required attention to the communication protocol used by various control system and other hardware vendors, attention to the needs of the system users who are operators, managers, engineers and maintenance personnel, and reconciling significant differences between control and monitoring systems. The El Segundo project is addressing these and other issues and producing guidelines for tight integration in subsequent power plant control retrofit projects (22).

The Next Step: The Roxboro Demonstration

This project with Carolina Power and Light will demonstrate the application of state-of-the-art automation technologies to improve heat rate and unit maneuverability throughout the load range and reduce forced outages and O&M costs. This will be achieved through full utilization of the power of a distributed control system and integration of monitoring, diagnostics, on-line testing, expert systems, and information management functions into a plant-wide automation system based on EPRI's Utility Communications Architecture (UCA). The program strategy is to utilize available state-of-the-art technologies and to develop the missing links essential to the realization of the project goals. The project is an alliance between EPRI, CP&L and an industry supplier that will advance integration technology.

STANDARDIZING THE POWER PLANT NETWORK

Collectively, these four demonstrations with other industry experience point towards the viability of a standard plant network that possesses these characteristics:

Figure 7

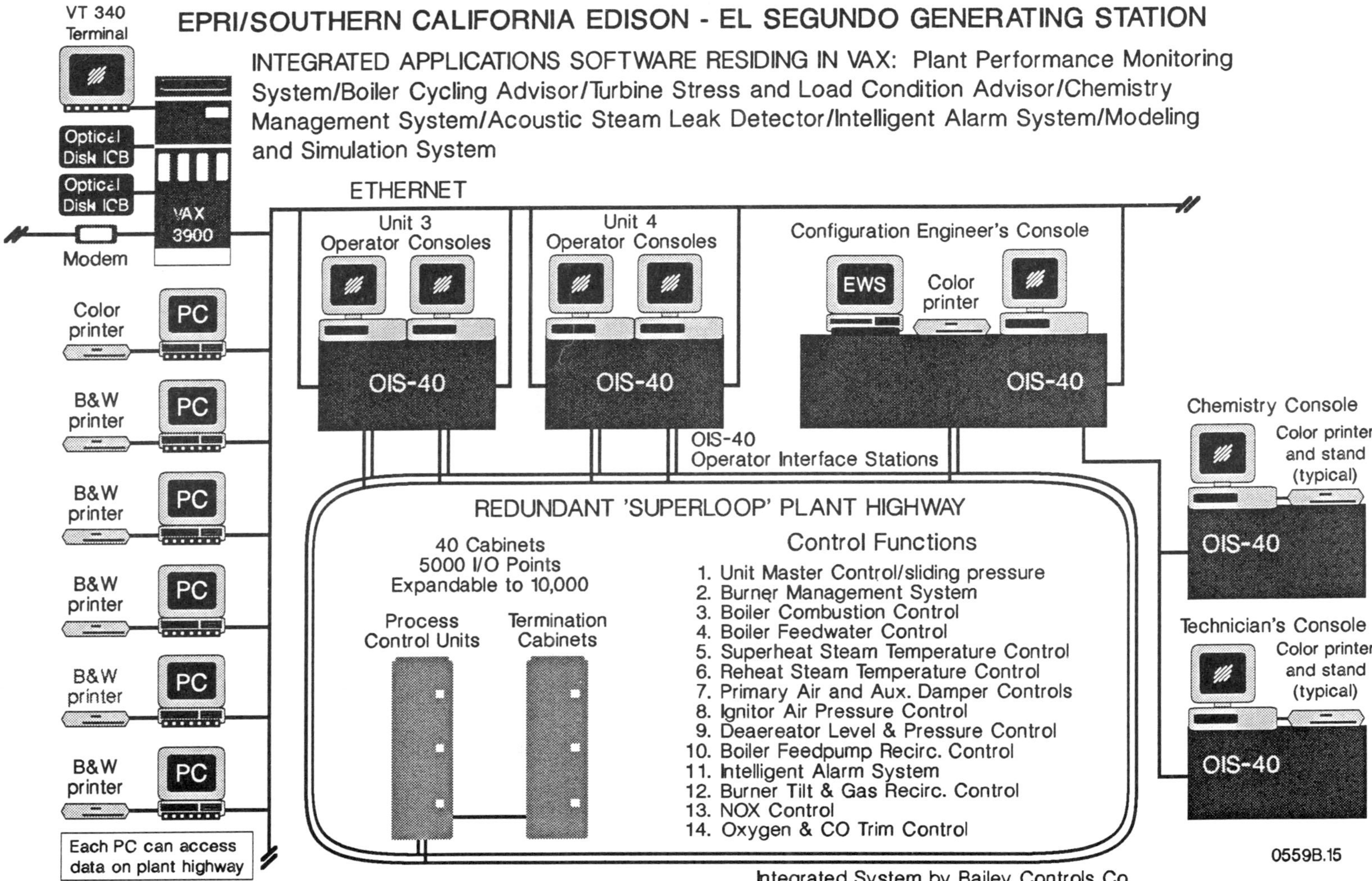

EPRI/SOUTHERN CALIFORNIA EDISON - EL SEGUNDO GENERATING STATION
INTEGRATED APPLICATIONS SOFTWARE RESIDING IN VAX: Plant Performance Monitoring System/Boiler Cycling Advisor/Turbine Stress and Load Condition Advisor/Chemistry Management System/Acoustic Steam Leak Detector/Intelligent Alarm System/Modeling and Simulation System
VT 340 Terminal
Optical Disk ICB
Optical Disk ICB
VAX 3900
Modem
ETHERNET
Color printer
PC
B&W printer
PC
B&W printer
PC
B&W printer
PC
B&W printer
PC
B&W printer
PC
Each PC can access data on plant highway
Unit 3 Operator Consoles
OIS-40
Unit 4 Operator Consoles
OIS-40
Configuration Engineer's Console
EWS
Color printer
OIS-40
OIS-40 Operator Interface Stations
Chemistry Console
Color printer and stand (typical)
OIS-40
Technician's Console
Color printer and stand (typical)
OIS-40
REDUNDANT 'SUPERLOOP' PLANT HIGHWAY
40 Cabinets
5000 I/O Points
Expandable to 10,000
Process Control Units
Termination Cabinets
Control Functions
1. Unit Master Control/sliding pressure
2. Burner Management System
3. Boiler Combustion Control
4. Boiler Feedwater Control
5. Superheat Steam Temperature Control
6. Reheat Steam Temperature Control
7. Primary Air and Aux. Damper Controls
8. Ignitor Air Pressure Control
9. Deaereator Level & Pressure Control
10. Boiler Feedpump Recirc. Control
11. Intelligent Alarm System
12. Burner Tilt & Gas Recirc. Control
13. NOX Control
14. Oxygen & CO Trim Control
0559B.15
Integrated System by Bailey Controls Co.

- Extensive use of digital distributed systems for control and diagnostic sensor data acquisition and display.

- Reduction and analysis of data to essential features as near as practical to the sensor using distributed processing.

- Work stations for detailed on-line analysis of plant mechanical, electrical, and thermal conditions.

- Maintenance planning computer functions directly linked to diagnostic database.

- A selection of core diagnostic technologies offered as a standard commercial package.

- On-site expert system assistance for fault diagnosis, root cause determination, and remedial O&M advise.

- A selection of automation technologies for improved operations.

THE INTEGRATED UTILITY NETWORK: OPTIMIZING THE PRODUCTIVITY OF UTILITY ASSETS

Although the focus of this paper deals with power plant networks, they are only one element of the overall "tapestry" of networks that serve an entire utility (18). A conceptual integrated utility network incorporating all elements is shown conceptually in Figure 8 (19). In this scheme, information is available and accessible, as needed and permitted by corporate policy, across the entire utility enterprise. This type of information availability offers dramatic potential for improved operations and productivity.

With the advent of new generations of computer-based information and automation technologies, many utilities have begun to install local and wide area networks (LANs and WANs) to support the technologies. However, as described earlier, integration of equipment by various vendors into networks is still a complex and costly task. While it may take many years for utilities to fully implement system-wide networking and to grow comfortable with its technical complexity and potential implications it may have on present utility organizational structures and operations, EPRI is charting a strategic course. Implementation of UCA and DAIS can be done in an evolutionary fashion starting with new or upgraded systems and interfacing to key existing systems as needed and as cost effective.

UCA Figure 8 **INTEGRATED UTILITY NETWORK**

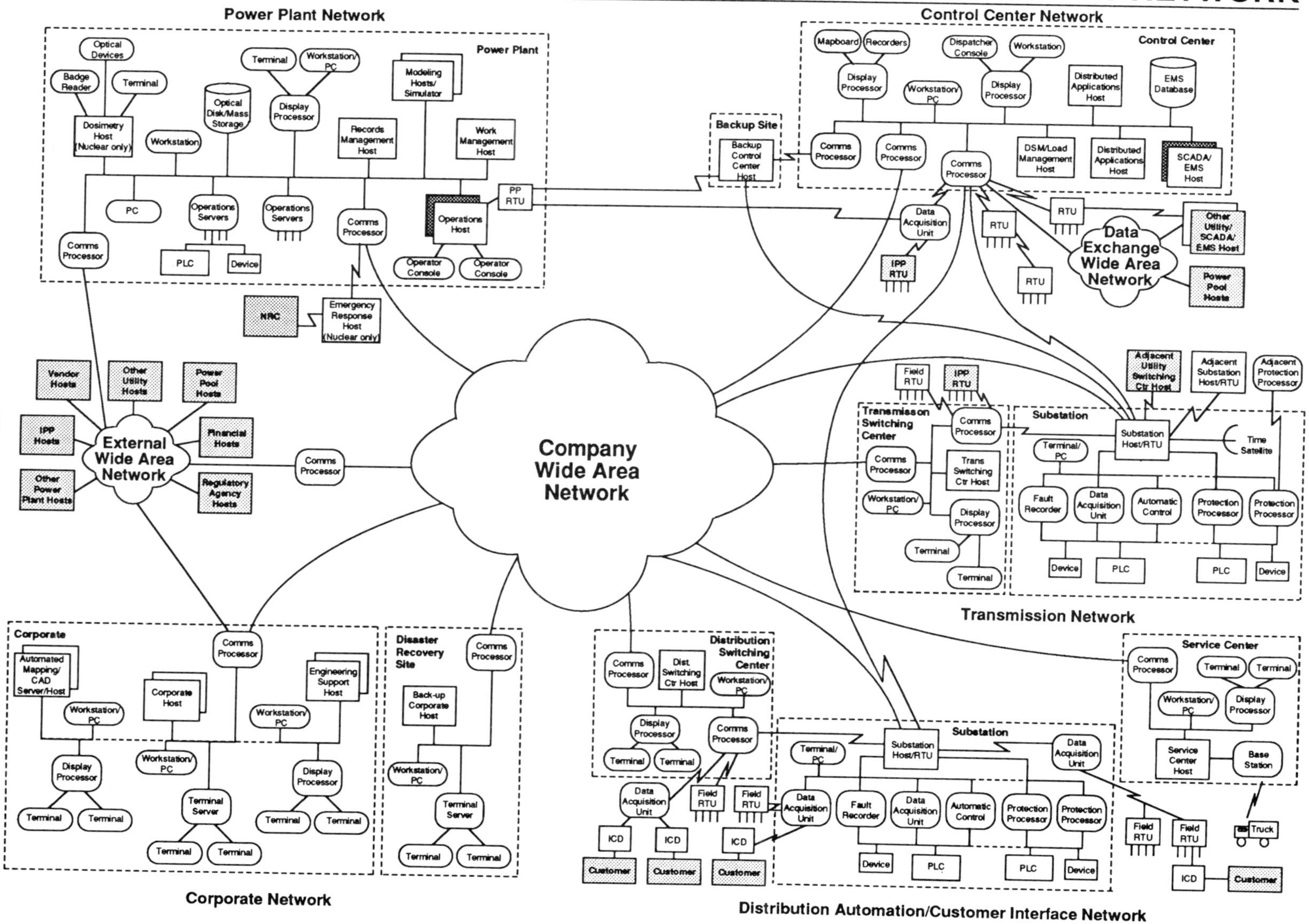

CONCLUSIONS

Networks will be the cutting edge of increasing asset productivity. At the plant level, equipment and performance diagnostic and monitoring capabilities will be networked with advanced controls to optimize the operation of existing and future power plants. With many existing plants required to run reliably into the indefinite future, advanced intelligence provided through the network will be a hedge against growing operational risks and uncertainties. EPRI plant demonstrations provide an available technology base and strong justification for a standard plant-wide information network that would be integrated with the control system and make extensive use of distributed digital processing. Networks and distributed computing are the enabling technologies for combining operations and maintenance — an emerging strategy EPRI has named "active O&M".

At the utility system level, networks will enable total asset management. Powerful asset management workstations, each tailored to the appropriate needs of the utility function served will access data wherever it resides on the system, and process it into information, which is then available to other network clients as needed for further processing. The plant manager will be able to analyze the return on investment for a rotor replacement based on aggregate system needs without the laborious iteration required today. Maintenance planners will have remaining life and condition information based on real-time monitoring for components across the system, and will plan for maintenance on a plant or system level.

Networks and distributed computing will effect all levels of utility operation, from the maintenance floor to the corporate offices. When communication integration becomes a reality, utilities will be able to maximize the use of their communication systems and optimize management of the wide range of their other assets. The power and flexibility of integrated networks will make possible innovations limited only by imagination.

ACKNOWLEDGEMENTS

The authors wish to acknowledge the contributions of Wade Malcolm and Bob Iveson of EPRI's Electrical Systems Division. These gentlemen are managing the Integrated Utility Communications initiative with the support of over 45 member utilities.

REFERENCES

1. A.F. Armor et. al.. "Managing Plant Assets for Profitability." Presented at EPRI Conference on Decision Support Methods for the Electric Power Industry, May 29-31, 1990, Cambridge, MA.

2. Verity, John W.. "Rethinking the Computer: With Superchips, the Network is the Computer." BusinessWeek, November 26, 1990, p. 116.

3. Michael P. Lukas. Distributed Control Systems. New York: Van Nostrand Reinhold Company, 1986, pp. 121-125.

4. Blanco, M., Armor, A. et. al.. "Diagnostic Monitoring of Power Plants: The Key to Predictive Maintenance." Presented at the EPRI 4th Incipient Failure Detection Conference, October, 1990, Philadelphia, PA.

5. Proceedings: Fossil Plant Heat Rate Improvement - 1981 Conference and Workshop . CS-2180. Palo Alto, CA: Electric Power Research Institute, December, 1981.

6. Proceedings: 1983 Fossil Plant Heat Rate Improvement Workshop. CS-4014. Palo Alto, CA: Electric Power Research Institute, May, 1985.

7. Proceedings: 1985 Heat-Rate Improvement Workshop. CS-4736. Palo Alto, CA: Electric Power Research Institute, September, 1986.

8. Proceedings: 1988 EPRI Heat Rate Improvement Conference. GS-6635. Palo Alto, CA: Electric Power Research Institute, January, 1990.

9. Proceedings: 1989 EPRI Heat-Rate Improvement Conference. GS-6985. Palo Alto, CA: Electric Power Research Institute, September, 1990.

10. Proceedings: First Incipient Failure Detection Conference. CS-2920. Palo Alto, CA: Electric Power Research Institute, March, 1983.

11. Proceedings: Second Incipient Failure Detection Conference. CS-4748. Palo Alto, CA: Electric Power Research Institute, October, 1986.

12. Proceedings: Third Incipient Failure Detection Conference. CS-5395. Palo Alto, CA: Electric Power Research Institute, August, 1988.

13. Proceedings: Fourth Incipient Failure Detection Conference. To Be Published. Palo Alto, CA: Electric Power Research Institute.

14. MARK I Performance Monitoring Products. GS/EL-5658. Palo Alto, CA: Electric Power Research Institute, September, 1989.

15. Monitoring & Diagnostic Center: An Overview of Operating Activities. Draft Interim Report, Electric Power Research Institute, December, 1990.

16. Beaty, Wayne. "Utilities told to expand DA/DSM involvement." Electric Light & Power, March, 1991, p. 24.

17. Proceedings: Integration of Utility Communication Systems Workshop. EL-5242. Palo Alto, CA: Electric Power Research Institute, August, 1987.

18. Utility Communications Architecture (UCA), Volume I: Functional Definition (Draft Report). RP2949-1. Palo Alto, CA: Electric Power Research Institute, September, 1990.

19. Utility Communications Architecture (UCA), Volume II: Communications Requirements (Draft Report). RP2949-1. Palo Alto, CA: Electric Power Research Institute, September, 1990.

20. Utility Communications Architecture (UCA), Volume IV: UCA Specification (Draft Report). RP2949-1. Palo Alto, CA: Electric Power Research Institute, September, 1990.

21. Database Access Integration Services (DAIS), Volume I: Background and Concept (Draft Report). RP2949-5. Palo Alto, CA: Electric Power Research Institute, February, 1990.

22. Preliminary Guidelines for Integrated Controls and Monitoring for Fossil Fuel Plants. GS-6868. Palo Alto, CA: Electric Power Research Institute, June, 1990.

©1991 - Paper #91-730
0074-056X/91/403-414/$0 + .50pp

Integration of Control and Data Acquisition In A Combined Cycle Generating Plant

Harry W. Forbes

Industry Specialist
Utility Sales and Marketing
The Foxboro Company
Foxboro, Massachusetts, USA

Abstract

This paper describes a control and data acquisition system which has been installed in a new single-train 214 MW combined cycle generating plant. One of the tasks faced by the control engineers in the project was to successfully integrate systems from different vendors in order to control a complex, highly interactive process. The importance of this integration was recognized early in the project, and a plan for systems integration was developed. The resulting integration plan is used here as a framework for description of the plant control and data acquisition environment.

The control "domains" within the plant are: 1) Gas Turbine control, 2) Steam Turbine control, 3) Balance of Plant control, 4) Unit-level coordinated control, and 5) Plant Master Sequence control.

Each of these domains is described. In addition, description is given of the Gateways between the turbine controllers and the distributed control system (DCS). Also the UNIX software environment for data acquisition is described as are several of the data acquisition software packages which were developed for use on the system.

Plant History

These units occupy space which formerly housed two 60 MW units. These two units were retired by the utility in 1982. In 1985 the utility began to study options for using this vacated space to meet the expanding demand of their electric customers. The company wanted to add two units with a nominal output of 200 MW each. In addition to the space constraints imposed by using the existing building, it was necessary that the new units have less thermal emission than the retired units, and acceptable flue gas quality. Finally, the utility wanted to be able to operate the units in either baseload mode or in a daily start/stop mode as midrange peaking capacity.

The evaluation made concluded that gas turbine combined-cycle units would be the best option for new generating capacity. At the same time, advances in gas turbine technology permitted design of a 200 MW unit with a single gas turbine/heat recovery steam generator train. The owner selected a General Electric STAG 107F design for the gas turbines, and retained GE as a turnkey system supplier for the new units.

Plant Equipment Configuration

Each unit's equipment consists of a single gas turbine, a single unfired, dual-pressure Heat Recovery Steam Generator (HRSG) with an integral deaerator, and a single flow, straight-condensing, reheat steam turbine (see Figure 1). There is no bypass for gas turbine exhaust, so the gas turbine and HRSG are always operated as a unit. The two units share an exhaust gas stack, fuel handling, and water treatment systems. A detailed description of the installed equipment can be found in [1].

All of the above plant design decisions were made in advance of the control system design. It was necessary to supply with the plant a control system that would both support ease of operation and the planned modes of operating the plant. There were several important considerations:

1) Both the gas turbine and the steam turbine packages included electronic control systems, with different human interfaces.

2) The design objective for rapid start-up capability dictated that all plant controls must be capable of coordinated operation.

3) This need for coordinated control operation was also required in order to ease the burden imposed on the plant operators by start-up and shutdown maneuvers.

4) A common and consistent human interface to plant controls was necessary for successful operation by the plant operating staff.

After a thorough evaluation of the systems available, a Foxboro Intelligent Automation Series control system was selected for the plant. Foxboro and GE control engineers worked together as the control system project team through the design, start-up, and commissioning phases of the project.

Control Domains and Function

For purposes of this discussion, let's define a control "domain" as a subset of the plant control system that is focused on a distinct part of the operating plant or on a unique control function. A control "function" is a method of control (continuous control, logic control, or sequence control). Each control domain will include all these control functions. The plant control system breaks logically into five domains:

1) Gas Turbine Control

2) Steam Turbine Control

3) Balance of Plant Control

4) Unit Coordinated Control

5) Plant Master Control

Figure 2 shows the connections between these domains of control and which systems perform each function.

Gas Turbine Control

This domain is controlled by a GE SPEEDTRONIC Mark IV system. The system includes redundant controllers, and is tied directly to gas turbine field sensors and actuators (about 200 hardwired points). The continuous control functions performed by the system include control of fuel flow,inlet guide vanes, and steam/water injection for control of NOX. The major logic control functions are equipment protection and alarm generation logic. Gas turbine auxiliaries operate under SPEEDTRONIC control sequences during turbine startup, shutdown, and cooldown. This system connects to the Distributed Control System (DCS) through a Gateway device and through a small set of hardwired signals.

Steam Turbine Control

The steam turbine control domain is run by a General Electric Digital Control & Monitoring (DCM) system. This system contains hardware very similar to the SPEEDTRONICS system. It also uses redundant controllers, and includes regulatory control, equipment protection logic, and equipment operating sequences. It connects directly to steam turbine field sensors and actuators (there are about 300 hardwired points). There is no hardwired connection between the DCM and the rest of the plant controls. The steam turbine uses a serial communication link to send measurement data to a DCS Gateway. All DCS commands to the steam turbine are passed via the same Gateway and serial link.

Balance of Plant Control

This control domain is managed by the I/A Series system. Its major control functions are the condensate and feedwater systems, the steam bypass system, steam temperature control, drum level control (for the HP drum, IP drum, and the deaerator), circulating water system, fuel gas system, chemical injection, and sampling system.

The I/O signals for these systems are assembled in a vendor-supplied termination cabinet mounted two levels below the unit control room. These cabinets contain 1500 hardwired points at present. From the termination cabinet, these field signals are fed to the DCS I/O modules using prefabricated cables. The I/O modules are actually programmable ladder-logic processors. Motor and valve interlocks are configured at the I/O card level. These ladder-logic algorithms execute every 5-10 milliseconds. Two redundant Fieldbus cables carry the I/O signals from the termination area to each control module. The controllers are located adjacent the control room (see Figure 3). There are 10 digital control modules for BOP functions. Control modules are located in the control room. Each of the control modules is a fault-tolerant pair of processors. These modules perform regulatory control, alarming, sequence-of-events reporting, and services for the operator interface units.

Unit Coordinated Control

This level of control resides in the I/A Series DCS. Its purpose is to coordinate the operation of the overall unit in order to control unit output. It manipulates the gas turbine load using RAISE/LOWER signals to the Speedtronic system, while the steam turbine is running a follow mode accepting all available steam from the HRSG. The load setpoint is generated by the operator (in MANUAL mode), by the Plant Master Control Sequence (during unit startup and shutdown), or by a remote unit dispatch center when the unit is on line (RDC mode). In addition to setting a load target for the unit, the operator may set high and low load limits as well as rate limits using the displays at the DCS CRTs. The control system sends two analog signals to the RDC center which represent the high and low limits for unit load. Three discrete signals which represent high load limit, low load limit, and RDC control status are also sent by the system to remote dispatch. The multiplexer for signals to and from RDC is supplied by the utility.

Plant Master Control

The DCS includes a Plant Master Control Sequence (PMC). The sequence resides in a fully fault-tolerant control processor, identical to the controllers which perform BOP control functions. While all control processors are capable of performing both regulatory and sequence control algorithms, the processor used for the PMC function includes only control sequences.

The PMC sequencer automatically brings the plant from a READY state to a SPINNING RESERVE state in less than 70 minutes. It will also move the plant from any initial condition to a STANDBY state. The sequence starts, stops, and controls major equipment within the plant. The PMC communicates directly with the Unit Coordinated Controls, BOP controls, HRSG controls, and the DCM system of the steam turbine. It is capable of controlling sequences in the DCS and of starting/stopping sequences which are performed by the Gas turbine and Steam turbine controllers.

Also provided as part of the PMC is a hierarchy of CRT screen displays. There is a master CRT display, and sets of displays for both the startup and shutdown sequences. These displays provide increasing levels of detail concerning the operation of the Auto Start/Stop functions, and allow the operator to place or release holds in the sequence. Also, some sequence conditions can be bypassed by the plant operator using the CRT screens. A typical CRT screen display is shown in Figure 5.

Human Interface Domain

Great pains were taken to make an excellent human interface for the operators. A hierarchy of graphic displays was designed. Control faceplates can be overlaid on the displays by "picking" sensitive areas of each display. Process alarm information is routed to both a CRT display and a printer in the operator workstations. In addition, alarms are saved by the DCS Historian for review. A separate CRT display and printer log all "system level" alarms. These are alarms which pertain to the DCS equipment status rather than the process.

Device Gateway Functions

Situated between the DCS and the dedicated controllers for the gas and steam turbines are two "Device Gateways". Simply stated, the purpose of the Gateways is to allow the DCS to send and receive information from the dedicated controllers so that (to the DCS and the plant operator) the remote data appears to be the same as DCS data. The difference between process data read via the Gateway and data hardwired to the DCS controller should be "transparent" at higher level functions, such as the Historian, alarms, and human interface.

The steam turbine's "DCM" system sends 100 analog and 560 discrete data points to the gateway every 2 seconds. The gateway sends 6 analog and 73 discrete signals to the DCM on a change-driven basis. The gas turbine governor ("SPEEDTRONIC Mark IV") system sends 90 analog and 346 discrete values every second to its gateway. As the Mark IV cannot accept commands over the communication link (at present), outputs from the DCS to the gas turbine are hardwired (there are 8). Typical of the data points received by the DCS are gas turbine exhaust temperatures, turbine Speed/Load Reference, and valve position signals. The turbine control systems receive commands from the DCS to control motor-operated valves, set turbine operating modes, and change speed/load/pressure setpoints.

The gateway provides a number of services for the DCS. These are:

Communication port management - Each Gateway has two communication links with the turbine governor. The status of the ports is monitored continuously. The gateway maintains timers for message transmission, reception, and idle time on each port. Alarms are generated in cases of port failures. These DCS system-level alarms are annunciated visually and audibly at the operator workstation as well as printed. The function also provides message and error accumulated counts for monitoring the channel status over longer periods.

Message error checking and identification - Data received from the turbine governor is tested for validity using a CRC-16 checksum that is transmitted as part of each message. Valid messages are then parsed to identify their content. Erroneous messages are discarded. If the gateway receives a series of erroneous messages, a port failure is declared, and the port will be taken off line.

Format conversion - Data received (or sent) between the governors and the DCS will be converted to the appropriate format. Scaling is applied if required.

Ownership of data - The DCS contains a global distributed database of process data. The database is "global" in that all data objects are referenced by a name that is unique within the entire DCS. Data access is via this unique name. The database is "distributed" in that many stations serve as owners of data objects. The gateway serves as the owner of those objects which are sent or received from the turbine governors. Each object is assigned a name, and other stations of the DCS know the objects only by name. DCS stations using this data (for example an operator workstation) do not know that an item of data is received from a "foreign" device, and can mix gateway and DCS data freely in order to achieve the desired human interface.

Alarming - In the DCS any control station can generate process alarms. The gateway generates alarms for data it receives from the turbines. This alarm functionality is identical to that available with the other DCS control stations.

Change notification - The DCS does not use "polling" in transferring data between its stations. All data updates are change-driven, using a configurable deadband. The gateway generates update messages for both the DCS and the governor. Changes in data values received from the governor cause the gateway to send update messages to other DCS stations. Similarly, if another DCS station changes a value which is being sent by the gateway to the turbine, the gateway will generate and send an update message to the turbine governor over its communication link with the turbine.

To summarize, the Gateways service the communication links and map remote data points into the DCS so that they behave as if they were hardwired into the DCS. This way no special treatment is needed for these points at higher level DCS control and engineering functions.

Data Acquisition Environment

In addition to the components shown in Figure 3, the DCS contains an environment for application program development. The programming environments which are provided with distributed control systems (DCS) have traditionally been highly proprietary. Most environments have supported either an extended Fortran and/or Basic language, but editors, compilers, assemblers, linkers, etc., vary widely. Also, the software "toolset" provided with these systems varies greatly. Some systems provide no tools. Others provide proprietary tools of uneven quality. The main reason for the deplorable toolsets on these systems is the relatively small number of systems which are in the installed base. Most DCS suppliers sell a few hundred systems per year. This is not enough volume to justify the development of an extensive software toolset. The only options available to the DCS user have been to either 1) bear with what is provided, 2) make your own tools, or 3) link the DCS to some minicomputer which carries its own toolset.

The emergence of UNIX as the predominant operating system has changed this situation. UNIX includes an excellent set of software tools, far superior to that which has existed previously on any Distributed Control product. This represents a great leap forward in what has been available to DCS programmers. They are no longer the "poor cousins" of the programming community. For example, the UNIX system included with this plant DCS supports 7 languages (C, Fortran 77, Pascal, assembly, awk, Bourne shell, and C shell). Most older DCS families support 1 or 2 languages at most. There are 4 text editors (ed, sed, ice and vi). There is an extensive set of library functions including support for: 1) UNIX system calls, 2) process data access, 3) mathematical functions, 4) materials properties, 5) relational database query (this also provides access to historical process data), 6) terminal support ("curses"), and 7) graphical workstation support. Finally, there are tools for source code control (SCCS), automatic software module generation (make), code generators (lex, yacc), code checkers and debuggers (lint, adb), and software performance profilers (monitor, prof). Figure 4 summarizes the DCS software tools by function.

The UNIX system's ability to be ported from one hardware environment to another is what makes this large toolset available. Another advantage of UNIX (and perhaps equally important) is its widespread use in the academic world. College graduates in engineering and computer science are already conversant in UNIX and C. This results in a huge increase in productivity. There is no need to "break in" new people for long periods, and skilled programmers are more willing to work on UNIX systems than proprietary systems. Finally, it is easier to get outside help. In the case of older proprietary software systems help came only from the supplier and perhaps a few small shops. With UNIX, many more people are already trained to pick up the tools and do useful work.

Several data acquisition functions were implemented for this project using the UNIX programming environment. Here again we see concrete examples of the advantage of a UNIX-based distributed control system. For example, the DCS operator workstations and the File Servers operate under UNIX. This allows functions to be run from either place. Some of the functions described below are designed for the operator, while others are strictly for engineers. Nevertheless, any function can be run from any operator workstation.

Another example of a UNIX advantage is that the DCS Historian stores process data into a full relational database management system (DBMS), Informix. Any Informix database can be accessed by library functions which are available in the UNIX programmer's environment. Most of the application programs below use these UNIX library functions to select and obtain historical data from the Historian, using functions from this library provided with the DBMS.

The UNIX applications configured for this project are:

Event-Triggered Plot - A CRT plotting function that creates multi- variable plots. The plots are time trends of the variables. The starting time of the trends is configured based on a time before and after a triggering event (such as generator synchronization or trip). Triggering events can be defined for any boolean variable in the DCS database.

Performance Monitoring - A function for Heat Rate calculations.

Crossplot - Another CRT plotting function, which generates X-Y plots of historical data.

Start-up/Shutdown Analysis - This function generates an analysis report of events which occur during a startup or shutdown of either turbine.

Reactive Capabilities - This function generates a realtime CRT display which shows the applicable generator excitation curves for the gas turbine, based on measured generator loads and hydrogen pressure. A "tick" on the display shows the operator the current operating point in realtime.

Combustion Monitor - This function generates 2-D and 3-D plots of turbine exhaust gas temperature. Both realtime and historical plots can be requested. The three dimensions plotted are temperature, time and position on the turbine exhaust annulus.

Parts History - This application uses DCS Historian data which originates in the gas turbine control system. The function maintains a database recording the history of key turbine parts. The parts tracked are those which have limited life and/or require frequent inspection.

Alarm/Event Report - The DCS Historian is also set up to collect DCS alarm messages. This function queries the alarm report database and generates frequency distributions and chronological reports for any defined interval of plant history.

Application Example - The Combustion Monitor

A typical application program is the "Combustion Monitor" function. This application is used to generate historical or real-time plots of gas turbine exhaust temperatures. These can be 2-dimensional or 3-dimensional plots of up to 34 temperature measurements. The 107F Gas Turbine has 27 exhaust thermocouples. These sensors are arranged in a ring around the turbine exhaust annulus. The readings of these sensors are passed to the DCS via Device Gateway every second. Each temperature measurement is logged by the DCS historian, and is available for use on CRT displays.

The Combustion Monitor allows the user to view this data in a number of ways. The simplest is a two-dimensional plot, as in Figure 6. The X-axis on the plot indicates the position of the thermocouple on the exhaust ring, and the Y-axis indicates temperature. The values of the data points are listed at the top of the display, as well as a few other key turbine measurements.

A three-dimensional plot can also be drawn at the user's request. A typical plot is shown in Figure 8. The three axes represent 1)time, 2)temperature, and 3)position of the reading on the turbine exhaust annulus. This plot can consist of either historical or realtime data. There are several Menu-pick items on the right side of the screen. Two are for grid display. The alphanumeric data posted at the top of the plot represents one data set from the plot. The location of this data set within the plot is chosen by the "caret" pick item on the right side of the display. The "?" pick is used to display the plot specifications. The pick labelled "REV" reverses the direction of the time axis. The "SEL" pick item is used to switch to a 2D plot of the selected data set.

At the very top of the display (again see Figure 8) are four "Menu-bar" items, which can be picked with a pointing device, such as a mouse or trackball. The function of the "Menu-bar" items are:

Range - This function allows the data to be plotted using a new scale for the Y-axis.

Grid - This item superimposes a grid over the plot.

New - This allows the user to specify a new CRT plot, again using a series of Dialog Boxes to specify the plot (see Figure 7 for a view of the dialog boxes).

Exit - This exits the Combustion Monitor.

Summary

A key aspect of the design of this plant control system was the ability to integrate different pieces of control equipment (turbine governors, DCS) so that a single integrated operator interface could be available. The ability of the each component in the system to perform ladder-logic, continuous control, and sequence control allows these different components to work together well. The incorporation of a UNIX operating system and programming environment greatly eased the creation of a set of application programs for the plant.

References

1) Combined-cycle in Virginia, International Power Generation, November, 1989

2) Steam Turbine Control RS-232 Computer Interface, GEI-100015, GE Drive Systems, Salem, VA

3) Foxboro Device Gateway User's Guide, I/A Series Part B0193CZ, Dated March 30, 1990

4) Plant Master Control Users's Manual, Rev 0, The Foxboro Company, Power Systems Engineering

5) The UNIX Programming Environment, Brian W. Kernighan and Rob Pike, Prentice-Hall, Inc., 1984

6) Combustion Monitor, User's Guide, Document CRS 983-015, Rev 1., The Foxboro Company, Power Systems Engineering

Acknowledgements

SPEEDTRONIC Mark IV and DCM are trademarks of General Electric Company
UNIX is a trademark of UNIX System Laboratories, Inc.
I/A Series is a trademark of The Foxboro Company
Informix is a trademark of Relational Database Systems, Inc.

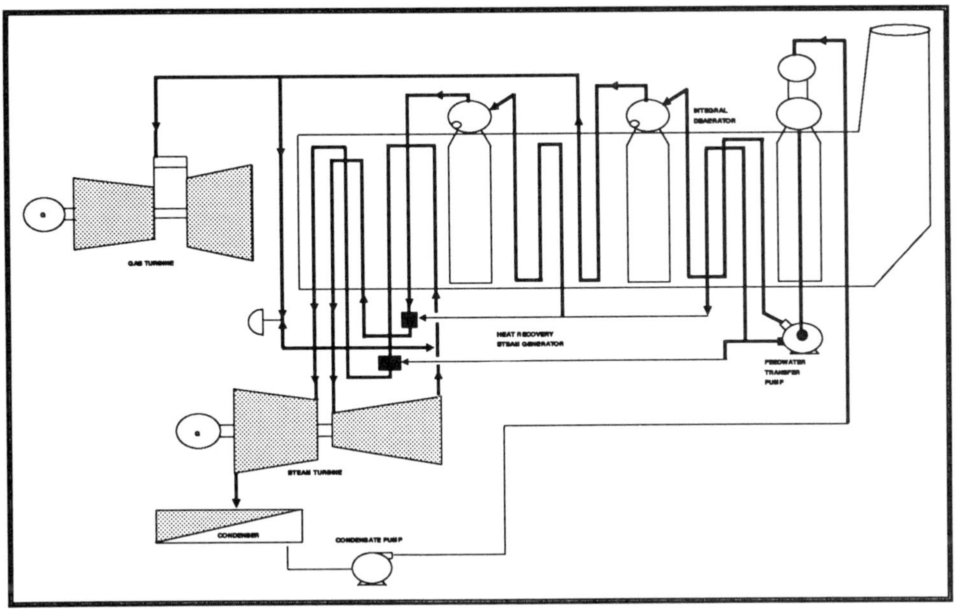

Figure 1 - Plant Equipment Configuration

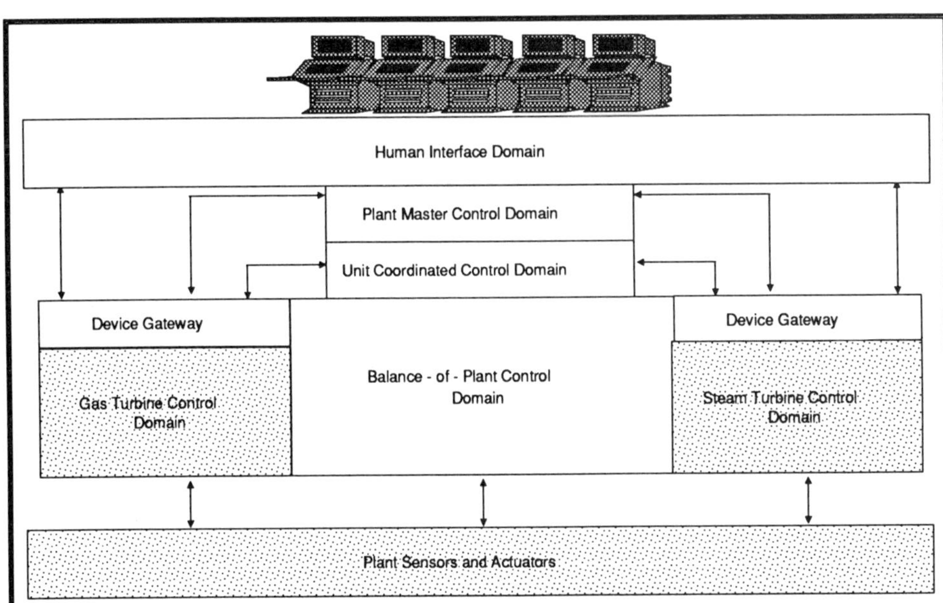

Figure 2 - Control Domain Diagram

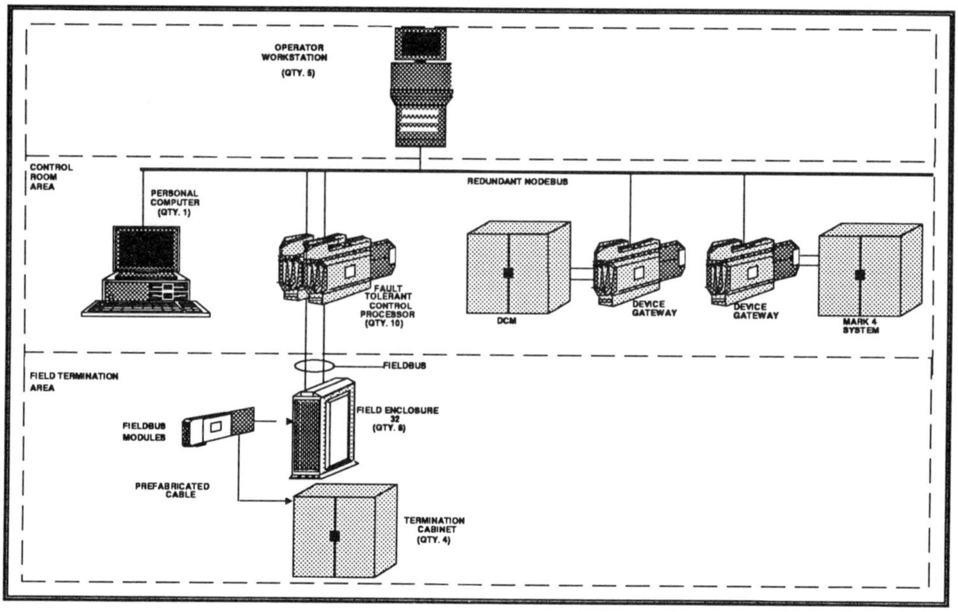

Figure 3 - Distributed Control System Architecture

Figures

Functions	Examples
DCS Programmer's Toolset	make, SCCS, vi, ed, ice, lint, adb, lex,yacc, prof, monitor
DCS Languages	C, Fortran 77, Pascal, ASM sh csh awk
DCS Library Functions	Process Data Access Historical Data Access Material Properties UNIX System Calls Workstation Support Terminal Support
DCS Hardware	Operator Workstation Personal Computer Controller Gateway Printer File Server

Figure 4 - Data Acquisition Environment Functions

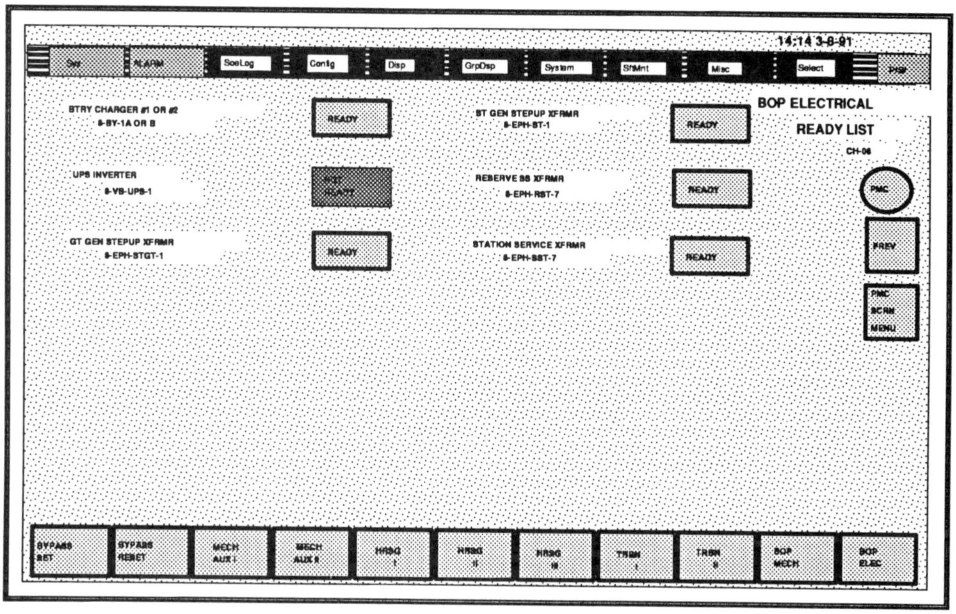

Figure 5 - Typical Plant Master Control CRT Display

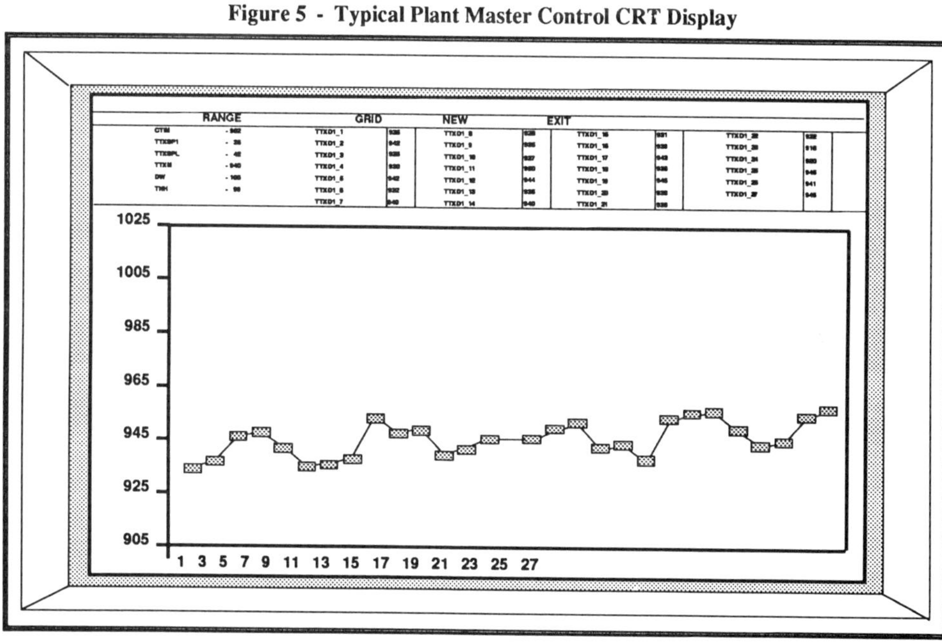

Figure 6 - Combustion Monitor Function 2D CRT Plot

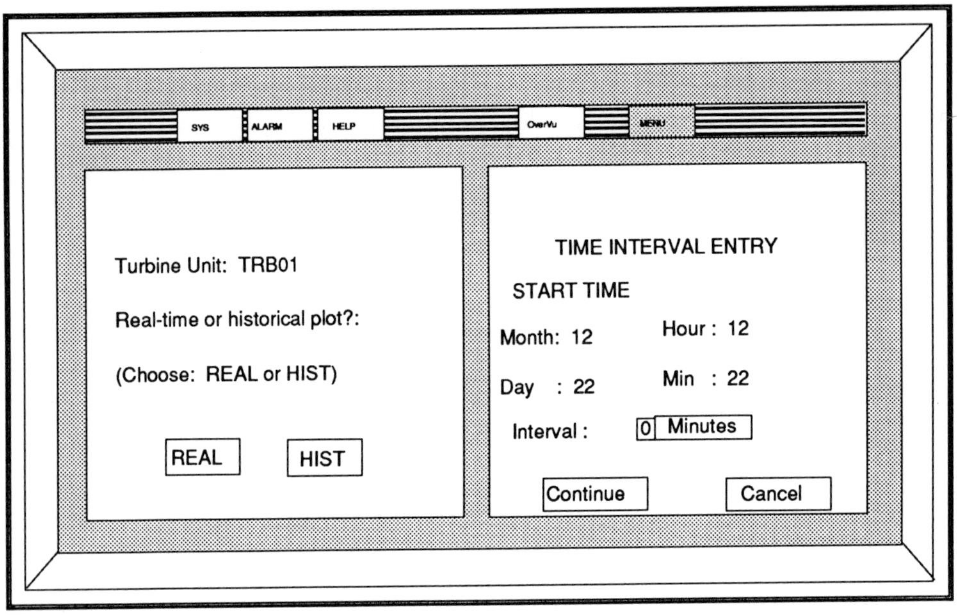

Figure 7 - Combustion Monitor Dialog Boxes

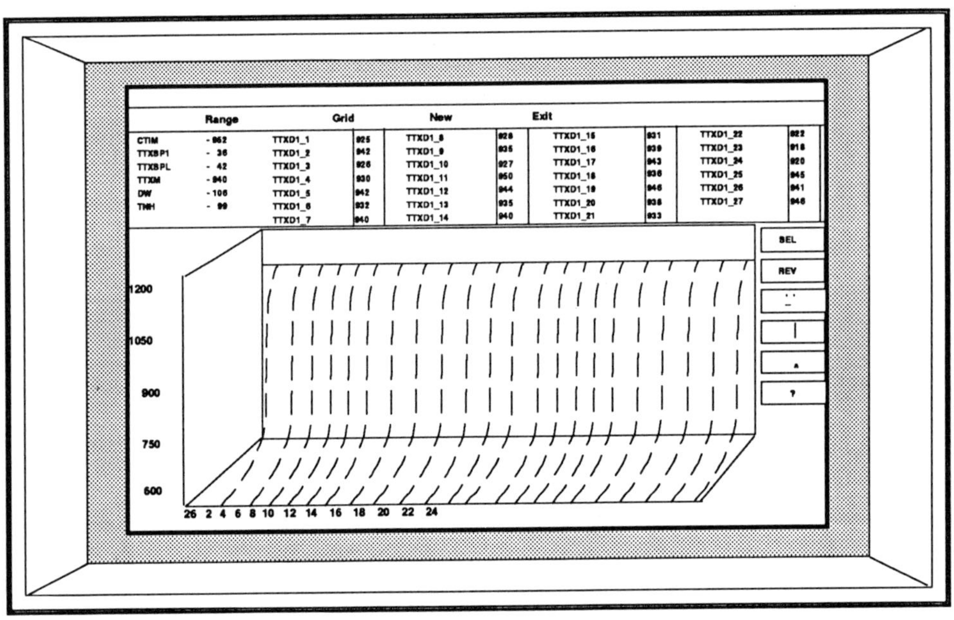

Figure 8 - Combustion Monitor 3D CRT Plot

©1991 - Paper #91-731
0074-056X/91/415-423/$0 + .50pp

AUTOMATED VERIFICATION OF SYSTEM CONFIGURATION*

W. H. Andrews, Jr.
S. P. Baker
A. V. Blalock
Development Engineers
Instrumentation and Controls Division
Oak Ridge National Laboratory**
Oak Ridge, Tennessee 37831-6285

ABSTRACT

Errors in field wiring can result in significant correction costs (if the errors are discovered prior to use), in erroneous or unusable data (if the errors are not discovered in time), or in serious accidents (if the errors corrupt critical data). Detailed field wiring checkout and rework are tedious and expensive, but they are essential steps in the quality assurance process for large, complex instrumentation and control systems.

A recent Oak Ridge National Laboratory (ORNL) development, the CONFiguration IDEnification System (CONFIDES) automates verification of field wiring. In CONFIDES, an identifier module is installed on or integrated into each component (e.g., sensor, actuator, cable, distribution panel) to be verified. Interrogator modules, controlled by a personal computer (PC), are installed at the connections of the field wiring to the inputs of the data acquisition and control system (DACS). Interrogator modules poll the components connected to each channel of the DACS and can determine the path taken by each channel's signal to or from the end device for that channel. The system will provide not only the identification (ID) code for the cables and patch panels in the path to a particular sensor or actuator but for individual cable conductor IDs as well. One version of the system uses existing signal wires for communications between CONFIDES modules. Another, more powerful version requires a single dedicated conductor in each cable. Both versions can operate with or without instrument power applied and neither interferes with the normal operation of the DACS. Identifier modules can provide a variety of information including status and calibration data.

*Research sponsored by the National Aeronautics and Space Administration, Kennedy Space Center, under interagency agreement 1926-B035-A1.

**Managed by Martin Marietta Energy Systems, Inc., for the U.S. Department of Energy under contract DE-AC05-84OR21400.

Key Words: Quality Assurance, QA, Automated, Configuration Identification, Field Wiring, Signal Wiring, Cabling, Module, Network, Smart Sensor, Smart Cable, Distributed Network, Verification, Identifier, Data Acquisition, Control System, Calibration, Termination Panel, CONFIDES.

INTRODUCTION

As we become increasingly reliant upon computers and other intelligent devices to control equipment and to automate processes, we also increase our dependance on electronic sensors and actuators and on interconnecting wiring. Fortunately, this trend is accompanied by a sharpened awareness of the issues of quality assurance, safety, and reliability.

A DACS comprises several components: the computer/controller, sensors and actuators, signal conditioning/multiplexing circuits, and signal transmission media. Various segments of the instrumentation, computer, and electronics industries are addressing the quality issues relating to these components. Redundancy, higher levels of embedded intelligence, computer-aided engineering tools, built-in test, and innovative design techniques are being applied to ensure quality and to increase reliability.

However, the problems of configuration control and field wiring integrity seem to have eluded such technically sophisticated approaches. DACS today can include thousands of sensors, actuators, intermediate termination points, and many cables consisting of miles of wires. Confidence in system configuration depends on the answers to three basic questions: What sensor or actuator is actually connected to which channel? What is the signal path from each sensor or actuator to its connection at the computer/controller? What is the integrity of that path? These questions are extremely important. Even though individual DACS components function reliably, the resulting data cannot be trusted unless interconnecting wiring is verifiably correct.

EXISTING TECHNOLOGY

The process control industry has begun to address the area of system configuration and integrity in a somewhat indirect manner. Multidrop communication links are available that simplify the interconnection of digital components and communications standards (e.g., RS-485 or BitBus) and promote interoperability by specifying hardware interfaces, data rates, electrical characteristics, and loading factors. Commercially available special-purpose transceivers assist in implementing many of these standards. All of these developments provide tools for attacking configuration control and wiring integrity problems, but they are insufficient alone. Automated intelligent supervision of these tools at the system level is needed before they can be brought to bear upon the configuration identification problem.

For several years, the industry has recognized the shortcomings of purely analog signal links to field devices. A new class of sensors with embedded intelligence is becoming available from process control equipment manufacturers. Most "smart sensors" still use traditional 4- to 20-mA analog current-loop signaling, but high-frequency digital signals can be superimposed for communication with hand-held programmers. This technique limits the use of smart sensors to low-bandwidth DACS. Because of the proprietary nature and incompatibility of the hardware and protocols in existing systems, smart sensors have yet to gain wide acceptance.

In response to these incompatibility problems, the Instrument Society of America SP-50 Committee is developing the FieldBus standard. This standard will specify signal transmission media, signal levels, and communication protocols for a multidrop serial digital data link for communication between field instruments. Committee members represent manufacturers, government, and end users. The issuance of the FieldBus standard has been delayed for several years while protocol issues are being resolved.

PREVIOUS WORK

During the 1970s, the Instrumentation and Controls Division of ORNL began to realize that an automated system for verifying the configuration of large research experiments could be extremely valuable. As the power of available computers increased, the size and complexity of the reactor safety experiments that they controlled grew and the number of the parameters to be monitored and controlled increased correspondingly. Systems with several thousand sensors and actuators were not uncommon, and configuration control became a major issue.

In the early 1980s, work began on sensor identifier technology, and after several approaches had been evaluated, the first demonstration system was built and tested in 1984. The Remote Sensor/Cable Identifier system, which received a 1987 IR-100 Award***, featured a single-board computer that interrogated passive "dumb" identifier devices via the sensor signal wiring. This system provides the identity code of the end device connected to each channel of a star-wired data acquisition system.

DESCRIPTION OF CONFIDES

CONFIDES expands the capabilities of previous developments by distributing intelligence to the remote identifier modules and is intended for retrofitting to existing DACSs or for integration into new DACSs.

CONFIDES has three basic components, as shown in Figure 1. A host computer and its resident software provide an operator interface and maintain the overall system database. Intelligent master modules provide the field wiring interface and intermediate storage for configuration data. Intelligent ID modules attached to (or integrated into) each cable, sensor, actuator, or other component to be identified respond to inquiries from their master modules by providing identification, status, and housekeeping data. CONFIDES operates with active or idle DACSs and does not interfere with normal DACS operation.

On power-up, a master module sends an **initialize** instruction down a dedicated data line to the first module in the serially connected chain of identifiers. This ID module responds by sending its unique ID code back to the host and placing itself in "transparent" mode so that subsequent commands from the master module are passed through. The next **initialize** command reaches the second ID module in the chain, which responds with its own code and then places itself in transparent mode. This process continues until all modules in the "daisy chain" have identified themselves. Thus, the master unit knows not only the ID codes for all modules connected to it but the order of their connection as well.

The master module assumes that all ID modules have been polled if a predetermined time increment passes with no response to the last **initialize** instruction. It then sends an **initialization complete** message to the ID modules. The ID modules then take themselves out of transparent mode and "listen" for messages from the master module.

***IR-100 Award formerly given by *Industrial Research* and now by *Research and Development*.

If an ID module receives a message addressed to the master module or to another ID module, it simply retransmits the message in the appropriate direction. If an ID module receives a message addressed to itself, it checks its own status or retrieves the required information from memory and responds accordingly.

When the master module has identified all ID modules connected to it, the master module determines which end device (sensor or actuator) is connected to each DACS input/output (I/O) channel. To do this, it excites the signal wires connected to each channel with a high-frequency pulse train and then interrogates the end-device ID modules (via the CONFIDES data line) to find which ones can detect the pulses on their signal lines.

Master modules can perform configuration updates at regular intervals to detect any changes that may occur. Upon instruction by the host computer, the master modules execute "configuration dumps" or perform other, more specialized tasks such as obtaining housekeeping information from an ID module, rechecking the status of a suspect ID module, or performing diagnostic tests.

Thus, the host computer can present system configuration information from the system database in several useful formats:

1. a graphical presentation of the system configuration or a more detailed description of a specified subsystem,
2. identification of the end device on any particular DACS I/O channel,
3. the path (including cable and junction box IDs) to any given end device, and
4. a list of all end devices whose data or control signals pass through a given cable.

ID modules can store other useful information such as in-service dates, calibration requirements, and descriptions of the type of component to which the module is attached. The end user will determine what "housekeeping" data to include. If calibration dates are included, for example, the user could instruct the system to poll all end devices and list all components that will require recalibration in the following month.

Each module contains a microcontroller, a communications interface, and power-conditioning circuitry. Each ID module has a failover switch to allow messages from other modules to pass through in the event of a local hardware or software failure (Figure 2a). Master modules and end-device modules have connections to selected DACS signal lines (Figure 2b). The serial data line is also used to distribute power, thus minimizing the number of cable conductors that must be dedicated to CONFIDES.

In the system under development, each master module can accommodate ≤30 ID modules. Usually, one master module will be required for each trunk cable (a multipair cable carrying signals from several sensors). In smaller systems, the host communicates with the master modules through standard serial ports. In large systems, the host is linked to the master modules through data concentrators over a commercially available local area network (LAN), as shown in Figure 3.

ID modules are designed around the Motorola 68HC05 high-speed complementary metal-oxide semiconductor (CMOS) microcontroller. This chip was selected as the CONFIDES module central processing unit (CPU) for its (1) low power consumption; (2) built-in serial communications interface; (3) large number of digital I/O lines; and (4) availability with internal PROM, EPROM, or EEPROM. Another feature of the chip, the high-speed serial peripheral interface, will be used to add "hot backup" redundancy to future versions of the CONFIDES modules. EEPROM chips are being used in the early development stages and will be used for applications in which field reconfiguration is desirable.

Modularity and maintainability are emphasized heavily in software development, which is using state transition techniques. ID module programming is in assembly language for speed and code compactness.

Design goals for the CONFIDES system include

1. module cost: less than that of the cable or end device to which the module is attached,
2. size: diameter equal to or less than a trunk cable connector so that cable ID modules can remain attached to the cables when they are pulled,
3. speed: 10 min to determine the configuration of a 6000-component system, and
4. reliability: CONFIDES will be transparent to the DACS and will not introduce new failure modes.

CONCLUSIONS

CONFIDES offers valuable features not now available and not promised by new industry standards currently under development. Its self-documenting autoconfiguration function will greatly reduce the manual labor required for cabling checkout, whether for troubleshooting or for quality assurance. Records generated by the system are useful for highly reliable archival data storage, for postincident analysis, and for tracking individual cable and component histories.

This technology should find use in many different industries–wherever there is a need to verify the integrity and configuration of a complex system of electrical devices, active or passive, "smart" or "dumb". Integration of certain CONFIDES features with the emerging FieldBus standard will be examined when it is issued.

FIGURE CAPTIONS

Figure 1 - Block Diagram of CONFIDES Modules and Host Personal Computer.

Figure 2a - Block Diagram of Cable Identification Module Showing Failover Switch.

Figure 2b - Block Diagram of Master Module Showing Connections to Data Acquisition and Control System.

Figure 3 - Block Diagram of Host Personal Computer Linked to CONFIDES Master Modules Through Data Concentrators over a Local Area Network, as Used in Large Systems.

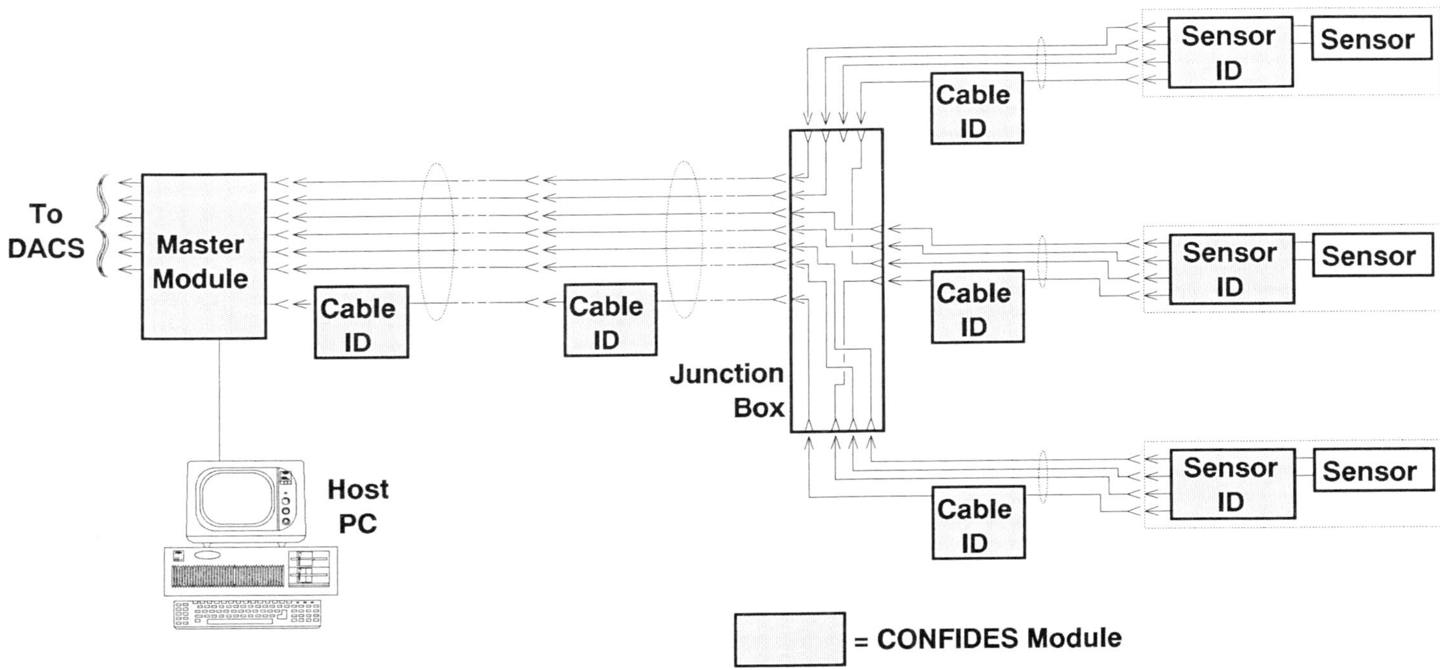

Figure 1: Block Diagram of Confides Modules and Host Personal Computer

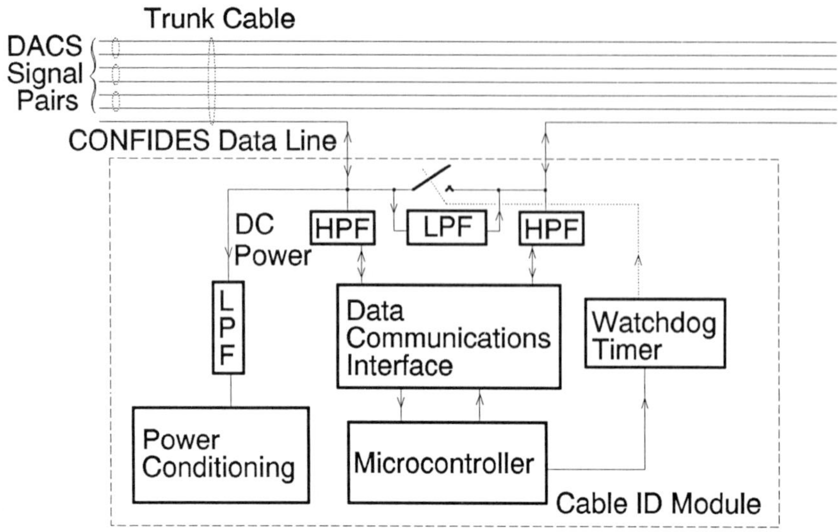

Figure 2a: Block Diagram of Cable Identification Module Showing Failover Switch

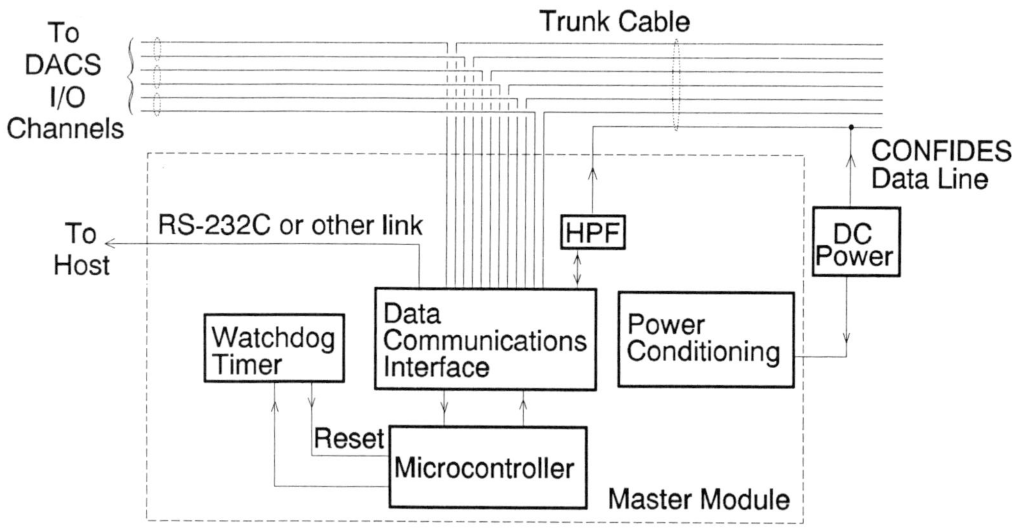

Figure 2b: Block Diagram of Master Module Showing Connections to Data Acquisition and Control System

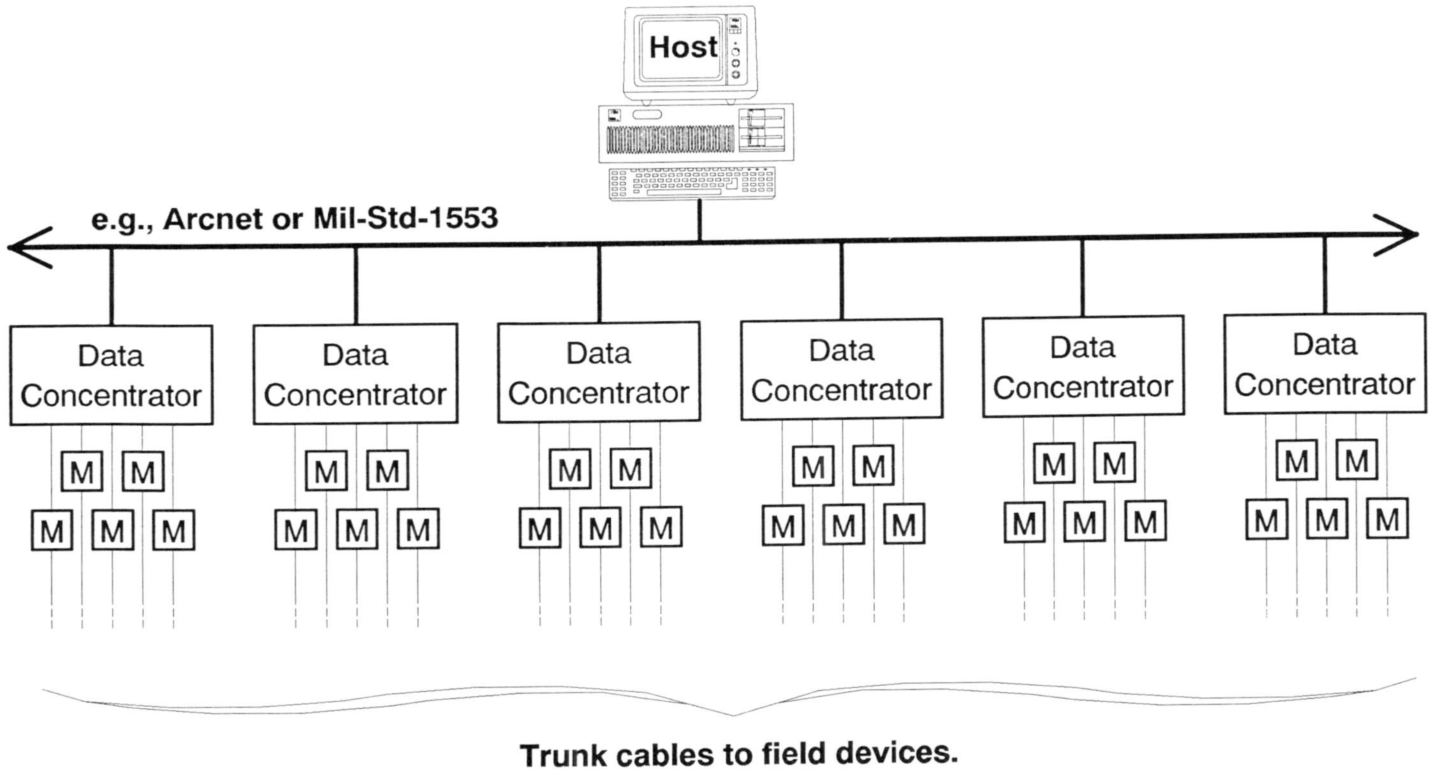

Figure 3: Block Diagram of Host Computer Linked to CONFIDES Master Modules Through Data Concentrators over a Local Area Network, as Used in Large Systems

0074-056X/91/425-450/$0 +.50pp

THE RELIABLE ON-LINE ANALYSIS OF BOILER CONDENSATE WATER

Eric F. Mooney

Tytronics Inc., 224 Calvary Street, P.O. Box 590,
Waltham, MA 02254-0590.

ABSTRACT

The direct measurement for the leakage of heat-exchange fluids, such as the Dowtherms, into boiler condensate water will be described using a simple fiber optics based photometer.

Making use of the fluidics capability of the FPA 800 series of process analyzers, the direct photometric measurement of copper, silica, phosphate and trace sulfate will be described. All these measurements are made using a fiber optics probe photometer.

A further adaption of the same FPA 800 analyzer permits the reliable on-line measurement of very low levels of hardness in boiler condensate water using a colorimetric titration method which, for maximum accuracy, utilizes measurement at two or more wavelengths.

INTRODUCTION

The care of boilers in the power industry and most industrial plant sites has often been neglected. This neglect is not on account of a lack of understanding of the requirement to maintain the quality of the recycled condensed water. Invariable it has been due to the inability to maintain analyzers in both a working and fully calibrated condition.

The difficulty in repairing or recalibrating some of these analyzers has required fully trained and highly skilled technicians and analysts. In consequence, there has been a tendency to take a grab sample when an upset condition is "suspect" and hope that no extensive damage has been done or corrosion started to occur within the boiler system.

Some simple, reliable continuous analyzers requiring the absolute minimum of maintenance and easily maintain when necessary by qualified but not necessarily highly skilled instrument technicians has been required by the industry.

DIRECT PHOTOMETRIC MEASUREMENTS.

The direct photometric measurement is an ideal solution as the minimum of a sample system is required. In addition, unless there are very large temperature swings in the process stream, no temperature control is required. The measurement is continuous and will immediately detect any break through of the component.

Cations such as Fe^{3+} and Cu^{2+} and polyatomic anions, such as PO_4^{3-} and SiO_3^{2-}, exhibit absorption bands in the electromagnetic radiation region. At the low concentrations of interest the absorption is too weak to provide reliable analytical data and hence are not amenable to direct photometric measurement.

One of the common group of materials used in the heat exchange systems are the Dowtherms [1]. The Dowtherms are eutectic mixtures of diphenyl and diphenyloxide derivatives and four mixtures are commonly used: Dowtherm J is probably the most common but Dowtherm A, Dowtherm G, Dowtherm Q, Dowtherm LF and Dowtherm HT all have specialized applications.

All these mixtures, however, show one common property in that, being aromatic compounds, all show very characteristic absorption spectra in the ultraviolet (UV) region and hence may be directly measured, see Figure One. The absorption is sufficiently strong that between 0 - 1 ppm of Dowtherm J, for example, may be measured for full scale of measurement of the analyzer; more commonly a range of 0 - 3 ppm is employed as the reasonable detectable limit

[1] Dowtherm is a registered trade mark of the Dow Chemical Co.

is 0.15 ppm. A 25 cm. pathlength cell is normally used for this measurement range; it should be remembered that pure water does not absorb in the UV region. Hence the use of UV spectroscopy, where applicable, is an ideal means of measuring components in water.

Other heat exchange fluids also exhibit UV absorption. Most of these other fluids do not exhibit such strong characteristic absorptions as that of the Dowtherm family. Hence, because of the high sensitivity obtained, the immediate detection of the on-set of leakage in the heat exchange system is easier when using the Dowtherms. Thus, with an appropriate detection system, the onset of leakage of Dowtherm into the system is immediately detected. This will be more readily appreciated when we compare Figure One with the spectrum, Figure Two, of Paratherm[2].

Another application of the use of photometric measurement is that of the direct measurement of the color of water. There are many color standards but one of the most useful for water analysis is that of the APHA (American Public Health Authority) scale (1) which is also the basis of the ASTM method (2). The APHA scale is based upon a well defined potassium chloroplatinate standard which

---

2 Paratherm is a product of the Paratherm Corporation.

essentially measures the degree of "yellowness" of the sample. The same chloroplatinate standard is used in the ISO standard (3) and are referred to as Hazen units (10 Hazen units = 10 APHA).

At an APHA color of 10 it is almost impossible to clearly see the "yellowness" of the sample while a process UV analyzer, properly set up, is able to measure 0 - 10 APHA for full scale with a repeatability of better than +/- 0.2 units and more usually better than +/- 0.1 unit, as shown in Figure Three.

Although the color of boiler condensate water is frequently not routinely measured, it should be recognized that the onset of an increase in color may give warning of a deterioration in the quality of the water. This could well be related to undesirable electrolytic reactions occurring in the condenser pipes.

INSTALLATION OF THE ANALYZER

The majority of the conventional process UV analyzers are rather large as a stable optical bench is required. Using a 50 cm. path-length cell (which is required for the measurement of Dowtherm), the optical bench may be 4 - 5 ft. long. This often presents a significant problem for the mounting of a unit in an already crowded work area.

The advent of the use of fiber optics has certainly changed this;

the photonics unit is 22" x 16" x 6" and the cell may be built in many configurations - from pipe-line cells (Figure Four) with slip flanges, to cylindrical cells with stubs for mounting parallel to a larger pipe-line (Figure Five), to the more conventional cells having 1/4" compression tube fittings. The latter are normally used when a sample conditioning system is required. The fiber optics bundles conduct the light from and back to the photonics unit. The fiber optics bundle may be supported in conventional cable trays or with flexible conduit; all providing a much simpler installation than for the conventional analyzers.

COLORIMETRIC ANALYSIS.

The trace levels of cations and anions required to be measured in water may be measured using well established colorimetric methods. These methods all require the addition of a reagent, or reagents, which reacts specifically with the ions in question to form a characteristic color. These methods have been developed in the laboratory and are used routinely for such analysis.

The analysis is, however, time consuming, requires a reasonable level of skill for reliable results, is costly in manpower and, more importantly, the results are often obtained too late to take immediate remedial corrective actions.

Using the simple photonics unit of the FPA 1000/2000 series described above with the well established fluidics handling system of the proven on-line process titrimeter, the FPA 300, the on-line process colorimeter, the FPA 800, and the colorimetric titrator, the FPA 400, have been developed.

The sample is circulated in a fast sample loop and, when designated (the frequency of analysis is menu selected), the control sample valve, Figure Six, opens and sample flows into the cell. The sequence of filling and washing is again menu selected to ensure that a representative and a reproducible volume of sample is collected and captured (4). One or two (depending upon the number of reagents to be added) FMI dosing pumps then add precise volumes of the reagent(s) to the stirred sample and the color is allowed to develop.

The absorbance is measured and the concentration of the component is displayed. The relative intensity at two wavelengths is measured. One wavelength is centered at the absorbance maximum of the colored complex being measured while the second wavelength is centered where there is normally no absorbance. For example, for the chromium(VI)-diphenylcarbazide complex a distinct maximum is observed at 540 nm, Figure Seven, which is selected for the measurement wavelength and a convenient reference wavelength would be at 750 nm, well outside the absorption band. The calibration curve for 0 to 0.5 ppm. Cr(VI) is shown in Figure Eight.

The use of the two wavelengths means that the Beer's-Lambert law may be used without the necessity to re-zero. The Beer's-Lambert law relates the absorbance to the ratio of incident to transmitted light. The incident radiation maybe taken as proportional to the output of the reference detector (since no absorbance should occur at this wavelength) and the transmitted radiation is given directly by the output of the measure detector.

$$\text{Thus, } A = \log_{10} I_o / I = \log_{10} V_r / V_m$$

where, A = absorbance and V_r and V_m are the output signals from the two detectors (Reference and measure).

The absorbance, A, is related to the concentration of the complex being measured by the expression:-

$$A = \frac{a.b.c.d.10}{MW}$$

where, a = molar absorbance, b = pathlength,

c = concentration, d = density of sample,

MW = mol.wt. of colored complex.

for a specific colorimetric measurement a, b, d and MW are constant, hence,

$$A = \underline{k}.c.$$

The factor, <u>k</u>, is determined by preparing calibration curve and for most process applications, where the absorbance, A, is kept below 0.4 or 0.5, the calibration curve is linear and measurement of a zero and a span value is usually more than adequate for verifying the calibration of the analyzer.

For practical purposes the standard pump (Figure Six) is actuated periodically (time interval again menu selectable) and calibration, certainly for compliance purposes, checked every shift or as infrequently as once a week or monthly. For the more sensitive measurements, eg. 0 to 250 ppb. of iron, the calibration should be checked on changing the reagents as the age of these may effect the color development. It must also be borne in mind that some of the calibration solutions are not particularly stable and often a manual calibration check using more stable optical calibration filters is sufficient.

INSTRUMENT MAINTENANCE.

With the increasing number of dedicated process analyzers being put on-line the load on trained and qualified instrument technicians is for ever increasing. Such real-time analysis in imperative for accurate, reliable and sensible control. The distributed control systems being used for plant optimization are absolutely dependent upon having the real-time reliable and accurate analytical data.

The process analyzers must, therefore, be reliable and require the minimum of maintenance. The analyzers should have built-in diagnostics to monitor various parameters, eg. in the FPA 400 and 800, lamp failure, detector failure or insufficient energy being returned to the detector, power rails out of range or complete failure, detector block temperature out of range, etc.

With both the FPA 400 and 800 process analyzers, as with any process analyzer using wet chemistry, the reagents must be topped up periodically - the frequency is dependent upon the time interval between the analysis but typically the reagents will need replenishing once every one or two weeks. No propriety reagents are used in either the FPA 400 or 800 and the solutions may be prepared on site or there are companies which supply the solutions for these standard measurements. It is estimated that, apart from the replenishing of reagent solutions, the average time spent on maintenance is 1 to 2 hours per quarter. The fluidics of the FPA 400 and 800 are essentially identical to the FPA 300 process potentiometric titrator and this has been available for some years and the maintenance is thus well documented.

COLORIMETRIC MEASUREMENTS OF IMPORTANCE ON THE POWER PLANT.

Copper.

Some recent reports (5) have drawn attention to the importance of copper as a factor affecting the corrosion in boilers and the factors which affect the increase of copper concentration. The current guidelines of the American Boiler Manufacturers Association and ASME (6) states that the copper content should be less than 15 ppb in boilers operating up to 1000 psig and less in boilers operating at higher pressures.

The range of 0 to 100 ppb of copper may be conveniently measured using the bathocuproine reagent. There are currently no installations measuring at this low level. The range of 0 to 2 ppm. is measured on effluent streams from VCM plants. The FPA 801 has the capability of capturing a larger volume, hence more micrograms of copper are available (in the captured sample) for the color development.

Copper is present in many locations where it is in contact with the boiler feedwater, steam or condensate as indicated in Table One.

On a steam generation plant it might not be necessary to measure the copper concentration as regularly as on a waste water stream and once an hour or even less frequently may be sufficient. This provides a quicker warning of the on-set of corrosion without having to remember to take a sample for analysis to the laboratory. The cost of the analysis may not provide a fast return on investment. It does lighten the burden on the laboratory to enable the more routine work to be done which cannot be conveniently made on-line, such as the analysis of solid fuels.

TABLE One. Common Locations of Copper Alloys in Boiler Feedwater, Steam and Condensate Cycle (reproduced by permission of Betz Laboratories, Inc.)

Pulp and Paper Industry	Hydrocarbon Processing Indust.	Utility/Cogen.& Gen. Manufact.
Pulp and Paper Dryers (Ross, Flakt, Yankee, etc.)	Feedwater Heat Exchangers	Copper Jacketed Vessels
Paper Machine Dryer Coils	Condensate/Feed-waters Pump Impellers	Turbine Condensers
Condensate/Feedwater Pump Impellers		
Paper Machine Pocket Ventilation	Condensers	Feedwater Heat Exchangers
Sweetwater Condensers	Refrigeration Units	Pump Impellers
Turbine Condensers	Sweetwater Condensers	
Feedwater Heat Exchangers		

The pay-back is, however, very quick when the on set of corrosion is detected and saves a very expensive breakdown and loss of plant production.

Phosphate and Silica.

Some attention has been paid to the copper analysis as this is one element which is not routinely measured on line in power plants.

The measurement of phosphate and silica are also readily made on-line using well established colorimetric procedures. Phosphates are measured by making the molybdophosphate complex which is then reduced with stannous chloride to develop the intensely colored molybdenum blue color. This method is better than using the simple yellow vanadomolybdophosphate complex as interference from silica is avoided. Silica is similarly measured by the formation of the complex heteropoly acid. Ammonium molybdate reacts with silica, and any phosphates present, to form the corresponding molybdosilicic and molybdophosphoric acids. Oxalic acid is also added to destroy the molybdophosphoric acid and the yellow color of the remaining molybdosilicic acid is a direct measure of the silica content of the water. Both methods are capable of measuring the normal concentrations in the range of 0 to 30 ppm (0 to 30 mg/l).

Iron.

Iron may be measured in the range of 0 to 4 ppm and higher by using the 1,10-phenanthroline reagent. The lower levels, eg. of 0 to 250 ppb, are measured using the ferrozine (7) reagent.

Other Cations and Anions.

There are many reagents which give specific color reactions with a whole range of ions, which have been described in the literature over the years; thus many different analyses are possible using the same basic type of analyzer. Note: each FPA 800 series is set up to do one specific analysis only as in almost every case specific reagents and definite conditions are required for the color development. The maintenance is, however, greatly assisted as the only variation in the hardware is the selection of the filters used for the selection of the measurement and reference wavelengths. The variations in the reagents is but part of the standard chemistry being used.

Hardness.

The total hardness in the range of 0 to 2 ppm or higher, may be measured using the complexometric titration with EDTA (ethylene-diamine tetraacetic acid disodium salt) and an appropriate indicator, eg. Erichrome Black T (EBT). The analyzer, the FPA 400, is very similar to the FPA 800 shown in Figure Six, except only one FMI pump is used for the titrant. A second peristaltic pump is used to add the appropriate buffer for the measurement of either calcium hardness or total hardness. The titration is made differentially to detect the inflection where the Ca/Mg-EBT complex no longer exists and the EDTA-EBT complex just appears. This latter complex absorbs at 650 nm and using the 810 nm wavelength as a reference, where neither complex exists, the presence of excess EDTA is immediately detected.

PERFORMANCE.

The majority of our studies have been related to the waste water analysis. Such data is often of a sensitive nature and may not be readily published. The laboratory work on synthetic samples has demonstrated both the linearity and reproducibility. One case of copper analysis may be used to illustrate the stability and reproducibility.

Prior to the installation of one trace copper analyzer in the field a large sample was collected and run in the laboratory overnight. The sample was circulated using a pump and the analysis was performed automatically and unattended once every 30 minutes. The results are shown in Table Two which clearly demonstrate the reproducibility obtained. This particular analyzer has been operating for over six months. No maintenance, other than the replenishing of the reagents, has been carried out.

THE FUTURE.

Many of the colorimetric methods and the titration procedures are very old and one has to go back some decades to look at the original work. It is perhaps unfortunate that the modern day instrumentation such as AA, ICP, etc., none of which are ideally suited for on-line analysis, has accelerated the loss of interest in extending the development these proven classical methods of analysis.

At the present time the majority of papers on colorimetric analysis are being published by the Peoples Republic of China. A whole range of new reagents, having greater sensitivity and greater specificity, are being developed by these Chinese workers (8) for use in colorimetric analysis. These will enable a wide range of metals and non-metallic ions to be measured which had previously not been easy or impossible to determine specifically. Thus, one might predict that it is becoming possible to measure almost every metallic and non-metallic cations and also the simple and polyatomic anions using colorimetric or titrimetric procedures.

CONCLUSION.

The use of the photometric measurement and microprocessor controlled system, enabling routine on-line analysis to be made both simply and with the minimum of maintenance, together with the use of well established and proven chemistry provides for the simple and reliable on-line analytical measurements required by the power industry.

As briefly demonstrated the industry now has a range of on-line analyzers using common parts to measure the components of interest in the industry, either directly, colorimetrically or titrimetrically, depending upon the component to be measured.
This both simplifies the maintenance and reduces the spare parts required and thus ensuring more reliable and trouble free operation of the plant.

REFERENCES.

1. American Public Health Association.

2. American Standard Testing Materials, ASTM, D-365.

3. ISO Standard, UDC 66:535.661.4.

4. M. Platt, US Patent Nos. 4,910,151 & 4,911,191.

5. Paul Burgmayer, Technical paper 309, CORROSION/89, New Orleans, LA, April 1989.

6. Cited in Capability Profile 005, Betz Laboratories, Inc., publication 1988.

7. Stookey, L.L., Anal. Chem., **42**, 779 (1970).

8. Wang S., et al., Fenxi Huaxue, **15**(10), 920 (1987) and many subsequent papers.

TABLE TWO.

An overnight run on the effluent from a plant, copper in ppm levels.

------ AUTO ANALYSIS ------

3:13 PM 12-19-90

SAMPLE=	0.84	@	15:15	19-Dec-90
SAMPLE=	0.89	@	16:24	19-Dec-90
SAMPLE=	0.86	@	16:54	19-Dec-90
SAMPLE=	0.91	@	17:24	19-Dec-90
SAMPLE=	0.89	@	17:54	19-Dec-90
SAMPLE=	0.85	@	18:24	19-Dec-90
SAMPLE=	0.88	@	18:54	19-Dec-90
SAMPLE=	0.88	@	19:24	19-Dec-90
SAMPLE=	0.86	@	19:54	19-Dec-90
SAMPLE=	0.84	@	20:24	19-Dec-90
SAMPLE=	0.88	@	20:54	19-Dec-90
SAMPLE=	0.86	@	21:24	19-Dec-90
SAMPLE=	0.90	@	21:54	19-Dec-90
SAMPLE=	0.87	@	22:24	19-Dec-90
SAMPLE=	0.81	@	22:54	19-Dec-90
SAMPLE=	0.83	@	23:24	19-Dec-90
SAMPLE=	0.85	@	23:54	19-Dec-90
SAMPLE=	0.85	@	0.24	20-Dec-90
SAMPLE=	0.88	@	0.54	20-Dec-90
SAMPLE=	0.86	@	1.24	20-Dec-90
SAMPLE=	0.86	@	1.54	20-Dec-90
SAMPLE=	0.86	@	2.24	20-Dec-90
SAMPLE=	0.87	@	2.54	20-Dec-90
SAMPLE=	0.84	@	3.24	20-Dec-90
SAMPLE=	0.87	@	3.54	20-Dec-90
SAMPLE=	0.87	@	4.24	20-Dec-90
SAMPLE=	0.82	@	4.54	20-Dec-90
SAMPLE=	0.87	@	5.24	20-Dec-90
SAMPLE=	0.85	@	5.54	20-Dec-90
SAMPLE=	0.89	@	6.24	20-Dec-90
SAMPLE=	0.90	@	6.54	20-Dec-90
SAMPLE=	0.90	@	7:24	20-Dec-90

Mean = 0.87 ppm

Std.Dev. 0.024

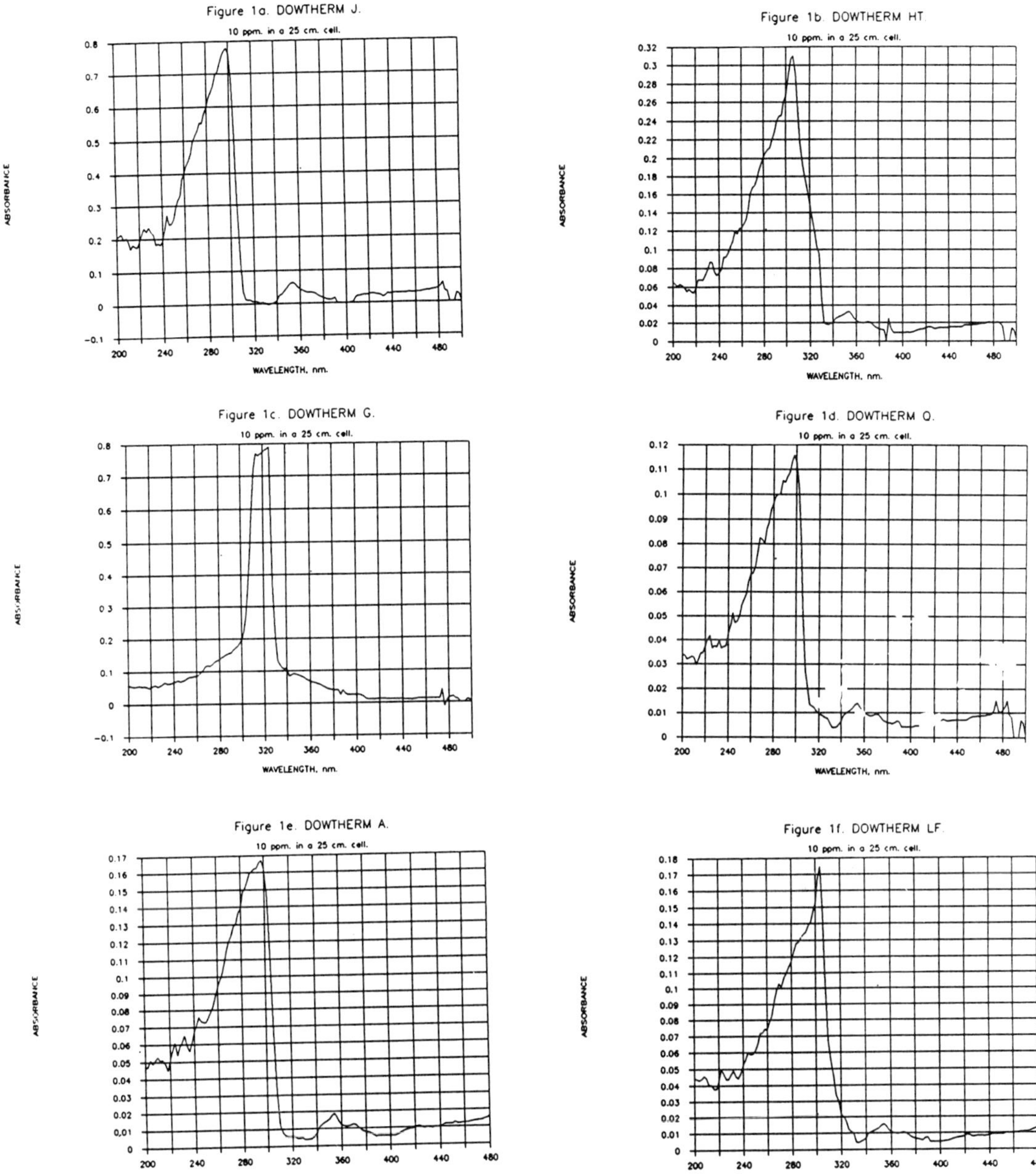

Figure One. UV spectra of Dowtherm Products.

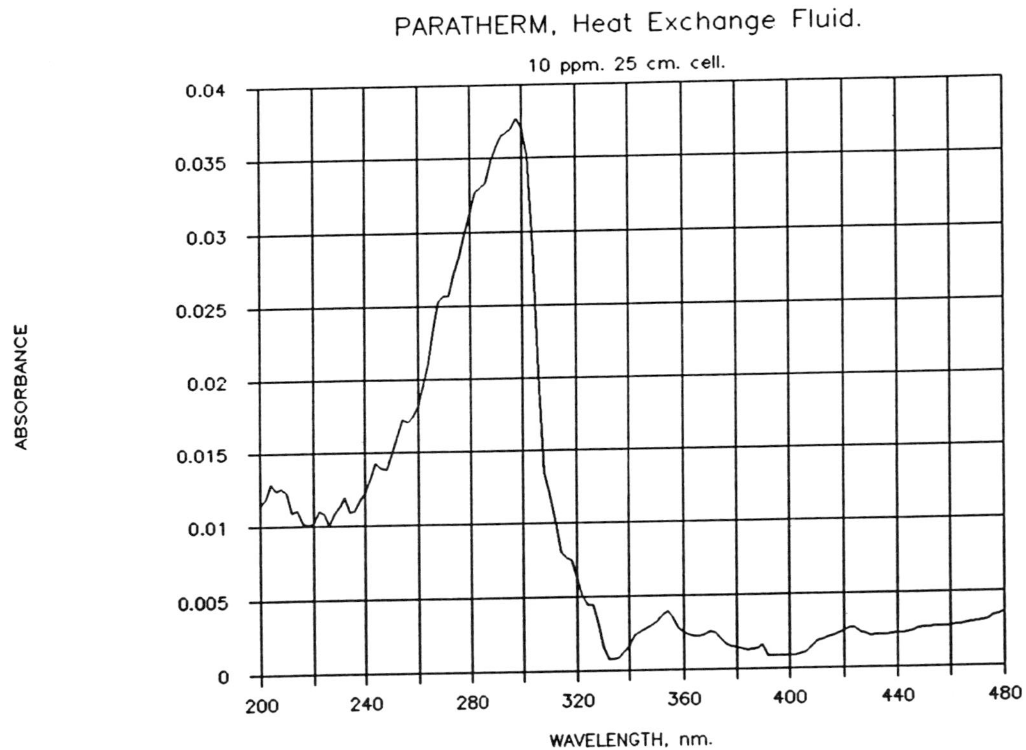

Figure Two. UV spectrum of Paratherm.

Figure Three.

Calibration curve 0 to 10 APHA.

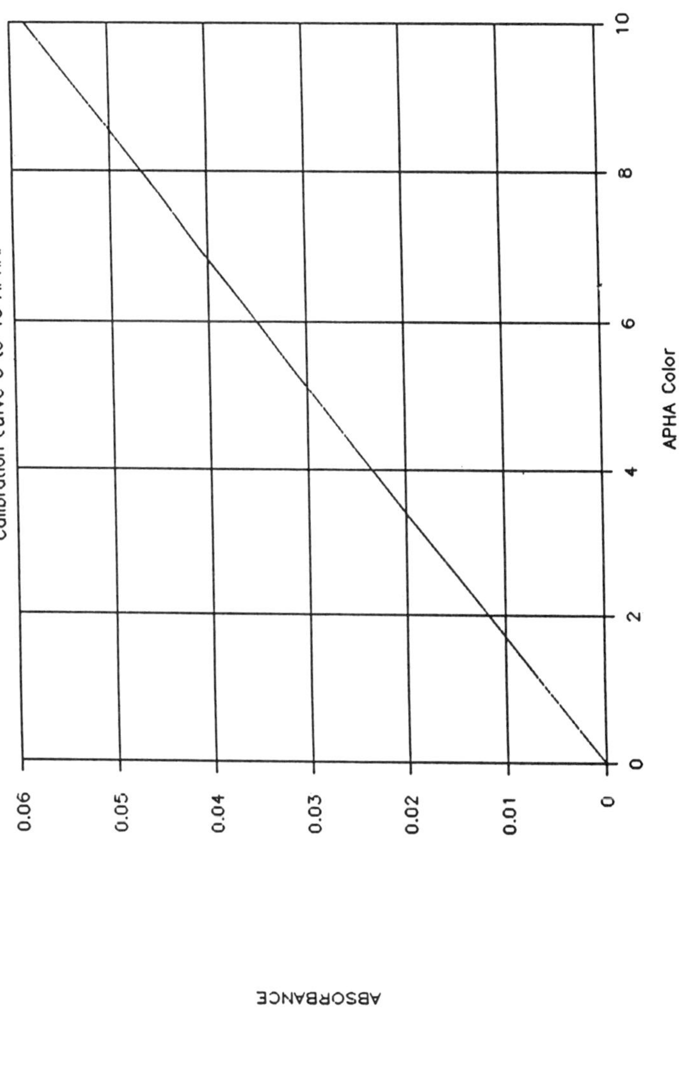

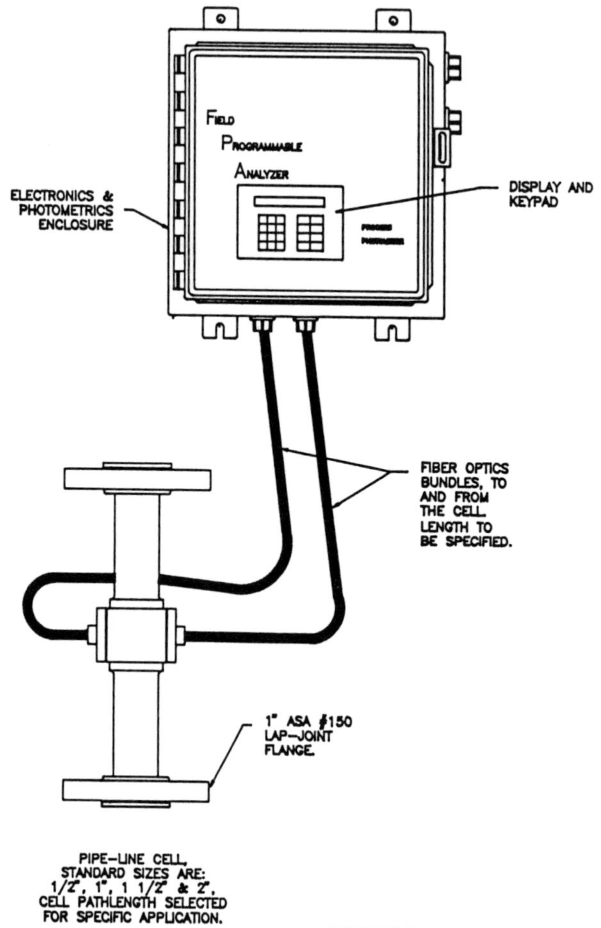

FIGURE FOUR.
PROCESS UV ANALYZER
WITH A PIPE-LINE CELL.

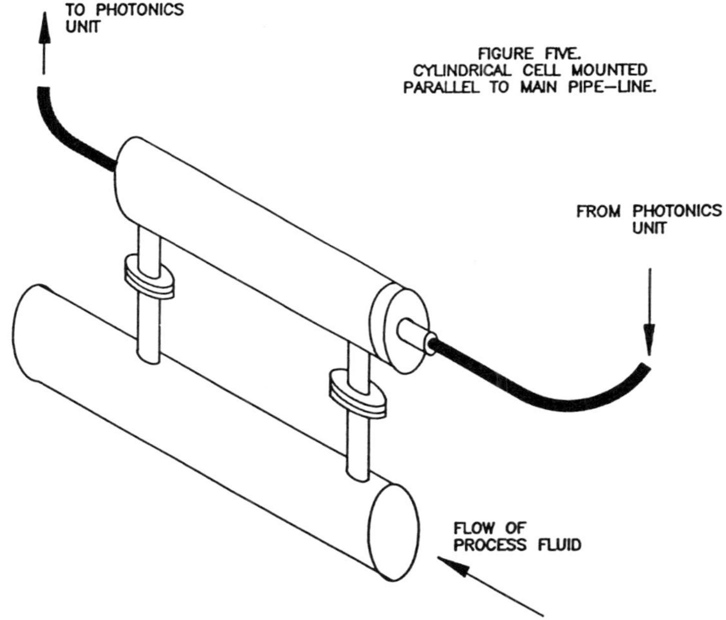

FIGURE FIVE.
CYLINDRICAL CELL MOUNTED PARALLEL TO MAIN PIPE-LINE.

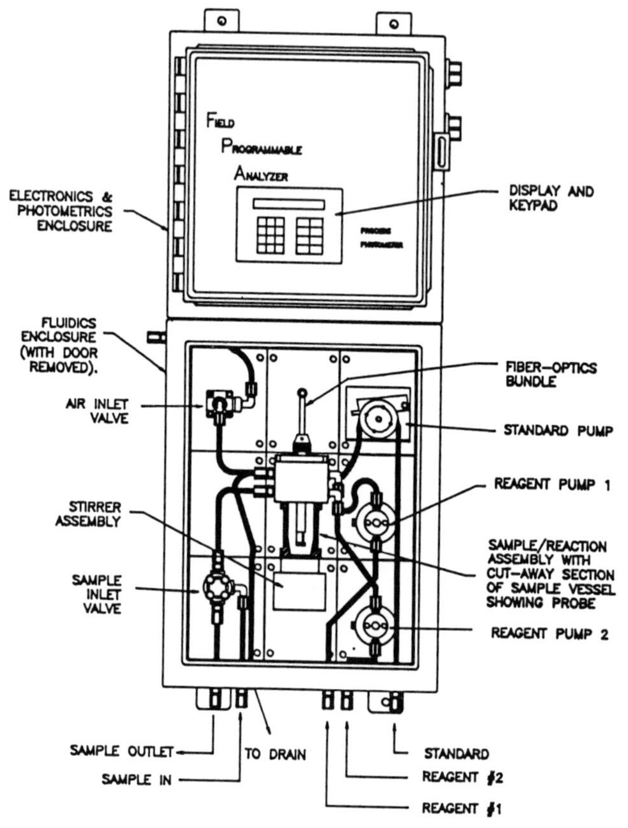

FIGURE SIX.
THE FPA 800
PROCESS COLORIMETER.

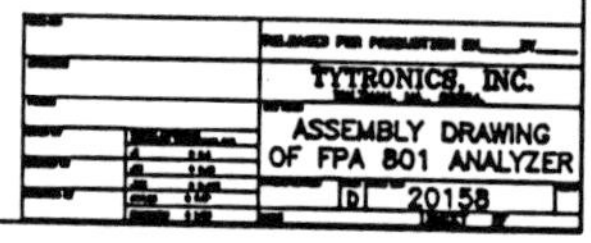

Figure Seven.

Spectra of Cr(VI)–diphenylcarbazide.

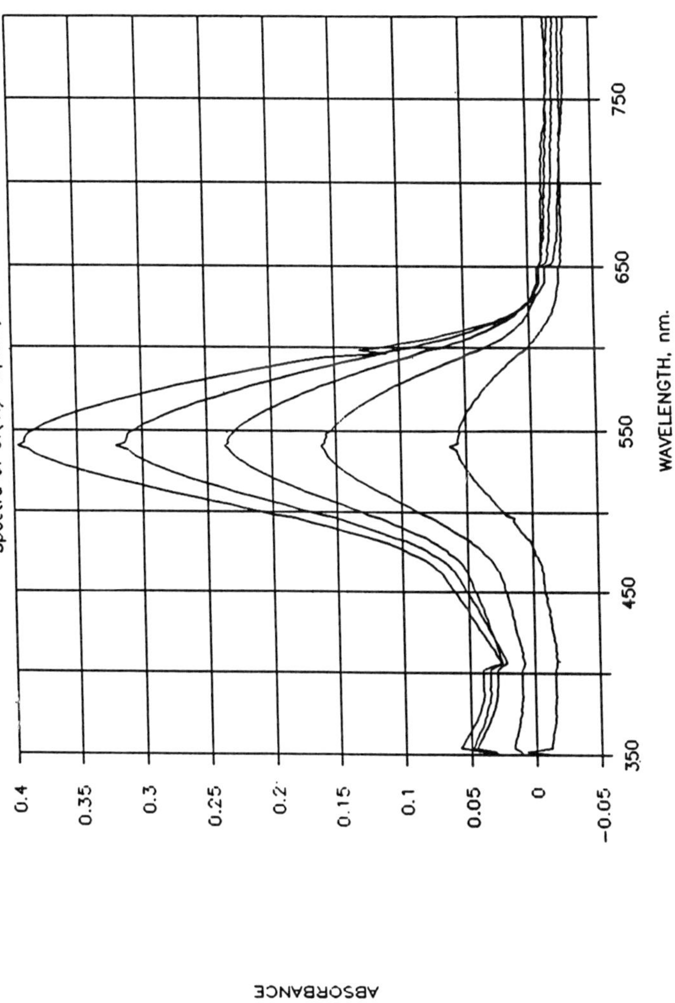

Figure Eight.

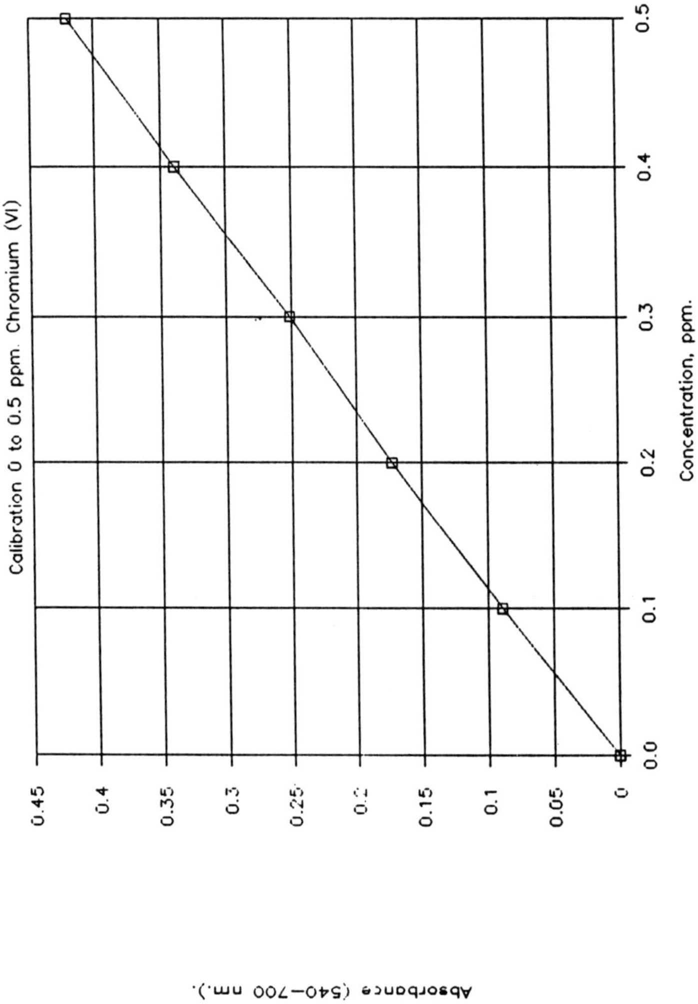

©1991 - Paper #91-733
0074-056X/91/451-459/$0 + .50pp

ADVANCES IN HIGH PURITY WATER MONITORING IN THE POWER INDUSTRY

David M. Gray
Leeds & Northrup
A Unit of General Signal

ABSTRACT

It is estimated that about fifty percent of unplanned power plant outages are caused by corrosion.(1) In fact, corrosion can be a determining factor in the life of a plant. A major preventive maintenance tool in combating corrosion is careful steam cycle chemistry control within EPRI guidelines. But the measurement of such basic parameters as specific and cation conductivity, pH, sodium ion and dissolved oxygen have been fraught with accuracy and reliability problems. Provided here is a tutorial discussion of the recent progress made in on-line analytical instrumentation to solve these problems. Microprocessor capabilities for temperature compensation and calibration as well as new measuring techniques and sensor developments have brought a higher level of performance, often warranting upgrade of monitoring instrumentation.

Key words: High purity water analysis, heat cycle chemistry monitoring, on-line analysis.

INSTRUMENT UPGRADE JUSTIFICATION

As the business of power generation and distribution becomes more competitive, preservation of generation capital equipment over a long operating life receives higher management priority. Long term, and in many cases, short term economics point in the direction of improved on-line measurement and control of cycle chemistry. The EPRI Cycle Chemistry Guidelines open with the statement that: "Any costs...should be evaluated in terms of the value of potential benefits, such as increased unit availability and reduced fuel, replacement power, maintenance, and component replacement costs. A high-level management commitment can provide the support necessary to realize these benefits."(1) In most plants, closer cycle chemistry monitoring and control with current analyzer technology can yield benefits in every area cited.

Total costs of all corrosion in the utility industry have been estimated at $3.5 billion annually, with a very substantial share on the steam cycle side. Steam generators or boilers, turbines, piping and condensers all suffer scaling and corrosion with unknown and/or uncontrolled excursions outside the optimum conditions. Neglecting to keep up with state-of-the-art monitoring instrumentation to protect these expensive components is indeed penny wise and pound foolish thinking. Described here are the reasons why it is more sensible than ever to upgrade on-line instrumentation.

At the same time economic pressures for life extension have mounted and understanding of the mechanisms of corrosion and scaling have matured, instrumentation has dramatically improved. Especially in the area of temperature compensation, measurement accuracy has been boosted very significantly. In some installations where expensive but unreliable sample chillers fail to deliver samples to the sensors at the required 25C, it is clearly more economical to replace the analyzers with current temperature compensation technology than to revamp and maintain the chiller system. New sampling system designs depend on this highly accurate instrument compensation in place of secondary cooling to achieve the same result.

A further advantage of some microprocessor-based instruments is a wide range capability and inclusion of all field-selectable software algorithms in a single model. Spare parts inventory can be dramatically simplified and reduced. Maintenance and troubleshooting time can be reduced with off-line diagnostics to aid system installation and checkout. On-line self-diagnostics assist greatly in locating faults in the sensors and analyzers.

CONDUCTIVITY TEMPERATURE COMPENSATION

Conductivity is the most reliable, simplest, and lowest cost of the pure water measurements. It is consistently on-line and therefore is the first line of defense in detecting ionic contamination. However, very sizeable temperature compensation errors with older instrumentation have prevented a full realization of its potential.

Conductivity is a direct measure of contaminant concentration but it is also directly affected by temperature. This is the reason water chemistry specifications require that conductivity values be referenced to 25°C. The temperature coefficient of most conventional samples is on the order of 2% of conductivity reading per degree C. This is because ionic contaminants become more mobile--carry a charge faster and conduct better--at higher temperatures. In pure water samples, the few ions present are mostly due to the ionization of water itself into hydrogen and hydroxide ions. The conductivity is low but the temperature coefficient is increased. The hydrogen and hydroxide ions are more mobile and they also increase in concentration due to the greater ionization of water at higher temperature. The combined effects of ion mobility and ionization in pure water yields a temperature coefficient near 5% per degree C. The conductivity versus temperature relationship is illustrated by the solid lines in Figure 1 for this transition range from pure water to 20 µS/cm.

Compensation for this variable coefficient in the past was attempted by various means, all involving compromises. As a result, many power plants required sample temperature control to 25°C. However, this control has been quite expensive and not always precise or reliable. With a coefficient as high as 5%/°C, temperature control even to ±1°C leaves considerable uncertainty, when it is working, and much more when it fails. Thus temperature compensation takes on a real importance.

General Electric, for their boiling water nuclear plants, required close sample temperature control for on-line measurement and developed an equation for compensating off-line measurements. This equation matches the curves in Figure 1 over a moderate temperature range around 25°C. It uses uncompensated conductivity and temperature values to calculate conductivity referenced to 25°C. With the advent of microprocessor technology it was natural to build in the G.E.-type equation to provide on-line compensation.

In fact, by that time there were significant improvements made on the G.E. equation to enhance its accuracy over the full range given in Figure 1. Today, quality conductivity instruments designed for high purity measurements implement this technique.(2)

However, a major limitation is that the G.E.-type equation describes the properties of boiling water reactor samples which have no treatment chemicals added. That is, it assumes that any contaminants are neutral salts with the temperature properties of sodium chloride. While this is a good assumption for BWR nuclear plant samples, it is definitely not correct for other kinds of power plants.

Fossil fuel and pressurized water nuclear plants invariably use treatment chemicals such as ammonia, morpholine, etc. These raise the pH to reduce iron corrosion and result in an increased hydroxide ion concentration. The temperature properties of a basic or acidic sample are considerably different from those of sodium chloride. This is because the extra hydroxide and hydrogen ions contributed by the base or acid directly influence the ionization of water. This influence is seen over temperature in Figure 1 which provides curves for the bases ammonia and morpholine as well as for sodium chloride. Morpholine exhibits different temperature properties from ammonia because its degree of ionization changes considerably with temperature whereas ammonia ionization does not. Thus two more algorithms in addition to the G.E. equation are needed to provide compensation for treated water. These have been included in microprocessor-based instruments which require only a switch setting or software command to select which temperature compensation algroithm is to be used.(3)

CATION CONDUCTIVITY MEASUREMENT/COMPENSATION

Conductivity is a non-specific measurement and responds to the treatment chemicals as well as the corrosive contaminants. Since the treatment chemicals are normally at much higher concentrations than contaminants, a direct conductivity measurement (often termed specific conductivity) indicates primarily the level of treatment, not the level of contamination. The cation conductivity technique passes the sample through a small cation exchange cartridge to remove any ammonia or amines. With the treatment materials removed, the conductivity measurement becomes much more sensitive to the corrosive contaminants. In addition, any neutral salt contaminants are converted by the cation exchange cartridge into acids which have higher conductivity than their respective salts. Thus cation conductivity measurement enhances sensitivity to contaminants by two mechanisms: removal of treatment chemicals and conversion of corrosive salts to acids.

As noted previously, trace acids have temperature properties different than salts. It has been found that the ionization of most acids typical of cation conductivity samples does not change significantly with temperature. In fact, the temperature properties are very close to those of ammonia such that the same temperature compensation algorithm may be applicable and the dotted curves in Figure 1 represent acids or ammonia. Again, field selection of the temperature compensation algorithm to match the application enables a single model of analyzer and cell to make all heat cycle measurements.

CONDUCTIVITY LEADWIRE ERRORS

Conductivity measurement is made by applying an AC voltage to the cell electrodes and determining the resulting conductance. AC must be used to prevent electrode polarization, electrolysis, etc. However, AC signals are vulnerable to capacitance errors. Distant separation of cells and analyzers, such as in makeup water treatment areas or occasionally in direct hotwell cell installations, creates a long path of parallel conductors in the connecting leadwire. Those parallel conductors have a significant capacitance, especially in pure water applications where the measured resistance is high. The result with conventional instruments is a positive error caused by AC leakage through leadwire capacitance adding to the measured conductivity.

A development available on some analyzers is a special driven shield wiring technique. One of the cell conductors is run on coaxial cable while the other cell and temperature sensor leads are carried by a separate 3-conductor cable which can be run in the same conduit or tray. This method has been found to allow cell/analyzer separation up to 1000 feet (305 m) while still meeting accuracy specifications. Previously, a 100 foot (30 m) separation was marginal in similar situations.

Another leadwire problem is its resistance. For pure water, the measured resistance is high enough that the leadwire contribution is usually insignificant in the conductivity measurement itself. However, depending on the type of temperature compensation circuit, there is frequently a problem in that area. Analog conductivity measurements depend on matching the temperature coefficient of the water with a thermistor network which is a part of the analog measuring circuit, to yield a compensated signal. That type of circuit typically used a few hundred ohm compensator. Thus the leadwire resistance could easily be significant.

Microprocessor-based instruments use a temperature sensor in the cell but perform the compensation within the analyzer by calculation. This allows for flexibility in selecting the specialized temperature compensation algorithms noted above. It also allows the use of an especially precise temperature sensor. Temperature measurement accuracy, usually to $\pm 0.1^{\circ}C$, is important since the temperature coefficient can be as high as $5\%/^{\circ}C$.

Some instruments use traditional resistance temperature detectors which require precise calibration in the installed measurement loop to account for the leadwire resistance. To eliminate the need for this exacting startup calibration, some instrument designs incorporate a precision thermistor which has a higher resistance and temperature coefficient and achieves the same accuracy without adjustment.(3) It also eliminates the possibility of misadjustment.

In conclusion, the installed accuracy of conductivity measurement depends heavily on accurate temperature sensing, appropriate temperature compensation and adequate signal handling. Very significant improvements in each of these areas has been made in recent years. In addition, improvements in determining and correcting for small variations in cell constant have been implemented.(4) As a result of progress in all of these areas, installed accuracy of conductivity measurement has been improved by more than an order of magnitude in many applications.

pH

The measurement of pH in absolutely pure water is unnecessary. However, when high purity is compromised, pH provides a supplementary indication of acid or base contamination which can be valuable in trying to identify the source of the problem. Also, with alkaline water treatment in power plants, pH is a required backup to conductivity to assure that unintentional contaminants are not present.

pH measurement in pure water is difficult. The low conductivity of the water, the variability of reference electrodes, the vulnerability to contamination and temperature compensation requirements provide many possible snags. However, addressing each of these potential problems adequately can bring success.

To measure pH, measuring and reference electrodes must have a conductive path between them to complete the measuring circuit. Pure water has barely enough conductivity to provide this. When such a poorly conductive fluid flows through non-conductive tubing, it generates a static charge or streaming potential. This potential is added to the millivolt level pH signal between electrodes. The net effect is a flow dependent, noisy pH reading. A necessary precaution is to use a grounded stainless steel flow chamber with low flowrate of 100-200 mL/min. This shields the electrodes with a uniform ground and prevents build up of any charges.

On-line reference electrodes utilize permanently filled electrolyte and junctions to minimize maintenance requirements. However, they exhibit a few tenths pH shift in going from calibrating buffer solutions to high purity samples due to variable reference electrode junction potentials. A specialized high purity calibration procedure can correct for this.(5)

The temperature effects on pH measurement are not always fully understood. The signal from pH electrodes is proportional to temperature as well as pH. They produce more millivolts per pH at higher temperatures and this effect must be compensated, either by an automatic temperature compensator immersed in the sample or by the manual adjustment on a typical lab pH meter. Virtually all pH instruments provide this electrode temperature compensation. However, electrode temperature compensation does not correct for actual solution pH changes with temperature.

In most applications, there is negligible pH change over the temperature ranges experienced and no compensation beyond that described above is needed. In natural, cooling, service and waste waters there are sufficient dissolved buffering minerals to hold the pH fairly constant with temperature variations. But in deionized water or ammonia-treated water, the lack of buffering allows water the freedom to increase in ionization as described for conductivity measurement. The result is significant temperature coefficients of approximately -0.017 pH/^{o}C for pure water and -0.032 pH/^{o}C for 0.25 to 1 ppm ammonia-treated water, as illustrated in Figure 2. This is the reason water chemistry specifications require that pH be referenced to 25°C--so that pH variations can be attributed solely to treatment levels, not to temperature. Since this effect is a property of the sample, its compensation must be optional and adjustable. Current high purity pH instrumentation now includes adjustable solution temperature compensation in addition to electrode temperature compensation.

A further benefit of microprocessor-based pH instrumentation is time saved in routine calibration. The normal two point standardize and slope (zero and span) adjustments of analog instruments require that the standardize pH value be near the electrical zero point of the measuring circuit, usually 7 pH, in order to prevent zero/span interaction. For power plant samples, slope adjustment is then made above that point, at 9 or 10 pH. Small deviations from the zero point cause interaction that must be adjusted out by repeated calibrations between the points. Thus a back and forth, trial and error routine is sometimes required.

With microprocessor-controlled calibration, any two pH values may be used and the interaction is automatically calculated out. The two points of calibration are firmly locked in with no need to repeat. Further, some on-line instruments now have tables of standard pH buffer solution values vs. temperature in memory to provide the features of automatic buffer recognition and automatic calibration. This can prevent operator errors of selecting the wrong buffer or neglecting the variation of buffer pH value with temperature.

High purity pH, while still a difficult measurement, has gained substantially in reliability. There is now a much greater understanding of solutions to the troublesome areas of streaming potentials, reference junction potentials, solution temperature compensation, calibration, etc.

SODIUM ION

On-line measurement of sodium ion is usually more sensitive than conductivity in detecting contamination. Sodium is a part of virtually any kind of contamination and is the first positive ion to break through a cation exchanger. Its measurement can be made at one-tenth part per billion levels with good sensitivity. This is due to the ion-selective electrode technique which yields an electrical response proportional to the logarithm of concentration. The signal has excellent resolution at low concentrations--like the scales of a slide rule toward the left end. Whether the instrumentation has a multi-decade log scale or a digital display with autoranging, the wide range ppb readout is well-supported by the electrode signal.

Instrumentation is very similar to that for pH except that the sodium electrode selectivity for sodium ion is not as good as the pH electrode selectivity for hydrogen ion. The interference of hydrogen ion on sodium measurement usually requires addition of a basic reagent to raise the pH (suppress the hydrogen ion concentration.) However, a unique electrode with especially high selectivity for sodium enables measurement down to 1 ppb in heat cycle applications where pH is above 8.8 without using a reagent.(6) The sample conditions above the diagonal line in Figure 3 represent the area of sodium response of this electrode. Measurements going below 1 ppb experience increasing positive errors, always in the safe direction--indicating more sodium than the actual amount. Typical alarm points at the higher 2 to 5 ppb levels are maintained with good accuracy.

For sample conditions below the diagonal line, ammonia must be added to raise the pH above the line before measurement. For accurate measurements below 1 ppb, ammonium ion interference must be prevented by the use of an amine to raise the pH. An amine forms a larger cation which will not interfere like ammonia. It is also a stronger base which suppresses the ionization of any existing ammonia already in the sample. For research applications with 0.1 ppb level measurement requirements such reagents as dimethylamine, monoethylamine, or morpholine are used.

Along with the sample conditioning and sensing, progress in the analyzer part of the measurement is considerable. With the wide ranges of measurement to cover normal and upset conditions, autoranging linear displays and outputs have been developed. Other options include logarithmic and bi-linear output scaling to be compatible with desired recorder scales. Some analyzers provide the capability for automatic unattended calibration with internal timers and relay contacts to activate solenoid valves which in turn direct calibrating standards to the electrodes. A stable reading triggers completion of the calibration and return to process measurement.

Sample stream switching is a function which can lower the per point cost of analysis. Internal timers and relay contacts in the sodium ion analyzer can activate stream selection valves for a time-shared type of operation. Up to 10 minutes per point may be required to achieve stability. This limits the frequency of stream sampling and the number of streams that can be multiplexed. Three-way solenoid valves are recommended to allow samples to continue to flow (to drain) even when not being measured. This assures a fresh sample at the analyzer when each stream is finally selected. Additional contacts in the analyzer are needed to identify which stream is being sampled. An output signal hold function assures that only stabilized values are recorded, to aid chart legibility.

The extremely good sensitivity of sodium ion measurement has the tradeoff of higher maintenance than the somewhat less sensitive conductivity or cation conductivity measurement. This tradeoff must be evaluated in terms of the required speed of contaminant detection balanced with the capabilities of operating personnel. The advances described here generally reduce maintenance requirements.

DISSOLVED OXYGEN

Dissolved oxygen (D O) measurement is required to assure operation at the low ppb levels that will minimize oxidation corrosion. D O is sensed with an electrochemical probe separated by a gas permeable membrane. Maintenance of these probes has required considerable time and dexterity in rejuvenating an oxidized anode, refilling electrolyte, and replacing the critical membrane.

Some recent D O probe developments have simplified maintenance by employing membrane and electrode modules with straightforward installation. A considerably more reliable measurement has resulted. A further patented technique eliminates internal maintenance completely with a sealed probe design.(7) It measures at equilibrium with no anode oxidation and no net internal reaction limiting the life of the probe. The equilibrium design allows the use of a thicker permanent membrane since the signal does not depend on a continuous consumption of oxygen obtained by diffusion through the membrane. It has the added advantage of flow independent response. Conventional D O probes create a continuous flow of oxygen through the membrane and tend to deplete the oxygen concentration at the membrane surface at low sample flowrates. The patented equilibrium technique operates with no oxygen flow through the membrane except with changes in sample concentration. Thus oxygen is not depleted from the sample even under zero flow conditions.

CONCLUSION

The overall effect of on-line analytical progress has been to maximize accuracy and reliability of these measurements while reducing sample conditioning and maintenance requirements. As more plants reap the benefits of these advances to obtain better cycle chemistry control, we expect to see corrosion loss statistics improve.

REFERENCES

1. "Interim Consensus Guidelines on Fossil Plant Cycle Chemistry," CS-4629, Electric Power Research Institute, Palo Alto, CA, June, 1986.

2. Gray, D., Tenney, A., "Improved Conductivity/Resistivity Temperature Compensation for High Purity Water," Ultrapure Water, Voorhees, NJ, Vol. 3, No. 4, July/August 1986, pp. 27-30.

3. "Two-Cell Ratio Conductivity/Resistivity Analyzers," Leeds & Northup Publication C2.2215-DS, 1989.

4. Gray, D., Shaffer, E., "High Purity Conductivity Cell Standardization," EPRI/NIST International Conference on Measuring Waterborne Trace Substances, Baltimore, MD, August 1990.

5. Gray, D., "Continuous pH Measurement in High Purity Water," Ultrapure Water, Littleton, CO, Vol. 6, No. 5, July/August 1989, pp. 38-44.

6. "Sodium Ion Measuring System," Leeds & Northrup Publication C2.1114-DS, 1990.

7. "Dissolved Oxygen Analyzer," Leeds & Northrup Publication C2.3114-DS, 1989.

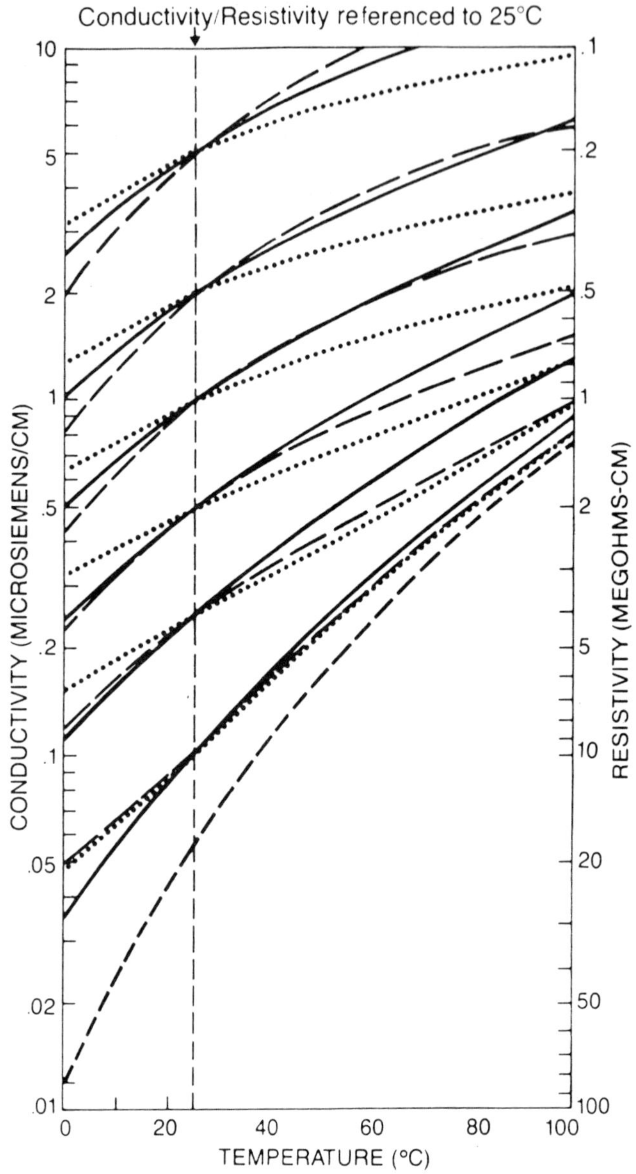

Figure 1 -- Conductivity Temperature Effects in High Purity Water, with traces of sodium chloride (solid), acids or ammonia (dotted), morpholine (dashed), pure water (bottom).

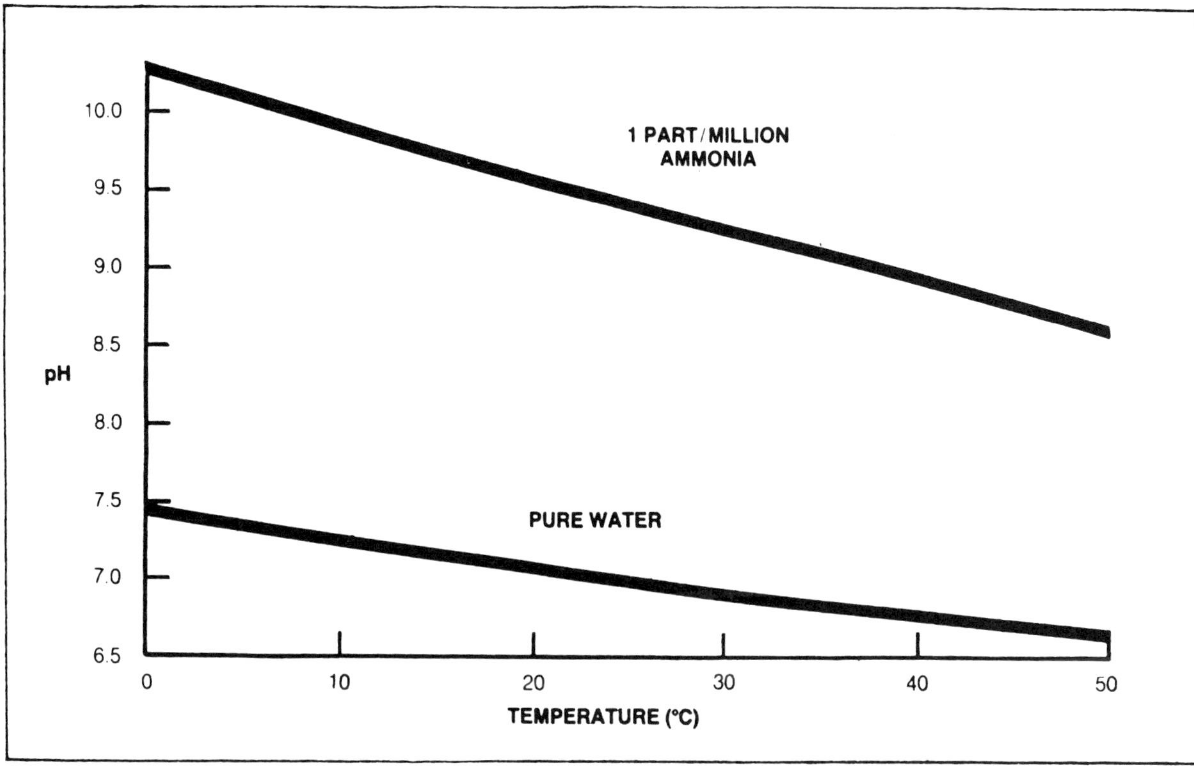

Figure 2 -- pH Temperature Effects on Pure Water, compensated only by solution temperature compensation

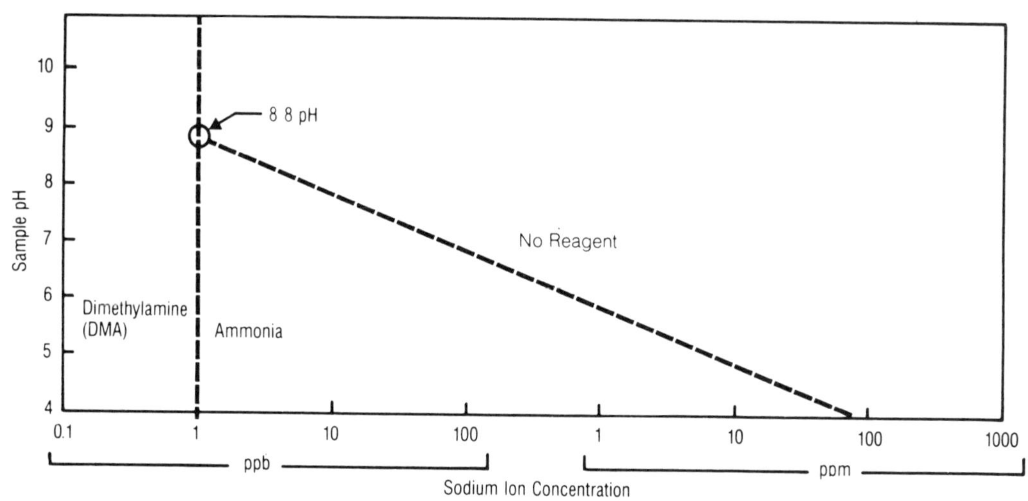

Figure 3 -- Selectivity of Sodium Ion Electrode, identifying reagent requirements for sample conditions (L&N electrode 003639)

©1991 - Paper #91-734
0074-056X/91/461-481/$0 + .50pp

"PORTABLE APPLICATIONS" - HELPING INFORMATION SYSTEMS TO BECOME AS ADAPTABLE AS PLANT CONTROL SYSTEMS

DeLynden Lersch, P.E.
Stone & Webster Engineering Corporation
245 Summer Street
Boston, Massachusetts 02210

ABSTRACT

In terms of ease of use, responsiveness and portability, process control systems have always led plant information and business systems. Now sophisticated new tools (relational and object oriented data base management systems, the Electric Power Research Institute's Plant Information Network (PIN), high speed data networks, distributed CAD, optical scanning and retrieval systems) are improving this situation.

This paper will discuss the successful implementation of such technologies in three facilities to support Plant Configuration Management and outline the increased potentials for blending such systems with plant control systems in the future.

INTRODUCTION

There are 110 nuclear-power plants currently in commercial operation in the United States, with another 11 in various stages of design, construction, and licensing. An additional 106 units are deferred. Fossil-fueled power plants (coal, oil, and gas) number in excess of 430. Uncounted heavy industrial installations, manufacturing plants, petrochemical refineries, and water treatment systems are also in operation, or under construction or renovation. All of these facilities share a common need for an information management system that effectively and efficiently captures, retains, and retrieves the voluminous information generated during the life of a facility.

Architects, engineers, and constructors view a facility through the information they generate or collect, process, and store during engineering, design, and construction processes. Vast amounts of information involved with the physical (design) and dimensional characteristics of the facility must be manipulated, along with data from complex engineering and design calculations in areas such as structural analysis, thermodynamics, fluid flow, pipe stress analysis, and process simulation; resource allocation and leveling in project management; and the tracking, archiving, and retrieval of myriad procedural, procurement, contractual, and correspondence documents.

Owners and operators of these facilities face the additional need for information related to the physical, as-built, and design-basis characteristics of the facility, as well as to the operations-related procedural, statistical, administrative, training and health-physics records. Programmed maintenance-based frequency of repair, reliability data and spare parts inventories, and regulatory and licensing documentation are generated from an assortment of the other existing information. It has been estimated that personnel involved in such efforts spend in excess of 50 percent of their time gathering the information necessary to support their assigned tasks, i.e., in "getting ready to do work."

Today, computers are essential to successful information management, from the preconceptual design phase through plant decommissioning. Recent advances in human interface tools, (such as software which supports user views of data, 3-D digital models, and optically scanned images) are being employed to decrease the time spent "getting ready to do work" associated with the sheer bulk of paper and amount of complex work handled by diverse plant groups throughout the lifetime of a facility. These tools are also being deployed to help identify and protect the plant's confirmed configuration baseline. More importantly, many of these systems use emerging standards for computing and communication, as well as flexible architecture to enhance the portability of applications. While not nearly as modular as plant process control systems, such systems are much more adaptable than ever.

CONFIGURATION BASELINE AND CONFIGURATION CONTROL

Although significantly less challenging to design than the real-time process control systems, there is a similarity in the need to relate information among dissimilar objects and situations.

The established or confirmed configuration baseline consists of validated or verified system, component, and document information, along with the following relationships:

1. System to component
2. System to features/parameters
3. Features/parameters to components
4. Features/parameters to documents
5. Components to documents
6. System to documents

Figure 1 shows the relationships among design basis, design criteria, components, critical features/parameters, documents, and training courses.

Configuration control comprises the procedures that outline the related work-authorization approvals and work methods, as well as the control of the configuration baseline information to maintain a system(s) design basis and criteria. Effective control of the information requires reviewing the activities associated with the process of configuration control, i.e., change control, Figure 2 displays the change-control process, which includes the following activities:

1. Identification of Change/Modification A change or modification to a structure, system, or component can be identified as either regulatory or non-regulatory.

2. Justification for the Change/Modification Justification for the change or modification can be categorized as follows: for personnel safety; for public safety; to improve/maintain or extend on-line performance; to reduce operating and maintenance (O&M) costs; or to improve plant capacity. Each justification can have a weight value assigned to prioritize the change/ modification.

   During this activity, a work authorization (change notice) should be prepared that would authorize the expenditure of hours to conceptualize the effects of the change. Upon approval, the work authorization will be assigned a number for work-hour expenditure and collection. The work authorization number can be documented in the accounting ledgers, allowing it to be traced or audited, as required.

3. Potential Impact on Existing Design Basis (System or Systems) The potential impact to the existing design basis should be assessed from a discipline interaction and system interface perspective. The assessment would provide the engineer with a comprehensive list of items for review in evaluating the technical impact of the change or modification, as well as the associated scope of work.

   The obvious changes that could potentially impact the existing design basis of a system occur during design changes and/or modifications. Other potential impacts can occur during O&M changes. For example, an operational change of equipment setpoints without an appropriate technical review or a maintenance change of component replacement with non-like kind, including the replacement of a manufacturer's equivalent, should be formally evaluated as to potential impact on the system's existing design. O&M changes that have a potential impact on the existing design basis of a system should proceed through the overall configuration-control process.

4. Technical Review of Potential Impact of the Change/ Modification A technical review is conducted to ensure that the change/modification does not degrade the existing design basis from a discipline interaction or a system interface perspective prior to authorization.

5. Detailed Scope of the Change/Modification One or more detailed scopes may exist depending on the number of technical solutions.

6. Cost/Schedule Analysis of the Change/Modification An optimal resolution of the change/modification will be provided based on the technical, cost, and schedule impact.

7. Change-Control Review Committee The change-control review committee should consist of operations, maintenance, and engineering managers or their designee(s). This committee can review and decide upon approval of a change/modification based on a thorough technical, cost, and schedule assessment.

8. Obtain Funding The owner of the facility must then obtain funding for the change/modification.

9. Detailed Engineering/Design and Procurement Detailed engineering/design and procurement is the effort associated with analyzing and communicating the change- or modification-related information to the organization responsible for physical implementation.

10. Physical Implementation of Change/Modification Physical implementation consists of the installation, quality control acceptance, testing, and operating acceptance.

11. Close-out Change/Modification Package Upon completion of the change/modification, all the affected documents must be revised to reflect the change/modification; e.g., procedures (O&M), technical specifications, FSARs, as-built drawings, and training courses.

The change-control process does not end with the closeout of the change/modification package. The data must be integrated into one central and accessible location to provide the plant with an accurate model and record of design, engineering, maintenance, construction, change/modification, and quality-related information.

CONTROLLING CONFIGURATION INFORMATION

Building a fully automated integrated configuration-control system (ICCS) is a realistic goal today because of recent advances in data base, CAD, and hardware technology. An ICCS provides plant personnel with useful, flexible, and cost-effective information tools. The integral parts of a fully automated ICCS include a rational data base management system (RDBMS), CAD system, cost/schedule system, and documentation management system, all of which can be linked through distributed data-processing technology and wide area networks (WAN) and local area networks (LAN). Given the number of documents associated with a nuclear facility and the frequency of changes, an optical laser disk storage/retrieval device and optical scanners become essential peripherals to the documentation management system. Figure 3 shows an ICCS to illustrate how such a system would support configuration control. The heart of the ICCS is the RDBMS, which should always be used to assess the potential impacts of any changes (training, maintenance, operations, etc.) to the design basis of a system(s).

A computerized system implemented with an RDBMS represents a practical approach to controlling configuration information. The flexible architecture of an RDBMS makes it particularly useful to utilities with multiple plants and diverse naming conventions. Structured Query Language (SQL)-based systems allow easier customization of data logic and functions to answer the specific requirements of each user. The ability to construct new views based on the derived relationship between entities is a key feature of the RDBMS. SQL provides a simple and consistent means for defining, querying, maintaining, and protecting the contents of the relational data base. Such a system can often be interfaced with existing computer applications/project systems and non-computerized procedures.

These systems can be driven either by menus or Graphical User Interfaces (GUI), allowing users to move through the system by the selection of information options from menu panels or pop-up windows. Informational HELP messages and HELP text can be made available to support infrequent or untrained users. In addition, such systems can provide multiple levels of security and password protection.

Relational Data Base Management System

An RDBMS allows for the association of information about equipment, documents, commitments, activities, and people in a single integrated data base. It is the means by which information is shared among the diverse plant user community. The simplicity of the relational data model organizes all data into simple table-like structures, where all access and data manipulation is done by value rather than position or location.

The data base is composed of a collection of relations or tables. The tables represent the physical storage of the records in the data base. Each table has relational records with all the unique attributes or data elements necessary to define the ID for one record. An RDBMS offers relational views of selected data elements from other records in the data base. SQL allows the user to create tables, store or change information, and retrieve information in desired tables, and also assists in maintaining the data base. This is accomplished in simple, English-like commands. Because SQL is relatively easy to learn, training time and costs are greatly reduced.

An RDBMS enables a project to determine the status of any design-change control, procedure, or correspondence document, as well of all the documents it affects and is affected by. Equipment, vendor documents, calculations, or work activities can be associated with a particular document. A maintained index of activities requiring physical work, administrative action, or project response can be achieved, along with an index of calculations and licensing documents with all of their associated document and equipment information. Design and qualification information related to a specific piece of equipment can be tracked. Spare-parts information, including stock identification, can be tracked to support maintenance and procurement activities. Training and skills information tracking, project assignments and scheduling, project authorization and hour tracking, and health-physics tracking also can be provided.

A system such as the Residual Heat Removal (RHR) has a design basis, design criteria, functional requirements, etc. In addition, the heat load is transferred to an interfacing system. With the capabilities of the RDBMS, the RHR systems engineer, through predefined or on-line queries, would be able to access data on interfacing systems, which already exist, without manually searching or developing his/her own files. Additional queries that could be made from the same data base could include the identification of all pumps by a certain vendor that are within the facility, design criteria, associated documentation, and licensing commitments. Data could be extracted as desired and used in root-cause analysis or trending by organizations outside of the operations staff. This would be extremely useful should a proposed modification be in the process of evaluation by engineering or if maintenance were considering to change out a component for one of increased efficiency.

In addition, RDBMS would provide the ICCS the capability to identify the intricate relationships between design basis, design requirements, documents, features/parameters, components, and system information, thereby providing required information to facility personnel including operations, engineering, and maintenance.

CAD System

With its 2-D drafting and 3-D modeling capabilities, CAD is another integral part of the ICCS. The ability of CAD equipment to share information among disciplines, readily illustrate alternatives, organize project information, and standardize presentations makes

CAD an important addition to the change/modification and the design processes. More sophisticated CAD equipment can recognize data, associate it with symbols and other data, and, most importantly, extract this data for further processing on other computer systems. Many CAD systems can accept raster scans of old manually drafted drawings, accommodate design changes with minimal impact to schedule, and rapidly generate new hybrid raster/vector drawings.

Discipline drawings can be produced on CAD and by combining data from two or more disciplines, composite general arrangement drawings can be created. Drawings are modeled at full scale on CAD. Commands are available that allow the disciplines to choose the scale at which the drawings will ultimately be plotted and, if necessary, to revise the scale factor. CAD libraries of symbols and drawings allow for fast and easy modifications of existing drawings, as well as the creation of new drawings.

CAD can be used as a display device for equipment data, and as a fundamental part of the engineering change and operations processes. Drawing information captured and stored would be available to engineers and operating personnel prior to the initiation of an activity. Changes and retrofits would be easily and quickly accommodated on the drawings or in the model as individual components can be moved as entities. The computer program also would check for any interferences, as the equipment is being located or modified. Similarly, the accessibility to components for operating and maintaining the equipment can be determined. CAD can demonstrate actual plant conditions, train employees, and serve as a librarian for the visual retrieval of the component, and associated documents and activities. Additionally, CAD would provide the interface to the data base by providing and/or retrieving attribute information for each component.

For several years, sophisticated CAD equipment has offered engineers the ability to develop a 3-D digital model, i.e., to describe with a design the shape, volume, and placement in space of components, and then to associate descriptive information to each component. CAD modeling offers tremendous possibilities for supporting applications that require accuracy to the actual facility. For example, the accuracy of P&IDs, electrical schematics, and logic diagrams are important to the maintenance troubleshooter. Nuclear maintenance is complex, and requires extensive cooperation and input from departments other than the maintenance department. With a computerized model, selected portions of the plant are modeled in the computer as 3-D objects. This permits components to be individually selected and located in the plant. Standard drawings can be made by creating slices of views of appropriate models either horizontally (for plants), vertically (for sections), or obliquely (for isometrics).

CAD would provide the ICCS the capability to: ensure that the design basis and design criteria are maintained when developing the change/modification; allow the user to work from the required view, not a predefined view; identify interferences and demonstrate the physical viability of the change/modification before activities begin; and simulate modifications for optimal solutions that fully support the engineering/design and implementation activities associated with the change-control process.

Documentation Management System

Another integral part of an ICCS is a documentation management system with text, figure, and table change-control features that provide full text-editing and formatting functions along with a concurrent annotation capability. This allows a revised document to be electronically reviewed simultaneously by various users without loss of configuration control. Change-control features can be used to highlight changes, and automatically maintain a comprehensive "audit trail" on a document and its various changes. This can provide a detailed record of when and by whom each change was created, modified, or frozen.

Since so very many of the documents associated with power plant operations tend to be hand written, the ability to handle images is an essential part of a documentation management system. Optical scanning and disk storage/retrieval devices provide the ability to store, transmit and retrieve documents identified as being essential for configuration control. Tables 1 through 4 are "rule of thumb" listings of the parameters which can be used to size imaging systems for a typical nuclear or fossil facility (Note: costs shown are in 1990 dollars).

This part of the integrated system is used by plant personnel in assessing the technical impact of a change. It provides plant personnel with easy access to plant documents/information required to perform such a review. The system would also be used during the engineering/design, implementation, and closeout phases of a change by revising, updating, or creating documents that would communicate the change to the various responsible organizations. To date, most large imaging have been implemented only in nuclear facilities. However, small systems, targeted primarily for maintenance procedures, are presently being considered by fossil plant owners.

Cost/Schedule System

The cost/schedule system would provide the ICCS with the required dates for those tasks associated with a change/modification that are reflected in an activity/work tracking module in the RDBMS. The activity/work tracking module would track discrete tasks associated with a work authorization through the impact assessment and detailed scope activities associated with a change/modification. It would also track through the detailed engineering/design and procurement phase, the physical implementation phase, and finally, the closeout change package phase to the required schedule dates based on the number of changes/modifications assigned to an outage. This is predicated on the change/modification justification, to which a weigh value is assigned that established its priority, the scheduled outage windows, and the required versus available resources. In addition, a link from a facility information system (time collection system) would provide actual hours to the cost/schedule system on a work authorization number basis.

ICCS APPLICATION EXAMPLE

The following scenario describes just one of the unlimited applications of an ICCS. It is a Saturday, and the second shift maintenance crew has just come to work. Bill, the second shift supervisor, checks with the first shift supervisor and is given no indication of any potential problems. Bill is expecting a quiet night in which he will be able to review the work orders in the outage backlog file for the plant's upcoming planned outage.

Since, as we know, pumps have a mind of their own, a motor-driven, single-stage pump suddenly fails and shuts the system down. After tagging out the pump, Bill goes to the RDBMS to review the vendor manual's procedures listing to obtain the cross-reference code for the disassembly procedure which is in the documentation management system. Bill calls up the full disassembly procedure on the documentation management system because the maintenance activity will affect at least four crafts and Bill wants to get it right the first time. Bill reads that the mechanical craft isolates the system, the electrical craft racks out the breaker on the motor, and a millwright is responsible for taking the pump apart. Bill then prints out those sections of the manual he deems necessary to assist in the pump disassembly. Bill reviews on CAD various views of the vendor diagrams to ensure the disassembly goes smoothly and selects several views that will assist the crafts.

After the pump has been torn down, Bill is informed that the wear rings, seals, and bearings are worn, and furthermore, there is evidence of a crack in the impeller. Going back to the RDBMS, Bill finds the name and telephone number of the original equipment manufacturer (OEM), whom he calls. Bill explains the situation to the OEM service manager, who tells him to replace, not repair, the impeller. Bill retrieves vendor drawings from the CAD system, and the list of required replacement parts from the RDBMS. After obtaining the parts, including a new impeller, the impeller is replaced and the pump reassembled. (Had the pump not failed on Bill's shift and been kept running by operations, the pump could have been repaired in the upcoming planned outage. Then, the cost/schedule system of the ICCS would have been used to determine the activities associated with the repair for proper scheduling into the outage.)

Bill tests the pump to its baseline performance curve (which is the last surveillance test curve on the pump and maintained in the RDBMS). Since the flow rate is less, Bill checks it against technical specification requirements that he obtains from the RDBMS. There is a deviation from the last surveillance performance curve, and the pump does not meet technical specification requirements for flow. An attempt is made to reverify alignment and other assembly clearances and fits to determine the cause of failure. Subsequent retesting shows no improvement. Bill enters the new data and produces a copy of his findings to give to the plant engineering staff to determine the corrective action required, which could include the purchase of a new pump.

After engineering reviews the findings submitted by Bill, it is determined that a new pump is required. Prior to procurement, engineering uses the data in the RDBMS to assess if the new pump could have a potential impact to the existing design basis from a discipline interaction and system interface perspective, and ensure that it would not degrade the existing design basis. After determining that the new pump is acceptable (replacement with like kind), the existing pump is removed, and the new one put in place and tested to final acceptance.

The systems comprising the ICCS are updated to include such entries as the new pump's specifications, vendor manual, performance standards, and diagrams; parts and tooling requirements; manufacturer's and supplier's information; maintenance history on the old pump (including the time to repair and the downtime, response time, active maintenance, supply time, and administrative time, etc.), (Bill, by the way, spends time during his next shift documenting the need for a better preventive maintenance program.)

EPRI AND THE PIN

Of course the desire to relate information from different sources is neither a new one nor unique to plant configuration control. Professional organizations such as the ISA, IEEE, ASME and others have worked with manufacturers and utility leaders to develop standards to improve the ability to interface among dissimilar equipment.

In addition, the Electric Power Research Institute (EPRI) has for the past seven years sponsored a Computer-Aided Engineering (CAE) Guidelines Project to provide a foundation on which to improve the management of generating plant data. The results of this work was initially published as EPRI report NP-5159 "Guidelines for Specifying Integrated Computer-Aided Engineering Applications for Electric Power Plants" in May 1987. This report presented a comprehensive, high-level model of the work activities and underlying data elements that make up the generating plant life cycle. The 440 work activities were identified, 140 activities were investigated and documented, 299 entities and their 2500 associated attributes were catlogued. The report also provided instructions for using the plant data model, called the Plant Information Network (PIN), in developing power plant applications.

The PIN was used by organizations such as Stone & Webster to model activities at several nuclear facilities. The PIN logic and naming conventions have been programmed into and are referenced as part of the Stone*WORKS* family of products.

In 1990, EPRI authorized the work to move this data model from its paper report format into a computer-aided software engineering (CASE) tool to facilitate the enhancement and implementation of the PIN within individual utilities/plants. The model is now available from EPRI in Knowledgeware's Information Engineering Workbench format and utility updates and sample projects will be made available in this format in the future.

DEVELOPING A FULLY AUTOMATED INTEGRATED CONFIGURATION-CONTROL SYSTEM

The integration and use of nongraphical information systems to serve and supplement CAD are the foundation of the ICCS. Direct display and access to information associated with CAD documents but not included in CAD data bases are practical with the installation of high-speed WANs/LANs and the implementation of an SQL-based RDBMS.

Table 5 presents some specific examples of data generated, used, and required throughout each phase of a facility's life cycle.

Merging an RDBMS with CAD revolutionizes the formalization and communication of this data. Linking the tools allows plant personnel to retrieve, view, and (if authorized) manipulate or revise data supplied by document clerks and purchasing agents and by engineering, construction, inspection, maintenance, and operations personnel. Depending on terminal authorization levels, the drawing or model can be retrieved, viewed, or updated by accessing the plant data already keyed into the information, or the information system can be revised via the CAD station.

Stone & Webster has developed two ICCS foundations: the Integrated Plant Database Model (IPDM) and the Automated Configuration Control & Equipment Support System (ACCESS). IPDM and ACCESS are implemented in an RDBMS, and can accommodate 2-D and 3-D graphics. IPDM is a relational data model implemented in IBM's DB2. IPDM is supported by Dassult System's CATIA for modeling in 3-D solids and for producing 2-D drawings. ACCESS is implemented in ORACLE and in DB2, and is highly portable to a number of computers including DEC, Prime, Data General, Hewlett Packard, and IBM. ACCESS can be integrated with CAD facilities such as Intergraph, Calma, and

AutoCAD through system links and data transfers. Both products are part of Stone & Webster's Stone*WORKS* family of software.

Stone & Webster has also implemented the electronic data communications networks necessary to support the additional technology. An RDBMS links to and from an optical disk/storage. Both Token-Ring and TCP/IP networks have been used in such efforts. Optical scanning systems from vendors such as Optigraphics, FileNet, Laser View, Wang and IBM's Image Plus can be supported by these Networks.

For many organizations, the most difficult part of developing an ICCS is developing a plan for action, and obtaining senior management support. Improvements in the information tools which support configuration management can be implemented in several ways. Our experience has shown the most successful is to stage the improvements in a series of discrete, but related activities which may be phased in over a period of 1 to 4 years. These activities should be related to: 1) a decrease in operating non-manual payroll expenses (primarily associated with labor as error reduction); 2) a migration from obsolete computer hardware or programmer intensive software; and/or 3) the introduction of improved human interface tools (graphics, images, expert systems).

A Strategic Information Resource Plan (SIRP) is useful in portraying the logical connection between the improvement in seemingly unrelated activities (for example, the automation of procedure publication and the establishment of a work order tracking system which references the same procedures). Many tools to aid in the presentation of an SIRP have become popular in the past few years - two of the most worthwhile have been powerful computer-aided software engineering tools (such as Excelerator and Information Engineering Workbench) and EPRI's Plant Information Network (PIN).

Individuals planning improvements to utility information systems will find EPRI's PIN to be especially useful. As part of NP-5159, EPRI-sponsored groups developed entity-relationship diagrams and identified many of the redundancies and interrelationships associated with the cradle-to-grave work processes in an electricity generating facility (Note: For additional information contact Warren Bilanin, EPRI, P.O. Box 10412, Palo Alto, CA 94303 or Terry Molloy, Pacific Gas & Electric Company, One California Street, Room F-734B, San Francisco, CA 94106).

CONCLUSION

As can be noted in the above application example, the integration of information in one accessible location and on-line access to that information can save hours and thus, dollars. The benefits of an ICCS include control of the configuration baseline information to maintain a system(s) design basis and design criteria; enhanced control of work and minimization of rework; increased productivity; reduced downtime cost; improved plant performance (reliability); improved communication between disciplines; increased data accuracy; improved plant availability; better scheduling of resources; and achievement of a higher accuracy of quality management decisions.

It has been estimated that between $2-10 million/year could be saved by automation of plant activities and integration of plant data bases used to support those activities. These savings include the decreased cost of performing the activity and the cost avoidance of decreasing forced outages (1).

Using our example of Bill's night with the failed pump, reductions were achieved in time spent to locate the procedure, to determine the replacement parts, in obtaining parts not required, in pump disassembly and reassembly (by not deviating from the procedure process), and in producing drawings in the appropriate view. If one considers that a nuclear unit has a cost of downtime equivalent to $500,000 per day, and assumes that the pump's contribution to the system's unavailability is as follows: forced outage rate = .05%, lost days per year = .2, and cost per year = $100,000; a conservative estimate of 14.5 hours saved (in the application example) equates to a potential savings of $302,282 in lost downtime. Future savings could be realized since the new pump was assessed and determined to be in compliance with the design basis and criteria.

Most owners already own most of the hardware/software that fulfills many of the diverse aspects of configuration control. The focus now should be on implementing flexible applications and communications products that best suit the needs of non-technical users, plant managers, operators, designers, and maintenance personnel alike.

REFERENCES

1. R.Colley, Managing Power Plant Computer Communications and Information System Technologies. EPRI Seminar on Data Acquisition, Control, and Communication, San Diego, CA, February 23-25, 1988.

2. D.R.Lersch, W.B. Dodson, and P.W. Chevrier. Configuration Management Database Systems. EPRI Seminar on Data Acquisition, Control, and Communication, San Diego, CA, February 23-25, 1988.

Table 1

Compression Ratios

Size	Ratio
A–size	9:1 to 18:1
B–size	13:1 to 28:1
C–size	13:1 to 30:1
D–size	16:1 to 30:1
E–size	21:1 to 34:1

Table 2

Data Volume

- Number of bytes (compressed)/various size drawings

A–size	25,000 to 54,000
B–size	33,000 to 70,000
C–size	62,000 to 140,000
D–size	124,000 to 235,000
E–size	222,000 to 360,000

Table 3

Documents Processed

- Typical nuclear plant
  - 5,000 per hour
  - 40,000 per day
  - 800,000 per month
  - 10,000,000 per year
- Hardware/software costs $1,000,000
  - Hardware is 60% of cost

Table 4

Documents Processed

- Typical 300–450 MW fossil plant
  - 450 per hour
  - 3,000 per day
  - 60,000 per month
  - 720,000 per year
- Hardware/software costs $200,000–$250,000
  - Hardware is 70% of cost

Table 5

EXAMPLES OF DATA RELATED TO VARIOUS PHASES OF A FACILITY'S LIFE CYCLE

PHASE	DATA	DESCRIPTION
Engineering	Design Basis	Requirements to which the system must perform.
	Design Criteria	Requirements to implement the design basis.
	System Requirements	System performance requirements in support of the design basis and design criteria.
	Documentation	Drawings, specifications, calculations, technical data (summary data sheets for systems and/or components), etc.
Operations	Technical Specifications	Limiting condition for operation and surveillance.
	Operating Logs	Performance data such as pressure, temperature, flow, etc.
	Tagging	Requirements for isolation of systems/components.
	System Chemistry	Requirements and sample results.
Maintenance	Maintenance Requirements	Identification of tasks including frequency, hours, etc.
	Scheduling	Determination of implementation of maintenance activities.
	Spare Parts	Insurance of proper inventory of required parts.
	Work Orders	Issue authorization to perform work based on schedule and task requirements. Findings and descriptions of work performed may be included.
	Documentation	Drawings, specifications, procedure, bills of materials, etc.
Testing/Inspection	Test/Inspection Requirements	Identification of tests/inspections including frequency, actions requiring a test or inspection, etc.
	Scheduling	Determine implementation of predetermined tasks.
	Documentation	Drawings; specifications; procedures; reports that provide data concerning date, results, required actions, etc.
Health Physics	Personnel Exposure	Data related to specific exposures and overall integrated exposures for personnel.
	Survey Data	Dose rates of areas and/or components.
	Survey Maps	Visual location of survey data.
	Radiation Work Permits	Documents defining work to be performed, survey data, protection requirements, etc.
	Radwaste	Data identifying volumes, types, radionuclides analysis, storage, etc.
	Shielding Analysis	Based on survey data to provide analysis for the use of shielding personnel.

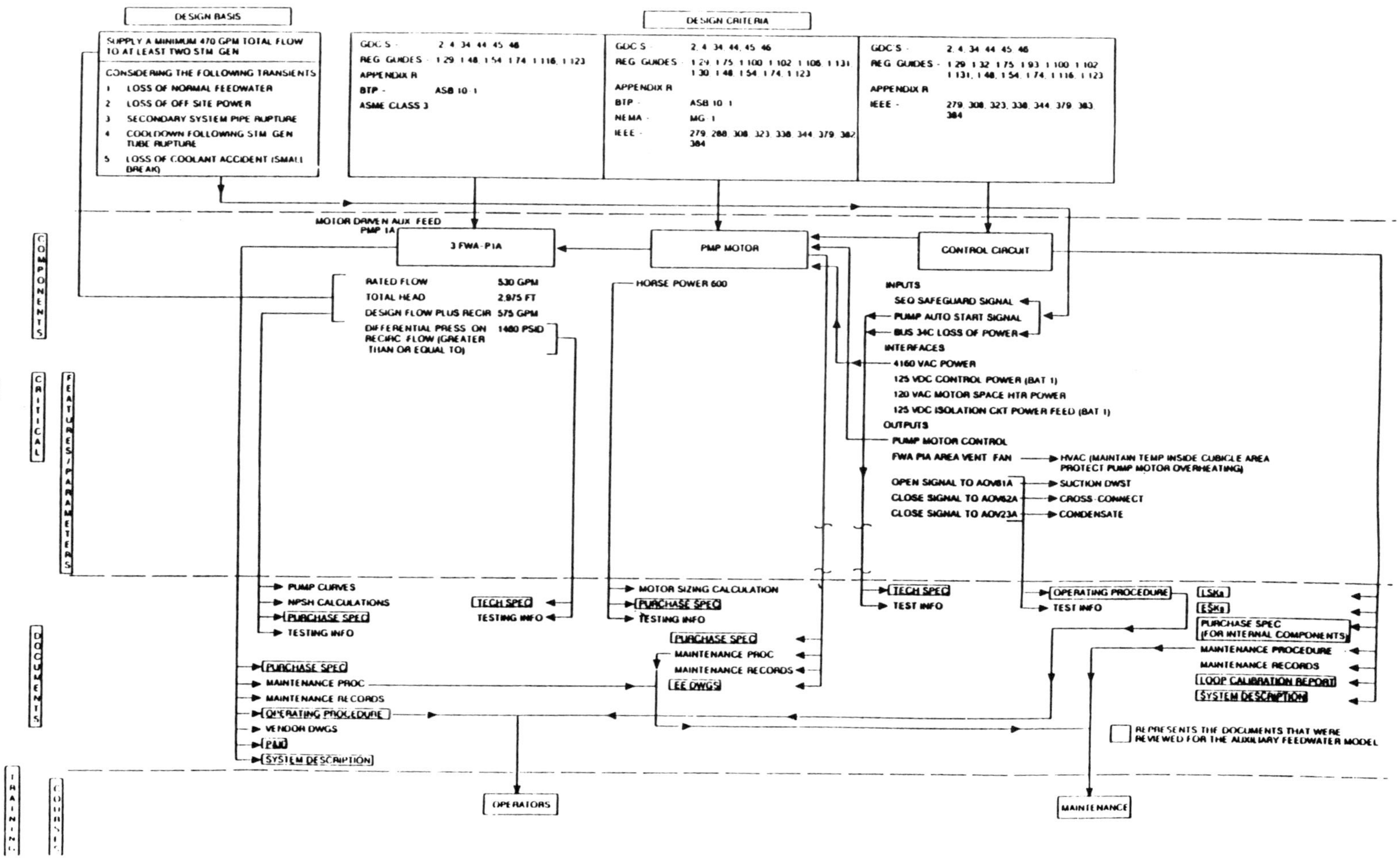

Figure 1. Example of Auxiliary Feedwater System Relationships

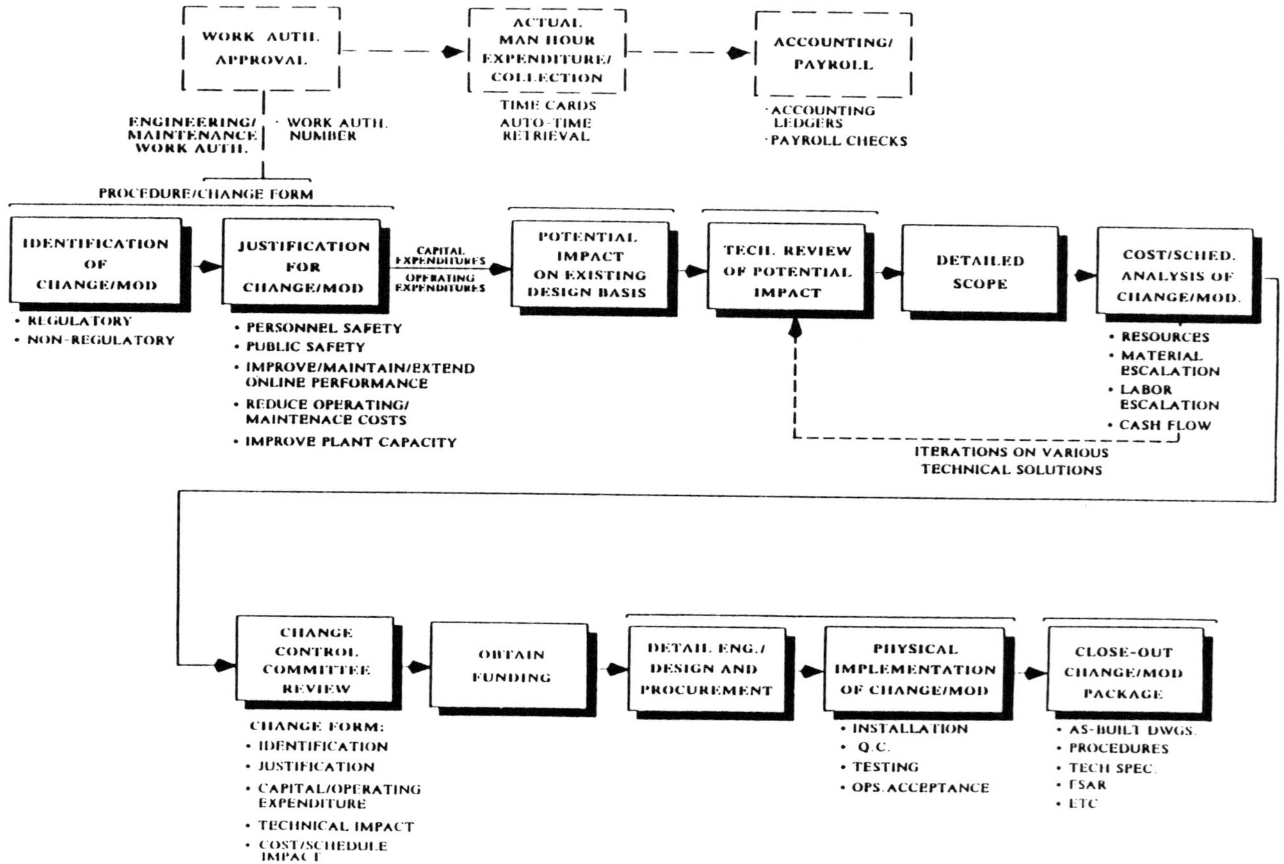

Figure 2. Change Control Process

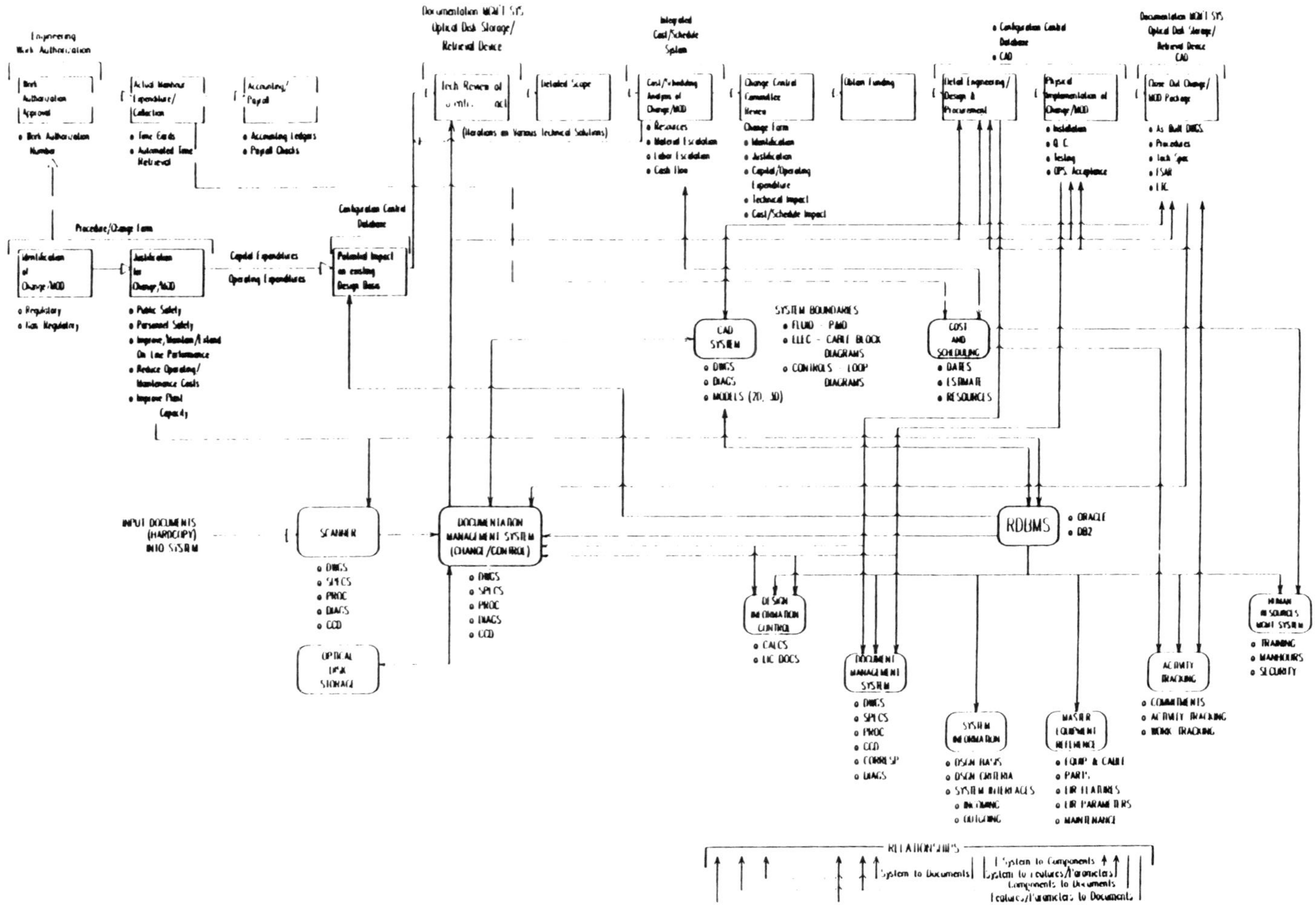

Figure 3. Typical Integrated Configuration-control Process

© 1991 - Paper #91-735
0074-056X/91/483-488/$0 + .50pp

IMPLEMENTING ON-LINE REAL-TIME EXPERT SYSTEMS

Robert L. Moore and Howard P. Rosenof
Gensym Corporation, 125 CambridgePark Drive
Cambridge, MA, 02140

ABSTRACT

The combination of Real-Time systems technology with Expert Systems has allowed applications in such areas as monitoring, diagnostics, control, and optimization, where expert human knowledge is put into a computer and applied on-line to improve operations. An object-oriented design allows large applications. The experience of several installations is discussed, including nuclear plant diagnostics and network control.

Key words: Artificial intelligence, Expert system, Real-time

INTRODUCTION

Advances in object-oriented expert systems have allowed new methodologies to be used in the construction of large-scale systems (1). Among the new aspects are knowledge representations that allow graphical constructions which are understood by an expert system, and which automatically become a deep knowledge representation of the structure of interactions of objects.

The object-oriented representation allows definition of problem behavior to be expressed generically, applying to whole classes of objects, and the understanding of interaction structure by the expert system then allows diagnosis to proceed from the object instance for which the problem behavior is detected, looking upstream for problem sources or downstream for effects.

Another new aspect is the definition of knowledge of rules, procedures, formulas, dynamic models and other attributes in structured natural language.

The combination of these methodologies, together with advances in reasoning paradigms, have together allowed large scale complex real-time expert systems to be implemented. Several examples will be discussed, each with thousands of objects or other complex characteristics.

APPLICATION EXAMPLES

Nuclear plant diagnosis.

An application using G2 for safety assessment and post-trip guidance at the Forsmark site in Sweden has been described (2,3). The Forsmarkverket nuclear power plant is the largest continuously operating plant in Sweden, providing 20% of Sweden's total domestic electrical energy consumption. The plant comprises three BWR units (boiling water reactors), each delivering 1000 megawatts.

Forsmarkverket has established a methodology for documenting safety systems, called CSF (Critical Safety Functions), based on process descriptions, logical safety functions, and alarm levels. A comprehensive CSF description of the plant has been worked out. The development work was subcontracted by Forsmark to IFE Halden of Norway, an applied research institute belonging to an international OECD group of nuclear energy research establishments.

The G2 real-time expert system, currently comprising well above 2000 rules, runs on a dedicated HP 9000/834, while the operator interface resides on a separate HP 9000/340 running Picasso, a graphics system developed by IFE Halden. More than 130 separate process mimic diagrams have been developed. Two Norsk Data ND 560 minicomputers provide real and simulated process data, and the whole set-up is linked together via an Ethernet.

Development of the application and all related software modules were completed in the summer of 1990, and the project has now entered the verification and validation phase. It will undergo comprehensive and stringent testing, and will be exposed to shift engineers and supervisors before being placed on-line.

Power distribution

An application in the power distribution area has been described by Pacific Gas & Electric in California (4). Currently in San Francisco, a group within PG&E is putting together a new operations station, to advise, assist, and diagnose faults and breakdowns at substation sites and in the power distribution network. The project name is IOPADS (Intelligent Operational Planning Aids for Distribution Systems). Running on a Sun Sparcstation, G2 Interfaces to an Oracle database which holds all the parts and line section information pertaining to the distribution circuits, and to the main SCADA system which monitors and controls the live distribution network.

By allowing faults to be monitored, and then overlaid on the distribution network, a large amount of information can be manipulated. IOPADS provides information about abnormal conditions, and advice on remedial operator actions. It allows the operations staff to dynamically simulate conditions and load transfers and consider various solutions before implementing a final solution.

Other diagnostic applications

The largest on-line nuclear plant diagnostic application, involving networks of G2 real-time expert systems and based on advanced diagnostic methodologies, involving multiple nuclear sites and organizations, has been underway for several years in Japan, but has not been published yet.

A long term project has been underway by Ontario Hydro (5), and the implementation of a demonstration using G2 is proceeding (6).

Applications in other areas.

A configuration management system (CMS) has been developed to address issues of efficiency, operational safety, and environmental compliance (7). The CMS is a complete model of the physical plant, implemented as a combination of computer databases and engineering drawings, along with procedures for maintenance of the model. The goal is to have the model accurately represent the plant "as-built," by tracking design changes. Providing the operator with an accurate "as-built" plant model, combined with timely real-time data, minimizes the risk of operator mistakes and maximizes speed of response to abnormal events.

The use of G2 for transient stability analysis, including simulation, interpretation, diagnosis, and selection of remedial measures to correct problems has been discussed (8).

Several companies are using G2 for dynamic scheduling applications (9). In these applications, the expert system knowledge is used to optimize resource allocation and to reschedule when resources are changed or when demands change. A related use could be to dynamically determine shutdown schedules, for example, in the event of partial malfunctions.

Published applications in the process industries, which illustrate methodologies which would be useful in the power industry, are also of interest. The use of object-oriented techniques to diagnose questionable measurements (10,11) helps operators to learn about problems before they reach the alarm stage. Mertz describes how the measurements across an entire plant are examined using generic rules applied to object classes (e,g. all temperature sensors, all pressure sensors, all control loops), so a type of aberrant behavior is continuously looked-for, and if the behavior is detected, then the operator is advised as to which of the sensors is behaving in the aberrant way.

Another use, combining dynamic simulation with the graphics and natural language rule expression, assists in capturing the knowledge of experts (12). In this installation, multiple networked G2's were used in concurrent knowledge base definition and testing, with multiple operators observing and commenting on the knowledge expression and behavior. Then the interactive editing of the knowledge base was used to improve and validate the knowledge.

METHODOLOGIES OF DEVELOPMENT

The availability of graphics workstations has permitted the implementation of a partially graphical domain definition. In a typical application, which might be a power plant, the material flows and process units form an interactive network. By representing these as objects, in the object-oriented programming sense, and by letting the graphical connectivity or proximity define the possible interactions, a significant part of the domain definition is done graphically. The cloning and connecting of objects defines the interaction network part of the application. This is both faster than programming, and easier to modify and understand.

An object is a convenient item to represent structure, graphic image, and other aspects. However the developer may wish to change the object class, adding attributes or behaviors for example. This may be desired even for a domain that is populated with objects already. Thus the methodology to allow editing of object classes, defining new attributes, changing existing attributes, and modifying behavior was designed into the software. An object class can be changed, and the already-existing instances will be changed in conformity with this.

To assist in the creation and editing of objects, and the definition of rules and other forms of knowledge, a structured natural language editor is used. A truly natural language human interface would be ideal. However this is not yet available. The structured natural language human interface is a step in this direction. Essentially the permitted language is limited in syntax and grammar to that understood by the system. Within that context, the human designer freely expresses the rules, procedures, models and other items of knowledge (Figure 1).

The behavior of objects may be, and often is, dynamic. Thus the integration was made of the well-known state variable representation. These models are expressed in natural differential equation form:

> d/dt(the pressure of any reactor)= ...etc.

Typical examples of dynamic behavior are material and energy balances, chemical reactions and equations of motion.

So now, as objects are cloned and connected, an interactive network of dynamic objects is instantiated, and the expert system can use this for simulation or for model-based reasoning. The interacting network of dynamic objects represents the "deep knowledge" of the domain. This is typically the knowledge possessed by the engineers. In the currently implemented systems, this engineer's knowledge is generally augmented with the knowledge possessed by operators or others who do not reason in terms of differential equations. Such knowledge may be expressed as rules, which may refer to behavior over time:

> if the rate of change of the temperature of any reactor during the last 10 seconds > 5 and the temperature of the reactor < 470 then conclude that the reactor-condition is unstable and invoke diagnostic rules for the process equipment connected at the output of the reactor.

Rules and other forms of knowledge are typically expressed generically, as shown in the above examples. This allows the knowledge to look for types of behavior. In one published application (11), 150 generic rules look for aberrant sensor behavior across a domain of 5000 objects including 1200 sensors in a chemical plant. Thus the generic form of knowledge provides a significant economy of representation, which is easier to implement and easier to maintain.

Another design concept, illustrated in the above example, is the propagation of the reasoning via connected objects. The graphical connections on the workstation screen are "live", so that "invoking rules for the objects connected", as in the above example, operates correctly even as connections are established or broken-off. The reasoning about connected objects allows the expert system to look for causes of problems, and to start looking for effects earlier, before alarm limits are reached.

The above concept, reasoning about connected objects, forms part of the methodology to distinguish between two common faults; aberrant behavior of domain objects and invalid measurements. A real process behavior problem often causes consequential behavior in connected objects, the presence or absence of which can be collaborative evidence for the diagnosis. An invalid sensor reading might be inferred by the lack of collaborative evidence of downstream process misbehavior.

The above described integration of graphics with object-oriented design, generic expressions of object behavior, and using connectedness of objects, is a powerful new and currently unique technology that has effectively opened the way to large applications.

SUMMARY

The combination of technologies in the G2 Real-Time Expert System, of object-oriented definition of structure and relationships, with graphic definition of object connectivity, together with generic definition of knowledge about behavior, integrated into a structured natural language development environment, has allowed large real-time expert system implementations. Examples in the diagnostic and operator decision support area are presented. As of this writing, over 600 such installations are underway or on-line, including more than 20 in the power industry.

REFERENCES

1. Moore, R.L., L.B. Hawkinson, M. Levin, A.G. Hofmann, B.L. Matthews, M.H. David. "Expert System Methodology for Real-Time Process Control." Proc. 10th World Congress on Automatic Control, IFAC, Munich, July 1987, vol. 6, pp 274-281.

2. Fjellheim, R. "Forsmark Nuclear Plant," Gensym Real-Times, Winter 1991 Issue.

3. Nilsen, S. "Experiences Made Using the Expert System Shell G2," Proc. 2nd International IEEE Conference on Tools for Artificial Intelligence 1990, Washington, D.C., November 6-9, 1990, pp. 520-529.

4. Mesa, B.; "G2 and the Power Industry", Gensym Real-Times, Winter 1991 Issue.

5. Chou, Q.B.; "Applying Expert System Concepts to Advanced Power Plant Control," Proc. ISA 1986, Paper #86-0710.

6. Chou, M.; Mensah, S., and Mylopolous, J.; "Expert Systems for Analysis and Control for Nuclear Power Plants, Ontario Hydro internal report, 1990.

7. Nevins, M.A.; "The Building of a Nuclear Power Knowledge-Based Control System", Proc. 1989 American Control Conference, vol 3 pp 2230-2232, Pittsburgh, June 21-23, 1989.

8. Hofmann, A. G.; "Power System Transient Analysis using the G2 Real-Time Expert System," Proc. European Simulation Multiconference p. 516-521, Nuremberg,Germany, June 10-13, 1990.

9. Stanley, G.; "The G2 Real-Time Expert System," Third International Conference on Expert Systems and the Leading Edge in Production and Operations Management, Hilton Head, S. Carolina, April 1990.

10. Rowan, D.A.; "On-Line Expert Systems in Process Industries," AI Expert, August 1989.

11. Mertz, G.E.; "Application of a Real-Time Expert System to a Monsanto Process Unit," Proc. Chemical Manufacturers Association, New Orleans, March 1990.

12. Braunstein, D.M., Brown, A.S., Drew, D.L.; "A Knowledge-Based System for Process Monitoring and Control in Propellant Manufacturing," Proc. ISA 1989, #89-0486, Philadelphia.

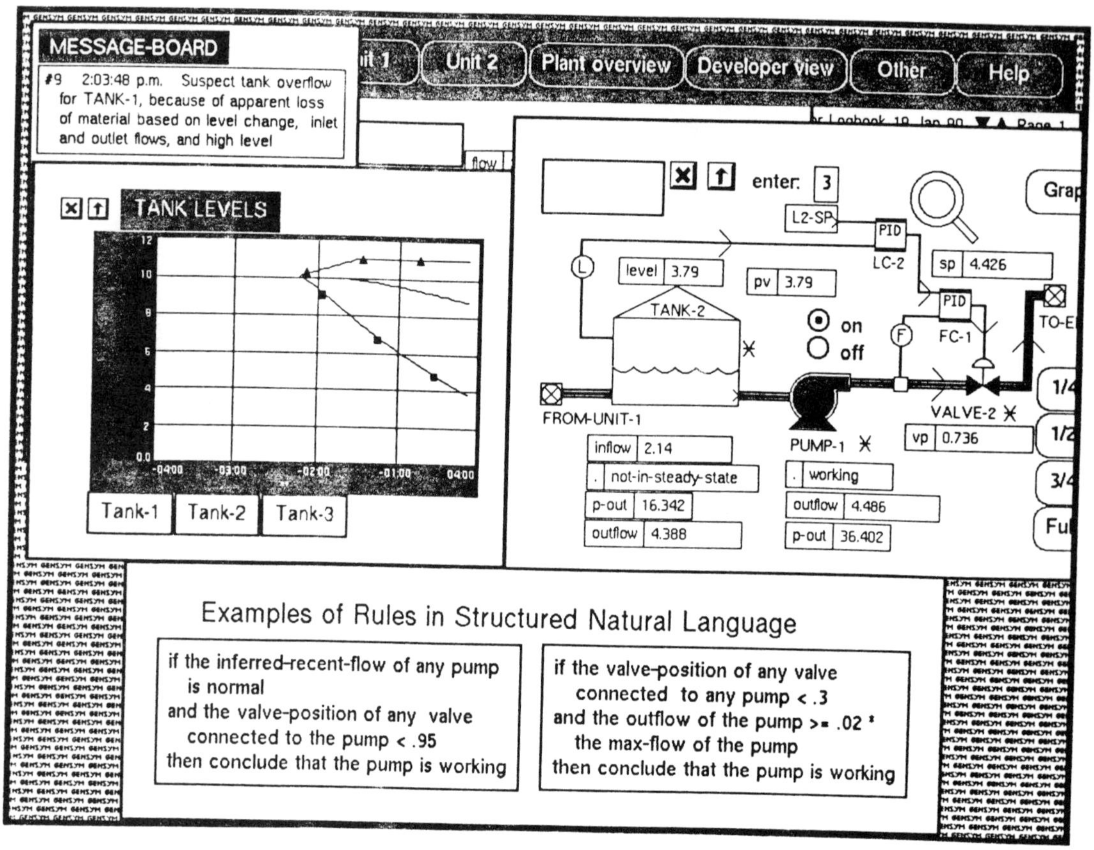

Fig. 1. **Structured natural language for building the knowledge base**

©1991 - Paper #91-736
0074-056X/91/489-499/$0 + .50pp

First Annual ISA/EPRI Joint Control & Automation Conference - POWID 34th Power Instrumentation Symposium

Human-Centered Automation of Testing, Surveillance and Maintenance

Siddharth C. Bhatt and Bill K. H. Sun
Electric Power Research Institute
3412 Hillview Avenue
Palo Alto, California, 94303
and
Joseph Thie
Consultant
12334 Bluff Shore Drive
Knoxville, Tennessee, 37922

1.0 Abstract

Manual surveillance and testing of instrumentation, control and protection systems at nuclear power plants involves system and human errors which can lead to substantial plant down time. Frequent manual testing can also contribute significantly to operation and maintenance cost. Automation technology offers potential for prudent applications at the power plant to reduce testing errors and cost. To help address the testing problems and to harness the benefit of automation application, input from utilities is obtained on suitable automation approaches. This input includes lessens from successful past experience at a few plants where some island of automation exist. The results are summarized as a set of specifications for semi-automatic testing. A human-centered automation methodology is proposed with the guidelines for optimal human/computer division of tasks given. Implementation obstacles for significant changes of testing practices are identified and methods acceptable to nuclear power plants for addressing these obstacles have been suggested.

2.0 Introduction

This paper describes a human-centered automation methodology to address the question: How can automation technology assist humans to effectively and efficiently perform testing, surveillance, and maintenance tasks for instrumentation, control and protection systems at power plants.

2.1 Significance of Testing, Surveillance and Maintenance

The power plant technical specifications require off-line and on-line functional tests, surveillance, and maintenance activities to ensure high availability and reliability of controls, protections, instrumentation, and other plant systems. The periodic and on-demand testing, surveillance, and maintenance tasks are required considering uncertainties in component failure rates, common mode failure rates, reduced redundancy during testing, operator errors during testing and component wear.

2.2 Problems Associated with Testing, Surveillance and Maintenance

The instrumentation, control and protection systems which are often referred to as the I&C systems of power plants are getting obsolete and difficult to test. The testing, surveillance and maintenance activities have led to both system and human errors. Some of the errors have resulted in unplanned scrams and shutdown of the power plants. The human burden for testing has been increasing due to additional regulatory requirements. Task executions related to testing, surveillance, and maintenance of I&C systems are manpower intensive and some tasks in radiation environment increase radiation exposure to plant I&C crews. Time-consuming manual testing, surveillance, and maintenance methods for I&C systems have been in use for the last two decades without any essential change. Typical problems experienced with current nuclear power plants are inadvertent scrams due to frequent testing, long outage time to perform repairs and maintenance, and excessive actuation of equipment contributing to wearout. The average rate of inadvertent scram due to I&C testing is about one scram per plant year of operation. The typical number of hours the plant is unavailable due to the inadvertent scram range from 12 to 24 hours. Approximately 40% of the human error related testing scrams can be attributed to I&C

testing in the Reactor Protection Systems or engineering safeguard systems. Some of the major systems involved are turbine, feedwater, recirculation, and logic systems. This is a growing problem due to aging equipment and associated errors during the testing, surveillance and maintenance activities.[1,2,3]

The problems associated with testing, surveillance, and maintenance tasks are well recognized by the industry as noted in the previous efforts [1, 2] involving surveys of I&C staff at the power plants. In the first of these surveys [1] , the responders ranked the worst problem to be in the documents and drawing areas followed by testability of equipment as the next problem area. The top problem identified in the other survey [2] was the inability of utility staff to obtain obsolete I&C systems replacement parts. Therefore, a careful evaluation and study of problems and processes related to I&C system testing, surveillance, and maintenance tasks are needed.

2.3 Automation Technology

Automation comes to mind as a form of technology that might be helpful to solve the present problems associated with the testing, surveillance, and maintenance tasks of power plants. This is because, in general, one finds trends toward automation when:

1) there are demands for better performance and safety;
2) there is increasing tedium and complexity;
3) technological advances have made automation methods more economical to implement.

Significant advances in instrumentation and computer controlled measurement technologies have occurred in the last decade. The cost to perform computer controlled testing has been steadily going down due to significant cost reduction of computer hardwares. Several opportunities exist to harness the microprocessor revolution in digital computers, smart instrumentation, and automation technologies to improve efficiency, productivity, and reliability of power plant operational and maintenance staff. Recognizing the potential, we look into the human-centered automation methodology and its application to solve power plant I&C system testing problems.

2.3.1 Factors to Consider for Level of Automation

It is important to consider what can be done to solve the problems associated with the I&C system testing, surveillance, and maintenance tasks. Of course, reduction in frequency of unnecessary testing and technical specification relief can help, but it is unlikely that such testing can be completely eliminated. Integration of testing, surveillance and maintenance functions can help reduce some human burdens by performing common tasks and sharing information.

Developing uniform utility standards for automation of testing and surveillance tasks can reduce operation and maintenance costs. First, we need to develop testing, surveillance, and maintenance automation methodology and tools to implement automation solutions effectively at the power plants.

The efficient execution of testing, surveillance and maintenance of a power plant is a complex and information intensive task which is currently accomplished through the manually initiated decision and test procedural actions. The level of automation of these tasks depend on many key factors, such as:

1) Economic return and cost benefit assessment
2) Radiation Exposure - REM reduction
3) Safety, reliability and security
4) Capital investment protection and assurance
5) Maintainability
6) User interface, required training and user acceptance
7) Sensors and instrumentation requirements
8) Environmental impacts
9) Accuracy
10) Repeatability
11) Operability
12) Flexibility/controllability
13) Robustness
14) Diagnostics

15) Repair and revalidation
16 Frequency of task
17) Response time required for completing the task
18) Potential impacts of the task's execution through automation

3.0 Human-Centered Automation Methodology

Studies of methodology to achieve balance between the automation and human operator interfacing with the automation has been an area of increasing interest [4,5,6] We need to recognize that automation in nuclear power plants is continuously increasing. We need to understand the capability of human and automation . We must develop and understand human performance data. Overall, we need to achieve a desired balance between human and automation. Both human and machine reliability are critically important to the overall system reliability for safe and economic operation of nuclear power plants. The human-centered automation methodology enables proper allocation of functions between human and machine throughout the process of design, construction, implementation, operation and maintenance of nuclear plant. The basis for human-centered automation methodology [5,6] is provided in the following discussions.

3.1 Some Useful Definitions

In order to describe the human-centered automation methodology, we use some common terms to avoid confusion [6].

Automation: Allocation or assignment of a human function to a machine.
Machine: A man-made process or device capable of performing tasks assigned.
Function: The specified role or purpose for which human or machine actions are performed.
Task: An action item or work or obligation to be performed to achieve a desired result.

3.2 Capabilities of Human and Automation

Both, human and automation have their strengths and weaknesses. Specifically, humans are best at certain capabilities whereas machines are best at other complementary capabilities.

3.2.1 Human Strengths and Weaknesses

Human are best at certain capabilities which we describe as human strength, such as:

1) Capable of being creative and flexible
2) Able to use knowledge, routines, and patterns to cope with unexpected or beyond-design-basis situations.
3) Recognize patterns within complex sets of information during fault and accident situations
4) Able to abstract useful information from noisy systems
5) Able to make decisions from incomplete information

Some of the human weakness are that humans make errors and human performance degrades under stress.

3.2.2 Automation Strengths and Weaknesses

Some of the strengths of machine automation are:

1) Speed of execution
2) Reliability and repeatability of processing information
3) Accuracy of results and processed information
4) Rapid response
5) Sustained performance
6) Able to handle large amounts, high rates and large capacity of information

Some of the weaknesses of automation are that machine performance generally cannot readily be modified, unless designed with flexibility, and automation fails in a discontinuous way.

3.3 Considerations for Integration of Human and Automation

There are several considerations we can point out for the integration of human and automation. Some of the major considerations are:

1) Integration design should match tasks with capabilities of assigned operator such as education, training, experience, etc.
2) Man-machine interface must follow human factor principles
3) Automatic devices should support human in adverse situations
4) Human involved with the use of automation must understand automation functions and performance during normal and failure modes.

3.4 Basic Principles of Human-Centered Automation

Careful integration of human and automation to complement strengths of both is the key to the human-centered automation methodology. The following are basic principles of human-centered automation:

1) Human cognitive strengths should be exploited by automation designers.
2) Automation should be used to improve safety and reliability by protecting from the variability of human.
3) Automation should start with prescriptive procedural functions.
4) Automation should reduce human cognitive overload.
5) Correct process of allocation of functions should be an integral part of the automation design.
6) Evaluations should consider the professional motivation and psychological well-being of the operator.

3.5 Allocation of Functions

The allocation of functions to human and automation is an iteration process. In order to perform the allocation of functions, several key elements should be considered. Some basic and essential elements are:

1) Knowledge of individual tasks and functions for safety and effective operations
2) Knowledge of capabilities and limitations of the human operators
3) Understanding of capabilities and limitations of the available technology for design, manufacturing and implementation
4) Criteria for determining the allocation process

Some of the influencing features for the allocation process are consideration of existing practices, feedback from experiences, regulatory requirements, feasibility, cost, technical climate and proven technology, policy matters, and cultural as well as social aspects.

3.6 Classification of Tasks

Overall, tasks can be grouped into four major classes. The first class of tasks are those which must be automated. The second class is that class which contains task which should be automated but does not meet the 'must' category. The third class are those tasks which should be done by humans. The fourth class are those tasks which should be shared between human and automation.

The first class of tasks which must be automated are those tasks which exceed the capacity or capability of human. These tasks require:

1) processing of large quantities of data
2) high accuracy
3) high repeatability
4) rapid performance, and
5) where the consequences of errors are severe

The second class of tasks are those which should be automated because they require:

1) lengthy execution or high consistency and accuracy
2) unnecessary exposure to risk to operators
3) tasks which result in boredom or monotony
4) tasks which automation offers more economic performance than human

The third class of tasks are those which should be done by human because they are:

1) tasks which, due to complexity, need heuristic or inferential knowledge
2) tasks which need flexibility due to lack and precise task specification
3) tasks in extreme fault or accident situations that require human flexibility and high level skills

The fourth class of tasks are those which should be shared between human and automation. Some examples are:

1) Operator Information Systems in which automation detects and annunciates plant conditions and humans make judgments, take decisions, and execute control
2) Computerized decision aids and monitoring and diagnostic expert systems in which machines extend the human capability.

For the fourth class of tasks, human must consider system integrity, validation and verification, and consequences of error.

3.7 Usefulness of Human-Centered Automation Methodology

A human-centered methodology, summarized in Figure 1, can be applied to design of new systems, existing plant modifications and upgrades, and examination of existing allocations of functions in systems. The potential users for this methodology are operators who develop requirements specifications for automation, technology designers who design automated machines, and researchers who refine the methodology. The iterative nature of the methodology and its application is clearer from Figure 1.

This methodology utilizes human strengths such as creativity, flexibility, ability to use knowledge, routines and patterns to cope with unexpected or beyond-design-basis situations, ability to recognize patterns with complex sets of information during faults and accident situations. Similarly, this methodology utilizes the digital automation strengths such as speed, reliability, repeatability, accuracy, rapid response, sustain performance and ability to handle large amount, high rate, and large capacity of measured and processed data.

The methodology supports development and implementation of automation of tasks such that it corrects the human errors and human performance degradation under stress. Similarly, the methodology helps to correct the automation weakness in its difficulty to modify or update during the life cycle and its failure in a discontinuous way such as to crash the system.

4.0 Application of Human-Centered Automation Methodology to Testing

A practical approach for accessing the usefulness of automation to I&C was adopted in this study. The basic information and advice on testing improvements was obtained from I&C experts at plants. Confidence in this approach is based on the belief that the ultimate users of any proposed testing aids would be the most knowledgeable in what is appropriate. Therefore a survey was performed to implement this approach.

The survey involved extensive contacts with, and visits to, I&C departments at a number of nuclear plants. Among them, the PWRs are Braidwood, Davis-Besse, Diablo Canyon, Kewaunee, Palo Verde, Prairie Island, and Zion, and BWRs are Dresden, Monticello and Nine Mile Point.

At each plant senior personnel in their I&C departments were consulted. All of them were well aware of its problems and some solutions. The judgmental and opinion aspects of this paper are from this valuable source of input.

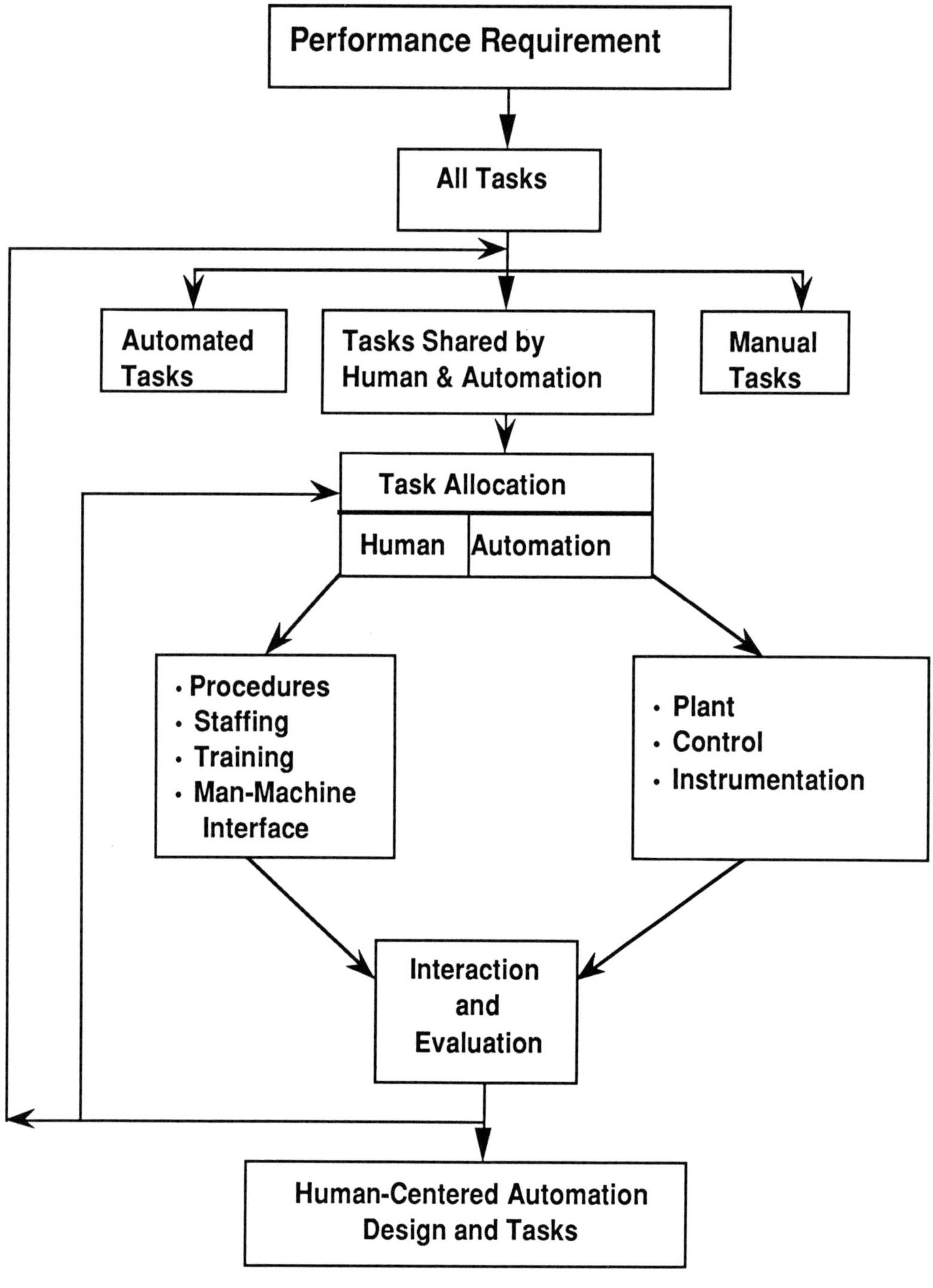

Figure 1: A Methodology to Design Human-Centered Automation

Table 1
Stages Of On-Line Computer Use In The Automation Of Testing

Stage	Computer Use	Signal Injection	Test Method Description
Manual	None	Manual	Paper and pencil data recording, meters, chart recorders, and oscilloscopes used.
Partial Automation	Data Acquisition	Manual	Computer data acquisition with interactive manual control.
Semi-automation	Full	Computer	Interactive manual control.
Full automation	Full	Computer	Very little human involvement.

Table 2
Practical Considerations for Optimal Partitioning of Human and Automation Activities in Testing

Item	Explanation
Human cognitive strengths	Perception, management, reasoning, verification, documentation, generalizing and judgment.
Automation strengths	Error reduction, accuracy, consistency, multiple complex operations, and insensitivity to extraneous factors.
Data processing and display	CPU and CRT capabilities.
Environment	Harsh environment, such as radiation.
Job satisfaction	A need to maintain personnel interest.
Understanding	Intelligent overseeing and not blind useage of a "black box"
Human backup	For failure of automated testing device.
Peripheral personnel	Exercising of mental processes by experienced observations
Costs and benefits	An optimal combination sought.
Human control	An appropriate extent of personnel being in charge.
Computer's conclusions	Information presented without any tendency to bias, or mislead the user.
Responsibility	Maintaining a technician's pride in test ownership.

4.1 Stages of Automation in Testing

To understand what might be meant by automation aids to surveillance and testing, various stages of computer utilization are listed in Table 1. Presently manual methods are primarily used at nuclear plants.

4.2 Existing Testing Automation in Nuclear Power Plants

Though not as progressive as other industries, nuclear plants already exhibit a number of inroads of automation into their testing activities. Specific examples of this follow.

1) A commercial portable automated calibration system (from Loveland Controls) is used by Commonwealth Edison for balance-of-plant instruments at their Dresden BWR.

2) At the Braidwood PWR, Commonwealth Edison initially developed and continues to use a special device called MESAC for computer-assisted reactor protection system testing

3) The newer solid state reactor protection systems from PWR and BWR vendors, already in use, have self-testing as a design feature.

4) Internationally, EdF has replaced manual testing of its reactor protection systems in their 34 older 900 MWe units by a mobile automated system called RPS tester interfacing to cabinets installed as a backfit and, in Japan there is use of computer-stored and computer-monitored procedures along with the computer's data acquisition capabilities.

4.3 Allocation of Human Participation in Testing

The methodology for achieving an appropriate allocation of man and computer is a well-developed subspecialty within human factors engineering [4]. An important question when introducing testing automation aids is how to optimize the remaining extent of participation by personnel. Recognition of this is evidenced by an extensive study by the International Atomic Energy Agency (IAEA) recently.[5,6]. Results from this study, along with corroborating inputs from plant I&C personnel, are given in Table 2.

4.4 Description of an Automated Testing Aid

The following listing originates from plant personnel desires in the area of semi-automation. Some items, marked with a star, have possibilities for additional uses beyond surveillance and testing. Each item is approximately allocated to typical performance requirement task to be primarily done by Automation, Human, or Shared between Human and Automation.

Structural Features

1. Portable and transportable like a laptop computer with expansion slots. (Shared between Human & Automation)

2. Modular functions and modular hardware. (Automation)

3. Facilitation for usage in contaminated areas. (Shared)

Input/Output

4. Screen and full-size keyboard as user interface. (Shared)

5. Versatile input data acquisition and output signal generation. (Automation)

6. Bar-code reading capabilities.* (Automation)

7. Test data entered digitally in the early chain of data handling. (Shared)

8. Printouts appropriate for audits and quality assurance required records. (Automation)

Interfacing Capabilities

9. Hard-wiring, with minimal design change effort. (Automation)

10. Appropriate technician involvement via screen messages. (Shared)

11. Standardization among users. (Automation)

12. Verbal communications-checking software and hardware. (Shared)

13. Procedure monitoring, step by step. (Shared)

14. Interfacing with various database systems. (Automation)

Software Capabilities

15. Easily reconfigurable for new options, data ranges, added modules, new uses, new conditions, changes in assumption, and changes in procedure. (Automation)

16. Information processing and interpreting capabilities.* (Automation)

17. Checking test permissibility.* (Shared)

Operational Features

18. Self-tests, self-diagnoses, and self-calibration. (Automation)

19. Procedures incorporation facilitated by toolkit software.* (Shared)

20. On-screen documents, drawings, and other useful aids.* (Shared)

21. On-line signal monitoring and data analyses capabilities.* (Automation)

22. Error-checking during test performance. (Shared)

23. Troubleshooting aids.* (Shared)

24 Accuracy as good as or better than manual (Automation)

Other Features

25. Favorable cost/benefit as well as being affordable. (Shared)

26. NRC acceptability methodology, especially verification and validation. (Shared)

27. A simple user-friendly device. (Shared)

28. Minimal dependence on vendor support. (Automation)

29. Documentation at levels for the user vs.the engineer. (Shared)

30. A startup and installation service. (Automation)

31. Users groups for exchanges and cost-sharing. (Manual or Shared)

32. Controls over access and changes to software and data. (Shared)

33. Training assistance during usage. (Automation)

34. Fully debugged and highly reliable. (Automation and Manual)

35. Manual backup tests always possible as an option. (Manual)

4.5 Cost and Benefits

According to I&C departments at plants, it is most important that any major changes in approaches to testing must be cost effective.This is because expenditures would likely come from discretionary funds where there is competition from other projects.

A list of cost factors for consideration are hardware, software, installation, customizing, modifying, database handling, training, overseeing, maintenance and updating. A special cost and deterrent is that associated with paperwork. Plants are willing to implement hardware to improve balance-of-plant equipment testing.However they are reluctant to backfit safety system tests because of engineering paperwork loads.

Benefits of automation aids fall into tangible and intangible categories. For the former there are reduced personnel time, consistency and accuracy, error reduction, reduced radiation exposure, less scram risk in the method, less time in an undesirable test mode, and sometimes less wear.Intangible benefits include improved image from fewer incidents, more information from data, better quality assurance, easier test scheduling, and peripheral non-test uses.

5.0 Conclusions

A human-centered automation methodology for testing and surveillance is proposed. The following general conclusions are made:

1. Needs for and advantages of automated testing aids are widely and unanimously recognized in plant I&C departments.

2. The cost and benefits are major considerations in justifying any test method changes in a plant. For use in safety systems, engineering paperwork associated with modifications and procedures dominate the cost and can be a significant deterrent.

3. To reduce cost by sharing by multiple plants agreeing on similar approaches is attractive.

4. Deterrents in making major changeovers, as from manual to automatic methods, can be ameliorated by gradual implementation: test methods would be changed at times of obsolescent equipment replacement, procedure revisions, regulations changes, corrective actions for incidents, etc.

5. A general set of specifications for a versatile portable aid to automated testing has been derived based on inputs from potential users at plants.

6. Diversity in plant instrument configurations and testing customs, combined with diversity in opinions, leads to requirements of flexibility and modularity if one automated test system is envisioned for use in multiple plants.

7. Care must be taken to make proper use of human participation when automation aids are introduced. Special guidelines for optimizing human involvement should be followed.

8. Usage of test automation by a plant should be a fully debugged application in an area of essentially guaranteed success with economic payback.

6.0 Acknowledgements

The authors are indebted for the advice and judgments received from members of the I&C departments from the following utilities : Arizona Public Service, Commonwealth Edison, Niagara Mohawk Power, Northern States Power, Pacific Gas and Electric, Toledo Edison, and Wisconsin Public Service.

7.0 References

[1] J. A. Thie, "Surveillance of Instrumentation Channels at Nuclear Power Plants", Volume 2: An Approach to Classifying Problems and Solutions, EPRI report NP-6067, June 1989.

[2] "Instrumentation and Control Strategies for Power Plants", EPRI NSAC 153, Vol. 1 - Notes on the I&C Workshop, August 1990 and Vol. 2 - Problem Definition and Recommendations, December 1990.

[3] "Integrated Instrumentation and Control Plan", EPRI Nuclear Power Division draft report, January 1990.

[4] H. E. Price, R. E. Maisano, and H. P. Van Cott, "The Allocation of Functions in Man machine Systems: A Perspective and Literature Review." DOE, Oak Ridge National Laboratories, June 1982, NUREG/CR-2623.

[5] "Balance between Automation and Human Actions in Nuclear Power Plant Operation," IAEA International Symposium, Munich, Germany, July 9-13, 1990.

[6] W. Bastl, J. Jenkinson, A. Kossilor, R. Olmstead, A. Oudiz, and B. Sun, "Balance Between Automation and Human Actions in Nuclear Power Plant Operation - Results of International Cooperation," IAEC International Symposium, Munich, Germany, July 9-13, 1990, Paper No. KAEA-SM-315/1.

©1991 - Paper #91-737
0074-056X/91/501-511/$0 + .50pp

NEURAL NETWORK APPLICATIONS TO MEASUREMENT CALIBRATION VERIFICATION IN POWER PLANTS

A. Ipakchi
M. Khadem
C. M. Peng
ABB Systems Control
2550 Walsh Avenue
Santa Clara, CA 95051

Robert W. Colley
Nuclear Power Division
Electric Power Research Institute
3412 Hillview Ave
Palo Alto, CA 94303

ABSTRACT

Nuclear power plant operations and safety rely on the proper operation of the plant monitoring, controls and protection systems. These systems in turn, depend on plant instrumentation for a correct reporting of equipments status and the thermofluid condition of the process. The incorrect reading can be due to instrument failure or drift in calibration. Plant instrumentation are therefore recalibrated on a periodic basis. This is a time consuming and costly operation.

This paper presents the preliminary results of an EPRI sponsored research project to develop a neural network-based analytical redundancy technique for instrument calibration reduction. The neural network will be used to predict the reading of a target instrument using readings from other dissimilar instruments. This technique is similar to the conventional on-line parameter identification and model-based observer methods.

A methodology for selecting required dissimilar instruments inter-related to the target instrument, and determining the neural network structure is presented. This methodology requires the development of a dynamic model of the processes based on the first principles of physics. An observability test was used to determine the inter-relationship between selected input variables and the target variable.

1 INTRODUCTION

Safe and efficient operation of nuclear power plant requires reliable and correct measurement readings. Erroneous or contradictory instrument signals may provide the operator with incorrect information, or cause plant equipment to operate in an undesirable condition. Such upsets can challenge plant safety and availability. It is reported that [1] 10% of total unplanned shutdowns are due to instrument failures.

A measurement error can be due to a drift in the instrument's calibration or due to the instrument failure. Errors due to calibration drifts are not easy to detect because they often correspond to a gradual degradation of the measurement device. Due to the critical role of accurate instrumentation readings in plant operation, there are extensive regulatory requirements for periodic calibration of the key instrumentation. Instrument calibration typically requires manual intervention, and is a time consuming operation. A significant portion of a nuclear power plant operation and maintenance budget is consumed by the instrument recalibration activities.

In addition, technician errors and other inadvertent actions during instrument calibration, can potentially cause upsets in control or protection systems, and lead to undesirable transients, equipment or reactor trips, and loss of generation. Therefore, strict procedures are employed to avoid occurrence of such undesirable conditions.

In nuclear plants, those measurements which are critical to safety are equipped with redundant instruments. Typically a triple redundancy is employed. Signal validation techniques are used to determine the most accurate reading from redundant measurements [2]. Many of these techniques, such as parity space [3], rely on software to determine the most accurate value. They are also capable of detecting a failure, or a significant drift, e.g., several percent. However, these techniques are not widely applied to reduce the required frequency of instrument calibration.

Not all plant measurements, however, have redundant instruments. Most parameters, including those related to performance monitoring have a single instrument. Therefore, periodic calibration test is currently the only means for detecting soft failures and instrument drifts on non-redundant instruments. With the availability of digital data acquisition and on-line computer systems, it is possible to generate redundant readings using analytical techniques. This is basically done by computing the estimated value of a given measurement using data from other dissimilar instruments. Once such redundancy is available, signal validation or other software-based data comparison techniques can be employed to detect soft failures, or calibration drifts.

This study is sponsored by EPRI, and was started in 1989 as an exploratory research activity. It was initiated to study analytical techniques for instrument calibration reduction based on sound mathematical and physical principles. The study has explored mathematical modeling and systems analysis techniques for development of on-line observers. This included methods for determination of dynamic coupling between dissimilar instruments. This paper presents activities underway to use these results in design and development of neural network-based predictors for plant measurement readings.

2 ANALYTICAL REDUNDANCY TECHNIQUES

Computer processing power has dramatically increased over the past decade, while computer hardware cost continues to decline. Aging plant computers are being replaced with new systems offering order of magnitude increased processing power. This has opened new opportunities for developing software-based application programs for the improvement of plant operations, safety, and availability. One such application is signal validation and signal failure detection based on analytical techniques [1,4,5,6].

The concept of predicting instrument readings from a set of dissimilar instruments using analytical redundancy technique is illustrated in Figure 1. A computer model is used to accurately represent the relationship between an output, a target instrument reading, and a set of model inputs, readings from other instruments. This model will be installed on-line, fed by current readings of its inputs, and will compute an estimated value of the target instrument reading under the current plant operating condition. The current readings of the target instrument is then compared against its computed estimated value, and any deviations are analyzed for detecting calibration drifts or instrument failures. This is a viable concept, if the model could provide sufficient accuracy, e.g., better than 1 %, over a wide range of operating conditions.

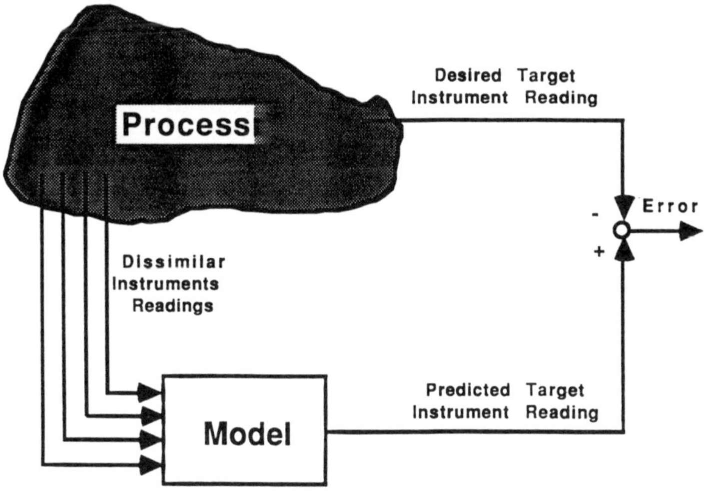

Figure 1: Layout of Analytical Redundancy Concept

Analytical redundancy may be based on first principle models, statistical models or neural network technology. First principle models include methods which use physical laws to develop mathematical descriptions of the physical system, i.e. differential equations that describe the physical processes. Such methods are more complicated to implement than statistical methods; however, they have the potential to perform over a wider range of system's operating conditions, and predict its dynamical behavior. The key limitation of this technique is a proper tuning and parameterization of the model to the process it is representing. On-line model identification techniques [7] have been successfully used for linear dynamical systems. However, most of power plant processes have non-linear characteristics, and on-line identification techniques for such systems are not well established.

Statistical methods [5] use correlation techniques to develop an input-output model using time history data. The advantage of the statistical methods is that no mathematical description of the process is needed and the implementation is relatively simple. The disadvantage is that they are based on a "curve fit" to process data, and depending on the conditions under which these data are collected, and the selection of the input variables, they may not accurately represent all of the modeled process characteristics. Most statistical techniques also ignore dynamical relationship between inputs and output, and are basically steady state models.

Neural network based techniques [8,9] have several attractive features for this applications. First they can model a highly non-liner input-output relationship. They can be trained, similar to statistical techniques, using process data. Eventhough training is computationally intensive, but once trained they can predict the output very quickly with a relatively little computation required.

In order for an analytical redundancy technique to be effective in detecting instrument failures and calibration drifts, it should meet the following requirements:

a. The target instrument, i.e., model output, should be observable [10] from the selected input dissimilar instruments. This means that the model inputs should collectively represent all the process parameters required to compute the output signal. Therefore, proper selection of required dissimilar instruments is critical.

   The observability test is a well defined concept in linear dynamical systems [7], and can be directly evaluated using the system equations.

b. The model should accurately represent the input-output relationship under various plant operating conditions. This includes plant modes of operation, and system configuration.

c. The model should account for any significant process dynamics. Eventhough nuclear plants mostly operate under steady state conditions, but the dynamical impact of process noise and control system regulation may be significant for this application.

d. Techniques should be available to parameterize the model to actual process conditions. This is especially important since plant equipment may not be operating at their designed characteristics due to aging or other process conditions. An automated model tuning capability should be available to allow periodic model re-parameterization (training), e.g., after a periodic instrument calibration or equipment maintenance activity.

An approach to obtaining information regarding one variable using measurements made on other variables is by means of an Observer [11]. There are two primary limitations of an observer based failure detection system. One is the difficulty of implementing an observer for high order systems such as a full nuclear plant model. The other is the possibility of generating inaccurate state estimates due to the inaccuracy of the assumed model.

The difficulty of implementing a high order observer in real time is due to the fact that the order of the observer is the same as the order of the model. A solution to this difficulty would be to obtain a lower order model that retains the essential features of the original model.

The problem of observer sensitivity to plant variations may be more difficult to solve. The linear model of a system is only an approximation to the actual non-linear system and there are continuing changes in the system parameters as the operating point changes. Special care must be taken to make the linear observers robust with respect to parameter variations, or non-linear observers must be used.

The conventional observability analysis technique results in a simple yes/no result while no relative measure of observability was provided. The observability properties of the state-space system tested by using Observability Gramians [15] can rank each output measurement on its ability to observe a certain set of states. Observability Gramians not only have better numerical properties than the observability matrices, but also can determine the observability of the states with respect to each sensor. This information can then be used to associate the outputs with previously determined states and control groupings to specify complete subsystem descriptions.

The observability Gramian also has its physical meaning. From the deterministic point of view, the "output energy" in the system is proportional to the Gramian. The diagonal matrix elements of the observability Gramian are a measure of the output energy available from the system and indicate information about the states of the system. The off-diagonal matrix elements of the observability Gramian are related to the inter-relationship of the states. This information can then be utilized to the weighting factor of the the artificial neural network for dissimilar instrument calibration use.

This paper present a systematic approach for addressing each of the above items. This approach is based on utilizing the knowledge of process physics, and applying mathematical modeling and systems analysis techniques. Eventhough the approach is generally applicable to other analytical redundancy techniques, we shall focus on neural networks here.

3 NEURAL NET DESIGN METHODOLOGY

The general approach for development of an on-line neural network model for instrument failure detection is illustrated in Figure 2.

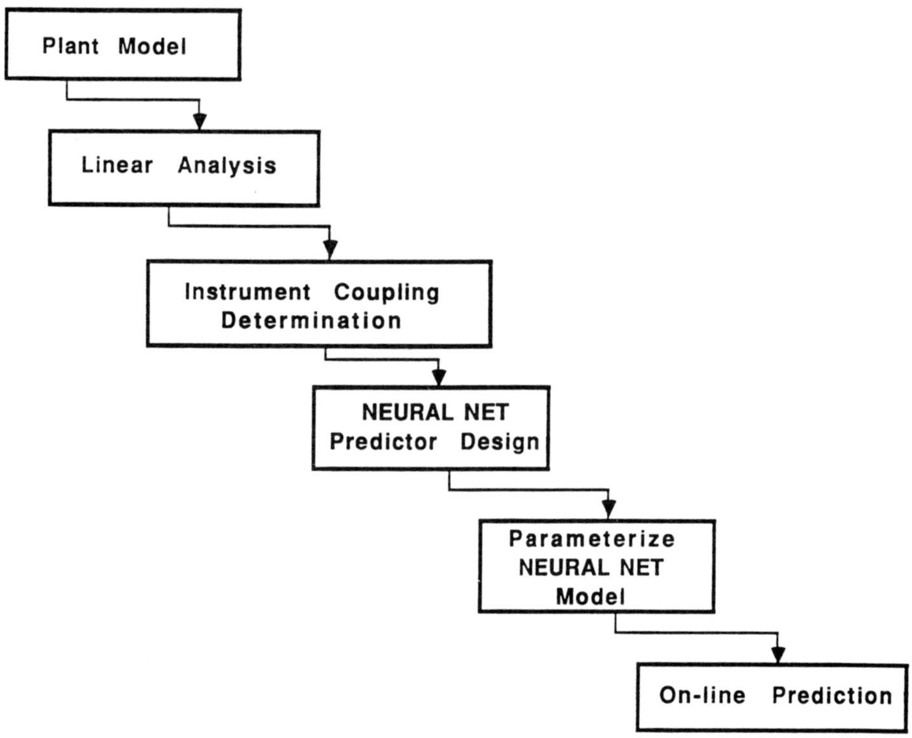

Figure 2: Neural Net Design Approach for an On-line Predictor

This approach starts with of a simulation model of the power plant processes. This model should be based on first principles of physics, and have sufficient fidelity to accurately represent the qualitative characteristics of the process. The model scope should cover the major plant instrumentation of interest. For a typical nuclear power plant this may result in a highly non-linear model with 100 or more state variables. Depending on the target instrument, it may be possible to use a subsystem model rather than a full plant representation. This depends on how decoupled is the subsystem of interest from the rest of the plant. Simulation packages, e.g., EPRI's Modular Modeling System [12] may be used for this purpose.

Since no easy computational method exists for directly determining the observability of non-linear systems, the model is linearized about an operating point. By analyzing the linear system equations several important information can be obtained. Eigen values are computed to provide information on system dynamic modes and time constants.

Next, as indicated by the block labeled "instrument coupling study" in Figure 2, an analysis is performed to determine the extent to which a functional relationship may exist between a target instrument, whose output is to be verified, and a reference set of dissimilar instruments. Using the observability test a minimum set of required dissimilar instruments are identified [10]. The model may be linearized for several different operating conditions, and this test be repeated to ensure a complete set is identified.

The degree of dynamical relationship between the inputs and the output should also be determined. This will simply indicate how much of input history is needed to correctly estimate the output signal. This step involves converting the system equations to modal form [7] using the eigen values and eigen vectors previously calculated.

The next step is to make use of this information to synthesize a neural network predictor that will produce a signal which can provide an independent check of the target instrument. The neural network structure is designed based on the observability test, required input signals, modal analysis, and the input time history requirement. Following this the parameters of the neural net will be optimized using plant or simulation data in conjunction with an optimization search such as backpropagation. Finally the effectiveness of the detection system will be tested, and evaluated through the use of actual plant data. In implementing this approach, one or more plant instruments, with a history of reported problems, such as feedwater flowrate would be selected.

Since this methodology requires mathematical models of the plant derived from first principles of physics, it is expected that the results could potentially have much greater validity than statistical results which have no firm basis for causality.

3.1 Neural Net Selection

Because of the complexity of the system and the presence of non-linearities, it is likely that there will be significant difficulties in using a linear observer as discussed in Section 2. This leads to the consideration of using a neural net to estimate the correct reading of one instrument from the reading of other dissimilar instruments. However, the concept of observability can be useful in providing insight into the proper selection of inputs to the neural net. The idea is that if a set of instruments could be used to produce a linear observer to accurately generate the reading of another instrument, then it should be possible to construct a neural net to accomplish the same result, and conversely, if it is not possible to construct a linear observer to achieve that result, then it is unlikely that a neural net would be successful.

A typical neural net consists of an interconnection of nodes starting at input nodes, and proceeding through one or more layers of hidden nodes and terminating at a set of output nodes. There is associated with each path a non-linear function, usually a sigmoidial function, and a weighting factor. Typically the weighting factors are determined during a series of training runs by searching for the set of values that minimizes an error functional that is related to the difference between the output of the neural net and the desired output. Usually a gradient search algorithm is used to make the adjustments in the weighting factors. Because neural net have a high degree of flexibility in describing functional relations between variables, even if those relationships are highly non-linear, the use of a neural net has good potential for success.

To apply this technique to the instrument calibration and failure detection problem, the output of the neural net would be the output of the sensor for which the estimated measurement is to be generated. The inputs would be the present and past outputs of sensors for which a functional relation will be established using the system model simulation and observability computation. It should be noted that the past output values are needed to represent the effects of the differential equations describing the system.

After training runs have been made to acquire the best neural net-based instrument reading predictor, the configuration will be "tuned" to actual plant data.

3.2 Dynamic Neural Net

A standard neural net may not explicitly represent dynamic relation that exists between variables as described by differential equations. To represent these dynamic relationships, the input to the net must include past values of the input variables. Thus for the instrument calibration problem, past values of the instruments readings (dissimilar instruments) to be used in the calibration process would be part of the neural net input vector.

One method of implementing the dynamic relationships between the output variable and input variables, is to use the standard neural nets and represent the discrete past values of each input variable as additional input to the net. One consideration in the structure design of the neural net, is the selection of discrete past values for each input variable. For example, for each of the input variables, how many and which past values from time "t-1", "t-2", ..., etc., should be included as additional input nodes. Linear and modal analysis of a dynamic system, help on the selection of past values of each of the input variables. linear analysis provides information on the time constants of the system. Each input variable is related to one or more of the system time constants. To find this relationship, the modal analysis provides contribution of each mode of the system (time constant) to each of the input variables. As to how many time interval should be considered, it is a matter of experience and testing with each individual dynamic system.

Another approach of implementing the dynamic relationships, is suggested by Werbos [13] in a procedure referred by him as "Backpropagation Through Time". In his approach, every hidden and output node of the net, receives input values at the two previous time intervals, from any of the nodes, including the input nodes. The input nodes in his approach also receives current values as well as the values at the two previous time intervals. Each input to any of the hidden and output nodes is associated with a weighting factor. The weighting factors are then obtained through a solution method similar to simple backpropagation method [13]. The solution is simply to calculate the weighting factors based on minimizing the square of the error function between actual and desired values of the output variable.

To successfully implement Werbos's backpropagation through time procedure the following considerations should be addressed:

1. The input output set must be properly selected. The output variable (target instrument) should be observable through the input variables (dissimilar instruments);

2. The neural net model (structure) should accurately represent the relationships between input variables and output variable. To accomplish this, the model should include the dynamics (input values for previous time intervals);

3. The input/output data sets from the system's dynamic processes should have the capability to parameterize the model during the supervised learning stage (i.e., obtain the node connections/equations and weighting factors);

4. The on-line implementation should have the capability of adapting to changing conditions.

3.3 Target Application

To demonstrate the application of the methodology, an instrument with problem in a nuclear power plant is targeted. Feedwater system in pressurized water reactors (PWRs) are reported to have difficulty in providing correct feedwater flowrate due to the change in physical characteristics of the venturi meter, thus requiring frequent recalibration [14]. The incorrect reading of the amount of feedwater flowrate due to drift in calibration, affects the feedwater control valve position which in turn causes improper change in steam and power generation. For cases where the feedwater flowrate reading is showing a value below its actual value, this problem can cause an unnecessary decrease in power generation. Feedwater flowrate parameter, is thus selected as the target instrument for this study.

To develop a neural net model for this application, as illustrated in Figure 1, one needs to properly select the input/output set such that the output is observable through the inputs. To perform an observability analysis, it is required to develop a dynamic model of the system under consideration (i.e., feedwater system) based on first principles. The EPRI Modular Modeling system (MMS) was selected for this purpose. The model includes the feedwater terrain from the condenser outlet to the inlet of the steam generator. Proper boundary conditions will be set at the two ends of the terrain.

Using observability and modal analysis for the simulation model of the feedwater system, the inter-relationship between the selected input variables (dissimilar instruments) and feedwater flowrate (target instrument) will be obtained.

The relationship thus obtained provides a guideline for selection of those input variables whose present and past values are coupled to the present value of the feedwater flowrate. The coupling will be ignored when falling below a predetermined threshold value.

A dynamic neural net model as discussed in previous section will be developed to use information obtained from observability analysis. Input/output data sets will be used to train the neural net and obtain the weighting factors. The data sets are either generated by the simulation model or obtained from actual plant data.

4 CONCLUSIONS

It has been proposed that a neural net could be used as part of a system to detect the need for recalibration of instruments by using the data from other instruments that are measuring different variables. In order to accomplish this effectively it is necessary that the inputs to the neural net have some significant functional relationship to the instrument whose calibration is in question. This functional relationship need not be an algebraic one, but might be a more complicated relationship involving the system dynamics. The necessary functional relation could be identified by using a procedure that involves linearization of the system of differential equations observability analysis. The observability Gramian provides the relative measure of the states with respect to each sensor such that the neural network weighting factor can be implemented based on this measure. This methodology provides information towards selection of a proper set of inputs and building a structure for the neural net predictor model, to be used for calibration of a target instrument using other dissimilar instruments.

5 REFERENCES

1) C.P. Tzanos, A Methodology of Plant-Wide Data Validation and Fault Identification, ANL/RAS 84-3, 1984.
2) Guidelines for Nuclear Plant Performance Data Acquisition, EPRI NP-3915, Volume 2, 1985.
3) On-line Power Plant Signal Validation Technique Utilizing Parity Space Representation and Analytical Redundancy, EPRI NP-2110, 1981.
4) Validation of Critical Signals for the Safety Parameter Display System, EPRI NP-5066M, April 1987.
5) C.M. Smith and R.C. Gonzalez, Automated Long-Term Surveillance of a Commercial Nuclear Power Plant, NUREG/CR-4577, 1987.
6) J.E. Mott, User's Guide for the System State Analyzer, Saratoga Engineering consultants, Inc., San Jose, CA, 1986.
7) Y. Takahashi, M.J. Rabins, and D.M. Auslander, Controls & Dynamics Systems, Addison-Wesley Publishing Co., 1970.
8) M. McCord Nelson and W.T. Illingworth, A Practical Guide to Neural Nets, Addison-Wesley Publishing Co., 1991.
9) T. Khanna, Foundations of Neural Networks, Addison-Wesley Publishing Co., 1990.
10) A. Ipakchi, M. Khadem, C.M. Peng, and F. Alexandro, Phase I Report, May 1990, (to be published as EPRI report).
11) D.G. Luenberger, "An introduction to Observers", IEEE Transactions On Automatic Control, AC-16, No. 6 (December 1971), P 596.
12) Modular Modeling System (MMS), EPRI CS/NP-3016, 1987.
13) Paul J. Werbos, "Backpropagation Through Time: What it Does and How to Do It", Proceedings of The IEEE, Vol. 78, No. 10, October 1990.
14) Robert W. Colley, Personal Communication, August 1990.
15) Kailath, T. "Linear Systems", Prentice-Hall, 1980.

©1991 - Paper #91-738
0074-056X/91/513-519/$0 + .50pp

Intelligent Control System Characteristics and Challenges

Siddharth C. Bhatt
Electric Power Research Institute
3412, Hillview Avenue, Palo Alto, California, 94303

1. Abstract

Advances in the areas of automatic controls, electronics, sensors & transmitters, signal processing, computer systems, process information analysis and decision making, artificial intelligence provide potential for intelligent approaches to the complex system controls through synthesizing Intelligent Control System, (ICS), concept. The ICS may be autonomous, human-aided or computer-aided, with the potential to yield higher level of system performance in terms of reliability, accommodation of significant changes in system parameters, and in providing a safe graceful system degradation. The ICS can utilize extensive computation to (1) reason about the current system state and detect changes in the system's internal state and environment, (2) use qualitative and quantitative models of system behavior by means of multiple sensor inputs, and (3) generate actions to modify system performance operation.

2. Introduction

There have been many successful applications and implementations of conventional control system concepts where the controller goals are essentially fixed and defined by the designer. While this approach has been adequate for controlling some systems, it is not sufficient for controlling complex control systems which require some form of automation and intelligent on-line decision making in the control objectives [1, 2]. This type of system has to select optimum goals based on the knowledge of the current controller state, background experience and knowledge of the process dynamics, and on other constraints and goals defined by the higher levels of control system [1]. Such systems may require the capability to reason under uncertainty which often need decision and action involving some imprecision. The control system may need to make decisions in an unstructured or fuzzy environment [3, 4]. Many practical complex processes have temporal and spatial non-linear behavior such as electric power plants and their transmission and distribution networks, military command, control and communication systems, and air traffic control systems which can potentially utilize some features of intelligent controls.

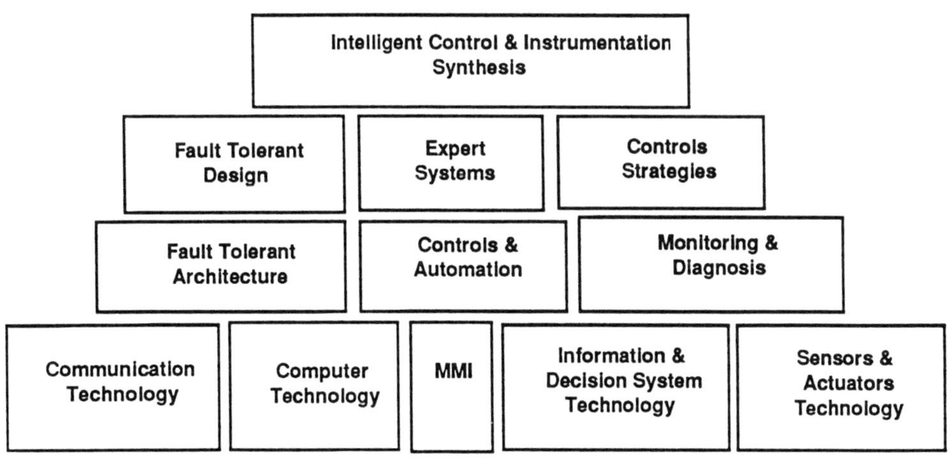

Figure 1 : Typical Advanced Technology Building Blocks for Developing an ICS System

Advances in the areas of computer technology, information processing and time-critical decision making technology, human-centered automation and control man-machine interface technology, smart sensors, signal processing and actuation systems, and electronics have provided potential for new approaches to the control of complex systems. A typical technology building blocks available for intelligent control system [5] are shown in Figure 1.

3.0 Characteristics of Intelligent Control System

Intelligent Control System concept is a relatively new area and therefore a limited precise definition of this concept is still in its infancy [6]. Thus, it is useful to describe intelligent control systems in terms of their attributes and characteristics [7].

3.1 Behavioral Attributes of Intelligent Control System

An intelligent control system utilizes dynamic process knowledge together with the recognition of the internal system state as well as external environmental changes from the sensors to achieve higher level system goals over time in the presence of uncertain, nonlinear, noisy environments with a high degree of autonomy and minimal human intervention [6, 7]. Thus, an intelligent control system exhibits such attributes as: high degree of autonomy, reasoning under uncertainty, performance in a goal seeking manner, higher level of abstraction, data fusion from multitude of sensors, learning and adaptation in a heterogeneous environment to perform timely and effective real-time decision making to carry out optimum actions to satisfy the over all controller goals.

Table 1 summarizes some of the key characteristics of an intelligent control system. They are, by no means, exhaustive characteristics but provide a general description of the key attributes.

3.2 Structural Attributes of Intelligent Control System

The intelligent control system can be developed utilizing typical technology building blocks of Figure 1, with variety of system architectures, most of which share common structural characteristics. There are several ways one can describe structural attributes of an intelligent control system. Two types of descriptions are given, one is in the form of traditional hierarchical form familiar to traditional control engineers, and another form is more generally used in the artificial intelligence field.

Table 1
Some Key Characteristics of Intelligent Control Systems

- Respond Autonomously to discrete symbolic and numeric commands
- Perform diagnosis and identification of changes in system structure or configuration
- Provide more responsiveness in operations through human-centered controls and automation
- Use sensor data and analytical redundacy to detect system faults and human errors
- Facilitate effective use of heterogeneous numerical and symbolic knowledge sources
- Decide on course of action based on sensor and process information knowledge
- Take on more of the functions currently performed by human operator
- Reconfigure a backup controller, if necessary, to respond to configuration changes.
- Ability to handle large number of sensors, process information in timely manner to make optimum decision to support higher level system goals
- Ability to perform data fusion of large number of sensors data
- Able to resolve conflicting data to reason under uncertainty
- Able to integrate distributed multiple intelligent subsystems operating concurrently and asynchronously
- Functional hierarchies may appear in organization due to the need for efficient processing of sensor data and decision making.
- Functional hierarchies may be associated with different time-scales of phenomena / signals and decision making / control

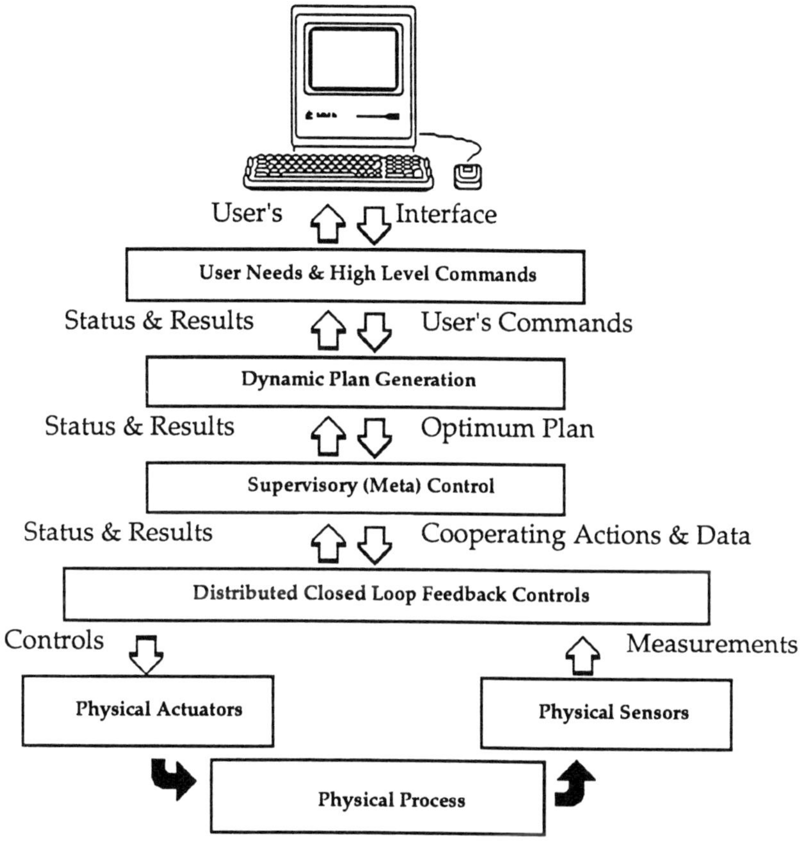

Figure 2
Hierarchical Structure of Intelligent Control System

3.2.1 Hierarchical Structural Attributes of Intelligent Control System

A schematic hierarchical architecture of intelligent control system is shown in Figure 2. Some of the features of the system are the high level interpretation of user's requirement, conversion of the user requirements into high level commands and dynamic goal setting supported by dynamic plan generator which in turn provides higher level control objectives to the supervisory controller.

The plan generator selects a control plan expected to meet requirements, evaluates prior run results, and modifies the plan to meet the objectives to optimize the performance.

The more common controller feature is the integrated distributed control structure of supervisory control supported by the distributed digital controllers which interface with the process through sensors and actuators. The supervisory control loads actions and data in control systems, monitors plan sequence progress, logs and displays data, detects and accommodates faults or abnormal events with alternate plans, and assists higher level user interface with upset recovery. The closed loop feedback controls tracks set-points, gains, and adaptive adjustments according to the plan, provides dynamic compensation for stable operation, monitors and manages low level scan, alarm, and protection limits, and assist higher level system in effective mode transition from normal to abnormal or transient operating conditions.

3.2.2 Artificial Intelligence Form of Structural Attributes of Intelligent Control System

The characteristics of intelligent control system can be grouped into three major organization layers found in most artificial intelligence reasoning systems: Representation, Inference, and Control layers [7].

The representation layer is concerned with identifying, describing, and maintaining the relevant data and knowledge essential to perform the intelligent control tasks.

The inference layer is focused on the reasoning process. The inference layer provides the mechanism that evaluates, interprets, and executes the data and knowledge in the intelligent control system to generate inferences or sequence of actions to be executed.

The control layer is responsible for monitoring and guiding the inference mechanism to ensure the proper handling of interrupts, time-deadlines, changes in goals, and states.

Table 2 summarizes some of the key attributes of the intelligent control system using artificial intelligence type of grouping of the system characteristics.

One of the key attributes of the intelligent control system is the capability to reason under time constraints. An ICS must provide real-time responses. To meet the real-time performance requirement, an ICS must be able to handle interrupts and reason within deadlines.

The ability to handle interrupts can be accomplished by storing the state of the reasoning system inference network current evaluation, current goal evaluation, and other related facts and related inferred data, attending and serving the time critical real-time interrupt, and re-entering the state taking into the account changes occuring during the interrupt which could have modified the original state. Techniques such as truth maintenance can be used for this incremental state revision. An ICS must have multi-tasking, multi-priority capability.

Table 2
Artificial Intelligence Attribute of ICS

Representation Layer Atrributes
• Able to handle both numerical and symbolic data • Able to describe heterogeneous data and models, user input, domain knowledge, and feedback information at different levels of abstractions
Inference Layer Atrributes
• Capable of reasoning by variety of means and techniques • Capable of using various models for reasoning e.g. Analytical models, Qualitative Models, Fuzzy Logic models with fuzzy production rules, knowledge based system utilizing production rules, and neural networks
Control Layer Atrributes
• Capable of Meta-reasoning including determining computing resource allocation, planning resources to meet hard and soft deadlines, able to handle interrupts • Capable of Learning including through strutural and parametric adaptation, rule generation, and rule modification • Capable of Monitoring and Reconfiguring with supervisory mode capable of monitoring goals, constraints satisfactions, and evaluating performance indices • Capable of state estimation, identification of normal, abnormal, and transient state changes • Capable of knowledge Base and model selection as a result of the estimated state of the system and desired future trajectory to achieve desired goals

An ICS must be able to reason within deadlines. The system must be able to select the appropriate model approximation suitable for the task and its time budget.

Table 3
Some of the Challenges Posed by ICS

• Verification and validation of hybrid ICS
• Scalability of ICS from a prototype to a full- size application usually two order of magnitude larger.
• Functional decomposition, modularity, hierarchy or topology, and heterogeneous knowledge representation of ICS design.
• Complexity of interactions among multiple agents including hardware as well as person in the control loop.
• Learning and adaptation such as parameteric and structural learning, self organization etc.
• Reasoning and planning in dynamic uncertain environments and synthesizing techniques such as rules, stochastic optimization, neural nets, fuzzy logic etc.
• Decoposition of high level human commands into lower level constraint and commands.
• Methods for obtaining knowledge (rules) from large amount of process data such as automatic generation of expert system rules from data.
• Methods of modifying knowledge (rules) as a result of past process measurements and experience such as responding to recurrent failures.
• Methods for automatically constructing knowledge bases such as synthesizing information from physical principles with validated process signals and meaurements. In other words machine learning
• Methods to gurantee fault tolerance, and techniques for the ICS to have capability for graceful degradation in the event of missing part of information or rules.

Furthermore, the system must determine which subset of the model can be used to achieve the goal within the deadline. This resource allocation process implics the availability of time-estimates for the execution of different sub-modules or subsets of the various available models.

Table 4
Some Benefit of ICS Applications

• Reduce human burden and error, and improve user's productivity through human-centered control products
• Increase autonomy and automation for various process control systems to reduce operation cost and improve performance
• Provide new functions to improve productivity of both human operator and the system
• Meet stringent safety, reliability, availability, and cost goals
• Able to solve complex tasks which were not solvable previously due to cost, reliability, and safety constraints

4.0 Challenges of Intelligent Control System Development and Implementation

The potential benefit of the use of the Intelligent Control System can be realized only after successfully resolving the challenges posed by the intelligent control system concept. The challenges posed by the ICS implementation are manyfold. Some of these challenges are summarized in Table 3.

Of course, there are numerous challenges to be met to make ICS application to modern complex problem a reality. The above list is an attempt to highlight some of the more important challenges.

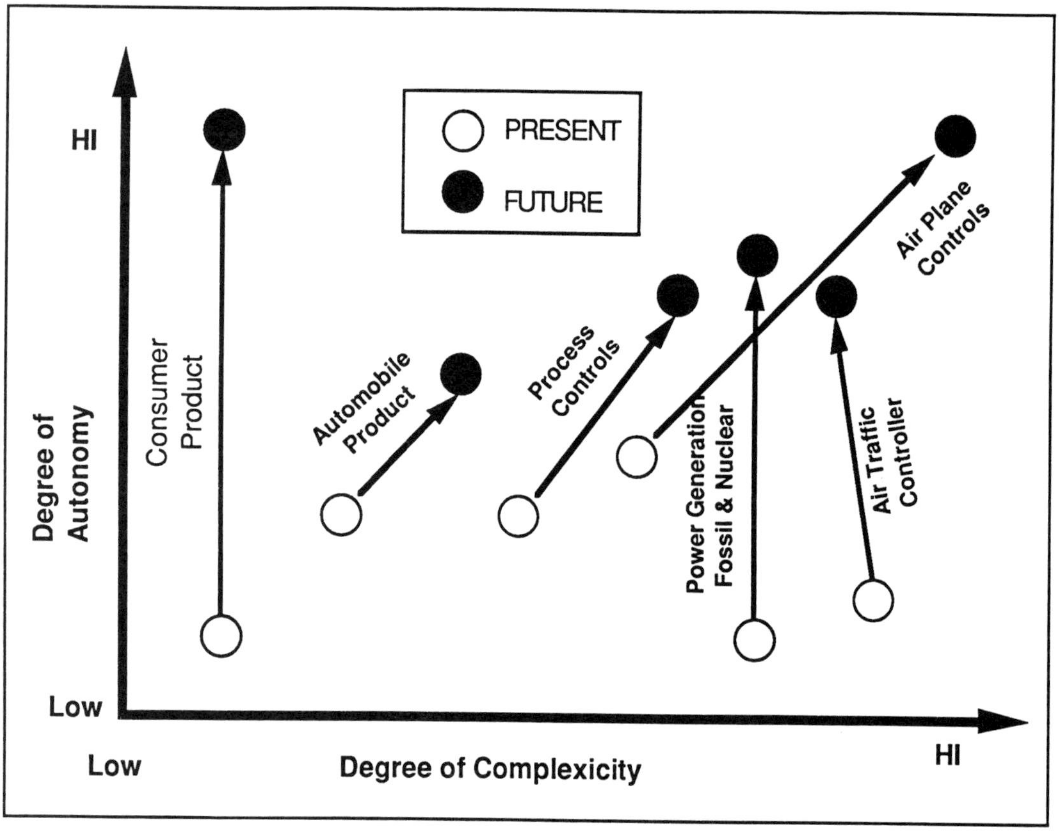

Figure 3 : Qualitative Trends in Complexity and Autonomy for Selected Intelligent Control Systems

5.0 Potential ICS Benefits and Applications

Some of the potential benefits of ICS are summarized in Table 4.

The potential ICS applications are expected in many fields. There are many characteristics of ICS involved in the applications. It is difficult to forecast progress of ICS applications in various fields but a qualitative trend can be developed.

The two primary features of ICS are the degree of autonomy of the control system operation and degree of complexity of the control system. One attempt to show a qualitative trend of future ICS applications in various fields was developed at the recent workshop [7] and it is shown in Figure 3. The predicted trend is increasing automation but relatively small reduction in the complexity of the future controls. From the end user point of view, it would be worthwhile to reduce the complexity of the large system through the use of automated intelligent modular subsystems and make the overall system design simpler and robust to reduce the risk and cost by making the verification and validation of the system less costly and more reliable. The trends shown in the Figure 3 reflect the views of technologists who participated at the workshop [7] but the true future trends will depend on the ability of ICS to deliver successful products which can make the end users more productive and the controls more reliable with attractive cost / benefit ratio. Clearly, the future ICS research, development and applications are exciting for all in the area of instrumentation and controls.

6.0 Conclusion

The Intelligent Control System is an emerging area in the field of instrumentation and control which is going to expand and provide exciting new developments in future research, development, and applications in commercial, industrial, and military fields. The ICS characteristics describe the new potentially powerful future controllers. The ICS challenges are many and encompass smart synthesis of various rapidly developing technologies.

7.0 Acknowledgements

The author participated at the joint EPRI / NSF workshop on Intelligent Control Systems held at Electric Power Research Institute, Palo Alto, California on October 15-16, 1990. Many ideas outlined in this paper came from the various participants at the workshop. Author gratefully acknowledges valuable contributions through group discussions by the participants of the joint EPRI/NSF workshop providing the crux of this paper. In particular, author thanks valuable inputs from Dr. Bill Sun, Dr. Joseph Naser, and Dr. John Maulbatch of EPRI during and after the EPRI/NSF workshop.

8.0 References

[1] Saridis, G., "Foundation of the Theory of Intelligent Controls", Proceedings of IEEE Workshop on Intelligent Control, Troy, N.Y. , August, 1985.

[2] Proceedings of IEEE Workshop, " Intelligent Motion Control", Istanbul, Turkey, August 19-22, 1990

[3] Proceedings of American Control Conference, Special Session, " Mystique of Intelligent Control", San Diego, California, May, 1990.

[4] Proceeding of International Joint Conference on Neural Networks, San Diego, California, June, 1990.

[5] Joint NSF/DOE Workshop on Assessment of European and Canadian Instrumentation and Control Technology, Washington D.C. January 31, 1991.

[6] Lawrence Livermore Report No. CONF-9007134, " Software Tools for Distributed Intelligent Control Systems", Lawrence Livermore National Laboratory, U.S. Army and DARPA Workshop, 1990.

[7] Draft Proceedings of Joint EPRI / NSF Workshop, "Intelligent Control Systems", Palo Alto, California, October 15-16, 1990.

©1991 - Paper #91-739
0074-056X/91/521-526/$0 + .50pp

Open System Architecture for Real-Time Controls and Automation

Vic Sohal
Lynx Real-Time Systems
16780 Lark Avenue, Los Gatos, California, 95030
and
Siddharth Bhatt
Electric Power Research Institute
3412, Hillview Avenue, Palo Alto, California, 94303

1.0 Abstract

The trend towards standardized computing architectures for software, or ``Open Systems" is unmistakable in any analysis of the growth of today's computer industry. One area that stands to benefit greatly from this trend is the real-time automation and controls industry. Real-Time controls have traditionally been developed and implemented using custom hardware and software. Performance, not portability was the key point in the design of these systems. This resulted in solutions that were difficult to implement, maintain and transfer. With the advent of Open Systems for real-time, such applications can be made portable and easily maintainable while actually increasing their performance. Open Systems make modular design, hardware independence, system verifications and validation, and over all quality control of such systems easier. Case studies of the use of Open Systems in process control are discussed.

2.0 The Relevance of Digital Controls for Power Plants

The utility industry has started the process of replacing older obsolete analog controls with newer digital control systems. The primary driving force for this change is the obsolescence of existing controls systems, the lack of spare parts, and the benefits of newer digital technologies which provide better price/performance and greater flexibility. This change over to digital controls is quite expensive and there are still a large number of analog controls in use at many power plants that will have to be replaced. Every control system change-over can cost a quarter of a million dollars or more. The older analog equipment had a long life cycle. By contrast, digital systems, and in particular computers, are evolving very rapidly and there is a new generation of computers every few years with vastly improved price/performance. The investment in control software is usually greater than that in the hardware. This ratio to software to hardware is also increasing.

2.1 The Trend Towards Open Systems

The basic problem of preserving software investment as one moves to newer hardware platforms is not unique to the utility industry. Many other users of computers and digital systems have started moving towards standard, open systems as opposed to proprietary systems to avoid this kind of technology obsolescence and the excessive cost of reimplementing software. Computer users in many industries, including many government agencies and major programs, have started to standardize on UNIX[1] as a solution to this problem.

2.2 Definition of an Open System

An ``Open System" is defined as a standardized computing architecture for software. Porting an application that is written for one CPU-specific implementation of an open system to another is easily or trivially accomplished. Although typically relegated to operating system kernel software, open systems frequently take the form of a combinations of an open kernel (such as UNIX or POSIX) with various auxiliary types of system software such as graphical window managers, and communications protocols.

2.3 A Short History of Open Systems

In the past, open systems were not very important to computer manufacturers, in fact, most manufacturers revelled in the fact that their operating system was completely different from that of their competitors. The point was made even more clearly by hardware enhancement of various operating system features. These differences were even used as selling points, with each manufacturer pointing out the features that his system had over that of his competitors. Since there was no standard computing architecture, applications writers had to spend more time per application porting it to each system architecture. This made applications software more expensive, and their release less frequent. This cycle of developing systems software to work specifically on proprietary hardware was interrupted, in short, by the arrival of the UNIX operating system. UNIX was developed at Bell Laboratories by Dennis Ritchie and Ken Thompson as a portable operating system. Initially, it was written in assembly for a DEC PDP-11, but was later ported to the language ``C" for portability. This was a major development because it made UNIX the only operating system that could run on many hardware architectures. At first, UNIX was used mostly at universities. As more and more people became aware of it, however, it spread to more systems. It soon became a de-facto standard. The advantages it offered were: it was relatively cheap, a large base of knowledge existed for it, and it was available on virtually all the major CPU architectures.

An important side note in the development of UNIX was the divergence of it into two factions, the BSD (Berkeley Standard Distribution) and SVID (System V Interface Definition). BSD represented an attempt by The University of California at Berkeley to add extensions to UNIX for networking, high-performance files, etc. For a long time it was actually the most popular version of UNIX even though it was not as heavily advertised. In recent times, however, it has fallen behind and even partially merged into SVID which has become popular by virtue of the large number of commercial companies supporting its development, as well as AT &T getting behind its commercialization. Other important developments are the formation of the OSF (Open Software Foundation) and the IEEE's POSIX [2] interface definition for portable operating systems. These are significant because they represent definitions of operating system interfaces (very similar if not exactly like UNIX) formed by committees of companies and users for use by everyone regardless if they are an AT\&T licensee. Of these,POSIX is actually the most universally accepted because of its adoption by the U.S. Government as a specification for software to be used by government agencies.

2.4 Why Open Systems are Important

Open Systems are important because they create an ``open" environment for the development of software. Users developing software for such systems do not need to worry about obsolescence and/or portability.They need only to concentrate on the application at hand and not on how well it will port to the myriad numbers of proprietary operating systems. This makes applications cheaper because the cost of development is amortized by their sale on a large number of different platforms. In general, open systems offer the following advantages:

- Vendor independence. Protection from technological obsolescence.
- Preservation of software investment.
- Availability of off-the-shelf software solutions from independent suppliers.
- Availability of trained programmers familiar with standard environments.
- Connectivity between dissimilar computers in a variety of environments (through the wealth of communications protocols available).

2.5 Examples of Open Systems

There are really only a few standards that have been adapted by the industry as general, open systems. They are listed below.

2.5.1 UNIX

Unix has been the primary motivator of open systems. UNIX was where the revolution started and it has kept going. No manufacturer today who builds a workstation can expect to compete in today's market unless UNIX is available on the hardware they make. The market has, to an extent, even wrested control of UNIX away from AT &T by creating a market for open systems and creating truly open standards such as POSIX available.

2.5.2 POSIX

POSIX [2] is the single most significant development in the arena of open systems because it represents an interface definition formed purely by committee of users and manufacturers rather than by a single company. The major parts of POSIX that are of importance to the controls industry are as follows:

- POSIX 1003.1 - This is the standard interface definition, it defines the basic operating system interface.
- POSIX 1003.4 - Real-time extensions to POSIX.
- POSIX 1003.4a -.Threads extensions to POSIX.
- POSIX 1003.2 - The utilities and the shell (i.e. the command interpreter).

2.6 The Role of Real-Time

A Real-Time System is defined as a system that has a completely deterministic response time to external events. In other words, the period of time which it takes the system to respond to an external event is completely predictable and known. Many times, systems that simply exhibit rapid response to external events are falsely billed as real-time systems. In order to qualify as a real-time system, a system must exhibit a predictable response interval to an external event regardless of what else is going on. Systems which exhibit rapid response *most of the time* do not qualify. Also, it is important to know what the *Worst Case Task Response Time* of a real-time system is. This is the most important time to designers of applications of such systems. It is defined as the longest period of time it will take for the highest priority process running in the system to respond to an external event, regardless of whatever else the system is doing (i.e. people are using it, other programs are running, etc).

Generally speaking, there are two types of real-time systems: *soft and hard.* Soft real-time system are systems in which a failure of the system to meet its deadlines for a given task is not considered disastrous. Examples of these are: transaction processing systems for banks, certain types of batch process control, etc. In a hard real-time system such a failure is considered disastrous and a failure of the system. Examples of hard real-time systems include: Power plant control, guidance systems for aircraft, flight simulators, etc.

2.7 Classes of Real-Time Operating Systems

Traditionally, real-time operating systems have been split into two classes, the modifications of large multiuser operating systems for real-time (i.e.real-time UNIX), and small, proprietary real-time kernels. This section talks about these systems as well as newer developments in high-performance real-time UNIX-compatible operating systems.

2.7.1 UNIX for Real-Time Applications

Although the benefits of UNIX are very attractive, standard UNIX is inappropriate for controls systems such as those used in the utility industry and other real-time applications. UNIX was designed as a general purpose time-sharing system and not as a real-time system. It introduces large unpredictable delays for high priority tasks that need to respond to external events. In spite of these limitations, the benefits of standard open systems for real-time are so compelling that many attempts have been made to address the demand for a real-time UNIX. None of them have been successful. The problem lies in the basic approach the designers of UNIX took in its creation. UNIX does not have a preemptive kernel, nor does it have any mechanisms for dealing with the large number of interrupting devices that occur in a real-time environment. It is also not ROMable. All of these disadvantages combine to make versions of real-time UNIX suitable only for soft real-time applications where the failure of the system to respond is not disastrous. Such systems typically have worst-cast task response times of 3-10 milliseconds or more. They also require additional hardware (i.e. multiprocessing CPUs, etc) to support even this level of response.

2.7.2 Proprietary Real-Time Kernels

The term proprietary" is a term that is an anathema to users of open systems. Unfortunately, this is exactly the recourse that many designers of real-time systems have had to use. Real-time programming with these systems is difficult, and requires expensive cross-development hardware and software. The primary difficulty with these systems is that they offer only a limited set of the functionality of more general-purpose operating systems like UNIX. The primary advantage of using these systems is their performance. They offer performance gains for real-time environments of up to an order of magnitude over UNIX modified for real-time. Worst case task response times are measured in hundreds of microseconds on such systems. All of the vendors for such systems (Wind River, Ready Systems, etc) have acknowledged the need for their systems to be complaint with some type of real-time standard and are hard at work modifying their systems to at least meet some of the POSIX standards.

2.7.3 Rewrites of UNIX for Real-Time

This is the latest development in real-time systems. A rewrite of UNIX should ideally provide adherence to all the current standards (POSIX,UNIX System V and BSD) as well as the performance typically associated with proprietary real-time kernels. Since this is a rather arduous task requiring the integration of two environments whose operations goals are orthogonal to each other, it has been achieved by only a few companies. This type of environment should be the ultimate goal of everyone involved with writing controls applications. It combines together the best elements of both approaches into a single unified approach. Developers can enjoy the wealth of software available for open systems while at the same time not compromising performance benefits necessary for their real-time applications to be successful.

The rewriting of UNIX and/or implementing open standards such as POSIX has become the goal of all the major players in the real-time kernel arena, but so far only one company, Lynx Real-Time Systems, has actually delivered a product (LynxOS) compliant with the current drafts of POSIX and UNIX.

2.8 ``Open Real-Time Systems" in the Controls Industry

Open software for the controls industry has requirements to which specific attention must be paid. These requirements are:

- Performance
- Reliability
- Backwards Compatibility
- Connectivity
- Graphics Support

2.8.1 Performance

This is of primary concern to developers. It is so much of a concern that custom applications are still developed using custom operating systems. Given the large investment involved in the implementation of a digital control system for a plant, developers try to squeeze as much out of existing systems as they can. Developing a custom real-time system for such applications can be very tedious indeed. The developer can never stop worrying about the real-time response of the system. It must always be considered and worst-case scenarios established for every new feature implemented. This makes new system very slow in coming out and updates to existing ones, a difficult proposition.

2.8.2 Reliability

Reliability is a concern not only in the development of software, but also in its execution. If a commercially available operating system is to be used, it should have adherence with a quality assurance standard. Adherence to such standards leads to more reliable software.

Compliance with existing quality assurance standards is a very desirable feature for a real-time operating system operating in a critical environment such as a nuclear power plant.

LynxOS [4], for example, has passed and has ongoing conformance with NASA's JSC [3] 31000 quality assurance specification space software. This specification has been devised by NASA as a means of maintaining highly reliable software for use in manned spacecraft. Other governmental organizations such as the FAA also have similar specifications for quality assurance. If an open system is to be used for critical control, this type of quality assurance is an absolute requirement. The operating system should also use the fault tolerant features provided in today's CPU architectures. Chief among these is the MMU (Memory Management Unit). Properly used, an MMU provides protection for each task running in the system. It keeps tasks from accidently writing into each other's memory space, and it also avoids system-wide memory fragmentation. The majority of commercial real-time kernels do not provide this feature because they were originally written for systems which did not have this feature

2.8.3 Backwards Compatibility

Software written for a huge application like a nuclear power plant should have some degree of backwards compatibility so that revisions of operating systems software will not take an inordinate amount of time. The industry-wide support of open systems solves this problem by providing an upgrade path for future developments. The standards are written with backwards compatibility in mind.

2.8.4 Connectivity

Given the size of most plants, distributed control is a necessity. A consistent means of communication between nodes of control is essential because of the different computer architectures that are involved. Support should be provided for standards such as Ethernet and TCP/IP as well as the newer fiber optic LANs.

2.8.5 Graphical Support

Support for the X window system is a must in today's open systems. The X window system was developed by a consortium of vendors at MIT to be a network-based client-server graphical interface standard. The source code for X is freely available for UNIX and it has been ported to many different implementations of it including rewrites of UNIX for real-time. Many of the proprietary kernel vendors also have X support. Graphical support for controls applications is important because it provides a user interface that is easier to learn and interpret. This has been proven with the success of graphical window interfaces in the Personal Computer world.

3.0 Conclusions

It is important to note that for a true open real-time system, all of the above elements must mesh together in such a way that a very high degree of real-time performance is realized at the same time as all of the other elements of the system (i.e. networking, graphics, etc) are working. The designers of the system should never stop thinking about what effect the additions they are making will do to the real-time performance of the system. Also, vendors of the real-time operating systems should always make available their system's performance in realistic, loaded scenarios. Many systems have been designed using unrealistic performance figures. Beginning work with such a system is a recipe for disaster.

4.0 REFERENCES

[1] System V Interface Definition - AT&T

[2] POSIX 1003.1, 1003.4D9, 1003.2 Specification Documents - Technical Committee on Operating Systems of the IEEE Computer Society.

[3] JSC 31000 Rev D. - Overview Space Station Projects and Requirements - National Aeronautics and Space Administration, Johnson Space Center, Houston, TX

[4] LynxOS User's Manual Vols 1 & 2 -
Lynx Real-Time Systems

© 1991 - Paper #91-740
0074-056X/91/527-534/$0 + .50pp

TWIN OAK STEAM ELECTRIC STATION PLANT DISTRIBUTED CONTROL SYSTEM

D.L. Wade
Sr. I&C Engineer
TU Electric
Dallas, Texas

D.C. Crow
Coordinator - Inst. & Controls
TU Electric
Dallas, Texas

R.H. Guerra
I&C Supervising Engr.
UE & C, Inc.
Philadelphia, Pennsylvania

ABSTRACT

Twin Oak Steam Electric Station will consist of two 750 MW lignite-fueled units. The engineering for the project was originally initiated in 1974. There have been numerous delays and suspensions of engineering on the project over the past 16 years for many different reasons. The primary reasons being load growth projections, economic factors, and repeal of the Gas Use Act. Before the last suspension of engineering at the end of 1987, efforts were underway to integrate the major control subsystems into a single plant-wide distributed control system which utilized state-of-the-art technology and equipment. Since restart of engineering on October 1, 1989, this effort has continued and been expanded to include reevaluation of the design to incorporate staffing and operating philosophies that are planned. This paper will detail the overall control complex conceptual design, the operating philosophies that are proposed, how the control system fits into the overall operating scheme, and some of the economics that were considered in the reconfiguration process.

INTRODUCTION

The Twin Oak Steam Electric Station and associated facilitites are located about 132 miles south of Dallas, Texas in Limestone and Robertson Counties near the town of Franklin. The steam electric station will consist of two lignite-fueled units rated at 750 MW each. The scheduled commercial operation dates are April 1, 1995 for Unit #1 and April 1, 1996 for Unit #2.

Engineering for this project was initiated in July 1974, when an agreement between United Engineers & Constructors, Inc. and Texas Utilities Services, Inc. was put into effect. The original trial operation dates were August 1, 1980 for Unit #1 and September 1, 1981 for Unit #2.

Over the past 16 years there have been numerous delays and suspension of engineering for many different reasons. The primary reasons being load growth projections, economic factors, and the repeal of the Gas Use Act. Prior to the last suspension of engineering, at the end of 1987, there had been a considerable effort to cancel the controls from the scope of supply of some of the major OEMs. The main intent was to integrate the major control systems into a single plant-wide distributed control system. This system was to be designated as the "Plant Distributed Control System" (PDCS) and would be comprised of the following systems:

1. Boiler Combustion Control System
2. Miscellaneous Control System
3. Auxiliary Boiler Control System (Combustion & FSSS)
4. Data Acquisition System (Boiler, Turbine, Generator, BOP)
5. Flue Gas Desulfurization Control System
6. Limestone Handling System (future)
7. Solid Waste Processing (future)

In addition the PDCS would have interfaces, for data acquisition, to the following systems:

1. Lignite Handling
2. Fly Ash Handling
3. Bottom Ash Handling
4. Fuel Supervisory Safeguard System (FSSS)
5. Condensate Polishing
6. Demineralizer
7. Limestone Stacker/Reclaimer
8. Lignite Feeders
9. Air Heater Leakage
10. Precipitator
11. Flue Gas Data Acquisition

As the engineering to incorporate this equipment into the control complex design began, it was not envisioned that there would be any drastic changes in layout or philosophy in the plant control system. In the Main Control Room there would be a Console with control switches and indicating lights for motor control and monitoring, manual/automatic (M/A) stations for regulatory control, edgewise indicators and recorders for system monitoring, and a window-box annunciator for annunciation. This typical arrangement would be present in the FGD Control Room, Limestone Handling Control Room, Solid Waste Processing Control Room, and Auxiliary Boiler Control Room.

Early in 1988 the Resource Plan again identified the need to delay Twin Oak construction and suspend all engineering activity until later. During this period the purchase orders for the Control Panels and the Annunciator System were cancelled because of the amount of changes that would be necessary to incorporate the PDCS into the design. Plans were to repurchase this equipment after initiation of an engineering restart. Engineering shutdown records were assembled documenting the status of design for the various plant systems and work on the Twin Oak design was stopped.

Engineering was resumed on the project on October 1, 1989. TU Electric began negotiations with United Engineers & Constructors (UE&C) to define the remaining engineering scope to complete the project and establish a firm price for the UE&C design effort required. During the scope definition phase it was decided that the control complex should be redesigned to include the following major changes:

1. The conventional control panels should be deleted and as much control and monitoring as possible, incorporated into the PDCS CRT-based operator's console.
2. Motor control and monitoring should be incorporated into the PDCS to allow initiation and monitoring via the CRT consoles.
3. The interface between the PDCS and the various PLC-based plant systems should be expanded to include control capability.
4. The conventional window-box annunciator should not be repurchased and all annunciation incorporated into the PDCS.
5. The turbine-generator control system should be modernized and integrated closely with the PDCS. This will allow for control and monitoring via the PDCS CRT-based operator's console.

In December 1989, Engineering presented, to a selected group from Operations, the status of the Twin Oak design and the philosophies that would be used to complete that design. A concern was raised that over the past few years there had been some restructuring of the Operation's Departments at the existing lignite plants that could impact how future units would be staffed. Because the layout of the control complex has a drastic affect on the staffing requirements of a plant, it was decided that a staffing plan should be formulated and this plan used as a basis to proceed with the design in this area.

OPERATING & STAFFING PHILOSOPHY

In February 1990, a meeting with Operations and Engineering was held to discuss the proposed staffing of Twin Oak and the changes required to incorporate this plan into the control complex design. Engineering explained that major areas of control were being provided in the following areas:

1. Main Control Room for boiler, turbine-generator and lignite handling.
2. Bottom Ash Control Bldg. for bottom ash and dewatering bins.
3. Water Treatment Bldg. for make-up demineralizer.
4. FGD Control Bldg. for precipitator, FGD system, and fly ash.
5. Limestone Slurry Prep. Control Bldg. for limestone slurry preparation, limestone handling, and limestone stacker/reclaimer.
6. Solid Waste Processing Bldg. for solid waste processing.
7. Auxiliary Boiler Control Bldg. for auxiliary boiler.

Operations proposed that two control areas be established in the plant. The first area would be designated as "Production" and would include control and monitoring of all systems directly involved in power production and requiring immediate operator response to prevent reduction in power output. The second area designated as "Support" would include control and monitoring of all systems that support power production and that do not need immediate operator response to prevent reduction in power output.

The Production area would have it's main control and monitoring performed from a CRT-based operator's console located in the Main Control Room. The following systems are included in this area:

1. Boiler Controls
2. FSSS
3. Data Acquisition
4. Flue Gas Desulfurization
5. Bottom Ash
6. Fly Ash
7. Economizer Ash
8. Switchyard
9. Demineralizer
10. Condensate Polishing
11. Auxiliary Boiler
12. Precipitator
13. Air Heater Leakage
14. Lignite Feeders

The Support area would have it's main control and monitoring performed from the Solids Waste Processing (SWP) Control Room located in the SWP Building. The following systems are included in this area:

1. Lignite Handling
2. Limestone Handling
3. Limestone Preparation
4. Limestone Stack/Reclaim
5. Solid Waste Processing
6. Dewatering Bins
7. Fly Ash Silo
8. Water Management
9. Railroad

The concept presented by Operations supported recent trends which were to reduce the operating staff required at a site and centralize control of plant subsytems. Therefore it was accepted as the basis for continuation of the design of the control complex.

DESIGN & BENEFITS

With a general agreement between Operations and Engineering on the direction that staffing and operation of the plant was proceeding, information was formalized for transmittal to the PDCS vendor to provide a revised proposal to meet the new requirements. Tabulation of the approximate input/output (I/O) count and type for each area was assembled to assist in this effort. The revised I/O for Unit 1 & Common and Unit 2 was estimated to be about 33,000 points. The total at the project shutdown in 1987 was approximately 21,000 points, an increase of about 57%. Refer to Figure 1 and Figure 2 for the tabulation of I/O by area.

The following requirements were presented as a basis to reconfigure the PDCS:

1. Two major areas of control will be established. One in the Main Control Room and the other in the Solid Waste Processing Building. Three backup areas of control will be established. One in the Auxiliary Boiler Control Bldg., one in the Water Treatment Bldg., and one in the FGD Control Bldg. Operator's and Engineer's consoles with touchscreen CRTs and peripherals will be provided as follows:

    A. Main Control Room - 10 CRTs per unit; 3 dot matrix printers per unit; and 1 ink jet printer per unit.
    B. Solid Waste Processing Bldg. - 8 CRTs for operator's console and 2 CRTs for engineer's console; 4 dot matrix printers; and 2 ink jet printers.
    C. Auxiliary Boiler Bldg. - 2 CRTs and 1 dot matrix printer.
    D. Water Treatment Bldg. - 2 CRTs; 1 dot matrix printer; and 1 ink jet printer.
    E. FGD Control Bldg. - 3 CRTs per unit; 1 dot matrix printer per unit; and 1 ink jet printer per unit.
    F. Lower Electronics Room - 3 CRTs per unit; 1 dot matrix printer per unit; and 1 ink jet printer per unit.
    G. Administration Bldg. - 1 CRT; 1 dot matrix printer; and 1 ink jet printer.

2. Conventional control boards will be deleted and all motor control will be initiated via the PDCS CRT-based operator's consoles.
3. Conventional window-box annunciator system will be deleted and all alarming will be via the PDCS CRT-based operator's consoles.
4. Control and monitoring of all PLC-based control systems will be initiated via the PDCS CRT-based operator's consoles.
5. Distribution of FBMs (field bus modules) local to medium voltage switchgear and 480 volt MCCs to reduce cable requirements will be provided.

The reconfiguration of the PDCS will provide the following benefits:

1. Deletion of conventional control boards and implementation of CRT-based operator consoles provides the ability and flexibility to make modifications to meet future operating needs without costly expenditures. Changes are easily accomplished with software changes instead of control board modifications.
2. Distribution of FBMs throughout the plant will reduce the amount of cable that is required because it will no longer be routed to central control board locations but will be terminated closer to the final control devices.
3. By limiting the main areas of control to two areas the number of operators required for control and monitoring of the plant will be reduced. However; backup control areas have been provided for the major subsystems to allow for the dedication of additional personnel to address specific problems.
4. Making the annunciator system an integral part of the PDCS eliminates the need for the conventional window-box annunciator system. The past design had all the signals that were terminated into the annunciator system also terminated into the PDCS, therefore the cable requirements will also be reduced.
5. The current design approach of the PDCS provides for future expandability and the necessary flexibility to accommodate changes for many years.
6. The interfacing of PLC-based controlled subsystems with the PDCS provides a common operator interface throughout the plant which reduces the need to train operators on a different operator interface for each system.

LOCATION	Analog Input				Analog Output	Digital Input (Dry Contact)			Digital Output		PLC Gateway (RS 232)				
	4-20 ma	T/C (mV)	RTD	0-1 ma	4-20 ma	DC	CUTOUT	PULSE	DC	AC	DI	DO	AI	AO	Trip Coil
Main Control / Electronics Room															
Integrated Control System (ICS)	475	75		75	350	500			200	400					
Miscellaneous Control System (BOP)	100	25		60	100	150			75	150					
Data Acquisition System (DAS)	300	325	150			1000	24	40							
o Sequence of Events (SOE)						256									
o Generator Stator	10	10	40												
o Boiler Water Wall		110													
Bottom Ash / Dewatering Bins (PLC)											450	600	20	20	
Flame Safety Safeguard System (PLC)											725	75			
Condensate Polishing System (PLC)											75	75	25	25	
Motor Control (Turbine Area)		70	40	70		410			80	230					80
FGD Control & Unit 1 Swgr. Bldg.															
Flue Gas Desulfurization System (FGD)	150	50		50	150	450			150	400					
Data Acquisition System (DAS)	100	100	30			425									
o FGD Temperature		75	50												
Precipitator Control System (PLC)											50		50		
Fly Ash Control System (PLC)											650	650	20	20	
Motor Control (FGD)		50	30	60		245			60	140					60
Solid Waste Processing Bldg.															
Solid Waste Processing System (SWP)	200	50		50	75	500			150	300					
Data Acquisition System (DAS)	200	200	50			800									
Lignite Handling System (PLC)											575	625	10	10	
Water Management System	100					100		100							
Limestone Stacker/Reclaimer (PLC)											25	25	10	10	
Motor Control (SWP Area)		70	35	75		560			65	470					70
Limestone Slurry Prep. Control Bldg.															
Limestone Slurry System	25	10			25	100			2,	100					
Data Acquisition System (DAS)	50	100				150									
Limestone Handling System	25	10			25	200			50	100					
Auxiliary Boiler															
Auxiliary Boiler Control	75	5			85	110		10	45	180					
Data Acquisition System (DAS)	50	100				150									
Water Treatment Bldg.															
Demineralizer Control System (PLC)											400	75	50	50	
Total - Unit No. 1 & Common	1860	1435	425	440	810	6106	24	150	900	2470	2950	2125	185	135	210

Figure 1 - PDCS Input and Output Summary for Unit No. 1 and Common

LOCATION	Analog Input				Analog Output	Digital Input (Dry Contact)			Digital Output		PLC Gateway (RS 232)				
	4-20 ma	T/C (mV)	RTD	0-1 ma	4-20 ma	DC	CUTOUT	PULSE	DC	AC	DI	DO	AI	AO	Trip Coil
Main Control / Electronics Room															
Integrated Control System (ICS)	475	75		75	350	500			200	400					
Miscellaneous Control System (BOP)	100	25		60	100	150			75	150					
Data Acquisition System (DAS)	300	325	150			1000	24	40							
o Sequence of Events (SOE)						256									
o Generator Stator	10	10	40												
o Boiler Water Wall		110													
Bottom Ash / Dewatering Bins (PLC)											450	600	20	20	
Flame Safety Safeguard System (PLC)											725	75			
Condensate Polishing System (PLC)											75	75	25	25	
Motor Control (Turbine Area)		50	30	50		375			60	215					60
FGD Control & Unit 1 Swgr. Bldg.															
Flue Gas Desulfurization System (FGD)	150	50		50	150	450			150	400					
Data Acquisition System (DAS)	100	100	30			425									
o FGD Temperature		75	50												
Precipitator Control System (PLC)											50		50		
Fly Ash Control System (PLC)											650	650	20	20	
Motor Control (FGD)		50	30	60		245			60	140					60
Solid Waste Processing Bldg.															
Solid Waste Processing System (SWP)															
Data Acquisition System (DAS)															
Lignite Handling System (PLC)															
Water Management System															
Limestone Stacker/Reclaimer (PLC)															
Motor Control (SWP Area)															
Limestone Slurry Prep. Control Bldg.															
Limestone Slurry System	25	10			25	100			2[illegible]	100					
Data Acquisition System (DAS)	50	100				150									
Limestone Handling System															
Auxiliary Boiler															
Auxiliary Boiler Control															
Data Acquisition System (DAS)															
Water Treatment Bldg.															
Demineralizer Control System (PLC)															
Total - Unit No. 2	1210	980	330	295	625	3651	24	40	570	1405	1950	1400	115	65	120

Figure 2 - PDCS Input and Output Summary for Unit No. 2

CONCLUSION

Although TU Electric has been retrofiting existing units with distributed control systems, they were never truly distributed because in most cases much of the DCS equipment was located into a centralized electronics room and/or control room. The Twin Oak design will, for the first time within the TU Electric system, actually distribute equipment throughout the plant to minimize cable requirements while maintaining the ability to provide centralized operator interface to the PDCS. Refer to Figure 3 for the layout of the PDCS and how equipment is to be distributed.

In summary the heart of the control and monitoring of the entire plant at Twin Oak will be accomplished by the PDCS. Integration of auxiliary systems to interface with the PDCS is continuing and the preliminary staffing plan is being incorporated into the design. The Production area will have its main point of control and monitoring from the Main Control Room with limited backup being provided in the FGD Control Room and the Auxiliary Boiler Building. The Support area will have its main point of control and monitoring from the Solid Waste Processing Control Room with limited backup being provided in the FGD Control Room. The control complex will take advantage of the latest technology and utilize the latest in state-of-the-art control equipment. Reliability as well as flexibility will continue to be important factors in the design related to this area of the plant.

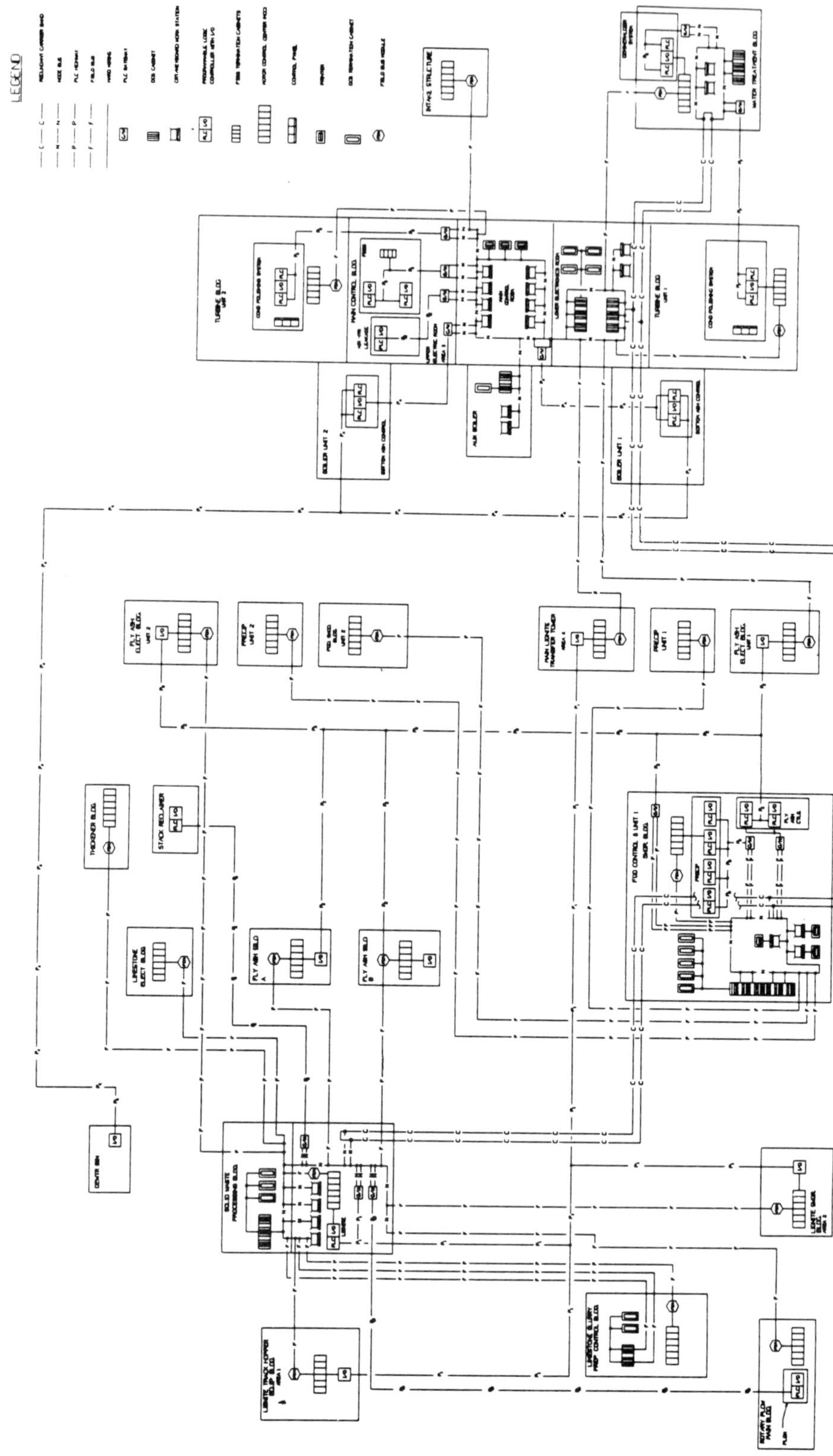

Figure 3 - Plant Distributed Control System Overview

©1991 - Paper #91-741
0074-056X/91/535-550/$0 + .50pp

TurbinAID: TURBINE ARTIFICIAL INTELLIGENCE DIAGNOSTICS

M. A. MORADIAN M. P. CHOW
R. L. OSBORNE M. A. JENKINS

WESTINGHOUSE ELECTRIC CORPORATION
POWER GENERATION TECHNOLOGY DIVISION
ORLANDO, FLORIDA

ABSTRACT

The Westinghouse Turbine Artificial Intelligence Diagnostics system or TurbinAID, can diagnose both thermodynamic and mechanical component anomalies within the turbine, and around the turbine cycle. Any monitoring system can detect that a variable is in an abnormal state, but TurbinAID can also indicate the cause, and provide recommended corrective action(s). The TurbinAID Expert Systems utilize multiple sensor and variable inputs, and their interdependencies in the generation of a diagnosis. The system performs sensor validation as part of the data acquisition scheme. The TurbinAID system has been in operation for several years. This paper describes the monitoring and diagnostic functions provided by TurbinAID, and how the utility industry both nuclear and fossil, can utilize the system to enhance unit operation.

1. INTRODUCTION

The need for providing the control room operators with the best information when he or she needs to make the critical operating decisions has never been greater. The utility industry has responded by installing large data gathering and display systems. This provides the operator with an overwhelming amount of data but does not interpret the meaning of the data, nor does it identify recommended actions or suggest priorities for corrective action. The information is especially important during times of start-up or unit excursions when many alarms can be present which compete for the operators immediate attention. In addition, it is also important to identify operator actions which will maximize plant efficiency on an hourly basis. All the desired elements that are not included in the data gathering system are included in a system called TurbinAID, i.e., Turbine Artificial Intelligence Diagnostics. "TurbinAID" is defined as a computer resident software program/package which is able to operate in any mainframe or control system computer. The TurbinAID overall objective is to provide the operator with the best diagnoses possible on the turbine generator island. The TurbinAID overall objective is to provide the operator with the best diagnoses possible on the turbine generator island.

TurbinAID is one of the three AI diagnostic products which are supported by a staff of engineers and scientist involved in diagnostics at the headquarters of the Westinghouse Power Generation Business Unit, in Orlando, Florida. The other two diagnostic products available are operator ChemAID, which is Chemistry Artificial Intelligence Diagnostics and, GenAID or Generator Artificial Intelligence Disgnostics.

The overall objective of TurbinAID is to oversee the operations of the Turbine Generator Island. The TurbinAID system can diagnose malfunctions in the operation of the HP turbine and IP turbine, in addition to the supporting systems such as Lube oil system and EH fluid system.

TurbinAID is comprised of three distinct components, Thermal Performance, Vibration Analysis, and Condition Analysis (diagnostics) refer to Figure 1. These three different components will be briefly discussed.

Thermal Performance:

The thermal performance component calculates overall thermal performance of the Turbine generator and identifies component efficiencies such as high pressure turbine efficiency, intermediate pressure turbine efficiency and/or LP turbine efficiency.

In addition the thermal performance component performs sensor validation and sensor substitution as a part of the diagnostic system. It identifies operator controllable parameters that the operator can control to obtain better Thermal Performance. The performance of the feedwater heaters are also calculated and monitored specifically with regard to calculating the drain cooler approach temperatures and terminal temperature differences for each heater.

Vibration Analysis:

The vibration analysis or the vibration analysis component looks at the mechanical condition of the rotating apparatus. Specifically, it looks at the phase angle, amplitudes, and other mechanical conditions that may exist on the rotating apparatus.

Component Condition Analysis:

The Diagnostic Component Condition Analysis, is really the brain or the "expert" part of the TurbinAID system. It interprets the measured or calculated variables in the Thermal Performance component or Vibration Analysis Component to diagnose conditions that may exist around the turbine. Lets assume the thermal performance component calculates that the turbine generator heat rate has increased by 200 BTU's. The component condition analysis goes through and identifies the cause of the increased heat rate. Specifically, if the calculated turbine generator heat rate increase has been due to cycle conditions, it identifies that the cycle is the cause of lower than expected thermal performance. And likewise, if there has been a drop in component efficiencies, the Component Condition Analysis, identifies the cause of the higher heat rate.

2. SYSTEM DESCRIPTION

TurbinAID (Turbine Artificial Intelligence Diagnostics) is an on-line steam turbine diagnostic system which utilizes both advanced data acquisition and computational systems (monitoring portion), and AI expert systems (diagnostics portion).

TurbinAID utilizes a Plant Data Center which is located in the control room at the power plant and serves to collect data, process it, and communicate with the Diagnostic Operations Center (DOC) in Orlando, Florida (fig. 2). The expert system at the DOC analyzes the data, and transmits the diagnoses and the associated recommendations back to the PDC where the information is displayed.

2.1 Plant Data Center (PDC) Functions

The main functions of the PDC are data collection and management, data processing, and display. Sensor signals are collected, converted to engineering units, and checked for quality and deviations from normal. The PDC computer calculates all the variables necessary to diagnose the condition of the turbine. A deadbanding technique is employed so that only data which changes significantly is transmitted to the DOC and stored at the PDC. Graphics showing turbine performance and diagnoses are provided for the plant operator.

2.2 Diagnostic Operations Center (DOC) Functions

The data transmitted to the DOC is analyzed automatically by a proprietary forward-chaining expert system named Process Diagnostic System (Lowenfeld, 1984) using a knowledge base created by turbine experts. Diagnostic results, ranked according to priority, are automatically transmitted from the DOC back to the PDC. Each diagnosis carries with it a confidence factor, or measure of belief that the condition exists. The DOC can also transmit recommendations to resolve abnormal conditions. The expert system operates continuously without human intervention. The diagnoses are also displayed on a screen at the DOC which is staffed around the clock by operators familiar with power plant operation.

2.3 SYSTEM HARDWARE

Prior to the TurbinAID installation a sensor audit is performed for each specific application to determine if sensor replacements or additions are necessary. TurbinAID uses existing plant sensors whenever possible. To reduce heat rate uncertainty, accurate and redundant sensors are used at key locations per ASME PTC6 test guidelines. A PTC6.1-type final feedwater flow nozzle is desirable to improve accuracy. TurbinAID uses sufficient instrumentation around the cycle to approach ASME test requirements. Additional sensors can be easily incorporated into TurbinAID, but this may not be cost effective for the utility. A generic TurbinAID sensor list for a fossil unit is shown in Table 1.

Sensor data may be acquired by a number of means, such as from the plant computer, from a data logger over an RS-232 link, or a separate I/O system. To reduce installation costs for existing units where many sensors may need to be added, Westinghouse has developed a SAM (Signal Acquisition Module) network in which the SAM modules can handle up to 20 sensors per module. Up to 50 SAM's can be mounted remotely in the plant and connected together by a two conductor cable that handles data transmission to the computer and supplies power for the SAM's.

The heart of the TurbinAID PDC is a DEC 3300 computer. The PDC also contains a color CRT and keyboard for operator interface and a color printer for hard copy data output.

At the DOC the expert system runs on a pair of DEC Vax 8530 super mini-computers. Various CRTs, printers and plotters are available for use by the DOC operators and the knowledge engineers.

3. THERMAL PERFORMANCE

All TurbinAID thermal performance calculations are performed at the PDC rather than at the DOC because the thermodynamic variables must be calculated based on current rather than deadbanded data. The PDC computer uses a Westinghouse-developed language called PDCCL (Plant Data Center Configuration Language).

Each sensor which affects the cycle heat balance calculation is automatically validated before use in the calculation, and a substitute value is used if the sensor is determined to be bad. This is done to prevent bad or drifted sensor readings from propagating errors through the thermal performance software. In this way the classic programming "garbage in -- garbage out" scenario is avoided. The first level of validation is checking the sensor quality as determined by the data acquisition system. This will catch any "hard" sensor failures. If the data acquisition system considers a sensor value as good, then it proceeds to a second level of validation, where "soft" sensor failures will be identified. Sensors whose accuracies are crucial to the heat balance calculations and locations with redundant sensors are subjected to an analytic 2-out-of-3 test. The three values being compared include the sensor being verified and two estimates. The estimates may be redundant sensor readings, functions of other validated sensors, functions of gross megawatts, etc. as appropriate. Less-important sensors are tested against a single estimate which itself is based on a validated reading. The test(s) compare the difference(s) between the sensor and its estimate(s) against a predetermined tolerance. Sensor values whose difference(s) meet the tolerance satisfy the test and are used in subsequent thermal calculations. If a sensor fails to meet the test tolerance and its estimates do, the average of the estimates is substituted for the sensor value and a digital variable is toggled to indicate that sensor has been determined to be bad. If neither the sensor nor the estimates meet the test tolerance, the average of all three is substituted for the sensor value and another digital is toggled to indicate that the final value is uncertain. TurbinAID performs this sensor validation/substitution automatically - no action is required by the plant operator.

Sensors are scanned at rates from once per second to once per minute depending on the response of the sensor and its use in the diagnostic system. Sensor readings are averaged over time periods as appropriate. Variables are calculated at

intervals from 1 minute (such as a pressure in final units of psia) to 30 minutes (such as heat rate).

Enthalpies, entropies, specific volumes, etc. are calculated from steam table functions based on 1967 ASME Steam Tables.

Once the steam properties have been calculated, the heat balance calculation can be performed. The calculation procedure follows ASME PTC6.1 methodology since feedwater flow is the primary flow measurement. The TurbinAID software calculates HP heater extraction flows based on heat and mass balances around each heater up to and including the deaerator. Flows and leakages throughout the turbine blade path are calculated up to the LP turbine inlet. TurbinAID calculates the gross turbine cycle heat rate as well as the efficiencies of the HP and IP turbines. Cycle component performance indices such as heater TTDs and DCTDs, hotwell subcooling, BFP turbine flow, etc., are determined. Selected variables are shown on graphics to the operator.

TurbinAID calculates LP efficiency in the following manner. LP turbine work is calculated from the total work (gross generator load plus losses) minus HP turbine work minus IP turbine work. The heat available to the LP turbine is known. Since the condensate flow is determined by a heat balance around the deaerator and the temperatures at the hotwell outlet and deaerator inlet are measured, the total heat gained across the LP heaters can be calculated. The total heat extracted out of the LP turbine can be approximated as a linear function of the total heat gained across the LP heaters. The LP exhaust flow can be determined approximately from LP inlet flow or from LP turbine extraction pressures. The only remaining unknown in the equation is the LP turbine exhaust enthalpy (used energy end point) and thus the LP turbine efficiency can be calculated.

TurbinAID calculates and displays the heat rate and fuel cost penalties (or savings) associated with operating at off-target parameters. The use of realistic, dynamic targets makes the Operator Controllable Parameters Screen and the Component Performance Parameters Screen more useful to the plant operators. These screens and the calculations behind them are discussed in more detail in the Operator Interface section of this paper.

TurbinAID monitors and diagnoses two turbine subsystems that many heat rate monitoring systems omit. These subsystems are the steam chest/valves system and the steam gland system. The heat rate can be improved by proper operation and alignment of the turbine control valves, and, to a lesser extent, the gland steam system.

The TurbinAID software goes on to calculate many variables that most heat rate monitoring systems do not. These variables are those which are not used to assist the operator in optimizing heat rate hour by hour, but will help the utility's bottom line by providing useful diagnostic information about degradation that can be used to plan and shorten outages and thus improve availability and heat rate on a long-term basis. These variables are those necessary to determine the condition of the turbine blade path, such as corrected pressures and flows (normalized to design inlet steam conditions), individual blade group efficiencies, blade group pressure ratios, and blade group flow factors, Blade path erosion, seal rubs, deposits, and blockage can be diagnosed only by considering all these variables and observing their changes over time. This is best handled by an expert system.

4. OPERATOR INTERFACE PARAMETERS

Color graphic displays are provided to inform the plant operator of the current state of the unit and to assist him in maintaining efficient operation.
A Controllable Parameters Screen (fig. 3) displays the current and target values of selected operator controllable parameters which affect heat rate. Heat rate and fuel cost effects of deviations from target values are displayed in dollars per shift and thousands of dollars per year. The controllable parameters include

throttle pressure and temperature, hot reheat temperature, back pressure, and superheat and reheat attemperating spray flows. Other controllable parameters outside of the turbine cycle which affect heat rate such as excess air, exit gas temperature, and station service are not presently included but can be added. This screen can assist the operator in optimizing the heat rate by minimizing controllable losses.

It is important that the target parameters displayed be realistic ones or the operator will lose confidence and interest in the system. Some heat rate monitoring systems use constants from contract heat balance values for the targets. These values may be unrealistic for most operating conditions. While it is important for the plant engineer to know about deviations from design, the operator only needs to know about deviations from achievable targets. Targets should be realistic at all times and at all loads as much as practically possible. For example, the contract back pressure for a particular unit may be 2.50 inHgA. This value may be realistic in the winter but absolutely unachievable in the summer when circulating water inlet temperature is high, and the operator may become discouraged by having this unrealistic target displayed. TurbinAID's target parameters are variables which are calculated to be as realistic as possible. The target back pressure, for example, is a multiple polynomial interpolation determined from circulating water inlet temperature, circulating water inlet flow, and condenser heat duty based on information supplied by the condenser manufacturer. The target throttle temperature is a function of load as specified by the boiler manufacturer, which includes the effects of boiler droop at low loads. In this way the Controllable Parameters Screen is as useful to the operator as possible. One of the advantages or TurbinAID's centralized system architecture is that targets can be changed at any time from the DOC to suit a particular plant's needs. A phone call or electronic mail message from the plant operator or engineer is all it takes to effect a change.

A Component Performance Parameters Screen (fig. 4) is also provided to keep track of longer-term equipment degradation which may adversely affect the heat rate. The Component Performance Parameters include HP, IP, and LP turbine efficiency, HP heater TTDs, BFP turbine flow, steam coil air heater flow, etc. Like the Controllable Parameters Screen, the heat rate and fuel cost effects of deviations between current and target values are displayed. The component performance targets are also variables rather than constants and are calculated to be as realistic as possible. The target HP turbine efficiency, for example, is calculated in both sequential valve mode (partial arc admission) and single valve mode (full arc admission). It may be based on data from recent performance tests or the current state of the unit. The target heater TTDs are functions of feedwater flow as specified by the heater manufacturer. Component performance parameter targets may be specified by the heater manufacturer. Component performance parameter targets may also be changed as needed. The Component Performance Parameters screen can aid the utility to justify and plan outages to work on components whose performance has deviated significantly from their targets.

The fuel cost, availability, and load factor used in the above "cost of deviation" calculations are operator inputs, which may be changed by the operator at any time.

Other schematic, P&ID-type screens are provided to display the performance of the turbines, heaters, condenser, and gland steam system. Data list and trending displays are also available.

Diagnostic results are displayed on a Diagnostic Summary Screen (fig. 5) which are listed in order of priority. Recommendations for diagnoses are also transmitted to the PDC when appropriate. The recommendations include a more detailed explanation of the condition, the consequences of inaction, and the recommended corrective measures.

5. VIBRATION ANALYSIS

Mechanical distress in steam turbine-generators often starts as an incipient condition which escalate to final malfunction because the condition was allowed to progress unchecked. The objective of vibration analysis is to identify incipient condition which can lead to more serious problems. Most incipient conditions cause changes in the vibration characteristics of a steam turbine-generator set. These changes can be detected by analyzing the vibration signature. This technique of vibration signature analysis has been used by the power industry for the last two decades and has proven effective in detecting these incipient conditions.

Vibration signature analysis is a key aspect of today's predictive maintenance program in utility companies. Westinghouse has incorporated this capability in the TurbinAID system. It was recognized during the early stages of the project that a front end data acquisition/processing system was required to perform accurate real-time vibration diagnostics. This includes the ability to digitize analog vibration signals, perform Fast Fourier Transform (FFT), and extract essential vibration information in real-time, not only during steady-state operation but also during startup and coastdown. A program was initiated to develop this hardware called Dynamic Data Analyzer (DDA).

5.1 Functional Description DDA

The DDA can be viewed as a real-time multi-channel spectrum analyzer. It takes analog vibration signals from existing Turbine Supervisory Instrument (TSI), converts the signals to digital form, transfers the data from the time domain to the frequency domain, and finally transmits the amplitude and phase information to a host computer. This information is then transmitted over a dedicated data line to the DOC in Orlando for use by the diagnostic expert systems.

The DDA differs from the traditional spectrum analyzer in the following areas:

- Digitizes analog signals based on a once-per-revolution signal
- Samples vibration signals up to 32 channels simultaneously
- Handles up to 32 channels during startup and coastdown with speed resolution less than 20 rpm
- Automatically captures the first critical speed, vibration vector at the first critical, and runout vector for each channel.

The DDA has the following functions:

Functions:

- Provides continuous vibration data acquisition and processing
- Operates under steady-state conditions as well as during coastdown and startup
- Transmits selected amplitude and phase to host computer
- No increase on data acquisition time between 8 and 32 channels
- Performs one FFT (1024 points) in 7 milliseconds
- Performs time synchronous averaging and frequency averaging
- Frequency resolution 1.25 Hz
- RPM resolution 20 rpm during coastdown
- Phase resolution within one degree
- Automatically captures critical speed and vibration vector at critical speed
- Automatically captures runout vectors

The DDA is designed to take analog signals from existing TSI, perform signal conditioning, digitize the signal using once-per-rev as reference, perform 32 channels FFT in less than a second, and send spectral parameters (selected amplitudes and phrases) to the host computer.

Each individual spectral parameter such as running speed amplitude and phase or 8th harmonic amplitude can be retrieved and plotted against time. Alarms can be set for each spectral parameter. Full spectrum (400 lines FFT) and time waveform data can be uploaded from the DDA to the Host computer through a single command. Bode and Waterfall plots can be reconstructed for the plant engineer to review.
The DDA combines with the present PDC not only for the monitoring information but also the diagnostic information such as critical speed, vibration vector at critical speed for each bearing, an unusual vibration peak at a certain frequency, etc. which are essential for early problem identification. This in turn would help plant personnel to reduce machine downtime and increase machine availability.

6.0 COMPONENT CONDITION ANALYSIS

Data from the Thermal Performance, and Vibration Analysis components of TurbinAID are utilized by the Expert System Knowledge Base to diagnose malfunctions internal to the steam turbine or turbine auxiliaries.

Generation of the TurbinAID Knowledge Base required Man-years of effort involving commitment of expert resources, research on each component system, interviewing domain experts, prototype testing, and loading the knowledge into the expert system shell. Also, the available sensors were reviewed to determine if additional sensors were required to accurately diagnose the desired malfunctions (refer to the Partial List of TurbinAID Sensors on Table 1, and Examples of Conditions Diagnosed by the TurbinAID System on Table 2).

The sophistication of the Expert System is the keystone of the TurbinAID system. For example, in order for the system to diagnose a cracked shaft, the system must recognize, the critical speed with a speed resolution of less than 20 rpm, the system damping at critical speed, the 1X vibration amplitude and phase during transient and rated speed, the 2X vibration amplitude and phase at one half of the critical speed, and the 3X vibration amplitude and phase during transient and rated speed. The system can also "memorize" this information from past tansients and utilize the information for on-line operation to differentiate among the conditions of shaft crack, shaft imbalance, and misalignment.

Use of the DDA in the TurbinAID System

The DDA can generate 32 FFTs in a second during steady-state operation and 32 FFTs within 20 rpm during coastdown. Data compression is required to ensure that only significant data is transmitted from the DDA to the PDC and to maximize the amount of short-term storage capabilities at the PDC. Turbine operating parameters such as MW, steam temperature and pressure, bearing temperature, oil film thickness, shaft center line position etc., are collected as well. Data from the PDC is transmitted to the DOC and processed automatically using AI Expert Systems. Processing is performed by a proprietary forward chaining expert system using a knowledge base created by knowledge engineers, field vibration specialists, and turbine designers. The incoming data is used to perform an incremental diagnosis based on current and previous machine conditions. Diagnostic results, stored according to priority, are automatically transmitted to the PDC for the plant operator's use as they emerge from the expert system.

Data is validated in the knowledge base (KB) first before it is propagated any further. This process is called sensor diagnosis and is an integral part of the knowledge base. The sensor diagnosis can be as simple as a range check or as complicated as detecting fluctuating vibration readings. Once the data is validated, it is used in the KB to generate baseline signature when the machine starts up. The current vibration signature is compared against baseline signature to generate "symptoms". Symptoms can be as simple as "the running speed amplitude increases with load" or as complicated as "the second harmonic amplitude increases as the inlet steam temperature decreases and the second harmonic amplitude peaks at a half of the first critical during startup". These symptoms are categorized and evaluated for their severities before they are used by the inference engine to reach conclusions. The conditions diagnosed by the KB along with the recommendations are forwarded to the plant for the operator to review.

Based on the data from the DDA, the TurbinAID can diagnose conditions which are related to subsynchronous vibration in addition to vibration signatures which are associated with common faults. It can also detect abnormal vibration signature which has not been seen in the past because of the availability of the full spectrum and time waveform data.

The ability of the TurbinAID to monitor and analyze vibration patterns easily allows utilities to detect developing faults early, help optimize maintenance and spare parts planning, and ensure safe, reliable plant operation.

This paper has placed a significant emphasis on the monitoring aspects of the thermodynamic portion of TurbinAID. Data acquisition, validation, calculation, and display are all monitoring functions. They are, however, necessary precursors to diagnostics. A monitoring system can exist without a back-end diagnostic system, but a diagnostic system cannot exist without a front-end monitoring system. A diagnostic system cannot be connected to a monitoring system without knowledge of how the monitoring system processes the data or the diagnostic system may draw false conclusions. The knowledge engineer writing the diagnostic expert system must know how variables are calculated, how often they are calculated, and how the calculations could break down if the expert system is to interpret those variables properly. He needs to know how particular sensors operate (and even malfunction) at particular locations to interpret sensor data properly. The monitoring and diagnostic functions of a total thermodynamic diagnostic system must be integrated "cradle to grave" for the system to work together properly. TurbinAID offers such an integrated diagnostic system.

TurbinAID can diagnose many conditions on the turbine ranging from those which help optimize heat rate to those which avert dangerous operating states to those which identify failed sensors. A sampling of diagnosable conditions is shown in Table 2.

For a more detailed description of the expert system diagnostics, refer to reference 4 (Chow, 1988).

7.0 CONCLUSION

TurbinAID can assist plant operators in otpimizing heat rate and improving availability through its monitoring functions and diagnostic capabilities. Turbine thermal performance calculations are an important and necessary part of the TurbinAID diagnostics system. Automatic sensor validation is required to minimize bad readings which could propagate errors throughout thermal performance software. Dynamic, realistic targets and training are crucial to the operator's acceptance and use of controllable parameters or component performance parameters displays.

1. A. F. Armor. "Expert Systems for Power Plants: The Floodgates are Opening." Power Engineering, July 1989, pp. 29-33.

2. S. Lowenfeld. "PDS: An Application of Artificial Intelligence Techniques to Process Diagnosis." Westinghouse Electric Corporation, Proprietary Document #84-1G1-ARTIN-R.1, 1984.

3. David L. Kenien and Joseph R. Nasa. "Operator Training as Part of a Heat Rate Improvement Program." 1988 EPRI Heat Rate Improvement Conference, May 10-12 1988, Richmond, VA.

4. M. P. Chow, et al. "Initial Experience with an On Line Steam Turbine Expert Diagnostic System." American Society of Mechanical Engineers, Steam Turbines in Power Generation, PWR-Vol 3, September 1988, pp. 227-231.

Table 1

Partial List of TurbinAID Sensors

PRESSURES	TEMPERATURES
Throttle *	Throttle *
First stage	Cold reheat *
Cold reheat *	Hot reheat *
Hot reheat *	Crossover
Crossover	Heater extractions at turbine
Heater extractions at turbine	HP heater feedwater inlets
Back pressure *	HP heater feedwater inlets
Gland steam header	Final feedwater *
Gland steam condenser vacuum	BFP suction
Main oil pump discharge	Deaerator condensate inlet
Bearing oil header	Condensate pump suction
Bearing oil header	Condenser circulating water inlet
	Bearing metal
	Bearing oil drain
	First stage steam
FLOWS LP exhaust hood	
Final feedwater *	
Superheat spray	
Reheat spray	
HP-IP gland leakoff	MISCELLANEOUS
Gland spillover	
BFP turbine steam	Dual vibration problems (each brng)
Steam coil air heater extraction	Keyphasor signal
Makeup	Rotor speed
Condenser circulating water inlet	Governor valve positions
	Throttle valve positions
	Single valve operation contact
	Megawatts

* Indicates redundant sensors.

Table 2

Examples of Conditions Diagnosed

Gland Steam Sealing System

1. HP/IP 18 psia gland seals are wearing.
2. Wearing gland seals on HP/IP between pressure chambers.
3. Gland condenser blast gate valve misadjusted.
4. Gland spillover station open at low header pressure.
5. CR gland supply station open at high header pressure.
6. Open HP and/or auxiliary gland supply bypass valve.
7. Insufficient desuperheating of LP gland steam supply.
8. Gland condenser vacuum is surging.

Steam Chest and Valves

9. DEH valve management system malfunction.
10. Broken governor valve stem detected.
11. Governor valves #1 and #2 are not moving in unison.
12. Lift on governor valve #4 is incorrect for single valve mode.
13. Position error on throttle valve #2.
14. Possible scale buildup on right hand governor valve stems.
15. Throttle valve #1 is sticking.
16. Communication failure from MVP card to throttle valve #1.
17. Valve #4 is not fully open in sequential mode - Seq 3.

Inlet Steam Conditions

18. Inadequate superheater attemperating flow.
19. Inlet features subject to creep damage.
20. Inlet features being damaged by abnormal throttle dT.
21. Reheat attemperating flow not functioning properly.
22. Possible creep damage of IP turbine is imminent.
23. Excessive throttle-reheat dT - lower reheat temperature.
24. Excessive throttle pressure is damaging inlet features.

Feedwater Heaters

25. Exceeding allowable load with heaters out of service.
26. Heater tube leak detected.
27. Heater air locked.
28. Heater fouling detected.

Turbine Steam Path

29. Erosion of nozzle block detected.
30. Deposits detected on second HP blade group.
31. Seal rub indicated in LP governor end first blade group.
32. LP is overloaded due to excess flow.
33. IP extraction bushing is leaking.

Rotating Parts

34. Cracked rotor shaft.
35. Loss of material from rotor.
36. Damaged bearing/journal.
37. Abnormal bearing clearance.
38. Coupling misalignment.
39. Turbine seal rub or internal rub.
40. Turbine rotor thermal distortion.
41. Thrust bearing front/rear face damage.
42. Improper generator frame foot loading.
43. Generator gland seal ring rub.

09:47:23 DD/MM/YY

PARAMETER	UNITS	CURRENT VALUE	TARGET VALUE	HEAT RATE DEVIATION (BTU/KWH)
THROTTLE PRESSURE	PSIG	2420	2400	–2
THROTTLE TEMP.	DEGF	994	1000	9
REHEAT TEMP.	DEGH	977	995	22
BACK PRESSURE	INHGA	2.66	2.52	9
SUPERHEAT SPRAY	KPPH	104	75	3
REHEAT SPRAY	KPPH	0	0	0

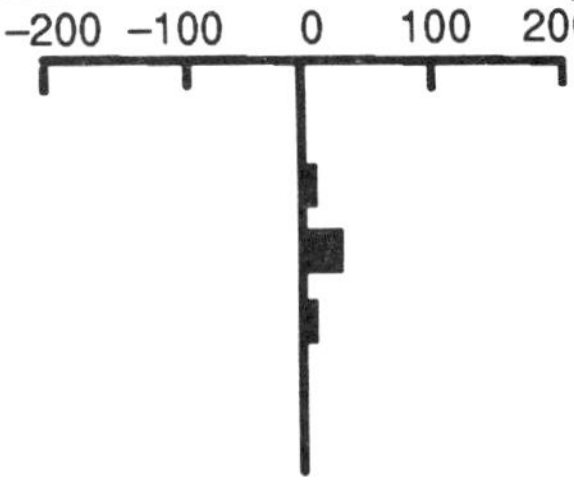

HEAT RATE ANALYSIS (BTU/KWH)	
CURRENT	8825
TARGET	7904
DEVIATION	41

COST OF HEAT RATE DEVIATION	
COST PER SHIFT (DOLLARS)	279
COST PER YEAR (K-DOLLARS)	238

(BASED ON 1.93 DOLLARS / MBTU, .95 AV, .85 LF)

Figure 3. Controllable Parameters Screen

09:47:23 DD/MM/YY

PARAMETER	UNITS	CURRENT VALUE	TARGET VALUE	HT RT DEVIATION (BTU/KWH)	ANNUAL COST (K- DOLLARS)
HP EFFICIENCY	PCT	85.4	86.0	10	58
IP EFFICIENCY	PCT	93.3	92.4	–15	–87
LP EFFICIENCY	PCT	89.2	88.5	–29	–166
HEATER 8 TTD	BEGF	–3.7	–5.0	2	12
HEATER 7 TTD	DEGF	–2.2	–1.0	–1	–4
BFPT FLOW	KPPH	124	112	19	103
AIR HEATER FLOW	KPPH	0	0	0	0

(ANNUAL COST BASED ON 1.93 DOLLARS / MBTU, .95 AV, .85 LF)

Figure 4. Component Performance Parameters Screen

709.6
MW

DIAGNOSTIC SUMMARY

MM/ DD/ YY
HH:MM:SS

PL1

5 DIAGNOSES ON MAINTENANCE LIST

NO.	CF	REC. AVAIL.	DESCRIPTION
1	0.55	Y	INSUFFICIENT DESUPERHEATING OF LP GLAND STEAM SUPPLY
2	0.92	Y	ABNORMAL BEARING CLEARANCE, BEARING #5
3	0.63	Y	EXCESSIVE MOISTURE CARRYOVER FROM MSR B
4	0.92	Y	GLAND HEADER SPILLOVER VALVE IS MISADJUSTED

MOVE CURSOR TO HIGHLIGHT DIAGNOSIS OF INTEREST AND PRESS F3 OR F4.

1	0.55	Y	INSUFFICIENT DESUPERHEATING OF LP GLAND STEAM SUPPLY
2	0.92	Y	ABNORMAL BEARING CLEARANCE, BEARING #5
3	0.63	Y	EXCESSIVE MOISTURE CARRYOVER FROM MSR B

F1 PDCMAIL	F2 DIAGNOSIS	F3 GUIDANCE	F4 ACTION				

Figure 5. Diagnostic Summary Screen

0074-056X/91/551-558/$0 + .50pp

Plant Control Technologies for the 90's

by
H.I.Sternberg and G.O.Provins

presented at the
I.S.A. Session of Advanced Control

Abstract

This paper focuses on the transisition from centralized analog control to distributed digital control that has taken place in the last fifteen years. The capability of Distributed Control Systems (DCS) to integrate both continuous and sequential (Logic state) process control in parallel with embedded data acquisition functions is discussed. Advancements in both the internal processing capabilities of these DCS's and their ability to tightly couple general purpose computer systems and their networks will be highlighted in reference to the development of a plant wide infrastructure for control and information.

A perspective on the major features that distributed technology offers will be introduced and an attempt will be made to extrapolate these techniques and to integrate them with both new and emerging technologies to postulate potential trends for control and information systems in the future. The authors view of the evolution of DCS into the enterprise wide Strategic Process Management System (SPMS) is discussed.

Historical Perspective

In the late 1970's and early 1980's trade magazines in the control industry were replete with articles comparing various control technologies and architectures. For a brief period of time, pneumatic, electric analog and mainframe digital control systems were simultaneously offered as solutions to various control requirements in the industry. There were centralized systems and distributed systems, and even a few variations calling for distributed fragments of centralized architectures (the so-called "islands of automation" approach). Each of these designs, for a period of time, was reported to the marketplace as the "wave of the future".

Electronic analog control systems were midstep in an evolution from discrete solid state transistor based models through the introduction of integrated analog circuits like linear and differential amplifiers. Split verses unitized architectures (whether or not to physically separate the control station from the control loop) and power redundancy schemes (auctioneering verses failover) were hot topics in the power industry.

At the same time, in the area of logic state control (sometimes called sequential or open loop control) the majority of applications were still accomplished with relay logic. There were some logic control systems available that used modules fabricated with discrete solid state components and even some flirtation with pneumatic based "fluidics" for state control in unique environments but the overwhelming majority of sequential state control was accomplished with electro-mechnical relays.

Before the introduction of automatic data collection and record keeping systems (Data Loggers), the functions of a plant wide management information system consisted of a collection of clipboards or an array of hand held walkie-talkies. From the automation of this clipboard function, simple data logger systems had grown into separate stand alone Data Acquisition computer systems. The introduction of computers in the plant to accomplish record keeping and the report generation function was a precursor of what would become a deluge of computer generated information in the mid 80's and sophisticated decision support systems in the early 90's. Back in the early 1980's the closest thing to a decision support function was having the home phone numbers of your supervisor and plant manager!

About this time a few Direct Digital Control (DDC) systems using a mainframe computer to accomplish the integrated process state and continuous control proved not only the practicality and feasibility of combining these control types but the desirability. The combination of all of the control functions with the data acquisition functions integrated into a common man-machine interface provided evidence that integration could provide tangible operational benefits. Some installations using DDC mainframes based systems were able to achieve impressive results in terms of efficiency and unit heat rate. However, These achievements were only possible through the implementation of complex and costly redundant computer backup designs. There were benefits in having a unified system but the price of having an "all in one processor" system was the need to effectively duplicate the whole system. The notorious phrase, "no Single Point Of Failure" was probably formulated around this time and became the focal point of many system design decisions. Actually, redundant mainframe systems were even more complex (costly) then simply duplicating the combined data acquisition and control systems. The capacity to perform the switch over from primary to backup on-line inflated the complexity (cost) of these DDC systems by magnitudes.

Adding to the already high initial cost of these DDC systems was the additional long term cost incurred by the need for maintaining or modifying these systems. In many cases these activities required an on-site staff of technicians and programmers. The on-going explosion in digital electronics resulted in a technology roll-over in mainframe processors that was completely unacceptable. In many cases the selection, design, implementation and installation cycle of the system was longer then the product life cycle of the mainframe that the system was based on.

Beyond the selection of an individual new system, there was always the additional challenge of finding some way to interface all the new "stuff" to all the existing "stuff". Each new application or change became an opportunity to learn a whole new technology. Every plant was a unique mosaïc of systems and technologies and most individual units within each plant were unique to themselves. All in all it was a confusing, but exciting time to be in the controls industry. It may not have been the most productive way to run a business but it was exciting!

As the controls industry progressed into the early 80's the distributed digital control system went from an introduction in 1974-75 by one vendor, to the dominant control technology offered by all vendors by the early 80's. A comparison of some of the advantages these systems possessed beyond the capabilities of the other analog control predecessors is shown in Figure 1. Distributed Digital Control Systems (DCS's) eventually became the industry standard because they incorporated the simplicity of the "building block" technique of the earlier analog systems, with the power, flexibility, precision and repeatability of the digital DDC systems. The fact that, in many cases, they did this at a lower installed cost then conventional analog approaches did not reduce or impede this change over!

Centralized / Analog	Distributed Digital
Constant recalibration required	"Drift free" set and forget adjustments
Changes require re-wiring	Changes accomplished by software
Hard to expand past the initial scope	Easy incremental expansion
Difficult to trouble shoot and repair	Self diagnostic, on-line module swaping
Separate Data Acquisition required	Integrated Data Acquisition functions
Single point-of-failure concerns	Fault tolerance through partitioning, redundancy or both

Figure 1

The reliability of the distributed microprocessors was quantitatively increased not only by having each module executing it's own internal diagnostics but also by having a hierarchy of modules monitoring the modules below them. These holistic techniques have made the DCS system more fault tolerant and easier to work on then their predecessors. In most cases a system would diagnose itself! Because cost continued to decrease for higher and higher performance microprocessors, with greater and greater functionality, DCS control technology came to dominate all other forms of control for medium to large process control applications.

Distributed Control Systems Now and in the Future

This array of distributed microprocessor based control modules connected by high speed, data secure redundant communication networks had become the standard DCS environment of the early 1980's. One point of interest is that the word "distributed " which, of course, is the distinguishing word in the term Distributed Control System, is a feature that up to now has been rarely used in the power industry. The DCS's capability to support physical distribution was originally projected to be a major cost saving for its users. All DCS's use some form of "data highway" to connect individual areas of control to each other and to communicate between the control nodes and the operator interface devices in the system. Early estimates of the savings to be realized by reduced labor and material cost from locating DCS nodes geographically around the plant without having to individually wire each transmitter back to a central cable gathering room or "relay room " ranged from the significant to the ridiculous. To date the wiring cost savings have rarely been realized because most DCS installations in power plants have been retrofits of existing control systems which have already incurred the cost of installation of these cables to the central facility.

One can take the position that the term distributed refers to the logical distribution of the functionality between multiple processors. This is, of course, completely accurate, but logical distribution does not save the mountain of copper and man years of labor that physical distribution will when it is used. So it is not a very risky prediction to anticipate that this capability for physical distribution will become much more widely used as existing plants require new instrumentation or instrumentation cabling. New plants that are constructed will be designed around the "no cable tray" concept. The use of "wireless LAN.'s (Local Area Networks) where radio channels accomplish the basic information transport between the transmitter and the DCS for all but the most critical applications, is only a slightly more risky projection.

Technical Advances

Electronic communication speeds on the early DCS data highways ranged from 9.6 KBaud (9,600 bits per second) through about 500 KBaud (0.5 MBaud). During this same period microprocessors using 8 and 16 bit registers and having clock speeds of 1 or 2 MHz were considered the state of the art. Advances in communication technology have supported the development of secure, high speed DCS data highways. Communication speeds of 8 to 10 Mbaud (10,000,000 bits per second) are now possible for the modern DCS system. These data communication channels can now be accomplished using either electrical or fiber optic cable without incurring the dreaded "Single Point Of Failure". The 32 bit microprocessor using an 8, 10, 16 or even 32 MHz clock rate is now available in advance DCS modules. The integration of these high performance processors with new expanded memories into DCS modules now supports the ability to preprocess raw data into useful information at this lowest module level of the system.

Large quantities of calculation intense algorithms can now be utilized in real-time by executing in the multi-processing environment of the DCS. This ability, to accomplish the data-to-information conversion in real-time, within the DCS system now allows many functions that heretofore had only been available in a retrospective , after-the-fact, manner to now become on-line control parameters (Example: Statistical Quality Control [SQC] and Statistical Process Control [SPC] factors). The resultant information derived from these calculations can now be presented to the user through the embedded communication system and the Man Machine Interface (MMI). This parallel processing approach along with the expanding capacity for higher speed and higher capacity data highways provides a mechanism to support greater and greater high level functionalities within the DCS system. The combination of these mechanisms for handling large amounts of processing in a multiprocessor network connected through high speed and high capacity communication offers the only economically viable means to convert the retrospective "what could have been done" into the active "what can be done".

Data Expansion

The first digital systems were commissioned with what seemed at the time to be large numbers of control tags and data points. As processing power grew and the scope of DCS systems expanded to include data acquisition functions, intermediate values and results of performance analysis, unit optimization and other forms of supervisory data, the quantity of data has doubled and tripled. Early DCS tag counts were in the area of 2,000 to 3,000 points for a 600 Mw unit. The inclusion of greater DA scope and the ever expanding requirements of regulatory monitoring and reports have expanded this 600 Mw unit to now average 5,000 to 6,000 information points. The emerging trend to incorporate parallel on-line simulation of all or portions of the unit steam cycle along with the resultant cascade of data quantities will steadily expand this count to 9,000 or 10,000 data points. Early results of the on-going EPRI intelligent alarm system project, in which data points have doubled, coupled with the expected increase in reporting data for federal as well as state agency requirements, will result in data (tag) counts of at least 20,000 in the very near future.

High Level Functions

Another easily predictable factor for future DCS systems is the continued expansion of the system scope and the inclusion of greater high level (operational and business) management functionality. This will require a much wider range of data access capabilities then we have seen so far. Given todays world-wide problems and constantly changing economic structures, does the notion of a security event at the perimeter of the plant triggering a unit ramp down sound too far fetched?

An extrapolation of scenarios found in manufacturing situations into the utilities service venue might result in a situation where the ownership of fuel is by one entity and the custody of the power produced from that fuel is the property of a second entity. The Individual units within the power plant might even be owned by other totally separate corporate entities. The choice of which unit to select, for what portion of the demand, at a particular time, might well be an on-line matrix equation. Custody transfer could be defined as having occurred at the point of conversion. The bookkeeping for such an operation will require real-time information on fuel as it is used (type, quantity, BTU content and cost), the power as it is produced, and the unit efficiency at the time of conversion. All of this information will be required just to determine who made or spent how much money during a specific time frame. The concept that the three entities may be using different monetary standards with dynamic rates-of-exchange would complicate the calculations even more. These scenarios will require tight coupling of the process control to many non-traditional resources.

When considered collectively the inclusion of these high level functions into the DCS scope, actually form a new type of plant-wide Information and control system. This "new breed" of system is based on the use of multiple computational resources of various sizes and backgrounds operating in parallel.

Man Machine Interface

Many early DCS systems were installed with hardwired Hand / Auto stations (one on each control loop) as the primary operator control interface. These "hard" stations were used for both monitoring the process values and for the interaction of the operator in making setpoint changes. As the use of CRT based control consoles became more widespread, their ability to duplicate the station functions with "soft "CRT accessed control stations provided an alternative to the hardwired station. As access to control loops from CRT displays became more acceptable to operators, more and more loops were commissioned without the hardwired stations. This process took place over time as the operators and engineers developed greater confidence in these "glass stations". The hardwired Hand / Auto (H/A) Stations were removed entirely and additional consoles used to provide backup. This migration away from the hardwired stations to the CRT stations continued through the 80's up to a point today when many DCS systems have few if any of the hardwired stations and most have multiple CRT consoles for the operator control interaction.

As mentioned earlier, the amount of data and information about the process that is available in the DCS is rising rapidly and shows no sign of slowing down. At the focus of this vast array of data and information flowing into the control room, the operator becomes the single, most critical variable in the safe control of the productivity, availability and profitability of any process control operation. The human capacity for processing information is, at best, magnitudes lower then what is required to monitor and correlate this huge data stream as it funnels into the control room. In order for the operator to be able to succeed in this information packed environment, the operator's video display console must be fully integrated with the data acquisition and control system to the point of becoming the operator's trusted "Assistant".

Advancements in the understanding of how human beings acquire information, conceptualize status, make decisions and interact with machines is leading to the development of sophisticated Man Machine Interfaces (MMI). New techniques for alarm management, prioritization, information reduction and on-line "expert" system assistance are in the development and application stages. New methods of conveying information and meaning to the decision maker will be required . Windows, Icons, Mice, and Pointers (the so called WIMP interface) are only the precursor of what will be required. Speech recognition and expansion of the simple console "tones" into information packed "cords" or other forms of easily recognizable acoustic patterns will be required. (Yes, the plant of the future may well be described as purring like a kitten, literally!) Even computer voice generation may return in a usable form. This presupposes that a technology greatly more sophisticated then the talking cars of the mid 80's and the "Some-one is pre-vent-ting us from leav-ving the stat-ion" of Atlanta airport fame can be developed!

Summary

Distributed multiprocessor technology, higher speed secure communication networks and new concepts for data collection, real-time calculations, and embedded simulations and data transformation will continue to evolve into plantwide information systems. Increased information access will form the basis for any projection of the future of process controlling. Systems composed of distributed Computational Resources Units (CRU) will by virtue of their embedded high speed high capacity communication networks share the raw analog values and logic state conditions derived from the process, as well as calculated or composed information developed internally or imported from other sources. These functions are now practical even though individual CRU's are partitioned separately and may be geographically distributed throughout the plant.

Systems comprised of these distributed CRU's can now provide a new and powerful platform for the integration of both primary and supervisory controls with their associated MMI. Because of the close coupling of the individual CRU's to the process they can execute on-line, application specific performance monitoring or economic condition evaluation in time frames that now allow the intervention to effect the result. This means that the information necessary to support decision making for plant wide resource management can be accomplished at lower levels of the system and in real-time. This information can then be correlated within the network to be accessible to all levels of the operation. The availability of higher level tactical information in real-time to the operations personnel will be the minimum requirement in order to meet the new levels of efficiency required in the future

References:

1) The Role of Communication Networks in Strategic Process Management by T.J.Fink, G.O.Provins, and J.R.Rege presented at the International ISA Conference on Industrial Control; March 1990, BCCo TP90-5

2) The future of Process Control (A View from the Drivers Seat) by M.A.Keyes presented at the British Petroleum Conference on Control "Automation for the 90's" ; Oct. 1990, BCCo TP90-9

3) Distributed Computational Resources in the Modern Plant by G.O.Provins presented at the Discovery 88 Computer Form for the Edison Electric Institute; Sept 88

4) Distributed Control Systems (Their Evaluation and Design) by Dr. Michael P.Lukas published by Van Nostrand Reinhold Company Inc. NY 1986

©1991 - Paper #91-743
0074-056X/91/559-567/$0 + .50pp

COMPARISON OF TWO ADVANCED STEAM TEMPERATURE CONTROL ALGORITHMS FOR COAL FIRED BOILERS

James B. Riggs
Dept. Chemical Eng.
Texas Tech University
P.O. Box 4679
Lubbock, TX 79409

Keith Curtner
Wendy Foslien
Honeywell SSDC
3660 Technology Drive
Minneapolis, MN 55418

Arvind Agarwal
Murthy Divakaruni
EPRI
3412 Hillview Ave.
Palo Alto, CA 94304

ABSTRACT

A nonlinear process model based control (nonlinear PMBC) algorithm and a state variable control (SVC) algorithm have been tested on a medium fidelity simulation of a drum-type coal-fired boiler. A simple PI controller was also tested as a point of reference. The simulation was based upon the Cromby 2 model and was used to test each controllers for 10 different upset scenarios that included soot blowing, load ramps, and changes in burner tilts. The simulation had control valve hystersis, control valve dynamics, and temperature sensor dynamics added to it.

The PMBC controller had the best control performance yielding factors of 2 to 5 times reduction in steam temperature variability over the optimally tuned PI controller for the 10 upset scenarios. The SVC controller outperformed the PI controller in 8 out of 10 of the upset scenarios, but typical variability reductions of only 20% was observed.

INTRODUCTION

Steam temperature control is an important control problem for coal-fired utility boilers. Significant variations in the temperature of steam entering a turbine is known to reduce the useful life of a turbine. In addition, many times boilers are operated in a mode with reduced operating efficiency in order to obtain reliable steam temperature control.

Steam temperature control is coarsely controlled by adjusting burner tilts or changing the amount of recycled flue gas. Fine steam temperature control is affected by adjusting the flow rate of attemperator water spray injected between the primary and secondary superheater (Figure 1). As load changes are made, soot blowing activities, or any other operational changes require adjustments in both the coarse and fine steam temperature control. Ultimately the attemperator spray flow rate determines the variability in the steam temperature entering the turbine as long as the coarse control prevents the attemperator spray from becoming constrained.

This paper compares two advanced steam temperature control algorithms namely nonlinear process model based control and state variable control. A nonlinear dynamic model of a drum type coal-fired boiler has been used for testing of the algorithms under a variety of upset conditions.

DYNAMIC MODEL OF COAL-FIRED DRUM TYPE BOILER

The simulation is based on the Cromby 2 model developed by J.P McDonald, H.G. Kwatny and J.H. Spare of the Philadelphia Electric Company (McDonald et al, 1971). The Cromby 2 unit is a drum boiler with high- and low-pressure turbines and has a maximum capacity of 250 MW. The nonlinear model is based on equations derived from first principles along with curve-fit equations derived from the acceptance test for Cromby 2. The model includes boiler elements between the mills and low-pressure turbine, as shown in Figure 1.

The Cromby 2 model was selected because it covered the major elements of plants that were of interest in terms of superheater temperature, and it had been validated from 50 to 100 percent of load. The model is also well documented and available in the public domain.

The boiler simulation software consists of 14 separate modules, each representing a segment of the boiler system. The boiler is divided into the following segments:

- Feedwater valve and boiler feed pumps
- Downcomers
- Waterwalls
- Drum
- Primary superheater
- Superheater attemperator
- Secondary superheater
- Turbine valve
- High-pressure turbine
- Reheater attemperator
- Reheater
- Low-pressure turbine
- Mills
- Combustion furnace
- Superheater furnace
- Reheater furnace

It is important to note that the state equations in these routines are in the explicit form and thus require small step size integration.

Two general assumptions were made in the development of the model. First, it is assumed that the dynamics of the steam and flue gases could be neglected. This assumption was reasonable since the time constants of the steam and flue gas dynamics are much smaller than other process time constants. Second, it is assumed that the economizer and feedwater heater dynamics can be neglected. To account for changes in feedwater enthalpy, which occur with changes in feedwater flow, a curve fit was done on acceptance test data for Cromby 2.

To ensure that the FORTRAN code of the model was correct, steady-state values were compared to steady-state acceptance data for the Cromby 2 unit. This procedure was similar to that followed for the original validation of the model. Several issues involving limitations of the medium-fidelity model used must be addressed. First, it is important to note that this particular plant has some decoupling of temperatures due to the split-furnace design. Thus, steam temperature control is less of a problem than for a single-furnace plant. Also, because the model does not incorporate variations in coal composition and other time-varying parameters, we do not see all of the time-varying disturbances that may occur in a plant. In spite of these limitations, we can still assess whether the new algorithms offer a relative improvement over conventional control strategies. In order to further improve the realism of the simulation, hysteresis in the attemperator spray control valve, lag in the control valve, and lag in the temperature sensor was added to the Comby 2 model.

TEST SCENARIOS

A set of baseline tests were established in order to compare and evaluate the three control strategies (conventional PI- controls, SVC and PMBC). These baseline tests were selected as stringent tests of steam temperature control system performance. The tests were chosen by utility personnel with extensive plant controls and performance evaluation experience. The scenarios that follow include the interactive effects of controllers and physical elements in the boiler for events likely to occur in normal operation. In each of the following upset scenarios, the initial conditions are lined out for operation at 190 MW. In addition, in all cases the burner tilt is controlled by a PI controller and only the attemperator spray is controlled by each of the different algorithms.

(1) <u>Increase in MW load</u>. The load demand is increase at 5 MW/min until it reaches 230 MW at which point load demand remains constant.

(2) Removal of a set of burners. A set of burners in the furnace is removed.

(3) Drop in tilt position. The tilt of the burners is reduced by 5 degrees which reduces the heat transfer to the furnace wall.

(4) Load profile For this scenario, the load is ramped up at 5 MW/min until it reaches 230 MW and then it is ramped down at 5 MW/min until it reaches 190 MW where it is maintained for the duration of the test.

(5) Five-percent decrease in the fuel BTU content. For this upset, a step change reduction of 5% is assumed for the heating value of the coal being burned.

(6) One-half percent O_2 increase. The oxygen content of the injected air is increased 0.5% in the form of a pulse lasting approximately two minutes.

(7) Primary superheater soot blowing. The primary superheater soot blowing operation is modeled by a decrease in the ash thickness on the primary superheater tubes.

(8) Secondary superheater soot blowing. The secondary superheater soot blowing operation is modeled by a decrease in the ash thickness on the secondary superheater tubes.

(9) Furnace wall soot blowing-conduction change. The furnace wall soot blowing operation is modeled by decreasing ash thickness on the furnace wall.

(10) Furnace wall soot blowing-radiation changes. The furnace wall soot blowing operation is modeled by decreasing the radiation heat transfer coefficient for the furnace wall.

STATE VARIABLE CONTROLLER (SVC)

The SVC is a modified version of state variable control with estimation. The state variable control technique has been used successfully by European companies for several years (Laubli et al, 1979). The basic concept of state variable control is to use additional state information from the process (e.g., intermediate temperatures along the superheater) to provide a more accurate control action. Various temperatures along the length of the superheater are multiplied by individual gain factors ($f_1,f_2,...f_n$) and summed to create a control signal. The standard state variable control approach assumes that all of the desired states are available to the controller. However, providing the state information can be costly. The alternative is to develop an estimate of the states and use the estimated values to drive the controller (Figure 2).

The estimator used in the SVC consists of a series of first-order lags. Each of the lags has an associated gain and time constant. In the systems implemented by European companies, the gain of each of the lags is used to model the temperature rise across the superheater. In contrast, the SVC uses an overall gain to model the temperature rise. Thus, each of the individual lags has a fixed gain of 1.0 and a time constant that may vary as a result of changing operating conditions.

The SVC provides proportional action in response to major changes in load, temperature or other operating conditions. Because the SVC provides proportional action only, a steady-state error will exist in the absence of any other form of control. A PI controller is used to provide trim action and eliminate steady-state error.

NONLINEAR PROCESS-MODEL-BASED CONTROL (NONLINEAR PMBC)

Nonlinear process-model-based control (nonlinear PMBC) uses a nonlinear approximate process model directly for control purposes. The controller model does not have to be rigorous but should contain the major characteristics of the process. One type of nonlinear PMBC is Generic Model Control (GMC), developed by Lee and Sullivan (1988). To understand GMC, let us consider a SISO (single input/single output) system described by the following approximate model:

$$\frac{dy}{dt} = f(y,u,\underline{d},\underline{k}) \tag{1}$$

where y is the output variable, u is the manipulated variable, $\underline{d}$ is a vector of measured disturbance, and $\underline{k}$ is a vector of parameters. Assuming that y has a value of y_o, and it is desired to move the process from y_o to y_{sp} in some time interval τ. Equation 1 can be approximated using the forward difference approximation of the derivative:

$$\frac{y_{sp} - y_o}{\tau} = f(y_o, u, \underline{d}_o, \underline{k}) \quad (2)$$

Equation (2) can be solved directly to determine u, the control action, if one knows $\underline{d}_o$ and $\underline{k}$. The time interval (τ) is a tuning parameter. If τ is small, rapid response is required; if τ is large, a more sluggish response is required.

Since Equation (2) uses only an approximate model, using this control law will result in steady-state offset. To eliminate this offset, Lee and Sullivan added an integral term (analogous to a PI controller), resulting in the GMC control law given by Equation (3)

$$f(y_o, u, \underline{d}_o, \underline{k}) + K_1(y_o - y_{sp}) + K\int_0^1 (y - y_{sp})\,dt = 0 \quad (3)$$

Note that K_1 is equal to $1/\tau$ in Equation (2).

Consider the superheater section shown in Figure 3. Applying an unsteady-state energy balance around the secondary superheater and the attemperator (denoted by the dashed lines) results in.

$$M_m Cp_m \frac{dT_{sso}}{dt} = Q_{ss} + w_{pso} h(T_{pso}) + w_{sp} h_e - (w_{pso} + w_{sp}) h(T_{sso}) \quad (4)$$

Note that the steam and metal tube in the secondary superheater are lumped together in this model. Also, Q_{ss} is modeled using an overall heat transfer coefficient, h_T

$$Q_{ss} = h_T(T_g - T_s) \quad (5)$$

where T_s is the average steam temperature in the secondary superheater and T_g is the average flue gas temperature around the secondary superheater.

Each time the controller is called, a steady-state energy balance around the secondary superheater is used to calculate h_T, and this value is filtered with a first-order exponential filter to provide the value of h_T used by the controller. Also, it is assumed that T_g is constant and equal to 2400°F.

Next, the GMC control law (minus the integral term) is used to calculate the new attemperator spray rate using Equation (3):

$$\left[\frac{Q_{ss} + W_{pso} h(T_{pso}) + w_{sp} h_e - (w_{pso} + w_{sp}) h(T_{sso})}{Cp_m M_m}\right] = k_i(T_{sso} - T_{SP}) + k_3 \frac{dT_{sso}}{dt} = 0 \quad (6)$$

where $Q_{ss} = h_T(T_g - T_s)$. Since w_{sp} affects T_s, solution of this equation requires an iterative numerical procedure. The PMBC controller uses the secant method to find the solution to this equation on-line.

For the superheater system, the integral term in the GMC control law is not required because the on-line calculation of h_T will eliminate any process/model mismatch. However, a derivative term was added to the GMC control law to improve controller performance.

RESULTS

First each method was optimally tuned based upon the integral square error (ISE) for scenario number 1 (ramp up from 190 to 230 MW). Figure 4 shows these results along with the spray flow rate for each controller. With these controller settings, the remaining scenarios were run.

Figure 5 shows the average results of these tests for six different performance criteria normalized to the PI controller results. The six performance criteria were (1) integral square error (ISE), (2) integral time square error (ITSE), (3) integral absolute error (IAE), (4) integral time absolute error (ITAE), (5) maximum deviation from setpoint, and (6) integral of control action. The nonlinear PMBC controller out performed the PI and the SVC controller for each of the six performance criteria.

Figure 6 shows the IAE's for the three controller under the ten test scenarios. Note that the SVC controller provided about 20% reduction in IAE over the PI controller in most cases while the nonlinear PMBC controller provided a factor of 3 to 5 reduction in IAE over the PI controller.

Figure 7 compares the IAE for the PI controller and the nonlinear PMBC controller as a function of the proportional gain used by each controller. This shows that the nonlinear PMBC controller is also less sensitive to controller tuning than the PI controller.

CONCLUSIONS

A nonlinear dynamic model of a coal-fired drum-type boiler has been used to compare two advanced control algorithms to a conventional PI controller for steam temperature control. The nonlinear PMBC controller clearly out performed the SVC controller which offered only marginal advantages over the PI controller.

Because the nonlinear PMBC algorithm uses an approximate nonlinear model of the process, this algorithm is less sensitive to the selection of tuning constants, results in smoother use of the manipulated variable, can effectively handle a wide range of process upsets, and, as a result, reduced the variability in the steam temperature by a factor of three to five based upon the IAE for the ten upset scenarios tested.

NOMENCLATURE

Cp_m = heat capacity of the metal (characteristic of superheater)

h_p = the enthalpy of water used by the attemperator (calculated variable)

h_{pso} = steam enthalpy at temperature T_{pso} (calculated variable)

h_{sso} = steam enthalpy at temperature T_{sso} (calculated variable)

h_T = the effective heat transfer coeficient between the flue gas and the superheated steam (calculated value)

M_m = mass of metal in the secondary superheater (characteristic of superheater)

Q_{ss} = the heat transfer rate from the flue gas to the steam in the secondary superheater (estimated variable)

t = time

T_{sso} = outlet steam temperature from the secondary superheater (measured variable)

T_{pso} = the measured temperature of the steam leaving the primary superheater (measured variable)

w_{sp} = flow rate of water spray in the attemperator (manipulated variable)

w_{pso} = flow rate of steam out of the primary superheater (measured variable)

REFERENCES

1. F.Laubli, D. Le Febve, and R. Herzog, "Large Coal Fired Once-Through Boilers for Middle Load Operation, Part 4: Process Dynamics and Automatic Control," Sulzer Brothers Limited, 1979

2. P.L. Lee and G.R. Sullivan, "Generic Model Control," Computer and Chemical Engineering, No. 12,p.573, 1988.

3. J.P. McDonald, H.G. Kwatny, and J.H. Spare, "Nonlinear Model of a Reheat Boiler-Turbine Generator System," Philadelphia Electric Company, 1971

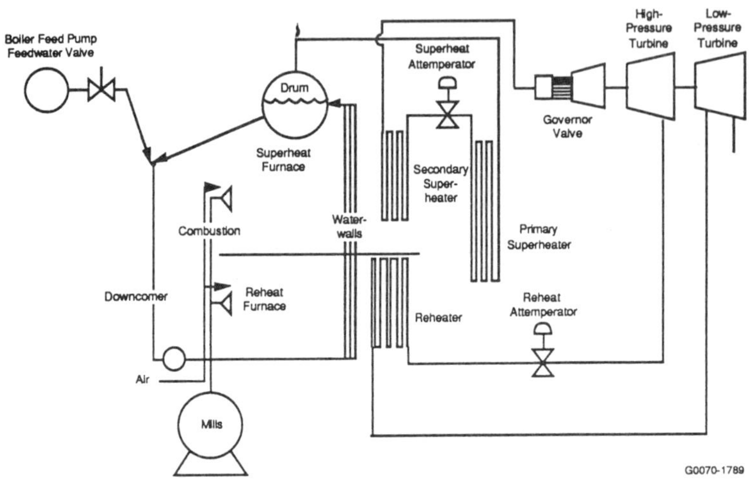

FIGURE 1. Schematic diagram of coal-fired drum-type boiler system.

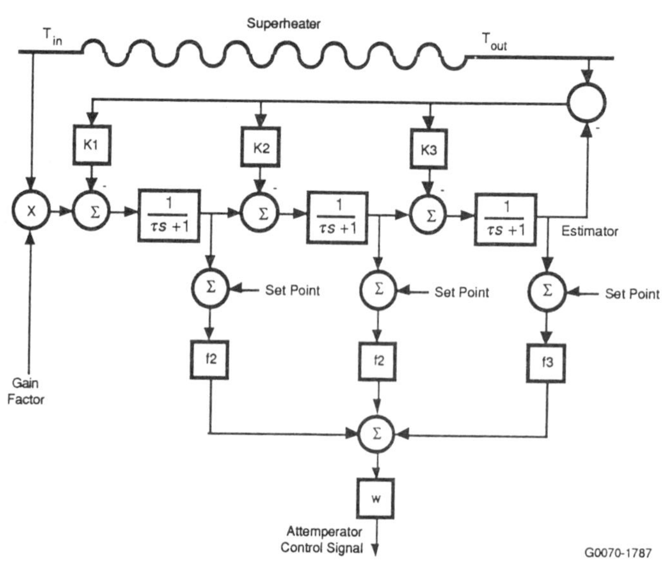

FIGURE 2. SVC controller schematic

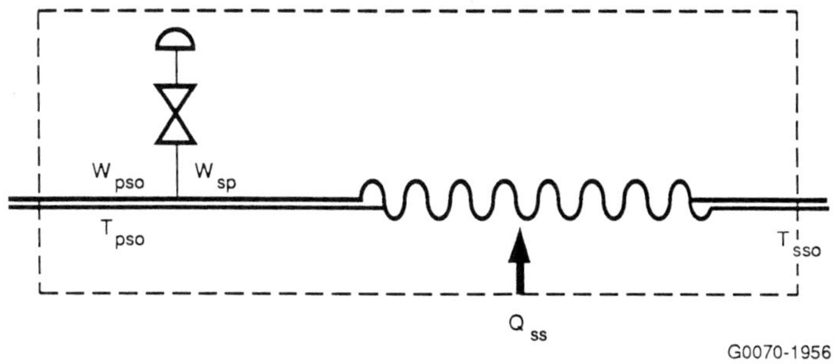

FIGURE 3. Superheater (Secondary) and attemperator section of the boiler

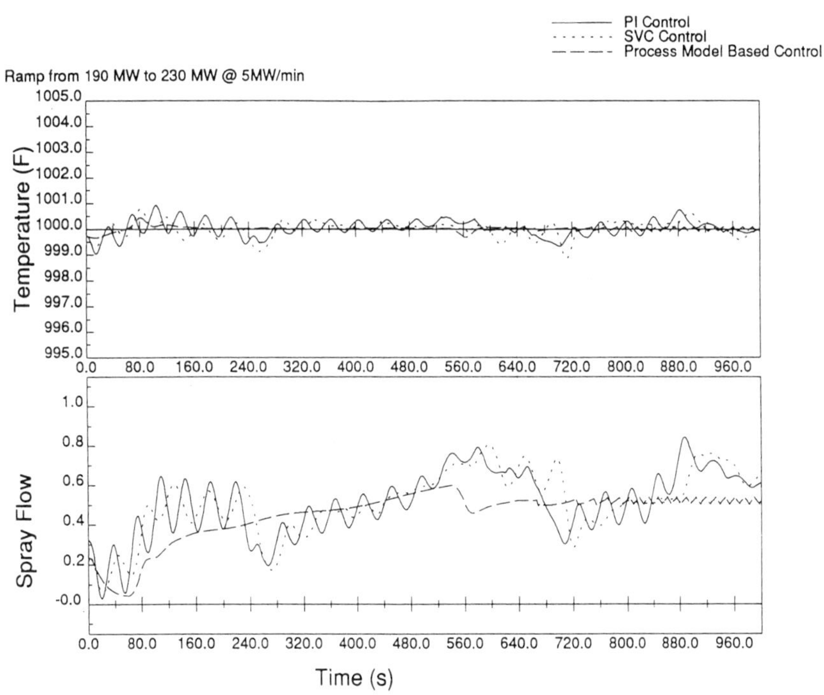

FIGURE 4. Control results for senario (1)

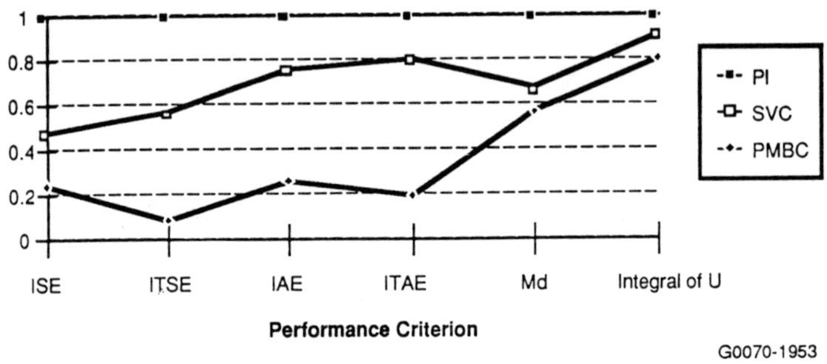

FIGURE 5. Average controller performance

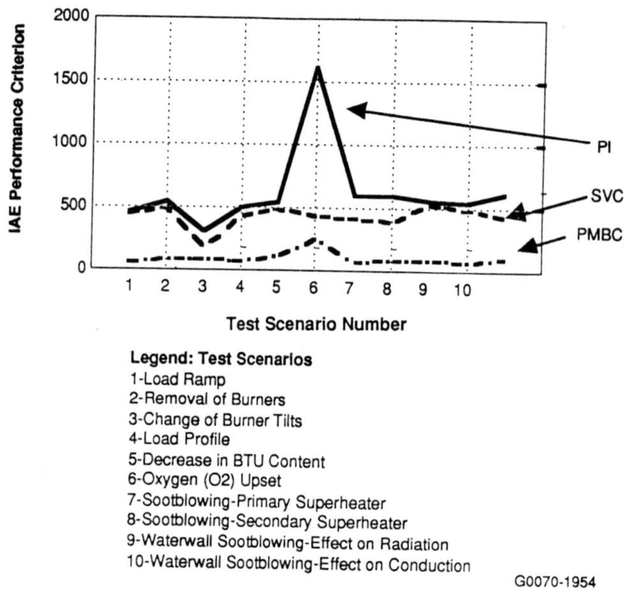

FIGURE 6. IAE for each senario and each controller

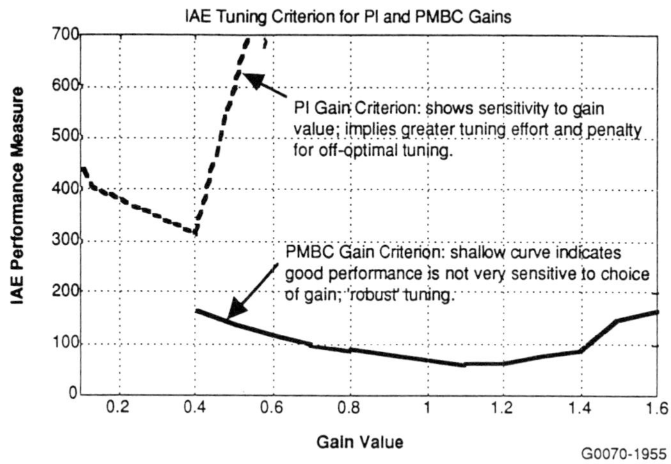

FIGURE 7. Tuning bands

APPENDIX I

ACKNOWLEDGEMENTS

1991 Executive Committee

President	Alan J. Robertson	Consultant
President-Elect Sec.	David N. Bishop	Chevron USA Inc.
Treasurer	C. Blair Ives, Jr.	C.B. Ives & Co., Inc.
Past President	William Calder	The Foxboro Co.

Department Vice Presidents

Education	Thomas J. Picard, Jr.	M.D. Duncan & Assoc.
Industries & Sciences	W. Gerald Wilbanks	Rust International Corp.
Strategic Planning	Roger L. Radeloff	SSOE, Inc.
Publications	Thomas J. Fisher	Lubrizol Corp.
Standards & Practices	John Rennie	Factory Mutual Research Corp.
Automation & Technology	Ronald H. Dieck	Pratt & Whitney

ISA Headquarters

Glenn F. Harvey
Executive Director

ISA Power Division Officers - 1990

Director	Robert N. Hubby	Leeds & Northrup
Director Elect	Milton B. Neher	Central & Southwest Services
Past Director	Edwin M. Good	Florida Power Corp.
Symposium General Chairaman	Edwin M. Good	Florida Power Corp.
Symposium Program Chairman	Paul L. Kenny	Forney International Inc.
Editorial Review Chairman	Roger L. Hull	Westinghouse Electric Corp.
Newsletter Editor	Eric H. Schaefer	The Foxboro Co.
Publicity Chairman	Ronald Hicks	Black & Veatch

Symposium Session Developers

Vision 2000	Murthy Divakaruni	EPRI
Fossil Power System Applications	Thomas W. Stevenson	Baltimore Gas & Electric Co.
Nuclear Power Systems Applications	Joseph Weiss	EPRI
Waste to Energy Applications	James Waung	Babcock & Wilcox
Nuclear Testing and Diagnostics	Dan Wilkinson	EPRI
Co-Gen/IPP/Combined Cycle Applications	Michael C. Annon	I&C Engineering Associates
Plant Wide Automation Concepts	Arvind Agarwal	EPRI
Advanced Instrumentation	Milton B. Neher	Central & Southwest Services
Intelligent Control	Siddarth Bhatt	EPRI
Advanced Control	Lewis Adams	TU Electric Co.

ISA Tampa Bay Section

President	Walter D. McMillan	Imc

Host Committee

Chairman	Howard K. Cuddy	Retired
Registration	Les Miller	State Instruments, Inc.
Spouses Program	Carol Good Pam Miller	

ISA District 3

Vice President	E.V. (Gene) Benedict, Jr.	Stone Mountain Applications

Symposium Luncheon Speakers

V.P. Power Operations	John A. Hancock	Florida Power Corp.
Chief Executive Officer	Dr. Jack Critchfield	Florida Progress Corp.

CALL FOR PAPERS

2ND ANNUAL JOINT CONTROLS AND INSTRUMENTATION CONFERENCE

Instrument Society of America/Electric Power Research Institute

JUNE 1-3, 1992
THE RITZ-CARLTON HOTEL
KANSAS CITY, MISSOURI

Power Industry Division (POWID) of Instrument Society of America and Electric Power Research Institute will consider original papers for presentation and publication for such topics as:

Control Systems
- Fossil/Nuclear Retrofits
- PLC Applications
- Digital Protection Systems
- Distributed Control Systems
- Validation and Verification

Instrumentation/Systems
- Test Reduction
- New Sensors
- Smart Sensors
- Sensor Applications
- Set Points-Nuclear

Control Room
- Upgrades/Modernizations
- MMI
- Operator Aids

International Experience
- Field Devices
- Control Strategies
- Operation
- Design/Maintenance

Expert Systems
- Operating Guidance
- Human Factors Concerns
- Smart Alarming
- Help Diagnostics

Data Communication
- Field Bus
- Plantwide Systems
- Standardization

Technical papers for the Joint ISA/EPRI Conference shall discuss - new and/or innovative techniques in the design and use of instrumentation and control systems. All papers must be original and will be published and copyrighted by ISA. All papers and presentations will be in English.

Authors will submit abstracts for review by the ISA POWID/EPRI Program Committee. Authors will be notified promptly of the Program Committees decision and provided with paper preparation guidelines for the full manuscript.

Abstracts MUST be submitted by October 14, 1991

Manuscripts MUST be submitted by December 16, 1991.

For further information and for "Intent to Present" forms for abstract, contact any of the following organizations or individuals:

General (All)
Meetings Department
Instrument Society of America
Post Office Box 12277
Research Triangle Park, NC 27709
Phone 919-549-8411
FAX 919-549-8288

Fossil
Mr. Cyrus Taft
Southern Company Services
Post Office Box 2625 Bin B464
42 Inverness Center Parkway
Birmingham, AL 35202
Phone 205-877-7424
FAX 205-868-5103

Nuclear
Mr. Joe Weiss
EPRI
Post Office Box 10412
Palo Alto, CA 94303
Phone 415-855-2751
FAX 415-855-2954

AUTHOR INDEX